Proceedings of the QMath12 Conference

Mathematical Results in Quantum Mechanics

Berlin, Germany 10 – 13 September, 2013

Editors

Pavel Exner
Nuclear Physics Institute, Academy of Sciences & Doppler Institute for Mathematical Physics and Applied Mathematics, Prague, Czechia

Wolfgang König
WIAS Berlin, Germany & Technische Universität Berlin, Germany

Hagen Neidhardt
WIAS Berlin, Germany

World Scientific

NEW JERSEY • LONDON • SINGAPORE • BEIJING • SHANGHAI • HONG KONG • TAIPEI • CHENNAI

Published by

World Scientific Publishing Co. Pte. Ltd.
5 Toh Tuck Link, Singapore 596224
USA office: 27 Warren Street, Suite 401-402, Hackensack, NJ 07601
UK office: 57 Shelton Street, Covent Garden, London WC2H 9HE

British Library Cataloguing-in-Publication Data
A catalogue record for this book is available from the British Library.

MATHEMATICAL RESULTS IN QUANTUM MECHANICS
Proceedings of the QMath12 Conference
(With DVD-ROM)

ISBN 978-981-4618-13-7

Printed in Singapore

Preface

A series of conferences always provides snapshots of the state of the art of a research field. However, every conference is also influenced by the local environment where it takes place. This is a factor which should be not underestimated even in our highly connected world of today. From this point of view it should be interesting and inspiring for a conference to return to a place where it took place many years previously. This gives the possibility to appreciate not only the scientific development of a research field but also the local changes of scientific and other nature in this place.

After two decades, the QMath series returned to the Berlin area. However, unlike in 1993, when QMath 5 took place in the village of Blossin near Berlin, the 2013 meeting QMath 12 was organized in the heart of the German capital, in a central building of the famous Humboldt University which means that the conference was not only closer to the daily but also to the scientific life of Berlin.

Like the previous eleven QMath events, the main purpose of QMath 12 was to display and to discuss the latest achievements in the mathematical treatment of quantum physics. This subject keeps presenting many challenges, as has been stressed many times, particularly in the proceedings of the previous meetings. With this respect it seems to be very interesting to compare the present conference with the one of twenty years ago.

As in Blossin the conference dealt with the traditional subject, the spectral theory of one-body and many-body Schrödinger and related operators, viewed from various angles. The majority of talks was devoted to this subject. In contrast to QMath 5 there was a new section on random operators and random matrices. At that time talks on stochastic spectral analysis and quantum chaos were of interest which were completely absent in 2013. Also new were contributions on relativistic quantum mechanics and quantum field theory, topics that have rapidly grown in the decades since Blossin. Completely new compared to all previous QMath conferences was a section on numerical methods in mathematical physics. This reflects the influence of the local environment. Indeed, numerical methods have been rapidly developed in the last twenty years in Berlin and can be now regarded as an important part of mathematical physics here.

This volume contains a vast selection of contributions of the speakers to the field in general, not necessarily describing their respective talks, but also giving a general view, a survey or describing just some particular new issue that was inspired by and raised in discussions during the meeting. All the speakers and also all the other participants were encouraged to contribute, and many did so with pleasure. In this way, the present volume conveys not only a significant part of the meeting contents, but also quite some of its scientific atmosphere.

Needless to say, no conference can be organized without competent people and a sufficient material support. We are pleased to thank those who made the meeting possible: the (other) members of the scientific board, the agency *event lab. GmbH* (Leipzig, Germany), which ran the meeting in an excellent manner, the section organizers, and the Humboldt–Universität zu Berlin, which made it possible to hold the conference sessions in the center of Berlin. Last but not least we gratefully acknowledge the financial support that we received from the Weierstrass Institute (WIAS), the German Research Foundation (DFG), the International Association of Mathematical Physics (IAMP), the DFG Research Center MATHEON, the publishers Birkhäuser, de Gruyter, Springer, AIP Publishing (and its *Journal of Mathematical Physics*), and Cambridge University Press (and its *Forum of Mathematics*) as well as the TNG Technology Consulting.

Enjoy reading!

Berlin and Prague, June 2014

Pavel Exner
Wolfgang König
Hagen Neidhardt

Committees

Scientific board

Volker Bach	TU Braunschweig, Germany
Michael Demuth	TU Clausthal, Germany
Pavel Exner	Doppler Institute Prague, Czech Republic
Wolfgang König	TU Berlin, Germany
Alexander Mielke	Humboldt–Universität zu Berlin, Germany
Hagen Neidhardt	Weierstraß–Institut Berlin, Germany
Reinhold Schneider	TU Berlin, Germany

Conference office

Susanne Jansen	event lab. GmbH, Leipzig, Germany
Jenny Kaftan	event lab. GmbH, Leipzig, Germany

Sections and organizers

Section A: ***Spectral theory of Schrödinger operators***
Michael Demuth, TU Clausthal, Germany
Pavel Exner, Doppler Institute Prague, Czech Republic
Hagen Neidhardt, WIAS Berlin, Germany

Section B: ***Spectra of random operators***
Wolfgang König, TU Berlin, Germany
Simone Warzel, TU München, Germany

Section C: ***Quantum field theory and relativistic quantum mechanics***
Marcel Griesemer, Universität Stuttgart, Germany

Section D: ***Interacting many body systems***
Volker Bach, TU Braunschweig, Germany

Section E: ***Numerical methods in mathematical physics***
Reinhold Schneider, TU Berlin, Germany

Contents

PART A

Plenary talks

A bound for the eigenvalue counting function for higher-order Krein Laplacians on open sets

F. Gesztesy*, M. Mitrea[§] and S. Sukhtaiev[‡]

Department of Mathematics, University of Missouri,
Columbia, MO 65211, USA
**E-mail: gesztesyf@missouri.edu*
[§]E-mail: mitream@missouri.edu
[‡]E-mail: sswfd@mail.missouri.edu

A. Laptev

Department of Mathematics, Imperial College London,
London SW7 2AZ, UK
E-mail: a.laptev@imperial.ac.uk

For an arbitrary nonempty, open set $\Omega \subset \mathbb{R}^n$, $n \in \mathbb{N}$, of finite (Euclidean) volume, we consider the minimally defined higher-order Laplacian $(-\Delta)^m|_{C_0^\infty(\Omega)}$, $m \in \mathbb{N}$, and its Krein–von Neumann extension $A_{K,\Omega,m}$ in $L^2(\Omega)$. With $N(\lambda, A_{K,\Omega,m})$, $\lambda > 0$, denoting the eigenvalue counting function corresponding to the strictly positive eigenvalues of $A_{K,\Omega,m}$, we derive the bound

$$N(\lambda, A_{K,\Omega,m}) \leqslant (2\pi)^{-n} v_n |\Omega| \{1 + [2m/(2m+n)]\}^{n/(2m)} \lambda^{n/(2m)}, \quad \lambda > 0,$$

where $v_n := \pi^{n/2}/\Gamma((n+2)/2)$ denotes the (Euclidean) volume of the unit ball in $\mathbb{R}^n$.

The proof relies on variational considerations and exploits the fundamental link between the Krein–von Neumann extension and an underlying (abstract) buckling problem.

Keywords: Krein Laplacian, eigenvalues, spectral analysis, Weyl asymptotics, buckling problem.

1. Introduction

To set the stage, suppose that S is a densely defined, symmetric, closed operator with nonzero deficiency indices in a separable complex Hilbert space $\mathcal{H}$ that satisfies

$$S \geqslant \varepsilon I_{\mathcal{H}} \text{ for some } \varepsilon > 0. \tag{1}$$

Then, according to M. Krein's celebrated 1947 paper [35], among all nonnegative self-adjoint extensions of S, there exist two distinguished ones, S_F, the Friedrichs extension of S and S_K, the Krein–von Neumann extension of S, which are, respectively, the largest and smallest such extension (in the sense of quadratic forms). In particular, a nonnegative self-adjoint operator $\widetilde{S}$ is a self-adjoint extension of S if and only if $\widetilde{S}$ satisfies

$$S_K \leqslant \widetilde{S} \leqslant S_F \tag{2}$$

(again, in the sense of quadratic forms).

An abstract version of [26, Proposition 1], presented in [6], describing the following intimate connection between the nonzero eigenvalues of S_K, and a suitable abstract buckling problem, can be summarized as follows:

$$\text{There exists } 0 \neq v_\lambda \in \operatorname{dom}(S_K) \text{ satisfying } S_K v_\lambda = \lambda v_\lambda, \quad \lambda \neq 0, \tag{3}$$

if and only if

$$\text{there exists a } 0 \neq u_\lambda \in \operatorname{dom}(S^*S) \text{ such that } S^*S u_\lambda = \lambda S u_\lambda, \tag{4}$$

and the solutions v_λ of (3) are in one-to-one correspondence with the solutions u_λ of (4) given by the pair of formulas

$$u_\lambda = (S_F)^{-1} S_K v_\lambda, \quad v_\lambda = \lambda^{-1} S u_\lambda. \tag{5}$$

As briefly recalled in Section 2, (4) represents an abstract buckling problem. The latter has been the key in all attempts to date in proving Weyl-type asymptotics for eigenvalues of S_K when S represents an elliptic partial differential operator in $L^2(\Omega)$. In fact, it is convenient to go one step further and replace the abstract buckling eigenvalue problem (4) by the variational formulation,

$$\begin{aligned} &\text{there exists } u_\lambda \in \operatorname{dom}(S)\backslash\{0\} \text{ such that} \\ &\mathfrak{a}(w, u_\lambda) = \lambda\, \mathfrak{b}(w, u_\lambda) \text{ for all } w \in \operatorname{dom}(S), \end{aligned} \tag{6}$$

where the symmetric forms $\mathfrak{a}$ and $\mathfrak{b}$ in $\mathcal{H}$ are defined by

$$\mathfrak{a}(f, g) := (Sf, Sg)_{\mathcal{H}}, \quad f, g \in \operatorname{dom}(\mathfrak{a}) := \operatorname{dom}(S), \tag{7}$$

$$\mathfrak{b}(f, g) := (f, Sg)_{\mathcal{H}}, \quad f, g \in \operatorname{dom}(\mathfrak{b}) := \operatorname{dom}(S). \tag{8}$$

In the present context of higher-order Krein Laplacians, the role of S will be played by the closure of the minimally defined operator in $L^2(\Omega)$,

$$A_{min,\Omega,m} := (-\Delta)^m, \quad \operatorname{dom}(A_{min,\Omega,m}) := C_0^\infty(\Omega), \tag{9}$$

under the assumption that $\emptyset \neq \Omega \subset \mathbb{R}^n$ has finite (Euclidean) volume ($|\Omega| < \infty$). This closure, $\overline{A_{min,\Omega,m}}$, is denoted by $A_{\Omega,m}$ and explicitly given by

$$A_{\Omega,m} = (-\Delta)^m, \quad \operatorname{dom}(A_{\Omega,m}) = \mathring{W}^{2m}(\Omega). \tag{10}$$

The Krein–von Neumann and Friedrichs extensions of $A_{\Omega,m}$ will then be denoted by $A_{K,\Omega,m}$ and $A_{F,\Omega,m}$, respectively. (To provide a quick example, we note that in the special case $m = n = 1$ and $\Omega = (a, b)$, $-\infty < a < b < \infty$, the boundary condition associated with $A_{K,(a,b),1}$ explicitly reads $v'(a) = v'(b) = [v(b) - v(a)]/(b-a)$, and that for $A_{F,(a,b),1}$ is of course the Dirichlet boundary condition $v(a) = v(b) = 0$.)

Since $A_{K,\Omega,m}$ has purely discrete spectrum in $(0,\infty)$ bounded away from zero by $\varepsilon > 0$ (cf. Theorem 2.3), let $\{\lambda_{K,\Omega,j}\}_{j\in\mathbb{N}} \subset (0,\infty)$ be the strictly positive eigenvalues of $A_{K,\Omega,m}$ enumerated in nondecreasing order, counting multiplicity, and let

$$N(\lambda, A_{K,\Omega,m}) := \#\{j \in \mathbb{N} \,|\, 0 < \lambda_{K,\Omega,j} < \lambda\}, \quad \lambda > 0, \tag{11}$$

be the eigenvalue distribution function for $A_{K,\Omega,m}$ (which takes into account only strictly positive eigenvalues of $A_{K,\Omega,m}$). The function $N(\,\cdot\,, A_{K,\Omega,m})$ is the principal object of this note. Similarly, $N(\lambda, A_{F,\Omega,m})$, $\lambda > 0$, denotes the eigenvalue counting function for $A_{F,\Omega,m}$.

In Section 2 we recall the basic abstract facts on the Friedrichs extension, S_F and the Krein–von Neumann extension S_K of a strictly positive, closed, symmetric operator S in a complex, separable Hilbert space $\mathcal{H}$ and describe the intimate link between the Krein–von Neumann extension and an underlying abstract buckling problem. Section 3 then focuses on the concrete case of higher-order Laplacians $(-\Delta)^m$, $m \in \mathbb{N}$, on open, finite (Euclidean) volume subsets $\Omega \subset \mathbb{R}^n$ (without imposing any constraints on Ω in the case where Ω is bounded) and derives the bound

$$N(\lambda, A_{K,\Omega,m}) \leqslant (2\pi)^{-n} v_n |\Omega| \{1 + [2m/(2m+n)]\}^{n/(2m)} \lambda^{n/(2m)}, \quad \lambda > 0, \tag{12}$$

where $v_n := \pi^{n/2}/\Gamma((n+2)/2)$ denotes the (Euclidean) volume of the unit ball in $\mathbb{R}^n$. We remark that the power law behavior $\lambda^{n/(2m)}$ coincides with the one in the known Weyl asymptotic behavior. This in itself is perhaps not surprising as it is *a priori* known that

$$N(\lambda, A_{K,\Omega,m}) \leqslant N(\lambda, A_{F,\Omega,m}), \quad \lambda > 0, \tag{13}$$

and $N(\lambda, A_{F,\Omega,m})$ is known to have the power law behavior $\lambda^{n/(2m)}$ (cf. (148), due to [36], which in turn extends the corresponding result in [38] in the case $m = 1$). We emphasize that (13) is not in conflict with variational eigenvalue estimates since $N(\lambda, A_{K,\Omega,m})$ only counts the strictly positive eigenvalues of $A_{K,\Omega,m}$ less than $\lambda > 0$ and hence avoids taking into account the (generally, infinite-dimensional) null space of $A_{K,\Omega,m}$. Rather than using known estimates for $N(\,\cdot\,, A_{F,\Omega,m})$ (cf., e.g., [11–16, 20, 21, 29, 30, 36, 38, 39, 42, 46–50, 54]), we will use the one-to-one correspondence of nonzero eigenvalues of $A_{K,\Omega,m}$ with the eigenvalues of its underlying buckling problem (cf. (3)–(5)) and estimate the eigenvalue counting function for the latter in Section 3. In our final Section 4 we briefly discuss the superiority of the buckling problem based bound (12) over the known estimates for $N(\,\cdot\,, A_{F,\Omega,m})$.

Since Weyl asymptotics for $N(\,\cdot\,, A_{K,\Omega,m})$ and $N(\,\cdot\,, A_{F,\Omega,m})$ are not considered in this paper we just refer to the monographs [37] and [51], but note that very detailed bibliographies on this subject appeared in [5] and [7]. At any rate, the best known result on Weyl asymptotics for $N(\,\cdot\,, A_{K,\Omega,m})$ to date is proven for bounded Lipschitz domains [9], whereas the estimate (12) assumes no regularity of Ω at all.

We conclude this introduction by summarizing the notation used in this paper. Throughout this paper, the symbol $\mathcal{H}$ is reserved to denote a separable complex Hilbert space with $(\,\cdot\,,\,\cdot\,)_{\mathcal{H}}$ the scalar product in $\mathcal{H}$ (linear in the second argument), and $I_{\mathcal{H}}$ the identity operator in $\mathcal{H}$. Next, let T be a linear operator mapping (a subspace of) a Banach space into another, with $\operatorname{dom}(T)$ and $\operatorname{ran}(T)$ denoting the domain and range of T. The closure of a closable operator S is denoted by $\overline{S}$. The kernel (null space) of T is denoted by $\ker(T)$. The spectrum, point spectrum (i.e., the set of eigenvalues), discrete spectrum, essential spectrum, and resolvent set of a closed linear operator in $\mathcal{H}$ will be denoted by $\sigma(\cdot)$, $\sigma_p(\cdot)$, $\sigma_d(\cdot)$, $\sigma_{ess}(\cdot)$, and $\rho(\cdot)$, respectively. The symbol $\mathrm{s}-\lim$ abbreviates the limit in the strong (i.e., pointwise) operator topology (we also use this symbol to describe strong limits in $\mathcal{H}$).

The Banach spaces of bounded and compact linear operators on $\mathcal{H}$ are denoted by $\mathcal{B}(\mathcal{H})$ and $\mathcal{B}_\infty(\mathcal{H})$, respectively. Similarly, the Schatten–von Neumann (trace) ideals will subsequently be denoted by $\mathcal{B}_p(\mathcal{H})$, $p \in (0,\infty)$. In addition, $U_1 \dotplus U_2$ denotes the direct sum of the subspaces U_1 and U_2 of a Banach space $\mathcal{X}$. Moreover, $\mathcal{X}_1 \hookrightarrow \mathcal{X}_2$ denotes the continuous embedding of the Banach space $\mathcal{X}_1$ into the Banach space $\mathcal{X}_2$.

The symbol $L^2(\Omega)$, with $\Omega \subseteq \mathbb{R}^n$ open, $n \in \mathbb{N}$, is a shortcut for $L^2(\Omega, d^n x)$, whenever the n-dimensional Lebesgue measure is understood. For brevity, the identity operator in $L^2(\Omega)$ will typically be denoted by I_Ω. The symbol $\mathcal{D}(\Omega)$ is reserved for the set of test functions $C_0^\infty(\Omega)$ on Ω, equipped with the standard inductive limit topology, and $\mathcal{D}'(\Omega)$ represents its dual space, the set of distributions in Ω. The cardinality of a set M is abbreviated by $\#(M)$. In addition, we define $\mathbb{N}_0 := \mathbb{N} \cup \{0\}$, so that $\mathbb{N}_0^n$ becomes the collection of all multi-indices with n components. As is customary, for each $\alpha = (\alpha_1, ..., \alpha_n) \in \mathbb{N}_0^n$ we denote by $|\alpha| := \alpha_1 + \cdots + \alpha_n$ the length of α, and set $\alpha! := \alpha_1! \cdots \alpha_n!$.

Moreover, $A \approx B$ signifies the existence of a finite constant $C \geqslant 1$, independent of the main parameters entering the quantities A, B, such that $C^{-1}A \leqslant B \leqslant CA$.

Finally, a notational comment: For obvious reasons, which have their roots in quantum mechanical applications, we will, with a slight abuse of notation, dub the expression $-\Delta = -\sum_{j=1}^n \partial_j^2$ (rather than Δ) as the "Laplacian" in this paper.

2. Basic Facts on the Krein–von Neumann Extension and the Associated Abstract Buckling Problem

In this preparatory section we recall the basic facts on the Krein–von Neumann extension of a strictly positive operator S in a complex, separable Hilbert space $\mathcal{H}$ and its associated abstract buckling problem as discussed in [5, 6]. For an extensive survey of this circle of ideas and an exhaustive list of references as well as pertinent historical comments we refer to [7].

To set the stage, we denote by S a linear, densely defined, symmetric (i.e., $S \subseteq S^*$), and closed operator in $\mathcal{H}$ throughout this section. We recall that S is called *nonnegative* provided $(f, Sf)_{\mathcal{H}} \geqslant 0$ for all $f \in \operatorname{dom}(S)$. The operator S is called *strictly positive*, if for some $\varepsilon > 0$ one has $(f, Sf)_{\mathcal{H}} \geqslant \varepsilon \|f\|_{\mathcal{H}}^2$ for all $f \in \operatorname{dom}(S)$; one then writes $S \geqslant \varepsilon I_{\mathcal{H}}$. Next, we recall that two nonnegative, self-adjoint operators A, B in $\mathcal{H}$ satisfy $A \leqslant B$ (in the sense of forms) if

$$\operatorname{dom}\big(B^{1/2}\big) \subset \operatorname{dom}\big(A^{1/2}\big) \tag{14}$$

and

$$\big(A^{1/2}f, A^{1/2}f\big)_{\mathcal{H}} \leqslant \big(B^{1/2}f, B^{1/2}f\big)_{\mathcal{H}}, \quad f \in \operatorname{dom}\big(B^{1/2}\big). \tag{15}$$

We also recall ([19, Section I.6], [31, Theorem VI.2.21]) that for A and B both self-adjoint and nonnegative in $\mathcal{H}$ one has

$$0 \leqslant A \leqslant B \ \text{ if and only if } \ (B + aI_{\mathcal{H}})^{-1} \leqslant (A + aI_{\mathcal{H}})^{-1} \ \text{ for all } \ a > 0. \tag{16}$$

Moreover, we note the useful fact that $\ker(A) = \ker(A^{1/2})$.

The following is a fundamental result to be found in M. Krein's celebrated 1947 paper [35] (cf. also Theorems 2 and 5–7 in the English summary on page 492):

Theorem 2.1. *Assume that S is a densely defined, closed, nonnegative operator in $\mathcal{H}$. Then, among all nonnegative self-adjoint extensions of S, there exist two distinguished ones, S_K and S_F, which are, respectively, the smallest and largest such extension (in the sense of (14)–(15)). Furthermore, a nonnegative self-adjoint operator $\widetilde{S}$ is a self-adjoint extension of S if and only if $\widetilde{S}$ satisfies*

$$S_K \leqslant \widetilde{S} \leqslant S_F. \tag{17}$$

In particular, (17) determines S_K and S_F uniquely. In addition, if $S \geqslant \varepsilon I_{\mathcal{H}}$ for some $\varepsilon > 0$, one has $S_F \geqslant \varepsilon I_{\mathcal{H}}$, and

$$\operatorname{dom}(S_F) = \operatorname{dom}(S) \dotplus (S_F)^{-1} \ker(S^*), \tag{18}$$

$$\operatorname{dom}(S_K) = \operatorname{dom}(S) \dotplus \ker(S^*), \tag{19}$$

$$\begin{aligned} \operatorname{dom}(S^*) &= \operatorname{dom}(S) \dotplus (S_F)^{-1} \ker(S^*) \dotplus \ker(S^*) \\ &= \operatorname{dom}(S_F) \dotplus \ker(S^*), \end{aligned} \tag{20}$$

and

$$\ker(S_K) = \ker\big((S_K)^{1/2}\big) = \ker(S^*) = \operatorname{ran}(S)^{\perp}. \tag{21}$$

One calls S_K the *Krein–von Neumann extension* of S and S_F the *Friedrichs extension* of S. We also recall that

$$S_F = S^*|_{\operatorname{dom}(S^*) \cap \operatorname{dom}((S_F)^{1/2})}. \tag{22}$$

Furthermore, if $S \geqslant \varepsilon I_{\mathcal{H}}$ then (19) implies

$$\ker(S_K) = \ker\big((S_K)^{1/2}\big) = \ker(S^*) = \operatorname{ran}(S)^{\perp}. \tag{23}$$

For abstract results regarding the parametrization of all nonnegative self-adjoint extensions of a given strictly positive, densely defined, symmetric operator we refer the reader to Krein [35], Višik [53], Birman [10], Grubb [24, 25], subsequent expositions due to Alonso and Simon [4], Faris [19, Sect. 15], and [27, Sect. 13.2], [52, Ch. 13], and Derkach and Malamud [18], Malamud [40], see also [23, Theorem 9.2].

Let us collect a basic assumption which will be imposed in the rest of this section.

Hypothesis 2.2. *Suppose that S is a densely defined, symmetric, closed operator with nonzero deficiency indices in $\mathcal{H}$ that satisfies $S \geqslant \varepsilon I_{\mathcal{H}}$ for some $\varepsilon > 0$.*

For subsequent purposes we note that under Hypothesis 2.2, one has

$$\dim\big(\ker(S^* - zI_{\mathcal{H}})\big) = \dim\big(\ker(S^*)\big), \quad z \in \mathbb{C}\backslash[\varepsilon, \infty). \tag{24}$$

We recall that two self-adjoint extensions S_1 and S_2 of S are called *relatively prime* (or *disjoint*) if $\operatorname{dom}(S_1) \cap \operatorname{dom}(S_2) = \operatorname{dom}(S)$. The following result will play a role later on (cf., e.g., [5, Lemma 2.8] for an elementary proof):

Lemma 2.1. *Suppose Hypothesis 2.2. Then the Friedrichs extension S_F and the Krein–von Neumann extension S_K of S are relatively prime, that is,*

$$\operatorname{dom}(S_F) \cap \operatorname{dom}(S_K) = \operatorname{dom}(S). \tag{25}$$

Next, we consider a self-adjoint operator T in $\mathcal{H}$ which is bounded from below, that is, $T \geqslant \alpha I_{\mathcal{H}}$ for some $\alpha \in \mathbb{R}$. We denote by $\{E_T(\lambda)\}_{\lambda \in \mathbb{R}}$ the family of strongly right-continuous spectral projections of T, and introduce for $-\infty \leqslant a < b$, as usual,

$$E_T\big((a,b)\big) = E_T(b_-) - E_T(a) \quad \text{and} \quad E_T(b_-) = \text{s} - \lim\nolimits_{\varepsilon \downarrow 0} E_T(b - \varepsilon). \tag{26}$$

In addition, we set

$$\mu_{T,j} := \inf\big\{\lambda \in \mathbb{R} \,\big|\, \dim(\operatorname{ran}(E_T((-\infty,\lambda)))) \geqslant j\big\}, \quad j \in \mathbb{N}. \tag{27}$$

Then, for fixed $k \in \mathbb{N}$, either:
(i) $\mu_{T,k}$ is the kth eigenvalue of T counting multiplicity below the bottom of the essential spectrum, $\sigma_{ess}(T)$, of T,
or,
(ii) $\mu_{T,k}$ is the bottom of the essential spectrum of T,

$$\mu_{T,k} = \inf\big\{\lambda \in \mathbb{R} \,\big|\, \lambda \in \sigma_{ess}(T)\big\}, \tag{28}$$

and in that case $\mu_{T,k+\ell} = \mu_{T,k}$, $\ell \in \mathbb{N}$, and there are at most $k-1$ eigenvalues (counting multiplicity) of T below $\mu_{T,k}$.

We now record a basic result of M. Krein [35] with an extension due to Alonso and Simon [4] and some additional results recently derived in [6]. For this purpose we introduce the *reduced Krein–von Neumann operator* $\widehat{S}_K$ in the Hilbert space

$$\widehat{\mathcal{H}} := \big(\ker(S^*)\big)^{\perp} = \big(\ker(S_K)\big)^{\perp} \tag{29}$$

by

$$\widehat{S}_K := P_{(\ker(S_K))^{\perp}} S_K|_{(\ker(S_K))^{\perp}}, \quad \mathrm{dom}(\widehat{S}_K) = \mathrm{dom} S_K \cap \widehat{\mathcal{H}}, \tag{30}$$

where $P_{(\ker(S_K))^{\perp}}$ denotes the orthogonal projection onto $(\ker(S_K))^{\perp}$. One then obtains

$$\big(\widehat{S}_K\big)^{-1} = P_{(\ker(S_K))^{\perp}} (S_F)^{-1}|_{(\ker(S_K))^{\perp}}, \tag{31}$$

a relation due to Krein [35, Theorem 26] (see also [40, Corollary 5]).

Theorem 2.3. *Suppose Hypothesis 2.2. Then*

$$\varepsilon \leqslant \mu_{S_F,j} \leqslant \mu_{\widehat{S}_K,j}, \quad j \in \mathbb{N}. \tag{32}$$

In particular, if the Friedrichs extension S_F of S has purely discrete spectrum, then, except possibly for $\lambda = 0$, the Krein–von Neumann extension S_K of S also has purely discrete spectrum in $(0, \infty)$, that is,

$$\sigma_{ess}(S_F) = \emptyset \text{ implies } \sigma_{ess}(S_K) \subseteq \{0\}. \tag{33}$$

In addition, if $p \in (0, \infty]$, then $(S_F - z_0 I_{\mathcal{H}})^{-1} \in \mathcal{B}_p(\mathcal{H})$ for some $z_0 \in \mathbb{C}\backslash[\varepsilon, \infty)$ implies

$$(S_K - zI_{\mathcal{H}})^{-1}|_{(\ker(S_K))^{\perp}} \in \mathcal{B}_p\big(\widehat{\mathcal{H}}\big) \text{ for all } z \in \mathbb{C}\backslash[\varepsilon, \infty). \tag{34}$$

In fact, the $\ell^p(\mathbb{N})$-based trace ideal $\mathcal{B}_p(\mathcal{H})$ $\big($resp., $\mathcal{B}_p(\widehat{\mathcal{H}})\big)$ of $\mathcal{B}(\mathcal{H})$ $\big($resp., $\mathcal{B}(\widehat{\mathcal{H}})\big)$ can be replaced by any two-sided symmetrically normed ideal of $\mathcal{B}(\mathcal{H})$ $\big($resp., $\mathcal{B}(\widehat{\mathcal{H}})\big)$.

We note that (33) is a classical result of Krein [35]. Apparently, (32) in the context of infinite deficiency indices was first proven by Alonso and Simon [4] by a somewhat different method. Relation (34) was proved in [6].

Assuming that S_F has purely discrete spectrum, let $\{\lambda_{K,j}\}_{j\in\mathbb{N}} \subset (0, \infty)$ be the strictly positive eigenvalues of S_K enumerated in nondecreasing order, counting multiplicity, and let

$$N(\lambda, S_K) := \#\{j \in \mathbb{N} \,|\, 0 < \lambda_{K,j} < \lambda\}, \quad \lambda > 0, \tag{35}$$

be the eigenvalue distribution function for S_K. Similarly, let $\{\lambda_{F,j}\}_{j\in\mathbb{N}} \subset (0, \infty)$ denote the eigenvalues of S_F, again enumerated in nondecreasing order, counting multiplicity, and by

$$N(\lambda, S_F) := \#\{j \in \mathbb{N} \,|\, \lambda_{F,j} < \lambda\}, \quad \lambda > 0, \tag{36}$$

the corresponding eigenvalue counting function for S_F. Then inequality (32) implies

$$N(\lambda, S_K) \leqslant N(\lambda, S_F), \quad \lambda > 0. \tag{37}$$

In particular, any estimate for the eigenvalue counting function for the Friedrichs extension S_F, in turn, yields one for the Krein–von Neumann extension S_K (focusing on strictly positive eigenvalues of S_K according to (35)). While this is a viable approach to estimate the eigenvalue counting function (35) for S_K, we will proceed along a different route in Section 3 and directly exploit the one-to-one correspondence between strictly positive eigenvalues of S_K and the eigenvalues of its underlying abstract buckling problem to be described next.

To describe the abstract buckling problem naturally associated with the Krein–von Neumann extension as described in [6], we start by introducing an abstract version of [26, Proposition 1] (see [6] for a proof):

Lemma 2.2. *Assume Hypothesis* 2.2 *and let* $\lambda \in \mathbb{C}\backslash\{0\}$. *Then there exists some* $f \in \operatorname{dom}(S_K)\backslash\{0\}$ *with*

$$S_K f = \lambda f \tag{38}$$

if and only if there exists $w \in \operatorname{dom}(S^*S)\backslash\{0\}$ *such that*

$$S^*Sw = \lambda Sw. \tag{39}$$

In fact, the solutions f *of* (38) *are in one-to-one correspondence with the solutions* w *of* (39) *as evidenced by the formulas*

$$w = (S_F)^{-1} S_K f, \tag{40}$$

$$f = \lambda^{-1} Sw. \tag{41}$$

Of course, since $S_K \geqslant 0$ *is self-adjoint, any* $\lambda \in \mathbb{C}\backslash\{0\}$ *in* (38) *and* (39) *necessarily satisfies* $\lambda \in (0, \infty)$.

It is the linear pencil eigenvalue problem $S^*Sw = \lambda Sw$ in (39) that we call the *abstract buckling problem* associated with the Krein–von Neumann extension S_K of S.

Next, we turn to a variational formulation of the correspondence between the inverse of the reduced Krein–von Neumann extension $\widehat{S}_K$ and the abstract buckling problem in terms of appropriate sesquilinear forms by following [32–34] in the elliptic PDE context. This will then lead to an even stronger connection between the Krein–von Neumann extension S_K of S and the associated abstract buckling eigenvalue problem (39), culminating in the unitary equivalence result in Theorem 2.4 below.

Given the operator S, we introduce the following symmetric forms in $\mathcal{H}$,

$$\mathfrak{a}(f,g) := (Sf, Sg)_{\mathcal{H}}, \quad f,g \in \operatorname{dom}(\mathfrak{a}) := \operatorname{dom}(S), \tag{42}$$

$$\mathfrak{b}(f,g) := (f, Sg)_{\mathcal{H}}, \quad f,g \in \operatorname{dom}(\mathfrak{b}) := \operatorname{dom}(S). \tag{43}$$

Then S being densely defined and closed implies that the sesquilinear form $\mathfrak{a}$ shares these properties, while $S \geqslant \varepsilon I_{\mathcal{H}}$ from Hypothesis 2.2 implies that $\mathfrak{a}$ is bounded from below, that is,

$$\mathfrak{a}(f, f) \geqslant \varepsilon^2 \|f\|_{\mathcal{H}}^2, \quad f \in \operatorname{dom}(S). \tag{44}$$

(The inequality (44) follows based on the assumption $S \geqslant \varepsilon I_{\mathcal{H}}$ by estimating $(Sf, Sg)_{\mathcal{H}} = \big([(S - \varepsilon I_{\mathcal{H}}) + \varepsilon I_{\mathcal{H}}]f, [(S - \varepsilon I_{\mathcal{H}}) + \varepsilon I_{\mathcal{H}}]g\big)_{\mathcal{H}}$ from below.)

Thus, one can introduce the Hilbert space

$$\mathcal{W} := \big(\operatorname{dom}(S), (\cdot, \cdot)_{\mathcal{W}}\big), \tag{45}$$

with associated scalar product

$$(f, g)_{\mathcal{W}} := \mathfrak{a}(f, g) = (Sf, Sg)_{\mathcal{H}}, \quad f, g \in \operatorname{dom}(S). \tag{46}$$

In addition, we note that $\iota_{\mathcal{W}} : \mathcal{W} \hookrightarrow \mathcal{H}$, the embedding operator of $\mathcal{W}$ into $\mathcal{H}$, is continuous due to $S \geqslant \varepsilon I_{\mathcal{H}}$. Hence, precise notation would be using

$$(w_1, w_2)_{\mathcal{W}} = \mathfrak{a}(\iota_{\mathcal{W}} w_1, \iota_{\mathcal{W}} w_2) = (S\iota_{\mathcal{W}} w_1, S\iota_{\mathcal{W}} w_2)_{\mathcal{H}}, \quad w_1, w_2 \in \mathcal{W}, \tag{47}$$

but in the interest of simplicity of notation we will omit the embedding operator $\iota_{\mathcal{W}}$ in the following.

With the sesquilinear forms $\mathfrak{a}$ and $\mathfrak{b}$ and the Hilbert space $\mathcal{W}$ as above, given $w_2 \in \mathcal{W}$, the map $\mathcal{W} \ni w_1 \mapsto (w_1, Sw_2)_{\mathcal{H}} \in \mathbb{C}$ is continuous. This allows us to define the operator Tw_2 as the unique element in $\mathcal{W}$ such that

$$(w_1, Tw_2)_{\mathcal{W}} = (w_1, Sw_2)_{\mathcal{H}} \text{ for all } w_1 \in \mathcal{W}. \tag{48}$$

This implies

$$\mathfrak{a}(w_1, Tw_2) = (w_1, Tw_2)_{\mathcal{W}} = (w_1, Sw_2)_{\mathcal{H}} = \mathfrak{b}(w_1, w_2) \tag{49}$$

for all $w_1, w_2 \in \mathcal{W}$. In addition, the operator T satisfies

$$0 \leqslant T = T^* \in \mathcal{B}(\mathcal{W}) \quad \text{and} \quad \|T\|_{\mathcal{B}(\mathcal{W})} \leqslant \varepsilon^{-1}. \tag{50}$$

We will call T the *abstract buckling problem operator* associated with the Krein–von Neumann extension S_K of S.

Next, recalling the notation $\widehat{\mathcal{H}} = \big(\ker(S^*)\big)^{\perp}$ (cf. (29)), we introduce the operator

$$\widehat{S} : \mathcal{W} \to \widehat{\mathcal{H}}, \quad w \mapsto Sw. \tag{51}$$

Clearly, $\operatorname{ran}\big(\widehat{S}\big) = \operatorname{ran}(S)$ and since $S \geqslant \varepsilon I_{\mathcal{H}}$ for some $\varepsilon > 0$ and S is closed in $\mathcal{H}$, $\operatorname{ran}(S)$ is also closed, and hence coincides with $\big(\ker(S^*)\big)^{\perp}$. This yields

$$\operatorname{ran}\big(\widehat{S}\big) = \operatorname{ran}(S) = \widehat{\mathcal{H}}. \tag{52}$$

In fact, it follows that $\widehat{S} \in \mathcal{B}(\mathcal{W}, \widehat{\mathcal{H}})$ maps $\mathcal{W}$ unitarily onto $\widehat{\mathcal{H}}$ (cf. [G]).

Continuing, we briefly recall the polar decomposition of S,

$$S = U_S|S|, \tag{53}$$

where, with $\varepsilon > 0$ as in Hypothesis 2.2,

$$|S| = (S^*S)^{1/2} \geqslant \varepsilon I_{\mathcal{H}} \text{ and } U_S \in \mathcal{B}\big(\mathcal{H}, \widehat{\mathcal{H}}\big) \text{ unitary.} \tag{54}$$

Then the principal unitary equivalence result proved in [6] reads as follows:

Theorem 2.4. *Assume Hypothesis 2.2. Then the inverse of the reduced Krein–von Neumann extension $\widehat{S}_K$ in $\widehat{\mathcal{H}}$ and the abstract buckling problem operator T in $\mathcal{W}$ are unitarily equivalent. Specifically,*

$$\big(\widehat{S}_K\big)^{-1} = \widehat{S}T\big(\widehat{S}\big)^{-1}. \tag{55}$$

In particular, the nonzero eigenvalues of S_K are reciprocals of the eigenvalues of T. Moreover, one has

$$\big(\widehat{S}_K\big)^{-1} = U_S\big[|S|^{-1}S|S|^{-1}\big](U_S)^{-1}, \tag{56}$$

where $U_S \in \mathcal{B}\big(\mathcal{H}, \widehat{\mathcal{H}}\big)$ is the unitary operator in the polar decomposition (53) of S and the operator $|S|^{-1}S|S|^{-1} \in \mathcal{B}(\mathcal{H})$ is self-adjoint and strictly positive in $\mathcal{H}$.

We emphasize that the unitary equivalence in (55) is independent of any spectral assumptions on S_K (such as the spectrum of S_K consists of eigenvalues only) and applies to the restrictions of S_K to its pure point, absolutely continuous, and singularly continuous spectral subspaces, respectively.

Equation (56) is motivated by rewriting the abstract linear pencil buckling eigenvalue problem (39), $S^*Sw = \lambda Sw$, $\lambda \in \mathbb{C}\backslash\{0\}$, in the form

$$|S|^{-1}Sw = (S^*S)^{-1/2}Sw = \lambda^{-1}(S^*S)^{1/2}w = \lambda^{-1}|S|w \tag{57}$$

and hence in the form of a standard eigenvalue problem

$$|S|^{-1}S|S|^{-1}v = \lambda^{-1}v, \quad \lambda \in \mathbb{C}\backslash\{0\}, \quad v := |S|w. \tag{58}$$

Again, self-adjointness and strict positivity of $|S|^{-1}S|S|^{-1}$ imply $\lambda \in (0, \infty)$.

We conclude this section with an elementary result recently noted in [9] that relates the nonzero eigenvalues of S_K directly with the sesquilinear forms $\mathfrak{a}$ and $\mathfrak{b}$:

Lemma 2.3. *Assume Hypothesis 2.2 and introduce*

$$\sigma_p(\mathfrak{a}, \mathfrak{b}) := \big\{\lambda \in \mathbb{C} \,\big|\, \text{there exists } g_\lambda \in \operatorname{dom}(S)\backslash\{0\} \text{ such that } \mathfrak{a}(f, g_\lambda) = \lambda\, \mathfrak{b}(f, g_\lambda), \quad f \in \operatorname{dom}(S)\big\}. \tag{59}$$

Then

$$\sigma_p(\mathfrak{a}, \mathfrak{b}) = \sigma_p(S_K)\backslash\{0\} \tag{60}$$

(counting multiplicity), in particular, $\sigma_p(\mathfrak{a}, \mathfrak{b}) \subset (0, \infty)$, *and* $g_\lambda \in \mathrm{dom}(S)\backslash\{0\}$ *in* (59) *actually satisfies*

$$g_\lambda \in \mathrm{dom}(S^*S), \quad S^*Sg_\lambda = \lambda S g_\lambda. \tag{61}$$

In addition,

$$\lambda \in \sigma_p(\mathfrak{a}, \mathfrak{b}) \text{ if and only if } \lambda^{-1} \in \sigma_p(T) \tag{62}$$

(counting multiplicity). Finally,

$$T \in \mathcal{B}_\infty(\mathcal{W}) \iff \big(\widehat{S}_K\big)^{-1} \in \mathcal{B}_\infty\big(\widehat{\mathcal{H}}\big) \iff \sigma_{ess}(S_K) \subseteq \{0\}, \tag{63}$$

and hence,

$$\sigma_p(\mathfrak{a}, \mathfrak{b}) = \sigma(S_K)\backslash\{0\} = \sigma_d(S_K)\backslash\{0\} \tag{64}$$

if (63) *holds. In particular, if one of* S_F *or* $|S|$ *has purely discrete spectrum (i.e.,* $\sigma_{ess}(S_F) = \emptyset$ *or* $\sigma_{ess}(|S|) = \emptyset$*), then* (63) *and* (64) *hold.*

Proof. We begin by noting that (44) and the fact that $\mathfrak{b}(f, f) \geqslant \varepsilon\|f\|^2_{\mathcal{H}}$ imply $\sigma_p(\mathfrak{a}, \mathfrak{b}) \subset (0, \infty)$. Moreover, using the fact that the self-adjoint operator in $\mathcal{H}$ uniquely associated with the form $\mathfrak{a}$ is given by S^*S (cf. [31, Example VI.2.13]), and that $\mathfrak{a}(f, g_\lambda) = \lambda\, \mathfrak{b}(f, g_\lambda) = \lambda(f, Sg_\lambda)_{\mathcal{H}}$, $f \in \mathrm{dom}(\mathfrak{a}) = \mathrm{dom}(S)$, the first representation theorem for quadratic forms (cf. [31, Theorem VI.2.1 (iii)]) implies (61). An application of Lemma 2.2 then yields (60). Relation (62) then follows from (60) and (55). The first equivalence in (63) again is a consequence of (55) and the fact that $\widehat{S}$ maps $\mathcal{W}$ unitarily onto $\widehat{\mathcal{H}}$; the second equivalence in (63) follows from (30). The final claim in Lemma 2.3 involving discrete spectra of S_F or $|S|$ is a consequence of (33) or (56) and the equivalence statements in (63). □

One notices that $f \in \mathrm{dom}(S)$ in the definition (59) of $\sigma_p(\mathfrak{a}, \mathfrak{b})$ can be replaced by $f \in C(S)$ for any (operator) core $C(S)$ for S (equivalently, by any form core for the form $\mathfrak{a}$).

3. An Upper Bound for the Eigenvalue Counting Function for Higher-Order Krein Laplacians on Finite Volume Domains

In this section we derive an upper bound for the eigenvalue counting function for higher-order Krein Laplacians on open, nonempty domains $\Omega \subset \mathbb{R}^n$ of finite (Euclidean) volume. In particular, no assumptions on the boundary of Ω will be made.

Before introducing the class of constant coefficient partial differential operators in $L^2(\Omega)$ at hand, we recall a few auxiliary facts to be used in the proof of Theorem 3.4.

Lemma 3.1. *Suppose that S is a densely defined, symmetric, closed operator in $\mathcal{H}$. Then $|S|$ and hence S is infinitesimally bounded with respect to S^*S, more precisely, one has*

$$\text{for all } \varepsilon > 0,\ \|Sf\|_{\mathcal{B}(\mathcal{H})} = \||S|f\|_{\mathcal{B}(\mathcal{H})} \leqslant \varepsilon \|S^*Sf\|_{\mathcal{H}}^2 + (4\varepsilon)^{-1}\|f\|_{\mathcal{H}}^2, \quad f \in \operatorname{dom}(S^*S). \tag{65}$$

In addition, S is relatively compact with respect to S^*S if $|S|$, or equivalently, S^*S, has compact resolvent. In particular,

$$\sigma_{ess}(S^*S - \lambda S) = \sigma_{ess}(S^*S), \quad \lambda \in \mathbb{R}. \tag{66}$$

Proof. Employing the polar decomposition of S, $S = U_S|S|$, where U_S is a partial isometry and $|S| = (S^*S)^{1/2}$ (cf. [31, Sect. VI.2.7]), one obtains

$$\|Sf\|_{\mathcal{B}(\mathcal{H})} = \||S|f\|_{\mathcal{B}(\mathcal{H})}, \quad f \in \operatorname{dom}(S) = \operatorname{dom}(|S|), \tag{67}$$

and hence the spectral theorem applied to $|S|$, together with the elementary inequality $\lambda \leqslant \varepsilon\lambda^2 + (4\varepsilon)^{-1}$, $\varepsilon > 0$, $\lambda \geqslant 0$, proves inequality (65).

The relative compactness assertion then follows from

$$S(S^*S + I_{\mathcal{H}})^{-1} = \Big[S\big(|S|^2 + I_{\mathcal{H}}\big)^{-1/2}\Big]\big(|S|^2 + I_{\mathcal{H}}\big)^{-1/2} \in \mathcal{B}_\infty(\mathcal{H}), \tag{68}$$

since $S\big(|S|^2 + I_{\mathcal{H}}\big)^{-1/2} \in \mathcal{B}(\mathcal{H})$ and $\big(|S|^2 + I_{\mathcal{H}}\big)^{-1/2} \in \mathcal{B}_\infty(\mathcal{H})$. □

Given a lower semibounded, self-adjoint operator $T \geqslant c_T I_{\mathcal{H}}$ in $\mathcal{H}$, we denote by q_T its uniquely associated form, that is,

$$\mathfrak{q}_T(f,g) = \big(|T|^{1/2}f, \operatorname{sgn}(T)|T|^{1/2}\big)_{\mathcal{H}}, \quad f,g \in \operatorname{dom}(\mathfrak{q}) = \operatorname{dom}\big(|T|^{1/2}\big), \tag{69}$$

and by $\{E_T(\lambda)\}_{\lambda\in\mathbb{R}}$ the family of spectral projections of T. We recall the following well-known variational characterization of dimensions of spectral projections $E_T([c_T,\mu))$, $\mu > c_T$.

Lemma 3.2. *Assume that $c_T I_{\mathcal{H}} \leqslant T$ is self-adjoint in $\mathcal{H}$ and $\mu > c_T$. Suppose that $\mathcal{F} \subset \operatorname{dom}\big(|T|^{1/2}\big)$ is a linear subspace such that*

$$\mathfrak{q}_T(f,f) < \mu\|f\|_{\mathcal{H}}^2, \quad f \in \mathcal{F}\backslash\{0\}. \tag{70}$$

Then,

$$\dim\big(\operatorname{ran}(E_T([c_T,\mu)))\big) = \sup_{\mathcal{F}\subset\operatorname{dom}(|T|^{1/2})} (\dim(\mathcal{F})). \tag{71}$$

We add the following elementary observation: Let $c \in \mathbb{R}$ and $B \geqslant cI_{\mathcal{H}}$ be a self-adjoint operator in $\mathcal{H}$, and introduce the sesquilinear form b in $\mathcal{H}$ associated with B via

$$\begin{aligned} &b(u,v) = \big((B - cI_{\mathcal{H}})^{1/2}u, (B - cI_{\mathcal{H}})^{1/2}v\big)_{\mathcal{H}} + c(u,v)_{\mathcal{H}}, \\ &u,v \in \operatorname{dom}(b) = \operatorname{dom}\big(|B|^{1/2}\big). \end{aligned} \tag{72}$$

Given B and b, one introduces the Hilbert space $\mathcal{H}_b \subseteq \mathcal{H}$ by

$$\begin{aligned}
\mathcal{H}_b &= \big(\mathrm{dom}\big(|B|^{1/2}\big), (\cdot,\cdot)_{\mathcal{H}_b}\big), \\
(u,v)_{\mathcal{H}_b} &= b(u,v) + (1-c)(u,v)_{\mathcal{H}} \\
&= \big((B - cI_{\mathcal{H}})^{1/2}u, (B - cI_{\mathcal{H}})^{1/2}v\big)_{\mathcal{H}} + (u,v)_{\mathcal{H}} \\
&= \big((B + (1-c)I_{\mathcal{H}})^{1/2}u, (B + (1-c)I_{\mathcal{H}})^{1/2}v\big)_{\mathcal{H}}.
\end{aligned} \tag{73}$$

One observes that

$$(B + (1-c)I_{\mathcal{H}})^{1/2} \colon \mathcal{H}_b \to \mathcal{H} \text{ is unitary.} \tag{74}$$

Lemma 3.3 (see, e.g., [22]). *Let $\mathcal{H}$, B, b, and $\mathcal{H}_b$ be as in (72)–(74). Then B has purely discrete spectrum, that is, $\sigma_{\mathrm{ess}}(B) = \emptyset$, if and only if $\mathcal{H}_b$ embeds compactly into $\mathcal{H}$.*

Next we turn to higher-order Laplacians $(-\Delta)^m$ in $L^2(\Omega)$ and hence introduce the following assumptions on $\Omega \subset \mathbb{R}^n$, $n \in \mathbb{N}$.

Hypothesis 3.1. *Let $n \in \mathbb{N}$ and assume that $\emptyset \neq \Omega \subset \mathbb{R}^n$ is an open set of finite (Euclidean) volume (denoted by $|\Omega| < \infty$).*

The above hypothesis is going to be relevant for the validity of the compact embedding of $\mathring{W}^1(\Omega)$ into $L^2(\Omega)$. Necessary and sufficient conditions for this compact embedding to hold in terms of appropriate capacities can be found, for instance, in [2], [3, Ch. 6], [41, Ch. 6]. Since we seek such an embedding for finite-volume do mains, and the precise statement appears to be difficult to discern from the existing literature on this subject, we decided to spell out the details for the convenience of the reader. In fact, for completeness, we will discuss a more general result in connection with L^p-based Sobolev spaces which we introduce next.

Suppose that $\emptyset \neq \Omega \subseteq \mathbb{R}^n$ is open and, for each $p \in [1,\infty]$ and $k \in \mathbb{N}$, define

$$W^{k,p}(\Omega) := \big\{u \in L^1_{\mathrm{loc}}(\Omega) \,\big|\, \partial^\alpha u \in L^p(\Omega) \text{ if } 0 \leqslant |\alpha| \leqslant k\big\}, \tag{75}$$

where the derivatives are taken in the sense of distributions. This becomes a Banach space when endowed with the natural norm,

$$\|u\|_{W^{k,p}(\Omega)} := \sum_{0 \leqslant |\alpha| \leqslant k} \|\partial^\alpha u\|_{L^p(\Omega)} \tag{76}$$

(see also (106), but for the purpose at hand we now prefer to use (76)). In the same setting we also consider the closed linear subspace $\mathring{W}^{k,p}(\Omega)$ of $W^{k,p}(\Omega)$ given by

$$\mathring{W}^{k,p}(\Omega) := \overline{C_0^\infty(\Omega)}^{W^{k,p}(\Omega)}. \tag{77}$$

A useful observation, seen directly from definitions, is that whenever $\alpha \in \mathbb{N}_0^n$ is such that $|\alpha| \leqslant k-1$ then

$$\begin{gathered}\partial^\alpha : \mathring{W}^{k,p}(\Omega) \longrightarrow \mathring{W}^{k-|\alpha|,p}(\Omega) \text{ is a well-defined,}\\ \text{linear, and bounded operator, with norm} \leqslant 1.\end{gathered} \tag{78}$$

Our goal is to prove the following compact embedding result.

Theorem 3.2. *Assume Hypothesis* 3.1 *and pick* $p \in [1,\infty]$. *Let* p_* *be an arbitrary number in* $[1,\infty)$ *if* $p \geqslant n$, *and suppose that* $p_* \in \big[1, \frac{np}{n-p}\big)$ *if* $1 \leqslant p < n$. *Then*

$$\mathring{W}^{1,p}(\Omega) \hookrightarrow L^{p_*}(\Omega) \ \textit{compactly.} \tag{79}$$

Proof. To start the proof, we note that since Ω has finite measure, the scale of Lebesgue spaces in Ω is nested. Specifically, Hölder's inequality implies

$$L^{q_1}(\Omega) \hookrightarrow L^{q_2}(\Omega) \text{ continuously if } 0 < q_2 \leqslant q_1 \leqslant \infty. \tag{80}$$

We continue by recalling a useful result from [17]. Given any $p_1 \in (1,\infty]$ and $p_2 \in [1,\infty)$, define the space

$$E^{p_1,p_2}(\Omega) := \big\{u \in L^{p_1}(\Omega) \,\big|\, \partial_j u \in L^{p_2}(\Omega) \ \text{for} \ 1 \leqslant j \leqslant n\big\}, \tag{81}$$

and equip it with the natural norm

$$\|u\|_{E^{p_1,p_2}(\Omega)} := \|u\|_{L^{p_1}(\Omega)} + \sum_{j=1}^n \|\partial_j u\|_{L^{p_2}(\Omega)}. \tag{82}$$

Then a particular version of [17, Theorem 2.2, p. 27] implies that under the current assumptions on Ω,

$$E^{p_1,p_2}(\Omega) \hookrightarrow L^{p_3}(\Omega) \text{ compactly, for each } p_3 \in [1,p_1). \tag{83}$$

Next, we denote by tilde the operator of extension by zero of functions defined in Ω to the entire Euclidean space $\mathbb{R}^n$. Since for every $\varphi \in C_0^\infty(\Omega)$ we have that $\widetilde{\varphi} \in C_0^\infty(\mathbb{R}^n) \subset W^{1,p}(\mathbb{R}^n)$ and $\|\widetilde{\varphi}\|_{W^{1,p}(\mathbb{R}^n)} = \|\varphi\|_{W^{1,p}(\Omega)}$, it follows from (77) that tilde induces a mapping

$$\mathring{W}^{1,p}(\Omega) \ni u \mapsto \widetilde{u} \in W^{1,p}(\mathbb{R}^n) \text{ which is a linear isometry.} \tag{84}$$

Bearing this in mind, in the case when $1 \leqslant p < n$, for each $u \in \mathring{W}^{1,p}(\Omega)$ we may use the classical Sobolev embedding theorem (in $\mathbb{R}^n$) in order to estimate

$$\|u\|_{L^{\frac{np}{n-p}}(\Omega)} = \|\widetilde{u}\|_{L^{\frac{np}{n-p}}(\mathbb{R}^n)} \leqslant C_{n,p}\|\widetilde{u}\|_{W^{1,p}(\mathbb{R}^n)} = C_{n,p}\|u\|_{W^{1,p}(\Omega)}, \tag{85}$$

for some finite constant $C_{n,p} > 0$. This proves that

$$\mathring{W}^{1,p}(\Omega) \hookrightarrow L^{\frac{np}{n-p}}(\Omega) \ \text{continuously, if} \ 1 \leqslant p < n. \tag{86}$$

We divide the remaining portion of the proof into two cases.

Case 1: *We claim that* (79) *holds when* $1 \leqslant p < n$ *and* $p_* \in \big[1, \frac{np}{n-p}\big)$. In this scenario, pick $p_1 \in \big(p_*, \frac{np}{n-p}\big]$ and $p_2 \in [1, p]$. This choice entails

$$E^{p_1,p_2}(\Omega) \hookrightarrow L^{p_*}(\Omega) \text{ compactly,} \tag{87}$$

by (83), and

$$\mathring{W}^{1,p}(\Omega) \hookrightarrow E^{p_1,p_2}(\Omega) \text{ continuously,} \tag{88}$$

by (86) and (suitable applications of) (80). Collectively, (87) and (88) yield (79) in this case.

Case 2: *We claim that* (79) *holds when* $n \leqslant p \leqslant \infty$ *and* $p_* \in [1, \infty)$. To justify this claim, pick an arbitrary $q \in [1, n)$. In particular, $q < p$, so (80) yields that on one hand,

$$\mathring{W}^{1,p}(\Omega) \hookrightarrow \mathring{W}^{1,q}(\Omega) \text{ continuously.} \tag{89}$$

On the other hand, by what has already proved in Case 1,

$$\mathring{W}^{1,q}(\Omega) \hookrightarrow L^{q_*}(\Omega) \text{ compactly, for each } q_* \in \big[1, \tfrac{nq}{n-q}\big). \tag{90}$$

Combining (89) with (90) and observing that $\frac{nq}{n-q} \nearrow \infty$ as $q \nearrow n$, one ultimately deduces that (79) holds in this case as well. □

Theorem 3.2 has two notable consequences, recorded below. The first such corollary deals with the following compactness result.

Corollary 3.1. *Assume Hypothesis* 3.1. *Then for each* $k \in \mathbb{N}$ *and* $p \in [1, \infty)$, *it follows that*

$$\mathring{W}^{k,p}(\Omega) \hookrightarrow L^p(\Omega) \text{ compactly.} \tag{91}$$

Proof. This is an immediate consequence of Theorem 3.2, keeping in mind that $\mathring{W}^{k,p}(\Omega) \hookrightarrow \mathring{W}^{1,p}(\Omega)$ continuously and that $p < \frac{np}{n-p}$ when $1 \leqslant p < n$. □

The second corollary of Theorem 3.2 alluded to earlier deals with a Poincaré-type inequality.

Corollary 3.2. *Assume Hypothesis* 3.1. *Then for each* $k \in \mathbb{N}$ *and* $p \in [1, \infty)$, *there exists a constant* $C \in (0, \infty)$ *with the property that the following Poincaré-type inequality holds:*

$$\sum_{0 \leqslant |\beta| \leqslant k-1} \|\partial^\beta u\|_{L^p(\Omega)} \leqslant C \sum_{|\alpha|=k} \|\partial^\alpha u\|_{L^p(\Omega)}, \quad u \in \mathring{W}^{k,p}(\Omega). \tag{92}$$

Proof. We shall prove (92) by induction on $k \in \mathbb{N}$.

Step 1: *We claim that* (92) *holds when* $k = 1$*, that is, there exists* $C \in (0, \infty)$ *such that*

$$\|u\|_{L^p(\Omega)} \leqslant C\|\nabla u\|_{[L^p(\Omega)]^n}, \quad u \in \mathring{W}^{1,p}(\Omega). \tag{93}$$

Seeking a contradiction, assume that there exists a sequence $\{u_j\}_{j\in\mathbb{N}} \subset \mathring{W}^{1,p}(\Omega)$ with the property that

$$\|u_j\|_{L^p(\Omega)} > j\,\|\nabla u_j\|_{[L^p(\Omega)]^n}, \quad j \in \mathbb{N}. \tag{94}$$

For each $j \in \mathbb{N}$ define

$$v_j := \frac{u_j}{\|u_j\|_{L^p(\Omega)}} \in \mathring{W}^{1,p}(\Omega). \tag{95}$$

Note that

$$\|v_j\|_{L^p(\Omega)} = 1 \text{ for every } j \in \mathbb{N}, \tag{96}$$

and $\nabla v_j = \nabla u_j / \|u_j\|_{L^p(\Omega)}$ which, in light of (94), implies

$$\|\nabla v_j\|_{L^p(\Omega)} = \frac{\|\nabla u_j\|_{L^p(\Omega)}}{\|u_j\|_{L^p(\Omega)}} < j^{-1} \text{ for every } j \in \mathbb{N}. \tag{97}$$

From (95)–(97) it follows that $\{v_j\}_{j\in\mathbb{N}}$ is a bounded sequence in $\mathring{W}^{1,p}(\Omega)$. Granted this fact, Corollary 3.1 applies and yields the existence of a strictly increasing sequence $\{j_\ell\}_{\ell\in\mathbb{N}} \subseteq \mathbb{N}$ along with some function $v \in L^p(\Omega)$ with the property that

$$v_{j_\ell} \underset{\ell\to\infty}{\longrightarrow} v \text{ in } L^p(\Omega). \tag{98}$$

As a consequence of this and (96) we deduce that

$$\|v\|_{L^p(\Omega)} = 1. \tag{99}$$

Next, we recall that tilde denotes the operator of extension by zero of functions defined in Ω to the entire Euclidean space $\mathbb{R}^n$. In particular, in the sense of distributions,

$$\partial_m \widetilde{w} = \widetilde{\partial_m w} \text{ for each } w \in \mathring{W}^{1,p}(\Omega) \text{ and } m \in \{1, \dots, n\}. \tag{100}$$

Note that $\widetilde{v_j} \in W^{1,p}(\mathbb{R}^n)$ for each $j \in \mathbb{N}$, $\widetilde{v} \in L^p(\mathbb{R}^n)$, and $\widetilde{v_{j_\ell}} \to \widetilde{v}$ in $L^p(\mathbb{R}^n)$ as $\ell \to \infty$. As such, for each test function $\phi \in C_0^\infty(\mathbb{R}^n)$ and each $m \in \{1, \dots, n\}$ we may write

$$\begin{aligned}
&\big|{}_{\mathcal{D}'(\mathbb{R}^n)}\langle \partial_m \widetilde{v}, \phi\rangle_{\mathcal{D}(\mathbb{R}^n)}\big| = \big|{}_{\mathcal{D}'(\mathbb{R}^n)}\langle \widetilde{v}, \partial_m \phi\rangle_{\mathcal{D}(\mathbb{R}^n)}\big| = \left|\int_{\mathbb{R}^n} \widetilde{v}(x)(\partial_m\phi)(x)\,d^n x\right| \\
&= \left|\lim_{\ell\to\infty}\int_{\mathbb{R}^n} \widetilde{v_{j_\ell}}(x)(\partial_m\phi)(x)\,d^n x\right| = \left|\lim_{\ell\to\infty}\int_{\mathbb{R}^n} (\widetilde{\partial_m v_{j_\ell}})(x)\phi(x)\,d^n x\right| \\
&\leqslant \limsup_{\ell\to\infty} \left\|(\widetilde{\partial_m v_{j_\ell}})\right\|_{L^p(\mathbb{R}^n)} \|\phi\|_{L^{p'}(\mathbb{R}^n)} \\
&\leqslant \|\phi\|_{L^{p'}(\mathbb{R}^n)} \limsup_{\ell\to\infty} \left\|\nabla v_{j_\ell}\right\|_{L^p(\Omega)} = 0,
\end{aligned} \tag{101}$$

by (100), Hölder's inequality (with p' denoting the conjugate exponent of p), and (97). Here, and elsewhere, ${}_{\mathcal{D}(\Omega)'}\langle\,\cdot\,,\,\cdot\,\rangle_{\mathcal{D}(\Omega)}$ is the standard distributional pairing, with $\mathcal{D}(\Omega) := C_0^\infty(\Omega)$ equipped with the usual inductive limit topology.

In turn, the estimate (101) proves (cf., e.g., [45, Ch. 2]) that there exists a constant $c \in \mathbb{R}$ such that $\widetilde{v} = c$ a.e. in $\mathbb{R}^n$. In fact, from (99) we see that, necessarily, $\widetilde{v} = |\Omega|^{-1/p}$ a.e. in $\mathbb{R}^n$ which then contradicts the fact that $\widetilde{v} = 0$ in $\mathbb{R}^n \setminus \Omega$, given that $|\mathbb{R}^n \setminus \Omega| = \infty$. This contradiction establishes (93) and finishes the proof of Step 1.

Step 2: *We claim that if* (92) *holds for some* $k \in \mathbb{N}$, *then its version written for* $k+1$ *in place of* k *is also true.* To see that this is the case, assume that k is as above and pick an arbitrary $u \in \mathring{W}^{k+1,p}(\Omega)$. Since for each $j \in \{1,\dots,n\}$, (78) implies that $\partial_j u \in \mathring{W}^{k,p}(\Omega)$, with $\|\partial_j u\|_{W^{k,p}(\Omega)} \leqslant \|u\|_{W^{k+1,p}(\Omega)}$, the induction hypothesis guarantees the existence of some $C \in (0,\infty)$ independent of u such that

$$\sum_{0\leqslant|\beta|\leqslant k-1} \|\partial^\beta(\partial_j u)\|_{L^p(\Omega)} \leqslant C \sum_{|\alpha|=k} \|\partial^\alpha(\partial_j u)\|_{L^p(\Omega)}. \tag{102}$$

Summing over $j \in \{1,\dots,n\}$ and adjusting notation then yields

$$\sum_{1\leqslant|\gamma|\leqslant k} \|\partial^\gamma u\|_{L^p(\Omega)} \leqslant C \sum_{|\alpha|=k+1} \|\partial^\alpha u\|_{L^p(\Omega)}, \tag{103}$$

for a possibly different constant $C \in (0,\infty)$ which is nonetheless independent of u. Together with (93) this proves that

$$\sum_{0\leqslant|\gamma|\leqslant k} \|\partial^\gamma u\|_{L^p(\Omega)} \leqslant C \sum_{|\alpha|=k+1} \|\partial^\alpha u\|_{L^p(\Omega)}, \quad u \in \mathring{W}^{k+1,p}(\Omega). \tag{104}$$

This completes the treatment of Step 2 and hence finishes the proof. □

In the remainder of the paper we are going to concern ourselves exclusively with the L^2-based Sobolev space $W^{k,2}(\Omega)$. As such, we agree to drop the dependence on the integrability exponent and simply write $W^k(\Omega)$. Hence,

$$W^k(\Omega) = \big\{u \in L^1_{\mathrm{loc}}(\Omega) \,\big|\, \partial^\alpha u \in L^2(\Omega) \text{ if } 0 \leqslant |\alpha| \leqslant k\big\}, \quad k \in \mathbb{N}_0, \tag{105}$$

with $\alpha \in \mathbb{N}_0^n$ and $\partial^\alpha u$ denoting weak derivatives of u. The space $W^k(\Omega)$ is endowed with the norm

$$\|u\|_{k,\Omega} = \sum_{0\leqslant|\alpha|\leqslant k} \big\|\partial^\alpha u\big\|_{L^2(\Omega)}, \quad u \in W^k(\Omega). \tag{106}$$

In addition, define

$$\mathring{W}^k(\Omega) = \overline{C_0^\infty(\Omega)}^{W^k(\Omega)}, \quad k \in \mathbb{N}_0, \tag{107}$$

and note that $\mathring{W}^k(\Omega)$ is a closed linear subspace of $W^k(\Omega)$. Granted Hypothesis 3.1, Corollary 3.2 then implies the Poincaré-type inequality

$$|||u|||_{\ell,\Omega} \leqslant C\, |||\, u\, |||_{k,\Omega}, \quad u \in \mathring{W}^p(\Omega),\ \ell \in \mathbb{N}_0,\ k \in \mathbb{N},\ \ell \leqslant k, \tag{108}$$

where we introduced the abbreviation

$$|||u|||_{k,\Omega} := \sum_{|\alpha|=k} \|\partial^\alpha u\|_{L^2(\Omega)}, \quad u \in W^k(\Omega),\ k \in \mathbb{N}_0. \tag{109}$$

Thus, $|||u|||_{k,\Omega}$, $u \in W^k(\Omega)$, represents an equivalent norm on $\mathring{W}^k(\Omega)$.

We proceed with the following useful identity:

Lemma 3.4. *Let $k \in \mathbb{N}$ and assume $\emptyset \neq \Omega \subseteq \mathbb{R}^n$ is open. Then,*

$$\sum_{|\alpha|=k} \frac{k!}{\alpha!} \int_\Omega \big|(\partial^\alpha \phi)(x)\big|^2 \, d^n x = \int_\Omega \overline{(\Delta^k \phi)(x)}\, \phi(x)\, d^n x, \quad \phi \in C_0^\infty(\Omega). \tag{110}$$

Proof. Pick an arbitrary $\phi \in C_0^\infty(\Omega)$. Using the fact that $\operatorname{supp}(\phi) \subset \Omega$ and employing the Plancherel identity, one obtains

$$\int_\Omega \overline{(\Delta^k \phi)(x)}\, \phi(x)\, d^n x = \int_{\mathbb{R}^n} \overline{(\Delta^k)\phi(x)}\, \phi(x)\, d^n x = \int_{\mathbb{R}^n} |\xi|^{2k} \big|\widehat{\phi}(\xi)\big|^2 \, d^n \xi. \tag{111}$$

Similarly,

$$\begin{aligned} \sum_{|\alpha|=k} \frac{k!}{\alpha!} \int_\Omega \big|(\partial^\alpha \phi)(x)\big|^2 \, d^n x &= \sum_{|\alpha|=k} \frac{k!}{\alpha!} \int_{\mathbb{R}^n} \big|(\partial^\alpha \phi)(x)\big|^2 \, d^n x \\ &= \sum_{|\alpha|=k} \frac{k!}{\alpha!} \int_{\mathbb{R}^n} \xi^{2\alpha} \big|\widehat{\phi}(\xi)\big|^2 \, d^n \xi. \end{aligned} \tag{112}$$

Since in general, $\Big(\sum_{j=1}^n x_j\Big)^N = \sum_{|\alpha|=N} \frac{|\alpha|!}{\alpha!} x^\alpha$, with $x := (x_1, ..., x_n)$, by the Multinomial Theorem, one concludes that

$$\sum_{|\alpha|=k} \frac{k!}{\alpha!} \xi^{2\alpha} = \Big(\sum_{j=1}^n \xi_j^2\Big)^k = |\xi|^{2k}, \quad \xi \in \mathbb{R}^n. \tag{113}$$

Therefore, using (111)–(113), we may write

$$\sum_{|\alpha|=k} \frac{k!}{\alpha!} \int_\Omega \big|(\partial^\alpha \phi)(x)\big|^2 \, d^n x = \int_{\mathbb{R}^n} |\xi|^{2k} |\widehat{\phi}(\xi)|^2 \, d^n \xi = \int_\Omega \overline{(\Delta^k \phi)(x)}\, \phi(x)\, d^n x, \tag{114}$$

completing the proof of (110). □

Lemma 3.4 is a key input for the next result.

Theorem 3.3. *Assume Hypothesis 3.1 and let $m \in \mathbb{N}$. Consider the minimal operator*

$$A_{min,\Omega,m} := (-\Delta)^m, \quad \operatorname{dom}(A_{min,\Omega,m}) := C_0^\infty(\Omega), \tag{115}$$

in $L^2(\Omega)$. *Then the closure of* $A_{min,\Omega,m}$ *in* $L^2(\Omega)$ *is given by*

$$A_{\Omega,m} = (-\Delta)^m, \quad \operatorname{dom}(A_{\Omega,m}) = \mathring{W}^{2m}(\Omega). \tag{116}$$

In addition, $A_{\Omega,m}$ *is a strictly positive operator, that is, there exists* $\varepsilon > 0$ *such that*

$$A_{\Omega,m} \geqslant \varepsilon I_\Omega. \tag{117}$$

Proof. Clearly $A_{min,\Omega,m}$ is symmetric and hence closable. Assuming $\phi \in C_0^\infty(\Omega)$, repeatedly integrating by parts and an application of Lemma 3.4 yield,

$$\begin{aligned}
\int_\Omega \big|((-\Delta)^m\phi)(x)\big|^2\, d^n x &= \int_\Omega \overline{((-\Delta)^m\phi)(x)}\,((-\Delta)^m\phi)(x)\, d^n x \\
&= \int_\Omega \overline{((-\Delta)^{2m}\phi)(x)}\,\phi(x)\, d^n x \\
&= \sum_{|\alpha|=2m} \frac{(2m)!}{\alpha!} \int_\Omega \big|(\partial^\alpha\phi)(x)\big|^2\, d^n x.
\end{aligned} \tag{118}$$

By density of $C_0^\infty(\Omega)$ in $\mathring{W}^{2m}(\Omega)$, identity (118) extends to

$$\int_\Omega \big|((-\Delta)^m u)(x)\big|^2\, d^n x = \sum_{|\alpha|=2m} \frac{(2m)!}{\alpha!} \int_\Omega \big|(\partial^\alpha u)(x)\big|^2\, d^n x, \quad u \in \mathring{W}^{2m}(\Omega). \tag{119}$$

Next, combining the Poincaré inequality (92) with (119) implies that for some constant $C_{m,\Omega} > 0$,

$$\int_\Omega \big|((-\Delta)^m u)(x)\big|^2\, d^n x \geqslant C_{m,\Omega} \sum_{0\leqslant|\beta|\leqslant 2m} \big\|\partial^\beta u\big\|^2_{L^2(\Omega)} \approx \|u\|^2_{m,\Omega}, \quad u \in \mathring{W}^{2m}(\Omega). \tag{120}$$

Finally, consider $\{f_j\}_{j\in\mathbb{N}} \subset \mathring{W}^{2m}(\Omega)$, $f, g \in L^2(\Omega)$, such that

$$\lim_{j\to\infty} \|f_j - f\|_{L^2(\Omega)} = 0 \text{ and } \lim_{j\to\infty} \big\|(-\Delta)^m f_j - g\big\|_{L^2(\Omega)} = 0. \tag{121}$$

Applying (120) to $u := (f_j - f_k) \in \mathring{W}^{2m}(\Omega)$, one infers for some $c_{m,\Omega} > 0$,

$$\int_\Omega \big|((-\Delta)^m (f_j - f_k)(x)\big|^2\, d^n x \geqslant c_{m,\Omega}\|f_j - f_k\|^2_{m,\Omega}, \quad j,k \in \mathbb{N}, \tag{122}$$

implying that actually, $\{f_j\}_{j\in\mathbb{N}}$ is a Cauchy sequence in $\mathring{W}^{2m}(\Omega)$. By completeness of the latter space one concludes that $f \in \mathring{W}^{2m}(\Omega)$. Taking arbitrary $\psi \in C_0^\infty(\Omega)$, one concludes that

$$\begin{aligned}
(g,\psi)_{L^2(\Omega)} &= {}_{\mathcal{D}(\Omega)'}\langle g,\psi\rangle_{\mathcal{D}(\Omega)} = \lim_{j\to\infty} {}_{\mathcal{D}(\Omega)'}\big\langle (-\Delta)^m f_j, \psi\big\rangle_{\mathcal{D}(\Omega)} \\
&= \lim_{j\to\infty} \int_\Omega \overline{f_j(x)}\,((-\Delta)^m\psi)(x)\, d^n x = \int_\Omega \overline{f(x)}\,((-\Delta)^m\psi)(x)\, d^n x \\
&= {}_{\mathcal{D}(\Omega)'}\big\langle (-\Delta)^m f, \psi\big\rangle_{\mathcal{D}(\Omega)}.
\end{aligned} \tag{123}$$

Hence, $g = (-\Delta)^m f$, implying closedness of $A_{\Omega,m}$. By the definition of $\mathring{W}^{2m}(\Omega)$ (cf. (107)), $A_{\Omega,m}$ is the closure of $A_{min,\Omega,m}$.

Strict positivity of $A_{min,\Omega,m}$, and hence that of $A_{\Omega,m}$, follows from (110) and the Poincaré-type inequalities (92). □

In the following we pick $m \in \mathbb{N}$ and denote by $A_{K,\Omega,m}$ and $A_{F,\Omega,m}$ the Krein and Friedrichs extension of $A_{\Omega,m}$ in $L^2(\Omega)$, respectively. By Corollary 3.1 (for $p = 2$), $\mathrm{dom}(A_{\Omega,m}) = \mathring{W}^{2m}(\Omega)$ embeds compactly into $L^2(\Omega)$ and hence by Lemma 3.3, the operator $A^*_{\Omega,m}A_{\Omega,m}$ has purely discrete spectrum, equivalently, the resolvent of $A^*_{\Omega,m}A_{\Omega,m}$ is compact, in particular,

$$\big(A^*_{\Omega,m}A_{\Omega,m}\big)^{-1} \in \mathcal{B}_\infty\big(L^2(\Omega)\big), \tag{124}$$

as the form associated with $A^*_{\Omega,m}A_{\Omega,m}$ is given by

$$\mathfrak{a}_{\Omega,m}(f,g) := (A_{\Omega,m}f, A_{\Omega,m}g)_{L^2(\Omega)}, \quad f,g \in \mathrm{dom}(\mathfrak{a}_{\Omega,m}) := \mathrm{dom}(A_{\Omega,m}). \tag{125}$$

Consequently, also

$$|A_{\Omega,m}|^{-1} = \big(A^*_{\Omega,m}A_{\Omega,m}\big)^{-1/2} \in \mathcal{B}_\infty\big(L^2(\Omega)\big), \tag{126}$$

implying

$$\big(\widehat{A}_{K,\Omega,m}\big)^{-1} \in \mathcal{B}_\infty\big(L^2(\Omega)\big) \tag{127}$$

by (56). Thus,

$$\sigma_{ess}(A_{K,\Omega,m}) \subseteq \{0\}. \tag{128}$$

Let $\{\lambda_{K,\Omega,j}\}_{j\in\mathbb{N}} \subset (0,\infty)$ be the strictly positive eigenvalues of $A_{K,\Omega,m}$ enumerated in nondecreasing order, counting multiplicity, and let

$$N(\lambda, A_{K,\Omega,m}) := \#\{j \in \mathbb{N} \,|\, 0 < \lambda_{K,\Omega,j} < \lambda\}, \quad \lambda > 0, \tag{129}$$

be the eigenvalue distribution function for $A_{K,\Omega,m}$. Recalling the standard notation

$$x_+ := \max(0,x), \quad x \in \mathbb{R}, \tag{130}$$

then $N(\,\cdot\,, A_{K,\Omega,m})$ permits the following estimate following the approach in [36].

Theorem 3.4. *Assume Hypothesis* 3.1 *and let* $m \in \mathbb{N}$. *Then one has the estimate,*

$$N(\lambda, A_{K,\Omega,m}) \leqslant (2\pi)^{-n} v_n |\Omega| \{1 + [2m/(2m+n)]\}^{n/(2m)} \lambda^{n/(2m)}, \quad \lambda > 0, \tag{131}$$

where $v_n := \pi^{n/2}/\Gamma((n+2)/2)$ *denotes the* (*Euclidean*) *volume of the unit ball in* $\mathbb{R}^n$ ($\Gamma(\cdot)$ *being the Gamma function, cf. [1, Sect. 6.1]*).

Proof. Following our abstract Section 2, we introduce in addition to the symmetric form $\mathfrak{a}_{\Omega,m}$ in $L^2(\Omega)$ (cf. (125)), the form

$$\mathfrak{b}_{\Omega,m}(f,g) := (f, A_{\Omega,m}g)_{L^2(\Omega)}, \quad f,g \in \mathrm{dom}(\mathfrak{b}_{\Omega,m}) := \mathrm{dom}(A_{\Omega,m}). \tag{132}$$

By Lemma 2.3, particularly, by (62), one concludes that

$$N(\lambda, A_{K,\Omega,m}) \leqslant \max\big(\dim\{f \in \mathrm{dom}(A_{\Omega,m}) \,\big|\, \mathfrak{a}_{\Omega,m}(f,f) - \lambda\, \mathfrak{b}_{\Omega,m}(f,f) < 0\}\big), \tag{133}$$

by also employing (64) and the fact that

$$\mathfrak{a}_{\Omega,m}(f_{K,\Omega,j}, f_{K,\Omega,j})) - \lambda\, \mathfrak{b}_{\Omega,m}(f_{K,\Omega,j}, f_{K,\Omega,j}) = (\lambda_{K,\Omega,j} - \lambda)\|f_{K,\Omega,j}\|^2_{L^2(\Omega)} < 0, \tag{134}$$

where $f_{K,\Omega,j} \in \mathrm{dom}(A_{\Omega,m})\backslash\{0\}$ additionally satisfies

$$\begin{aligned} &f_{K,\Omega,j} \in \mathrm{dom}(A^*_{\Omega,m} A_{\Omega,m}) \ \text{ and} \\ &A^*_{\Omega,m} A_{\Omega,m} f_{K,\Omega,j} = \lambda_{K,\Omega,j}\, A_{\Omega,m} f_{K,\Omega,j}. \end{aligned} \tag{135}$$

To further analyze (133) we now fix $\lambda \in (0,\infty)$ and introduce the auxiliary operator

$$\begin{aligned} &L_{\Omega,m,\lambda} := A^*_{\Omega,m} A_{\Omega,m} - \lambda\, A_{\Omega,m}, \\ &\mathrm{dom}(L_{\Omega,m,\lambda}) := \mathrm{dom}(A^*_{\Omega,m} A_{\Omega,m}). \end{aligned} \tag{136}$$

By Lemma 3.1, $L_{\Omega,m,\lambda}$ is self-adjoint, bounded from below, with purely discrete spectrum as its form domain

$$\mathrm{dom}\big(|L_{\Omega,m,\lambda}|^{1/2}\big) = \mathrm{dom}(A_{\Omega,m}) = \mathring{W}^{2m}(\Omega) \tag{137}$$

embeds compactly into $L^2(\Omega)$ by Corollary 3.1 (cf. Lemma 3.3). We will study the auxiliary eigenvalue problem,

$$L_{\Omega,m,\lambda}\varphi_j = \mu_j \varphi_j, \quad \varphi_j \in \mathrm{dom}(L_{\Omega,m,\lambda}), \tag{138}$$

where $\{\varphi_j\}_{j\in\mathbb{N}}$ represents an orthonormal basis of eigenfunctions in $L^2(\Omega)$ and for simplicity of notation we repeat the eigenvalues μ_j of $L_{\Omega,m,\lambda}$ according to their multiplicity, assuming φ_j to be linearly independent in the following. Since $\varphi_j \in \mathring{W}^{2m}(\Omega)$, we denote by

$$\widetilde{\varphi}_j(x) := \begin{cases} \varphi_j(x), & x \in \Omega, \\ 0, & x \in \mathbb{R}^n\backslash\Omega, \end{cases} \tag{139}$$

their zero-extension of φ_j to all of $\mathbb{R}^n$ and note that

$$\widetilde{\varphi}_j \in \mathring{W}^{2m}(\mathbb{R}^n), \quad \partial^\alpha \widetilde{\varphi}_j = \widetilde{\partial^\alpha \varphi_j}, \quad 0 \leqslant |\alpha| \leqslant 2m. \tag{140}$$

Next, given $\mu > 0$, one estimates

$$\mu^{-1} \sum_{\substack{j\in\mathbb{N} \\ \mu_j<\mu}} (\mu - \mu_j) \geqslant \mu^{-1} \sum_{\substack{j\in\mathbb{N}, \\ \mu_j<0,\, \mu_j<\mu}} (\mu - \mu_j) \geqslant \mu^{-1} \sum_{\substack{j\in\mathbb{N}, \\ \mu_j<0,\, \mu_j<\mu}} \mu = n_-(L_{\Omega,m,\lambda}), \tag{141}$$

where $n_-(L_{\Omega,m,\lambda})$ denotes the number of strictly negative eigenvalues of $L_{\Omega,m,\lambda}$. Combining, Lemma 3.2 and (133) one concludes that

$$\begin{aligned} N(\lambda, A_{K,\Omega,m}) &\leqslant \max\big(\dim\{f \in \mathrm{dom}(A_{min,\Omega,m}) \,\big|\, \mathfrak{a}_{\Omega,m}(f,f) - \lambda\, \mathfrak{b}_{\Omega,m}(f,f) < 0\}\big) \\ &= n_-(L_{\Omega,m,\lambda}) \leqslant \mu^{-1} \sum_{\substack{j\in\mathbb{N} \\ \mu_j<\mu}} (\mu - \mu_j) = \mu^{-1} \sum_{j\in\mathbb{N}} [\mu - \mu_j]_+, \quad \mu > 0. \end{aligned} \tag{142}$$

Next, we focus on estimating the right-hand side of (142).

$$
\begin{aligned}
N(\lambda, A_{K,\Omega,m}) &\leqslant \mu^{-1}\sum_{j\in\mathbb{N}}(\mu-\mu_j)_+ = \mu^{-1}\sum_{j\in\mathbb{N}}\big[(\varphi_j,(\mu-\mu_j)\varphi_j)_{L^2(\Omega)}\big]_+ \\
&= \mu^{-1}\sum_{j\in\mathbb{N}}\Big[\mu\|\varphi_j\|_{L^2(\Omega)}^2 - \big\|(-\Delta)^m\varphi_j\big\|_{L^2(\Omega)}^2 + \lambda\big(\varphi_j,(-\Delta)^m\varphi_j\big)_{L^2(\Omega)}\Big]_+ \\
&= \mu^{-1}\sum_{j\in\mathbb{N}}\Big[\mu\|\widetilde{\varphi}_j\|_{L^2(\mathbb{R}^n)}^2 - \big\|(-\Delta)^m\widetilde{\varphi}_j\big\|_{L^2(\mathbb{R}^n)}^2 + \lambda\big(\widetilde{\varphi}_j,(-\Delta)^m\widetilde{\varphi}_j\big)_{L^2(\mathbb{R}^n)}\Big]_+ \\
&= \mu^{-1}\sum_{j\in\mathbb{N}}\bigg[\int_{\mathbb{R}^n}\big[\mu-\big(|\xi|^{4m}-\lambda|\xi|^{2m}\big)\big]\big|\widehat{\widetilde{\varphi}}_j(\xi)\big|^2\,d^n\xi\bigg]_+ \\
&\leqslant \mu^{-1}\sum_{j\in\mathbb{N}}\int_{\mathbb{R}^n}\big[\mu-\big(|\xi|^{4m}-\lambda|\xi|^{2m}\big)\big]_+\big|\widehat{\widetilde{\varphi}}_j(\xi)\big|^2\,d^n\xi \\
&= \mu^{-1}\sum_{j\in\mathbb{N}}\int_{\mathbb{R}^n}\big[\mu-|\xi|^{4m}+\lambda|\xi|^{2m}\big]_+\big|\widehat{\widetilde{\varphi}}_j(\xi)\big|^2\,d^n\xi \\
&= \mu^{-1}\int_{\mathbb{R}^n}\big[\mu-|\xi|^{4m}+\lambda|\xi|^{2m}\big]_+\sum_{j\in\mathbb{N}}\big|\widehat{\widetilde{\varphi}}_j(\xi)\big|^2\,d^n\xi.
\end{aligned}
\tag{143}
$$

Here we used unitarity of the Fourier transform on $L^2(\mathbb{R}^n)$, the fact that $\big[\mu-|\xi|^{4m}+\lambda|\xi|^{2m}\big]_+$ has compact support (rendering the integral over a compact subset of $\mathbb{R}^n$), and the monotone convergence theorem in the final step.

Next, one observes that

$$
\begin{aligned}
\sum_{j\in\mathbb{N}}\big|\widehat{\widetilde{\varphi}}_j(\xi)\big|^2 &= (2\pi)^{-n}\sum_{j\in\mathbb{N}}\big|\big(e^{i\xi\cdot},\widetilde{\varphi}_j\big)_{L^2(\mathbb{R}^n)}\big|^2 = (2\pi)^{-n}\sum_{j\in\mathbb{N}}\big|\big(e^{i\xi\cdot},\varphi_j\big)_{L^2(\Omega)}\big|^2 \\
&= (2\pi)^{-n}\big\|e^{i\xi\cdot}\big\|_{L^2(\Omega)}^2 = (2\pi)^{-n}|\Omega|,
\end{aligned}
\tag{144}
$$

employing the fact that $\{\varphi_j\}_{j\in\mathbb{N}}$ represents an orthonormal basis in $L^2(\Omega)$.

Combining (143) and (144), introducing $\alpha = \lambda^{-2}\mu$, changing variables, $\xi = \lambda^{1/(2m)}\eta$, and taking the minimum with respect to $\alpha > 0$, proves the bound,

$$
N(\lambda, A_{K,\Omega,m}) \leqslant (2\pi)^{-n}|\Omega|\min_{\alpha>0}\bigg(\alpha^{-1}\int_{\mathbb{R}^n}\big[\alpha-|\xi|^{4m}+|\xi|^{2m}\big]_+ d^n\xi\bigg)\lambda^{n/(2m)}, \quad \lambda > 0. \tag{145}
$$

Explicitly computing the minimum over $\alpha > 0$ in (145)[a] finally yields the result (131). □

[a] Mark Ashbaugh generously provided us with the explicit value of the minimum in (145).

4. Comparisons with Other Bounds and Weyl Asymptotics

In our final section we briefly discuss the bound (131) on the eigenvalue counting function $N(\lambda, A_{K,\Omega,m})$.

For smooth, bounded domains $\Omega \subset \mathbb{R}^n$, and smooth lower-order coefficients (not necessarily constant), Weyl asymptotics for $N(\lambda, A_{K,\Omega,m})$ as $\lambda \to \infty$ was first derived by Grubb [26],

$$N(\lambda, A_{K,\Omega,m}) \underset{\lambda\to\infty}{=} (2\pi)^{-n} v_n |\Omega| \, \lambda^{n/(2m)} + O\big(\lambda^{(n-\theta)/(2m)}\big), \tag{146}$$

where $v_n := \pi^{n/2}/\Gamma((n+2)/2)$ denotes the (Euclidean) volume of the unit ball in $\mathbb{R}^n$ ($\Gamma(\,\cdot\,)$ being the Gamma function, cf. [1, Sect. 6.1]), and

$$\theta := \max\left\{\frac{1}{2} - \varepsilon \, , \, \frac{2m}{2m+n-1}\right\}, \text{ with } \varepsilon > 0 \text{ arbitrary.} \tag{147}$$

We also refer to [43, 44], and more recently, [28], where the authors derive a sharpening of the remainder in (146) to any $\theta < 1$. In the case $m = 1$, Weyl asymptotics for $N(\lambda, A_{K,\Omega,1})$ was derived in [5] for (bounded) quasi-convex domains, and most recently, in [9] for bounded Lipschitz domains.

The power law behavior $\lambda^{n/(2m)}$ of the estimate (131) for general domains governed by Hypothesis 3.1 (no smoothness of Ω being assumed at all in the case of bounded domains), coincides with that in the known Weyl asymptotics (146) and is of course consistent with the abstract estimate (37). In this connection we note that Weyl-type asymptotics and estimates for $N(\lambda, A_{F,\Omega,m})$, and hence upper bounds for $N(\lambda, A_{K,\Omega,m})$, without regularity assumptions on Ω can be found, for instance, in [11–16, 20, 21, 29, 30, 36, 38, 39, 42, 46–50, 54]. We mention, in particular, the bound for $N(\lambda, A_{F,\Omega,m})$ derived in [36] (extending earlier results in [38] in the case $m = 1$) which reads

$$N(\lambda, A_{F,\Omega,m}) \leqslant (2\pi)^{-n} v_n |\Omega| \, [1 + (2m/n)]^{n/(2m)} \lambda^{n/(2m)}, \quad \lambda > 0. \tag{148}$$

A comparison of (148) with our result (131),

$$N(\lambda, A_{K,\Omega,m}) \leqslant (2\pi)^{-n} v_n |\Omega| \, \{1 + [2m/(2m+n)]\}^{n/(2m)} \lambda^{n/(2m)}, \quad \lambda > 0, \tag{149}$$

clearly demonstrates the superiority of the buckling problem approach developed here over the bound obtained by combining the generally valid estimate (13) with (148) due to the extra term $2m$ in (149) as compared to (148).

Additional comparisons between the bound (149) and Weyl asymptotics, as well as an extension of our approach replacing $(-\Delta)^m$ by $(-\Delta + V)^m$, $m \in \mathbb{N}$, for an appropriate class of potentials $V \geqslant 0$ supported in $\overline{\Omega}$, will appear in [8].

Acknowledgments. We are grateful to Mark Ashbaugh and Sergey Naboko for very helpful discussions. Especially, we are indebted to Mark Ashbaugh for communicating to us the explicit value of the minimum in (145). We also thank the anonymous referee for a very careful reading of this manuscript, implying a number of improvements in Section 3. In particular, this prompted us to present the compact embedding result, Theorem 3.2.

F.G. and A.L. are indebted to all organizers of QMath12, and particularly, to Pavel Exner, Wolfgang König, and Hagen Neidhardt, for fostering a very stimulating atmosphere during the conference, leading to this collaboration.

Work of M. M. was partially supported by the Simons Foundation Grant # 281566 and by a University of Missouri Research Leave grant.

Bibliography

[1] M. Abramowitz and I. A. Stegun, *Handbook of Mathematical Functions*, Dover, New York, 1972.

[2] R. A. Adams, *Capacity and compact imbeddings*, J. Math. Mech. **19**, 923–929 (1970).

[3] R. A. Adams and J. J. F. Fournier, *Sobolev Spaces*, second edition, Academic Press, 2003.

[4] A. Alonso and B. Simon, *The Birman-Krein-Vishik theory of selfadjoint extensions of semibounded operators*, J. Operator Th. **4**, 251–270 (1980); Addenda: **6**, 407 (1981).

[5] M. S. Ashbaugh, F. Gesztesy, M. Mitrea, and G. Teschl, *Spectral theory for perturbed Krein Laplacians in nonsmooth domains*, Adv. Math. **223**, 1372–1467 (2010).

[6] M. S. Ashbaugh, F. Gesztesy, M. Mitrea, R. Shterenberg, and G. Teschl, *The Krein—von Neumann extension and its connection to an abstract buckling problem*, Math. Nachr. **283**, 165–179 (2010).

[7] M. S. Ashbaugh, F. Gesztesy, M. Mitrea, R. Shterenberg, and G. Teschl, *A survey on the Krein–von Neumann extension, the corresponding abstract buckling problem, and Weyl-type spectral asymptotics for perturbed Krein Laplacians in non smooth domains*, in *Mathematical Physics, Spectral Theory and Stochastic Analysis*, M. Demuth and W. Kirsch (eds.), Operator Theory: Advances and Applications, Vol. 232, Birkhäuser, Springer, Basel, 2013, pp. 1–106.

[8] M. S. Ashbaugh, F. Gesztesy, A. Laptev, M. Mitrea, and S. Sukhtaiev, in preparation.

[9] J. Behrndt, F. Gesztesy, T. Micheler, and M. Mitrea, *Sharp Boundary Trace Theory on Bounded Lipschitz Domains and Applications to the Krein–Von Neumann extension*, in preparation.

[10] M. Sh. Birman, *On the theory of self-adjoint extensions of positive definite operators*, Mat. Sbornik **38**, 431–450 (1956) (Russian).

[11] M. Sh. Birman and M. Z. Solomyak, *Leading term in the asymptotic spectral formula for "non-smooth" elliptic problems*, Funkcional. Anal. i Priložen **4**,

no. 4, 1–13 (1970) (Russian); Engl. transl. in Funct. Anal. Appl. **4**, 265–275 (1970).

[12] M. Sh. Birman and M. Z. Solomyak, *On the asymptotic spectrum of "nonsmooth" elliptic equations*, Funkcional. Anal. i Priložen **5**, no. 1, 69–70 (1971) (Russian); Engl. transl. in Funct. Anal. Appl. **5**, 56–57 (1971).

[13] M. Š. Birman and M. Z. Solomjak, *Spectral asymptotics of nonsmooth elliptic operators. I*, Trans. Moscow Math. Soc. **27**, 1–52 (1972).

[14] M. Š. Birman and M. Z. Solomjak, *Spectral asymptotics of nonsmooth elliptic operators. II*, Trans. Moscow Math. Soc. **28**, 1–32 (1973).

[15] M. Sh. Birman and M. Z. Solomyak, *Asymtotic behavior of the spectrum of differential equations*, Itogi Nauki i Tekhniki, Matematicheskii Analiz., **14**, 5–58 (1977) (Russian); Engl. transl. in J. Soviet Math. **12**, no. 3, 247–283 (1979).

[16] M. S. Birman and M. Z. Solomjak, *Quantitative Analysis in Sobolev Imbedding Theorems and Applications to Spectral Theory*, AMS Translations, Series 2, Vol. 114, Providence, RI, 1980, pp. 1–132.

[17] S.-K. Chua, S. Rodney, and R. L. Wheeden, *A compact embedding theorem for generalized Sobolev spaces*, Pacific J. Math. **265**, 17–57 (2013).

[18] V. A. Derkach and M. M. Malamud, *Generalized resolvents and the boundary value problems for Hermitian operators with gaps*, J. Funct. Anal. **95**, 1–95 (1991).

[19] W. G. Faris, *Self-Adjoint Operators*, Lecture Notes in Mathematics, Vol. 433, Springer, Berlin, 1975.

[20] L. Geisinger, *A short proof of Weyl's law for fractional differential operators*, preprint, arXiv:1309.1867.

[21] L. Geisinger, A. Laptev, and T. Weidl, *Geometrical versions of improved Berezin–Li–Yau inequalities*, J. Spectral Theory **1**, 87–109 (2011).

[22] F. Gesztesy and M. Mitrea, *Generalized Robin Laplacians and some remarks on a paper by Filonov on eigenvalue inequalities*, J. Diff. Eq. **247**, 2871–2896 (2009).

[23] F. Gesztesy and M. Mitrea, *A description of all self-adjoint extensions of the Laplacian and Krein-type resolvent formulas on non-smooth domains*, J. Analyse Math. **113**, 53–172 (2011).

[24] G. Grubb, *A characterization of the non-local boundary value problems associated with an elliptic operator*, Ann. Scuola Norm. Sup. Pisa (3), **22**, 425–513 (1968).

[25] G. Grubb, *Les problèmes aux limites généraux d'un opérateur elliptique, provenant de le théorie variationnelle*, Bull. Sci. Math. (2), **94**, 113–157 (1970).

[26] G. Grubb, *Spectral asymptotics for the "soft" selfadjoint extension of a symmetric elliptic differential operator*, J. Operator Th. **10**, 9–20 (1983).

[27] G. Grubb, *Distributions and Operators*, Graduate Texts in Mathematics, Vol. 252, Springer, New York, 2009.

[28] G. Grubb, *Krein like extensions and the lower boundedness problem for elliptic operators*, J. Diff. Eq. **252**, 852–885 (2012).

[29] E. M. Harrell II and L. Hermi, *Differential inequalities for Riesz means and Weyl-type bounds for eigenvalues*, J. Funct. Anal. **254**, 3173–3191 (2008).

[30] E. M. Harrell II and L. Hermi, *On Riesz means of eigenvalues*, Commun. Partial Diff. Eq. **36**, 1521–1543 (2011).

[31] T. Kato, *Perturbation Theory for Linear Operators*, corr. printing of the 2nd ed., Springer, Berlin, 1980.

[32] V. A. Kozlov, *Estimation of the remainder in a formula for the asymptotic behavior of the spectrum of nonsemibounded elliptic systems*, Vestnik Leningrad. Univ. Mat. Mekh. Astronom. **1979**, no 4., 112–113, 125 (Russian).

[33] V. A. Kozlov, *Estimates of the remainder in formulas for the asymptotic behavior of the spectrum for linear operator bundles*, Funktsional. Anal. i Prilozhen **17**, no. 2, 80–81 (1983). Engl. transl. in Funct. Anal. Appl. **17**, no. 2, 147–149 (1983).

[34] V. A. Kozlov, *Remainder estimates in spectral asymptotic formulas for linear operator pencils*, Linear and Nonlinear Partial Differential Equations. Spectral Asymptotic Behavior, pp. 34–56, Probl. Mat. Anal. **9**, Leningrad Univ., Leningrad, 1984; Engl. transl. in J. Sov. Math. **35**, 2180–2193 (1986).

[35] M. G. Krein, *The theory of self-adjoint extensions of semi-bounded Hermitian transformations and its applications. I*, Mat. Sbornik **20**, 431–495 (1947) (Russian).

[36] A. Laptev, *Dirichlet and Neumann eigenvalue problems on domains in Euclidean spaces*, J. Funct. Anal. **151**, 531–545 (1997).

[37] S. Levendorskiĭ, *Asymptotic Distribution of Eigenvalues of Differential Operators*, Kluwer, Dordrecht, 1990.

[38] P. Li and S.-T. Yau, *On the Schrödinger equation and the eigenvalue problem*, Commun. Math. Phys. **88**, 309–318 (1983).

[39] E. H. Lieb, *The number of bound states of one-body Schroedinger operators and the Weyl problem*, in *Geometry of the Laplace Operator*, R. Osserman and A. Weinstein (eds.), Proc. Symposia Pure Math., Vol. 36, Amer. Math. Soc., Providence, RI, 1980, pp. 241–252.

[40] M. M. Malamud, *Certain classes of extensions of a lacunary Hermitian operator*, Ukrainian Math. J. **44**, No. 2, 190–204 (1992).

[41] V. Maz'ya, *Sobolev Spaces with Applications to Elliptic Partial Differential Equations*, 2nd, rev. and augmented ed., Springer, Berlin, 2011.

[42] G. Métivier, *Valeurs propres de problèmes aux limites elliptiques irégulières*, Mém. Soc. Math. France **51–52**, 125–219 (1977).

[43] V. A. Mikhaĭlets, *Distribution of the eigenvalues of finite multiplicity of Neumann extensions of an elliptic operator*, Differentsial'nye Uravneniya **30**, 178–179 (1994) (Russian); Engl. transl. in Diff. Eq. **30**, 167–168 (1994).

[44] V. A. Mikhailets, *Discrete spectrum of the extreme nonnegative extension of the positive elliptic differential operator*, in *Proceedings of the Ukrainian Mathematical Congress–2001, Section 7, Nonlinear Analysis*, Kyiv, 2006, pp. 80–94.

[45] D. Mitrea, *Distributions, Partial Differential Equations, and Harmonic Analysis*, Springer, New York, 2013.

[46] Yu. Netrusov and Yu. Safarov, *Weyl asymptotic formula for the Laplacian on domains with rough boundaries*, Comm. Math. Phys. **253**, no. 2, 481–509 (2005).

[47] G. V. Rozenbljum, *On the distribution of eigenvalues of the first boundary value problem in unbounded regions*, Sov. Math. Dokl. **12**, 1539–1542 (1971).

[48] G. V. Rozenbljum, *On the eigenvalues of the first boundary value problem in unbounded domains*, Math. USSR Sb. **18**, 235–248 (1972).

[49] G. V. Rozenblyum, *Distribution of the discrete spectrum of singular differential operators*, Sov. Math. **20**, 63–71 (1976).

[50] Yu. Safarov, *Fourier Tauber theorems and applications*, J. Funct. Anal. **185**, 111–128 (2001).

[51] Yu. Safarov and D. Vassiliev, *The Asymptotic Distribution of Eigenvalues of Partial Differential Operators*, Transl. of Math. Monographs, Vol. 155, Amer. Math. Soc., Providence, RI, 1997.

[52] K. Schmüdgen, *Unbounded Self-Adjoint Operators on Hilbert Space*, Graduate Texts in Mathematics, Vol. 265, Springer, Dordrecht, 2012.

[53] M. L. Višik, *On general boundary problems for elliptic differential equations*, Trudy Moskov. Mat. Obsc. **1**, 187–246 (1952) (Russian); Engl. transl. in Amer. Math. Soc. Transl. (2), **24**, 107–172 (1963).

[54] T. Weidl, *Improved Berezin–Li–Yau inequalities with a remainder term*, in *Spectral Theory of Differential Operators. M. Sh. Birman 80th Anniversary Collection*, Adv. Math. Sci., Vol. 62, Amer Math. Soc. Transl. (2) **225** (2008), pp. 253–263.

Trace formulae for the asymptotic density of eigenvalue clusters for the perturbed Landau Hamiltonian

T. Lungenstrass* and G. D. Raikov†

Departamento de Matemáticas,
Pontificia Universidad Católica de Chile
Vicuña Mackenna 4860, Santiago, Chile
**E-mail: tlungens@mat.puc.cl*
†*E-mail: graikov@mat.puc.cl*
www.mat.puc.cl

This note is based on the results contained in [1] and [2]. We consider the Landau Hamiltonian perturbed by an electric potential V that decays at infinity. The spectrum of the perturbed operator consists of eigenvalue clusters which accumulate to the Landau levels. We distinguish two regimes: a short-range and a long-range one, and estimate the rate at which the eigenvalue clusters shrink to the Landau levels, as the energy increases. Further, we obtain an explicit description of the asymptotic density of eigenvalues within the qth cluster as $q \to \infty$. For short-range potentials this description is given in terms of the Radon transform of the potential V, whereas in the long-range regime we consider potentials $V(x) = \mathbb{V}(x) + O(|x|^{-\rho-\varepsilon})$, with $\mathbb{V}$ homogeneous of order $-\rho$, $\rho \in (0,1)$, and obtain a description in terms of the mean-value transform of $\mathbb{V}$.

2010 AMS Mathematics Subject Classification: 35P20, 35J10, 47G30, 81Q10.

Keywords: Perturbed Landau Hamiltonian, asymptotic density for eigenvalue clusters, radon transform, mean-value transform.

1. Introduction and Main Results

1.1. *The perturbed operator*

The Schrödinger operator

$$H_0 := (-i\nabla - A)^2,$$

self-adjoint in $L^2(\mathbb{R}^2)$, where $A := (-Bx_2/2, Bx_1/2)$, is called the Landau Hamiltonian. It describes a two-dimensional quantum particle subject to a constant magnetic field $B > 0$. Its spectrum consists of infinitely degenerate eigenvalues $\lambda_q := B(2q+1)$, $q \in \mathbb{Z}_+ := \{0, 1, 2, \ldots\}$, called *Landau levels.*

Let $V : \mathbb{R}^2 \to \mathbb{R}$ be continuous and bounded. On $\mathrm{Dom}(H_0)$ define the operator

$$H := H_0 + V,$$

self-adjoint in $L^2(\mathbb{R}^2)$. If V decays at infinity, then it is relatively compact with respect to H_0, which implies that the essential spectra of H and H_0 coincide. Therefore, $\sigma(H) \setminus \sigma(H_0)$ consists of eigenvalues of finite multiplicity which may only accumulate to the Landau levels.

In the sequel, we consider potentials which satisfy the following

Condition 1.1. The potential V belongs to $C(\mathbb{R}^2; \mathbb{R})$, and the estimate

$$|V(x)| \leq C(1+|x|)^{-\rho}, \quad x \in \mathbb{R}^2, \tag{1}$$

holds for some $\rho \in (0, \infty)$.

We will distinguish two different cases according to the decay rate of the potential. If V satisfies Cond. 1.1 for some $\rho > 1$, we will say that it is short-range. If, on the other hand, V verifies Cond. 1.1 for some $\rho \in (0, 1)$ but not for $\rho \geq 1$, we will classify it as being long-range.

In the case of long-range potentials, we will consider also some additional conditions on V. Set $\langle x \rangle = (1+|x|^2)^{1/2}$, $x \in \mathbb{R}^2$. For $\rho \in \mathbb{R}$ set

$$\mathcal{S}_1^{-\rho}(\mathbb{R}^2) := \left\{ u \in C^\infty(\mathbb{R}^2) \mid \sup_{x \in \mathbb{R}^2} \langle x \rangle^{\rho+|\alpha|} |D^\alpha u(x)| \leq c_\alpha, \ \alpha \in \mathbb{Z}_+^2 \right\}.$$

Also, write $u \in \mathcal{H}_{-\rho}^\sharp(\mathbb{R}^2)$ if $u \in C^\infty(\mathbb{R}^2 \setminus \{0\})$, is a homogeneous function of order $-\rho$.

Condition 1.2. The potential V belongs to $\mathcal{S}_1^{-\rho}(\mathbb{R}^2)$ for some $\rho \in (0, 1)$ and there exists $\mathbb{V} \in \mathcal{H}_{-\rho}^\sharp(\mathbb{R}^2)$ such that

$$|V(x) - \mathbb{V}(x)| \leq C|x|^{-\rho-\varepsilon}, \ x \in \mathbb{R}^2, \ |x| > 1, \tag{2}$$

with given constants C and $\varepsilon > 0$.

Notice that Cond. 1.2 implies Cond. 1.1 with $\rho \in (0, 1)$.

1.2. *Eigenvalue clusters*

Our first result concerns the location of the spectrum of H. It provides an estimate on the rate at which the discrete eigenvalues approach the qth Landau level, as q tends to infinity, and justifies the terminology "eigenvalue clusters".

Theorem 1.1.

(i) (Ref. [1]) *Assume that V satisfies Cond.* 1.1 *with $\rho > 1$. Then*

$$\sigma(H) \subset \bigcup_{q \in \mathbb{Z}_+} \left(\lambda_q - C_1 \lambda_q^{-1/2}, \lambda_q + C_1 \lambda_q^{-1/2}\right) \tag{3}$$

with a constant $C_1 > 0$ independent of q.

(ii) (Ref. [2]) *Assume that V satisfies Cond.* 1.1 *with $\rho \in (0, 1)$. Then*

$$\sigma(H) \subset \bigcup_{q \in \mathbb{Z}_+} \left(\lambda_q - C_2 \lambda_q^{-\rho/2}, \lambda_q + C_2 \lambda_q^{-\rho/2}\right) \tag{4}$$

with a constant $C_2 > 0$ independent of q.

Estimates (3) and (4) are sharp (see Sec. 4.1 and the remark after Prop. 3.5). This will also follow from our main results, Theorems 1.2 and 1.3.

1.3. *Asymptotic density of eigenvalues within clusters*

Our next goal is to give an asymptotic description of the distribution of eigenvalues within the qth cluster, as q tends to infinity.

1.3.1. *Short-range regime*

For $[\alpha, \beta] \subset \mathbb{R} \setminus \{0\}$ and $q \in \mathbb{Z}_+$ we define the eigenvalue counting measures by setting

$$\mu_q^{\text{short}}([\alpha, \beta]) := \sum_{\lambda_q + \alpha\lambda_q^{-1/2} \le \lambda \le \lambda_q + \beta\lambda_q^{-1/2}} \dim \operatorname{Ker}(H - \lambda).$$

Notice that the interval $[\alpha, \beta]$ is rescaled in accordance with Eq. (3). For large values of q, we have that

$$[\lambda_q + \alpha\lambda_q^{-1/2},\ \lambda_q + \beta\lambda_q^{-1/2}] \cap \sigma_{\text{ess}}(H) = \emptyset, \tag{5}$$

so that the above quantity is finite.

Note that if $\varphi \in C_0^\infty(\mathbb{R} \setminus \{0\})$, then

$$\int_{\mathbb{R}} \varphi(\lambda) d\mu_q^{\text{short}}(\lambda) = \operatorname{Tr} \varphi(\lambda_q^{1/2}(H - \lambda_q)). \tag{6}$$

Definition 1.1 (Radon transform). *Assume that V satisfies Cond.* 1.1 *with $\rho >$ 1, and define its Radon transform*

$$\tilde{V}(\omega, s) := \frac{1}{2\pi} \int_{\mathbb{R}} V(s\omega + t\omega^\perp) dt, \quad \omega = (\omega_1, \omega_2) \in \mathbb{S}^1,\ s \in \mathbb{R},$$

where $\omega^\perp = (-\omega_2, \omega_1) \in \mathbb{S}^1$.

Condition 1.1 with $\rho > 1$ entails the following decay property of the Radon transform:

$$|\tilde{V}(\omega, s)| \leq C(1 + |s|)^{1-\rho}, \quad \omega \in \mathbb{S}^1, \ s \in \mathbb{R}. \tag{7}$$

Theorem 1.2. (Ref. [1]) *Let V satisfy Cond.* 1.1 *for some $\rho > 1$. Then*

$$\lim_{q\to\infty} \lambda_q^{-1/2} \mathrm{Tr}\varphi(\lambda_q^{1/2}(H - \lambda_q)) = \frac{1}{2\pi} \int_{\mathbb{S}^1} \int_{\mathbb{R}} \varphi(B\tilde{V}(\omega, s))\, ds\, d\omega \tag{8}$$

for each $\varphi \in C_0^\infty(\mathbb{R} \setminus \{0\})$.

For $[\alpha, \beta] \subset \mathbb{R} \setminus \{0\}$ define the measure

$$\mu_\infty^{\mathrm{short}}([\alpha, \beta]) := \frac{1}{2\pi} \left| \tilde{V}^{-1}([B^{-1}\alpha, B^{-1}\beta]) \right|_{\mathbb{S}^1 \times \mathbb{R}},$$

$|\cdot|_{\mathbb{S}^1\times\mathbb{R}}$ being the Lebesgue measure on $\mathbb{S}^1 \times \mathbb{R}$. In terms of convergence of measures, Th. 1.2 is equivalent to

$$\lim_{q\to\infty} \lambda_q^{-1/2} \mu_q^{\mathrm{short}}([\alpha, \beta]) = \mu_\infty^{\mathrm{short}}([\alpha, \beta]), \tag{9}$$

for any α, β such that $\alpha\beta > 0$ and

$$\mu_\infty^{\mathrm{short}}(\{\alpha\}) = \mu_\infty^{\mathrm{short}}(\{\beta\}) = 0.$$

However, the above condition does not automatically hold. Indeed, the image of $C_0^\infty(\mathbb{R} \setminus \{0\})$ under the Radon transform is well-known (see, for instance [3], Th. 2.10). According to this description, if $\phi \in C_0^\infty(\mathbb{R})$ is an even, real-valued function, then $\tilde{V}(\omega, s) := \phi(s)$ is the Radon transform of some $V \in C_0^\infty(\mathbb{R} \setminus \{0\})$. Of course, if the derivative $\phi'(s)$ vanishes on an open interval, the corresponding measure has an atom.

1.3.2. *Long-range regime*

Similarly to the short-range case, we define for $[\alpha, \beta] \subset \mathbb{R} \setminus \{0\}$ and $q \in \mathbb{Z}_+$ the corresponding eigenvalue counting measures

$$\mu_q^{\mathrm{long}}([\alpha, \beta]) := \sum_{\lambda_q + \alpha\lambda_q^{-\rho/2} \leq \lambda \leq \lambda_q + \beta\lambda_q^{-\rho/2}} \dim \mathrm{Ker}(H - \lambda).$$

Definition 1.2 (Mean-value transform). *Assume that $u \in C(\mathbb{R}^2 \setminus \{0\})$, and define its mean-value transform*

$$\mathring{u}(x) := \frac{1}{2\pi} \int_{\mathbb{S}^1} u(x - \omega)\, d\omega, \quad x \in \mathbb{R}^2 \setminus \mathbb{S}^1.$$

We describe some elementary yet useful properties of the mean-value transform of functions belonging to given classes. If $u \in \mathcal{S}_1^{-\rho}(\mathbb{R}^2)$, $\rho \in (0,\infty)$, then its mean-value transform $\mathring{u}$ extends to a function $\mathring{u} \in \mathcal{S}_1^{-\rho}(\mathbb{R}^2)$. If $u \in \mathcal{H}^\sharp_{-\rho}(\mathbb{R}^2)$, $\rho \in (0,\infty)$, then $\eta\mathring{u} \in \mathcal{S}_1^{-\rho}$ provided that $\eta \in \mathcal{S}_1^0(\mathbb{R}^2)$ and $\operatorname{supp}\eta \cap \mathbb{S}^1 = \emptyset$. Moreover, if $\rho \in (0,1)$, then the mean-value transform of $u \in \mathcal{H}^\sharp_{-\rho}(\mathbb{R}^2)$ extends to a function $\mathring{u} \in C(\mathbb{R}^2)$. Finally, if $u \in \mathcal{H}^\sharp_{-\rho}(\mathbb{R}^2)$, $\rho \in (0,1)$, and $\mathring{u}(x) = 0$ for each $x \in \mathbb{R}^2$, then $u(x) = 0$ for each $x \in \mathbb{R}^2 \setminus \{0\}$.

Theorem 1.3. (Ref. [2]) *Let V satisfy Cond.* 1.2. *Then*

$$\lim_{q\to\infty} \lambda_q^{-1}\mathrm{Tr}\varphi(\lambda_q^{\rho/2}(H-\lambda_q)) = \frac{1}{2\pi B}\int_{\mathbb{R}^2} \varphi(B^\rho \mathring{\mathbb{V}}(x))\,dx \tag{10}$$

for each $\varphi \in C_0^\infty(\mathbb{R}\setminus\{0\})$.

Again, defining for $q \in \mathbb{Z}_+$ the measure

$$\mu_\infty^{\mathrm{long}}([\alpha,\beta]) := \frac{1}{2\pi B}\left|\mathring{V}^{-1}([B^{-\rho}\alpha, B^{-\rho}\beta])\right|_{\mathbb{R}^2}, \quad [\alpha,\beta] \subset \mathbb{R}\setminus\{0\},$$

with $|\cdot|_{\mathbb{R}^2}$ the Lebesgue measure on $\mathbb{R}^2$, we find that Eq. (10) is equivalent to

$$\lim_{q\to\infty} \lambda_q^{-1}\mu_q^{\mathrm{long}}([\alpha,\beta]) = \mu_\infty^{\mathrm{long}}([\alpha,\beta]), \tag{11}$$

for any α, β such that $\alpha\beta > 0$ and such that α and β are not atoms of the measure $\mu_\infty^{\mathrm{long}}$. We remark that if, for instance, $\mathbb{V}$ is radially symmetric, the corresponding measure has no atoms.

Remark 1.1. The case $\rho = 1$ differs from both the short-range and the long-range cases and will not be considered here. In fact, for, say, V asymptotically homogeneous of order -1, the Radon transform $\tilde{V}$ is not well defined. Further, since $\mathring{\mathbb{V}}$ can generically have a logarithmic singularity at $\mathbb{S}^1$, given $\mathbb{V} \in \mathcal{H}^\sharp_{-1}(\mathbb{R}^2)$, the support of the limiting measure $\mu_\infty^{\mathrm{long}}$ need not be compact, which would mean that (4) fails to hold in this case.

2. Semiclassical Interpretation

For $(x,\xi) \in T^*\mathbb{R}^2$, consider the Hamiltonian function

$$\mathcal{H}(x,\xi) := \left(\xi_1 + \frac{1}{2}Bx_2\right)^2 + \left(\xi_2 - \frac{1}{2}Bx_1\right)^2.$$

The projections of the orbits of the Hamiltonian flow of $\mathcal{H}$ onto the configuration space are circles of radius $\sqrt{E}/B$; here $E > 0$ is the energy corresponding to the orbit. The classical particles move around these circles with period $T_B = \pi/B$. The orbits can be parametrized by the energy $E > 0$ and the center $c \in \mathbb{R}^2$ of the circle.

Let $\gamma(c, E, t)$, $t \in [0, T_B)$, be the path in the configuration space corresponding to such an orbit. Set

$$\mathrm{Av}(V)(c, E) := \frac{1}{T_B} \int_0^{T_B} V(\gamma(c, E, t))dt.$$

Under Cond. 1.1 with $\rho > 1$, we have the following identity concerning the short-range case:

$$\frac{1}{2\pi} \lim_{E\to\infty} \frac{1}{E^{1/2}} \int_{\mathbb{R}^2} \varphi(E^{1/2}\mathrm{Av}(V)(c, E))\, B\, dc = \frac{1}{2\pi} \int_{\mathbb{S}^1} \int_{\mathbb{R}} \varphi(B\tilde{V}(\omega, s)) ds\, d\omega. \quad (12)$$

The basis of this calculation is the fact that as E tends to infinity, so does the radius $\sqrt{E}/B$ of the orbits. Hence, these orbits approximately look like straight lines on any compact domain of the configuration space. Under this consideration, Th. 1.2 can be restated as

$$\lim_{q\to\infty} \frac{1}{\lambda_q^{1/2}} \mathrm{Tr}\varphi(\lambda_q^{1/2}(H - \lambda_q)) = \frac{1}{2\pi} \lim_{E\to\infty} \frac{1}{E^{1/2}} \int_{\mathbb{R}^2} \varphi(E^{1/2}\mathrm{Av}(V)(c, E))\, B\, dc. \quad (13)$$

Notice that if we consider the set M_E of orbits of given energy $E > 0$, parametrized by $c \in \mathbb{R}^2$, the measure $B\, dc$ appearing in the right hand side of Eq. (13) coincides with the restriction of the Lebesgue measure on $\mathbb{R}^4$ onto the quotient of the constant-energy surface

$$\{(x, \xi) \in \mathbb{R}^4 \mid \mathcal{H}(x, \xi) = E\}$$

with respect to the flow of $\mathcal{H}$.

In an analogous fashion, in the long-range regime we have

$$\begin{aligned} \frac{1}{2\pi} \lim_{E\to\infty} \frac{1}{E} \int_{\mathbb{R}^2} \varphi(E^{\rho/2}\mathrm{Av}(V)(c, E))\, B\, dc &= \frac{1}{2\pi B} \int_{\mathbb{R}^2} \varphi(B^\rho \mathring{\mathbb{V}}(x)) dx \\ &= \lim_{q\to\infty} \frac{1}{\lambda_q} \mathrm{Tr}\, \varphi(\lambda_q^{\rho/2}(H - \lambda_q)), \quad (14) \end{aligned}$$

provided that Cond. 1.2 holds true.

Equations (13) and (14) are in agreement with the semiclassical intuition and can be interpreted in the spirit of the "averaging principle" for systems close to integrable ones. According to this principle, a good approximation is obtained if the original perturbation is replaced by its average along the orbits of the free dynamics (see for instance [4], section 52).

3. Berezin–Toeplitz Operators

For $q \in \mathbb{Z}_+$ denote by P_q the orthogonal projection onto $\mathrm{Ker}(H_0 - \lambda_q)$. In what follows, a key role will be played by the Berezin–Toeplitz operators $P_q V P_q$. This section is devoted to the study of these operators and their fundamental properties.

Introduce the harmonic oscillator

$$h := -\frac{d^2}{dx^2} + x^2,$$

self-adjoint in $L^2(\mathbb{R})$. It is well known that the spectrum of h is purely discrete and simple, and consists of the eigenvalues $2q+1$, $q \in \mathbb{Z}_+$. Denote by p_q the orthogonal projection onto $\operatorname{Ker}(h - (2q+1))$, $q \in \mathbb{Z}_+$. Set

$$\Psi_q(x,\xi) := \frac{(-1)^q}{\pi} \mathrm{L}_q(2(x^2+\xi^2))e^{-(x^2+\xi^2)}, \quad (x,\xi) \in \mathbb{R}^2, \ q \in \mathbb{Z}_+,$$

where

$$\mathrm{L}_q(t) := \frac{1}{q!} e^t \frac{d^q(t^q e^{-t})}{dt^q} = \sum_{k=0}^{q} \binom{q}{k} \frac{(-t)^k}{k!}, \quad t \in \mathbb{R},$$

are the Laguerre polynomials. Then $2\pi\Psi_q$ is the Weyl symbol of the operator p_q (see the next subsection).

3.1. *Weyl pseudodifferential operators*

Let $d \geq 1$. Denote by $\mathcal{S}(\mathbb{R}^d)$ the Schwartz class, and by $\mathcal{S}'(\mathbb{R}^d)$ its dual class. If $f \in \mathcal{S}(\mathbb{R}^d)$, then

$$\hat{f}(\xi) := (2\pi)^{-d/2} \int_{\mathbb{R}^d} e^{-ix\xi} f(x)dx, \quad \xi \in \mathbb{R}^d,$$

is the Fourier transform of f. Whenever necessary, we extend by duality the Fourier transform to $\mathcal{S}'(\mathbb{R}^d)$.
Let $\Gamma(\mathbb{R}^{2d})$, $d \geq 1$, denote the closure of the set of smooth functions $s \in C^\infty(\mathbb{R}^{2d})$, bounded with all their derivatives, under the norm

$$\|s\|_{\Gamma(\mathbb{R}^{2d})} := \sup_{\{\alpha,\beta \in \mathbb{Z}_+^d \ | \ |\alpha|,|\beta| \leq [\frac{d}{2}]+1\}} \sup_{(x,\xi)\in\mathbb{R}^{2d}} |\partial_x^\alpha \partial_\xi^\beta s(x,\xi)|.$$

Definition 3.1 (Weyl pseudodifferential operator). (Refs. [5, 6]) *Assume that $s \in \Gamma(\mathbb{R}^{2d})$. Then, the operator $\mathrm{Op}^{\mathrm{w}}(s)$ defined initially as a mapping from $\mathcal{S}(\mathbb{R}^d)$ into $\mathcal{S}'(\mathbb{R}^d)$ by*

$$(\mathrm{Op}^{\mathrm{w}}(s)u)(x) := (2\pi)^{-d} \int_{\mathbb{R}^d}\int_{\mathbb{R}^d} s\left(\frac{x+x'}{2}, \xi\right) e^{i(x-x')\cdot\xi} u(x')dx'd\xi, \quad x \in \mathbb{R}^d,$$

extends uniquely to an operator bounded in $L^2(\mathbb{R}^d)$, and there exists a constant c_0 independent of s such that

$$\|\mathrm{Op}^{\mathrm{w}}(s)\| \leq c_0 \|s\|_{\Gamma(\mathbb{R}^{2d})}. \tag{15}$$

The operator $\mathrm{Op}(s)$ *is called a* pseudodifferential operator *with Weyl symbol* s.

Assume that $s \in \mathcal{S}'(\mathbb{R}^{2d})$ verifies $\hat{s} \in L^1(\mathbb{R}^{2d})$. Then the operator $\mathrm{Op}^{\mathrm{w}}(s)$ extends to an operator bounded in $L^2(\mathbb{R}^d)$, and we have

$$\|\mathrm{Op}^{\mathrm{w}}(s)\| \leq (2\pi)^{-d} \|\hat{s}\|_{L^1(\mathbb{R}^{2d})} \tag{16}$$

(see [7], Lemma 18.6.1).

For $\ell \in [1, \infty)$, let S_ℓ be the Schatten–von Neumann class, and $\|\cdot\|_\ell$ its respective norm. Assume now that $s \in L^2(\mathbb{R}^{2d})$. Then, evidently, the operator defined above extends to a Hilbert–Schmidt operator in $L^2(\mathbb{R}^d)$, and we have

$$\|\mathrm{Op}^{\mathrm{w}}(s)\|_2^2 = (2\pi)^{-d} \int_{\mathbb{R}^{2d}} |s(x,\xi)|^2 \, dx \, d\xi = (2\pi)^{-d} \int_{\mathbb{R}^{2d}} |\hat{s}(x,\xi)|^2 \, dx \, d\xi. \tag{17}$$

Proposition 3.1. *Let $\kappa : \mathbb{R}^{2d} \mapsto \mathbb{R}^{2d}$, $d \geq 1$, be a linear symplectic transformation, $s_1 \in \Gamma(\mathbb{R}^{2d})$, and $s_2 := s_1 \circ \kappa$. Then, there exists a unitary operator $U : L^2(\mathbb{R}^{2d}) \mapsto L^2(\mathbb{R}^{2d})$ such that*

$$\mathrm{Op}^{\mathrm{w}}(s_2) = U^* \mathrm{Op}^{\mathrm{w}}(s_1) \, U. \tag{18}$$

The operator U is called the *metaplectic operator* corresponding to the linear symplectic transformation κ. There is a one-to-one correspondence between metaplectic operators and linear symplectic transformations, except for a factor of modulus one. Proposition 3.1 extends to a large class of not necessarily bounded operators. In particular, it holds for Weyl pseudodifferential operators with quadratic symbols (see [7], Th. 18.5.9, and [8], Ch. 7, Th. A.2, for further details).

3.2. *Metaplectic mapping of $P_q V P_q$*

For $(x_1, x_2, \xi_1, \xi_2) \in \mathbb{R}^4$, set

$$\varkappa_B(x_1, x_2, \xi_1, \xi_2) := \left(\frac{1}{\sqrt{B}}(x_1 - \xi_2), \frac{1}{\sqrt{B}}(\xi_1 - x_2), \frac{\sqrt{B}}{2}(\xi_1 + x_2), -\frac{\sqrt{B}}{2}(\xi_2 + x_1) \right).$$

The transformation $\varkappa_B$ is linear and symplectic, denote by $\mathcal{U}_B$ its corresponding metaplectic operator in $L^2(\mathbb{R}^2)$. Note that

$$\mathcal{H} \circ \varkappa_B(x_1, x_2, \xi_1, \xi_2) = B\left(\xi_1^2 + x_1^2\right), \quad (x_1, x_2, \xi_1, \xi_2) \in \mathbb{R}^4, \tag{19}$$

where $\mathcal{H}$ (defined in Sec. 2) is the Weyl symbol of H_0. Applying Prop. 3.1 we obtain the following unitary equivalence of operators:

Corollary 3.1.

(i) *We have*

$$\mathcal{U}_B^* \, H_0 \, \mathcal{U}_B = B(h \otimes I_{x_2}), \tag{20}$$

$$\mathcal{U}_B^* \, P_q \, \mathcal{U}_B = (p_q \otimes I_{x_2}), \quad q \in \mathbb{Z}_+. \tag{21}$$

(ii) If $V \in \Gamma(\mathbb{R}^2)$, $\rho \in [0, \infty)$, then

$$\mathcal{U}_B^* V \mathcal{U}_B = \mathrm{Op}^{\mathrm{w}}(\mathbf{V}_B), \tag{22}$$

where

$$\mathbf{V}_B(x_1, x_2, \xi_1, \xi_2) := V(B^{-1/2}(x_1 - \xi_2), B^{-1/2}(\xi_1 - x_2)), \quad (x_1, x_2, \xi_1, \xi_2) \in \mathbb{R}^4.$$

Furthermore, the operator $P_q V P_q$ is conveniently mapped by $\mathcal{U}_B$ into a pseudodifferential operator, as is shown by the next

Theorem 3.1. *For any $V \in L^1(\mathbb{R}^2) + L^\infty(\mathbb{R}^2)$ and $q \in \mathbb{Z}_+$, we have*

$$\mathcal{U}_B^* P_q V P_q \mathcal{U}_B = (p_q \otimes \mathrm{Op}^{\mathrm{w}}(V_B * \Psi_q)), \tag{23}$$

where

$$V_B(x_1, x_2) := V(-B^{-1/2} x_2, -B^{-1/2} x_1), \quad (x_1, x_2) \in \mathbb{R}^2.$$

3.3. *Analysis of* $\mathrm{Op}^{\mathrm{w}}(V_B * \Psi_q)$

In order to obtain information about the operator $\mathrm{Op}^{\mathrm{w}}(V_B * \Psi_q)$, we introduce a suitable approximation. Set $k > 0$ and define the distribution $\delta_k \in \mathcal{S}'(\mathbb{R}^2)$ by

$$\delta_k(\phi) := \frac{1}{2\pi} \int_0^{2\pi} \phi(k \cos\theta, k \sin\theta) d\theta, \quad \phi \in \mathcal{S}(\mathbb{R}^2).$$

Proposition 3.2. *Assume that $V \in \mathcal{S}_1^{-\rho}(\mathbb{R}^2)$ with $\rho > 0$. Then,*

$$\|\mathrm{Op}^{\mathrm{w}}(V_B * \Psi_q) - \mathrm{Op}^{\mathrm{w}}(V_B * \delta_{\sqrt{2q+1}})\|_2 = O(\lambda_q^{-3/4}), \quad q \to \infty. \tag{24}$$

Estimate (24) can be interpreted as a manifestation of the equipartition of the eigenfunctions of the harmonic oscillator h, i.e. the appropriate weak convergence as q tends to infinity of the Wigner function associated with the qth normalized eigenfunction of h to the measure invariant with respect to the classical flow (see [9–11] for related results).

Note that the convolution of a function $u \in \mathcal{S}_1^{-\rho}(\mathbb{R}^2)$ with δ_k corresponds to averaging it over circles of radius k, so that $u * \delta_1(x) = \mathring{u}(x)$, $x \in \mathbb{R}^2$. Applying Calderón–Vaillancourt-type estimates we obtain norm bounds for this operator in terms of the parameter k (see [1, 2] for further details).

Proposition 3.3. *Assume that $V \in \mathcal{S}_1^{-\rho}(\mathbb{R}^2)$ with $\rho > 0$. Then,*

$$\|\mathrm{Op}^{\mathrm{w}}(V_B * \delta_k)\| = \begin{cases} O(k^{-\rho}) & \text{if } \rho \in (0,1), \\ O(k^{-1} \ln k) & \text{if } \rho = 1, \\ O(k^{-1}) & \text{if } \rho > 1, \end{cases} \quad k \to \infty. \tag{25}$$

3.4. *Norm estimates for Berezin–Toeplitz operators*

Various norm estimates for Berezin–Toeplitz operators $P_q V P_q$ in Schatten–von Neumann spaces have been known for a long time. The following proposition contains estimates of $\|P_q V P_q\|_\ell$ by quantities independent of q.

Proposition 3.4. (Refs. [12, 13]) *Let* $V \in L^\ell(\mathbb{R}^2)$, $\ell \in [1, \infty)$. *Then, for each* $q \in \mathbb{Z}_+$ *we have* $P_q V P_q \in S_\ell$ *and*

$$\|P_q V P_q\|_\ell^\ell \leq \frac{B}{2\pi} \|V\|_{L^\ell(\mathbb{R}^2)}^\ell. \tag{26}$$

Estimates (26) are not sufficient for our purposes. In Cor. 3.2 below we obtain estimates of $\|P_q V P_q\|_\ell$, by quantities which decay as $q \to \infty$. The first step is to obtain appropriate estimates of the operator norms $\|P_q V P_q\|$, which is done in the following

Proposition 3.5. *Assume that* V *verifies Cond.* 1.1 *with* $\rho \in (0, \infty)$. *Then, there exists a constant* c_∞ *independent of* q *such that*

$$\|P_q V P_q\| \leq c_\infty \begin{cases} \lambda_q^{-\rho/2} & \text{if } \rho \in (0,1), \\ \lambda_q^{-1/2} |\ln \lambda_q| & \text{if } \rho = 1, \\ \lambda_q^{-1/2} & \text{if } \rho > 1, \end{cases} \qquad q \in \mathbb{Z}_+. \tag{27}$$

These estimates are obtained by approximating the unitarily equivalent operator $\mathrm{Op}^{\mathrm{w}}(V_B * \Psi_q)$ by $\mathrm{Op}^{\mathrm{w}}(V_B * \delta_{\sqrt{2q+1}})$ and considering Prop. 3.3.

Estimates (27) are sharp for $\rho \neq 1$. If $\{\phi_{k,q}\}_{k=-q}^{\infty}$ is the so-called *angular-momentum* orthonormal basis of the Hilbert space $P_q L^2(\mathbb{R}^2)$, $q \in \mathbb{Z}_+$, (see, for instance [14]) and χ_R is the characteristic function of a disk of radius $R > 0$ centered at the origin, then

$$\liminf_{q \to \infty} \lambda_q^{1/2} \langle \chi_R \phi_{0,q}, \phi_{0,q} \rangle > 0, \tag{28}$$

which implies the sharpness of estimate (27) for $\rho > 1$. Similarly, if $\rho \in (0, 1)$ we can show that

$$\liminf_{q \to \infty} \lambda_q^{1/2} \langle \langle \cdot \rangle^{-\rho} \phi_{-q,q}, \phi_{-q,q} \rangle > 0, \tag{29}$$

which entails the sharpness for $\rho \in (0, 1)$ (see [2, 15]). We do not know whether the estimate for $\rho = 1$ is sharp.

Corollary 3.2.

(*i*) *Assume that* V *satisfies Cond.* 1.1 *with* $\rho > 1$. *Then, for each* $\ell > 1/(\rho - 1)$ $P_q V P_q \in S_\ell$. *Moreover, there exists a constant* c_ℓ *such that*

$$\|P_q V P_q\|_\ell \le c_\ell \lambda_q^{\frac{1}{\ell}-\frac{1}{2}}, \quad q \in \mathbb{Z}_+. \tag{30}$$

(*ii*) *Assume that* V *satisfies Cond.* 1.1 *with* $\rho \in (0,1)$. *Then, for each* $\ell > 2/\rho$ $P_q V P_q \in S_\ell$. *Moreover, there exists a constant* c_ℓ *such that*

$$\|P_q V P_q\|_\ell \le c_\ell \lambda_q^{\frac{1}{\ell}-\frac{\rho}{2}} (1 + |\ln \lambda_q|)^{1/\ell}, \quad q \in \mathbb{Z}_+. \tag{31}$$

Equation (31) should be considered an *a priori* estimate, sufficient for the purposes of this exposition, but not necessarily sharp.

4. Proofs of the Main Results

In this section we outline the proofs of our main theorems stated in Sec. 1. More detailed proofs and calculations can be found in [1] for the short-range case, and in [2] for the long-range case.

4.1. *Cluster size estimates*

The proof of Th. 1.1 is largely a corollary of the norm estimates for Berezin–Toeplitz operators contained in Prop. 3.5 and the application of the Birman–Schwinger principle. Set $R_0(\lambda) := (H_0 - \lambda)^{-1}$, $\lambda \in \mathbb{C} \setminus \sigma(H_0)$. The operator $|V|^{1/2} R_0(\lambda) V^{1/2}$ can be estimated in norm by the expression

$$\||V|^{1/2} R_0(\lambda) V^{1/2}\| \le \sum_{k=q-m}^{q+m} |\lambda_k - \lambda|^{-1} \|P_k |V| P_k\| + \||V|^{1/2} \tilde{R}_0(\lambda; q, m) |V|^{1/2}\|, \tag{32}$$

where

$$\tilde{R}_0(\lambda; q, m) := R_0(\lambda) - \sum_{k=q-m}^{q+m} (\lambda_k - \lambda)^{-1} P_k.$$

Now, taking $m \in \mathbb{N}$ large enough that $\|V\|_{L^\infty(\mathbb{R}^2)} \le \lambda_m/2$ and

$$\tilde{C}_1 \lambda_q^{-1/2} < |\lambda_q - \lambda| \le B$$

(resp. $\tilde{C}_2 \lambda_q^{-\rho/2} < |\lambda_q - \lambda| \le B$), the above estimate ensures that the norm of the Birman–Schwinger operator is strictly less than one for q large enough.

4.2. *Short-range case*

4.2.1. *Reduction to a trace formula for Berezin–Toeplitz operators*

An application of the Stone–Weierstrass theorem allows us to consider polynomials instead of $C_0^\infty(\mathbb{R}\setminus\{0\})$ functions. Then, a Cauchy formula representation yields

Proposition 4.1. *Let $\ell > 1/(\rho-1)$. Then, for q large enough, the operators $(H-\lambda q)^\ell \chi_{(\lambda_q-B,\lambda_q+B)}(H)$ and $(P_qVP_q)^\ell$ belong to S_1 and*

$$\mathrm{Tr}\,(H-\lambda q)^\ell \chi_{(\lambda_q-B,\lambda_q+B)}(H) = \mathrm{Tr}\,(P_qVP_q)^\ell + o(\lambda_q^{-(\ell-1)/2}), \quad q\to\infty. \tag{33}$$

Once Eq. (33) is established, results in Secs. 3.3 and 3.4 allow us to consider the traces of the operators $\mathrm{Op}^{\mathrm{w}}(V_B * \delta_{\sqrt{2q+1}})^\ell$.

4.2.2. *Trace asymptotics*

Proposition 4.2. *Assume that V belongs to $C_0^\infty(\mathbb{R}^2)$. Then,*

$$\lim_{q\to\infty} \lambda_q^{(\ell-1)/2}\mathrm{Tr}\,(\mathrm{Op}^{\mathrm{w}}(V_B * \delta_{\sqrt{2q+1}}))^\ell = \frac{B^\ell}{2\pi}\int_{\mathbb{S}^1}\int_{\mathbb{R}} \tilde{V}(\omega,s)^\ell\, ds\, d\omega \tag{34}$$

for every integer $\ell \in \mathbb{N}$.

The proof of Prop. 4.2 relies chiefly on using the integral representation of the trace given by the composition formula for symbols of Weyl pseudodifferential operators. Thus,

$$\mathrm{Tr}\,(\mathrm{Op}^{\mathrm{w}}(t_k))^\ell = (2\pi)^{-\ell+1}\int_{\mathbb{R}^{2(\ell-1)}} \hat{t}_k(-\xi_{\ell-1})\hat{t}_k(\xi_{\ell-1}-\xi_{\ell-2})\dots$$
$$\dots\hat{t}_k(\xi_1)e^{\frac{i}{2}\sum_{j=2}^{\ell-1}\sigma(\xi_j,\xi_{j-1})}d\xi, \tag{35}$$

with $t_k := V_B * \delta_k$, so that

$$\hat{t}_k(\xi) = \frac{1}{2\pi}\hat{V}_B(\xi)\int_{\mathbb{S}^1} e^{-ik\omega\xi}d\omega, \quad \xi\in\mathbb{R}^2. \tag{36}$$

We obtain Eq. (34) by applying the stationary phase method.

Finally, a continuity argument allows us to extend the result, for $\ell > 1/(\rho-1)$, from $C_0^\infty(\mathbb{R}^2;\mathbb{R})$ to the set of functions verifying Cond. 1.1 with $\rho > 1$, which completes the proof of Th. 1.2.

4.3. *Long-range case*

Methodology used for the short-range case is not readily extensible to the long-range scenario. Conversely, most of the results and methods exposed below do not adapt well to the previous scenario.

4.3.1. *Reduction to a trace formula for Berezin–Toeplitz operators*

The first step in the proof of Th. 1.3 is the passage to Berezin–Toeplitz operators.

Proposition 4.3. *Assume that V verifies Cond. 1.2 for some $\rho \in (0,1)$. Then,*

$$\mathrm{Tr}\varphi(\lambda_q^{\rho/2}(H-\lambda_q)) = \mathrm{Tr}\varphi(\lambda_q^{\rho/2}P_qVP_q)) + o(\lambda_q), \quad q \to \infty. \tag{37}$$

Note that, in contrast to the short-range case, no approximation by polynomials is used, instead, we work directly with the function φ. Proposition 4.3 amounts essentially to approximating H by its Weinstein average $\langle H \rangle$.

Definition 4.1 (Weinstein average). *Assume that $V \in L^\infty$ and set*

$$\langle V \rangle := \frac{B}{\pi}\int_0^{\pi/B} e^{-itH_0}Ve^{itH_0}dt = \sum_{s\in\mathbb{Z}_+} P_sVP_s, \tag{38}$$

$$\langle H \rangle := H_0 + \langle V \rangle. \tag{39}$$

Note that a sufficient condition for the uniform convergence of the series defining $\langle V \rangle$ in Eq. (38), is that the norm of P_qVP_q tends to zero as q goes to infinity, which holds true according to Prop. 3.5.

Proposition 4.4. *Under the hypotheses of Cond. 1.2 we have*

$$\mathrm{Tr}\varphi(\lambda_q^{\rho/2}(H-\lambda_q)) = \mathrm{Tr}\varphi(\lambda_q^{\rho/2}(\langle H \rangle - \lambda_q)) + o(\lambda_q), \quad q \to \infty \tag{40}$$

for each $\varphi \in C_0^\infty(\mathbb{R} \setminus \{0\})$.

The main ingredients in the proof of this proposition are the Helffer–Sjöstrand formula used to express the difference of both operators in terms of the difference of their resolvents, and the Schur–Feshbach formula to estimate this difference accordingly. Namely, let $\tilde{\varphi} \in C_0^\infty(\mathbb{R}^2)$ be a suitable almost-analytic continuation of φ, let $\psi := \partial\tilde{\varphi}/\partial\bar{z}$ and set

$$\psi_q(x,y) := \lambda_q^{\rho/2}\psi(\lambda_q^{\rho/2}(x-\lambda_q), \lambda_q^{\rho/2}y), \quad q \in \mathbb{Z}_+,$$

so that

$$\varphi(\lambda_q^{\rho/2}(H-\lambda_q)) = \frac{1}{\pi}\int_{\mathbb{R}^2} \psi_q(x,y)(H-z)^{-1}\,dxdy, \quad z := x+iy \in \mathbb{C}. \tag{41}$$

Iterating the resolvent identity, we can write the above expression as

$$\varphi(\lambda_q^{\rho/2}(H-\lambda_q)) = \frac{(-1)^\ell}{\pi}\int_{\mathbb{R}^2} \psi_q(x,y)((H_0-z)^{-1}V)^\ell (H-z)^{-1}\, dxdy, \tag{42}$$

so that we are able to extract the terms which contribute to the main asymptotic term of the trace of the operator. Indeed, writing

$$((H_0-z)^{-1}V)^\ell = (\lambda_q - z)^{-\ell}(P_qV)^\ell + r_{q,\ell}(z), \tag{43}$$

and taking ℓ to be the smallest integer strictly greater than $\rho/2$, we find that the trace of the part of the operator corresponding to $r_{q,\ell}$ is of order $o(\lambda_q)$, as $q \to \infty$.

We analogously express the trace of the operator obtained by replacing H by $\langle H\rangle$ and we are left with estimating the trace of the difference

$$\frac{(-1)^\ell}{\pi}\int_{\mathbb{R}^2} \psi_q(x,y)(\lambda_q - z)^{-\ell}(P_qV)^\ell \left((H-z)^{-1} - P_q(\langle H\rangle - z)^{-1}\right) dxdy.$$

Application of the Schur–Feshbach formula and estimates in Sec. 3.4 imply that this quantity is of order $o(\lambda_q)$ as q tends to infinity.

Once the approximation by the Weinstein average is established, a simple argument relying on the fact that φ has a compact support shows that for large enough q, the trace of this operator reduces to that of a single Berezin–Toeplitz operator:

Proposition 4.5. *Assume that V verifies Cond.* 1.2. *Then, for each $\varphi \in C_0^\infty(\mathbb{R}\setminus\{0\})$ there exists $q_0 \in \mathbb{Z}_+$ such that*

$$\mathrm{Tr}\varphi(\lambda_q^{\rho/2}(\langle H\rangle - \lambda_q)) = \mathrm{Tr}\varphi(\lambda_q^{\rho/2}P_qVP_q), \quad q \geq q_0. \tag{44}$$

Finally, as q tends to infinity, V_B is integrated over larger and larger circles, so that it can be well approximated by its behavior far from the origin. This is expressed by the following result:

Proposition 4.6. *Assume that V verifies Cond.* 1.2. *Then,*

$$\mathrm{Tr}\varphi(\lambda_q^{\rho/2}P_qVP_q) = \mathrm{Tr}\varphi(\lambda_q^{\rho/2}\mathrm{Op}^{\mathrm{w}}(\mathbb{V}_B * \delta_{\sqrt{2q+1}})) + o(\lambda_q), \quad q\to\infty, \tag{45}$$

for each $\varphi \in C_0^\infty(\mathbb{R}\setminus\{0\})$.

The proof of this fact makes use of the technical assumption in Eq. (2), and is based on certain estimates of trace-class norms of operators of the type $\varphi(T+Q)-\varphi(T)$, where $\varphi \in C_0^\infty(\mathbb{R}\setminus\{0\})$ and T and Q are appropriate self-adjoint compact operators. The main tools used to obtain these trace-class estimates are the results of [16].

4.3.2. *Trace asymptotics*

Proposition 4.7. *Assume that V verifies Cond.* 1.2. *Then, we have that*

$$\lim_{q\to\infty} \lambda_q^{-1} \mathrm{Tr}\varphi(\lambda_q^{\rho/2}\mathrm{Op}^{\mathrm{w}}(\mathbb{V}_B * \delta_{\sqrt{2q+1}})) = \frac{1}{2\pi B}\int_{\mathbb{R}^2} \varphi(B^\rho \mathring{\mathbb{V}}(x))dx. \tag{46}$$

The proof of Prop. 4.7 relies on the unitary equivalence between the operator $\lambda_q^{\rho/2}\mathrm{Op}^{\mathrm{w}}(\mathbb{V}_B * \delta_{\sqrt{2q+1}})$ and the pseudodifferential operator whose Weyl symbol is given by

$$s_\hbar(x,\xi) := B^\rho \mathring{\mathbb{V}}_1(x,\hbar\xi), \quad (x,\xi) \in \mathbb{R}^2,$$

with $\hbar := (2q+1)^{-1}$. If the symbol were regular, standard semiclassical results would then imply Eq. (46) (see, for instance [8], Th. 9.6). However, $\mathring{\mathbb{V}}_1$ has a singularity at $\mathbb{S}^1$, so a suitable approximation by smooth symbols is required to establish the result. For this, analogous estimates to those given in Prop. 3.4 for weak Schatten–von Neumann classes are used and the desired result is finally obtained.

Acknowledgements

The authors would like to acknowledge the financial support of *Núcleo Científico de Física Matemática* RC120002, as well as that of the Chilean Science Foundation *Fondecyt* under Grant 1130591.

This text was partly written while the authors were visiting Institut de Mathématiques de Bordeaux at Université Bordeaux 1, gratitude for the warm hospitality is kindly due.

Bibliography

[1] A. Pushnitski, G. D. Raikov and C. Villegas-Blas, *Comm. Math. Phys.* **320**, 425 (2013).
[2] T. Lungenstrass and G. D. Raikov, *Ann. H. Poincaré* **15**, 1523 (2014).
[3] S. Helgason, *The Radon Transform*, 2 edn. (Birkhäuser, 1999).
[4] V. I. Arnold, *Mathematical methods of classical mechanics* (Springer-Verlag, 1989).
[5] A.-P. Calderón and R. Vaillancourt, *Proc. Nat. Acad. Sci. U.S.A.* **69**, 1185 (1972).
[6] H. O. Cordes, *J. Funct. Anal.* **18**, 115 (1975).
[7] L. Hörmander, *The analysis of linear partial differential operators. III Pseudo-differential operators*, Grundlehren der Mathematischen Wissenschaften [Fundamental Principles of Mathematical Sciences], Vol. 274 (Springer-Verlag, Berlin, 1994). Corrected reprint of the 1985 original.

[8] M. Dimassi and J. Sjöstrand, *Spectral asymptotics in the semi-classical limit*, London Mathematical Society Lecture Note Series, Vol. 268 (Cambridge University Press, Cambridge, 1999).
[9] A. Bouzouina and S. De Bièvre, *Comm. Math. Phys.* **178**, 83 (1996).
[10] Y. Colin de Verdière, *Comm. Math. Phys.* **102**, 497 (1985).
[11] S. Zelditch, *Duke Math. J.* **55**, 919 (1987).
[12] C. Fernández and G. Raikov, *Ann. Henri Poincaré* **5**, 381 (2004).
[13] G. D. Raikov, *Comm. Partial Differential Equations* **15**, 407 (1990).
[14] G. D. Raikov and S. Warzel, *Rev. Math. Phys.* **14**, 1051 (2002).
[15] E. Korotyaev and A. Pushnitski, *J. Funct. Anal.* **217**, 221 (2004).
[16] D. Potapov and F. Sukochev, *Acta Math.* **207**, 375 (2011).

On long range behaviour of van der Waals force

I. Anapolitanos

Department of Mathematics, University of Stuttgart, Stuttgart, Germany

I. M. Sigal

Department of Mathematics, University of Toronto, Toronto, Canada

We review our recent results [2] proving van der Waals-London's law of decay of the van der Waals force for a collection of neutral atoms at large separation and present the proof is a special case.

1. Introduction

Van der Waals force between atoms and molecules plays a fundamental role in quantum chemistry, physics and material sciences. For instance, it defines the chemical character of many organic compounds and enable geckos - which can hang on a glass surface using only one toe - to climb on sheer surfaces (see the entry "Van der Waals force" in Wikipedia). It explains why water condenses from vapor as well as many properties of molecular compounds, including crystal structures (e.g. the shapes of snowflakes), melting points, boiling points, heats of fusion and vaporization, surface tension, and densities. It forces gigantic molecules like enzymes, proteins, and DNA into the shapes required for biological activity (see Senese [31]).

A microscopic explanation of this force was given by F. London soon after the discovery of quantum mechanics and was one of its early triumphs. This heuristic explanation showed that this force has a universal behavior at large distances - it decays as the inverse sixth power of the distance between atoms. This behavior was confirmed in [24] by proving - through a sophisticated test function construction - an upper bound. In our recent result [2] we have given a complete proof of the van der Waals - London decay law for atoms. In this contribution, we review our results [2] and present the proof in a special case. This proof displays most of the key techniques of the proof of the general result.

Let $-e$ and m denote the electron charge and mass. Consider a system of M multielectron atoms which we call a molecule though we do not assume binding between atoms. In the units where $\hbar = 1$ the Hamiltonian of the system is

$$H_{\text{mol}} = \sum_{i=1}^{N}(-\frac{1}{2m}\Delta_{x_i} - \sum_{j=1}^{M}\frac{e^2 Z_j}{|x_i - y_j|}) + \sum_{i<j}^{1,N}\frac{e^2}{|x_i - x_j|} - \sum_{j=1}^{M}\frac{1}{2m_j}\Delta_{y_j} + \sum_{i<j}^{1,M}\frac{Z_i Z_j e^2}{|y_i - y_j|}.$$

Here N is the total number of electrons, $x_i, y_i \in \mathbb{R}^3$ denote the coordinates of the electrons and the nuclei, respectively, eZ_j is the charge of the j-th nucleus, m_j is the mass of the j-th nucleus and Δ_{x_j} is the Laplacian and gradient acting on the coordinate x_j. We consider a system of neutral atoms so we must have $\sum_{j=1}^{M} Z_j = N$. The operator H_{mol} acts on the subspace $\mathcal{H}_{\text{fermi}}^{\text{mol}}$ of the space $L^2(\mathbb{R}^{3(N+M)})$, which accounts for the fact that the electrons are identical particles and are fermions and therefore they obey the Fermi-Dirac statistics. Also, possibly, some of the nuclei are identical and obey either the Fermi-Dirac or Bose-Einstein statistics (more details are given below). For $M = 1$, H_{mol} is the Hamiltonian of the atom with the nucleus of charge $eZ = eZ_1$.

To define the interaction energy, let a be the decomposition of the molecule into neutral atoms and/or ions, each with its own nucleus, and let H_a be the sum of the corresponding atomic or ionic Hamiltonians (see the next section for precise definitions). We define the energy $E(\infty)$ of the system with the atoms or ions infinitely far from each other, as

$$E(\infty) = \min_a E_a, \qquad \text{where} \qquad E_a := \inf \sigma(H_a). \tag{1}$$

(Of course, one expects that $E(\infty)$ is obtained by taking the minimum over the atomic decompositions only, see the discussion below.) Furthermore, let $y = (y_1, ..., y_M)$ be the collection of the nuclear co-ordinates. The interaction (or cohesive) energy $W(y)$ between atoms in this system is defined as

$$W(y) := E(y) - E(\infty), \tag{2}$$

where $E(y)$ is the ground state energy of the system with positions of the nuclei fixed, i.e. the ground state energy of the Hamiltonian

$$H_N(y) = \sum_{i=1}^{N}(-\frac{1}{2m}\Delta_{x_i} - \sum_{j=1}^{M}\frac{e^2 Z_j}{|x_i - y_j|}) + \sum_{i<j}^{1,N}\frac{e^2}{|x_i - x_j|} + \sum_{i<j}^{1,M}\frac{Z_i Z_j e^2}{|y_i - y_j|}, \tag{3}$$

acting on the subspace $\mathcal{H}_{\text{fermi}}$ of the space $L^2(\mathbb{R}^{3N})$, which accounts for the Fermi-Dirac statistics of electrons.

The Hamiltonian (3) is called the Born-Oppenheimer Hamiltonian. It arises as the central object in the key technique in solving the eigenvalue problem for H_{mol}, which is called the Born-Oppenheimer approximation, and which plays an important role in quantum chemistry (for example minima of its ground state energy $E(y)$ determine shapes of molecules). In this approximation the ground state energy

$E(y)$ (or the energy of an excited state) of $H_N(y)$ is considered as the potential energy of the nuclear motion, which leads to the Hamiltonian

$$H_{\text{nucl}} := -\sum_{j=1}^{M} \frac{1}{2m_j}\Delta_{y_j} + E(y).$$

One expects that due to the fact that the ratio of the electron and nuclear mass being very small, the eigenvalues of H_{nucl} give a good approximation to the eigenvalues of H_{mol}. For rigorous results on the Born-Oppenheimer approximation see original articles [5, 13, 19, 20, 27] and the textbook [12].

One can define the interaction energy to any order in the electron to nuclei mass ratio (see the paragraph,"Beyond Born-Oppenheimer approximation" at the end of this introduction), but this will produce only correspondingly small corrections to our results.

It is expected, after van der Waals, that asymptotically $W(y)$ is a sum of pair interactions, W_{ij}, which are attractive and decay at infinity as $-|y_i - y_j|^{-6}$. More precisely, one expects that

$$W(y) = -\sum_{i<j}^{1,M} \frac{e^4\sigma_{ij}}{|y_i - y_j|^6} + O(\frac{e^4}{R^7}), \tag{4}$$

where σ_{ij} are positive constants depending only on the parameters of the pair of atoms i, j and

$$R = \min\{|y_i - y_j| : 1 \le i, j \le M, i \neq j\}, \tag{5}$$

provided that R is large enough. Here and in what follows the remainder $O(\dots)$ signifies the behaviour in y and might depend on N and M, see however the remark below.

Let $E_{m,n}$, $n \le Z_m$, denote the ground state energy of the ion with a nucleus of charge eZ_m and $Z_m - n$ electrons and $E_m = E_{m,0}$, the ground state energy of the atom corresponding to the m-th nucleus of charge eZ_m. We formulate a property of many body systems playing an important role below:

(E) $\quad \sum_{i=1}^M E_i < \sum_{i=1}^M E_{i,n_i}, \quad \forall (n_1, \dots n_M) : \sum_i n_i = 0, \; \sum_i |n_i| > 0.$

We discuss this property below. Here we mention only that, since the ground state energies depend on the underlying spaces on which the Hamiltonians are defined, Property (E) depends on the symmetry type (see below) and that experiments and numerical computations (see below) show that it holds for all elements for which it was tested, however, theoretically, it proven only for a system of several hydrogen atoms.

The fact that the electrons are identical particles is expressed in the property that $H_N(y)$ commutes with the permutations

$$H_N(y)T_\pi = T_\pi H_N(y), \quad \forall \pi \in S_N, \tag{6}$$

where S_N is the symmetric group (the group of permutations of N indices) and, for $\pi \in S_N$, T_π are the unitary operators given by

$$(T_\pi \Psi)(x_1, ..., x_N) = \Psi(x_{\pi^{-1}(1)}, x_{\pi^{-1}(2)}, ..., x_{\pi^{-1}(N)}). \tag{7}$$

To treat the spectral theory of $H_N(y)$, we have to restrict it to a subspace of a fixed symmetry type, i.e the subspace on which the representation of S_N is multiple to an irreducible one. The possible symmetry types are restricted by the property that electrons are fermions of spin $\frac{1}{2}$.

The simplest symmetry type for fermions of spin $\frac{1}{2}$ is the one corresponding to totally antisymmetric functions, or, what is the same, to the greatest possible spin. In this case the subspace $\mathcal{H}_{\text{fermi}}$ is the space $\mathcal{H}_A = \bigwedge_1^N L^2(\mathbb{R}^3)$ of purely antisymmetric functions and our analysis simplifies considerably. We formulate our main result in this special case.

Theorem 1.1 (van der Waals law; highest spin). *Assume that the Hamiltonian $H_N(y)$ acts on the space $\mathcal{H}_A = \bigwedge_1^N L^2(\mathbb{R}^3)$ and assume Condition (D) stated below. Then for large distances between atoms* (4) *holds for some constants $\sigma_{ij} > 0$ depending on the nature of the atoms i, j, if and only if Property (E) stated above holds.*

The general result, dealing with general symmetry types, can be found in our recent work [2]. To prove the latter result, we have to handle several simple but subtle issues related to the representation theory of the symmetric group S_N. These issues have also complicated our analysis considerably.

Now, we state Condition (D) of Theorem 1.1:

(D) For each atomic decomposition a, the ground state subspace of H_a, acting on functions antisymmetric w.r.to permutations of the particles in the same clusters, is one-dimensional.

One expects that for every symmetry type, the ground state subspace consists of a single copy of the irreducible representation of the symmetry group (which is one-dimensional in the present case), but proving this is an open problem.

Condition (D) is automatically satisfied if the statistics is not taken into account and, consequently, for a system of several hydrogen atoms, where the group of permutation of electrons inside atoms is trivial. This follows from the positivity improving property of $e^{-\beta H}, \beta > 0$, and from Perron-Frobenius theory (see for example [12, 28]).

Theorem 1.1 describes the van der Waals force at a pairwise large separation between the atoms. For intermediate distances, the van der Waals – London law is modified due to overlapping between electron clouds of the atoms and for small distances, the van der Waals forces are repulsive (the energy is positive) as follows from the rough estimate

$$H_N(y) \geq -C + \sum_{i<j}^{1,M} \frac{e^2 Z_i Z_j}{|y_i - y_j|},$$

for some constant C independent of y, implied by the bound $\frac{e^2 Z_m}{|x_n - y_m|} \leq -\alpha\Delta_{x_n} + \beta$, valid for any $\alpha > 0$ and a corresponding $\beta > 0$. Often the interaction energy for two atoms ($M = 2$) is modeled by the Lennard-Jones potential $W_{LJ}(y) = \frac{a}{|y_1-y_2|^{12}} - \frac{b}{|y_1-y_2|^6}$ or by the Buckingham potential $W_B(y) = e^{-c|y_1-y_2|} - \frac{d}{|y_1-y_2|^6}$ where the constants a, b, c, d are determined experimentally.

If the molecules have dipole moments, then one expects a third power law to be true. This can be proven by the techniques developed in this paper.

Now we address Property (E). The following statements are proven in Appendix A:

(a) Property (E) holds for a system of several hydrogen atoms and consequently so is the van der Waals law.

(b) Property (E) follows from the following property

(E′) For any two nuclei i and j, $i \neq j$, in our system and for any integers $m, n \geq 0, l > 0$ satisfying $m+l \leq Z_i$, we have the following energy inequality

$$E_{i,m} + E_{j,-n} < E_{i,m+l} + E_{j,\ n-l}.$$

(c) Property (E) for a system of hydrogen atoms with electron statistics follows from Property (E) without statistics.

The meaning of Property (E′) is that ionization energies of atoms are greater than the electron affinities, where, in a standard terminology, the n-th ionization energy ($n \geq 1$) of an atom is the energy required to remove an electron from its $n-1$-ion and the n-th electron affinity is the energy required to remove an electron from its $-n$ ion. The table below, taken from [25], gives first ionization energies and first electron affinities of atoms. It shows that the first ionization energies are always much larger than the first electron affinities.

It is experimentally verified that the higher (second third and so forth) ionization energies are bigger than the first and it is expected that the higher electron affinities affinities are lower or zero. (Only ions with at most two extra electrons – i.e. of the charges at most $-2e$ – are observed experimentally.

That nuclei can bound only finite number of electrons was proven in [29, 32] (with the bound $\leq 17Z$ on the number of the extra electrons). It was shown in [23], with some improvements by [8, 30], that asymptotically ions are neutral and in [21] that the maximal number of extra electrons $\leq Z + 1$. The latter bound was recently improved to $< 0.22Z + 3Z^{1/3}$ in [26]. Moreover, it was shown in [34, 35] that in the Hartree - Fock approximation the maximal number of extra electrons $\leq$ const.)

Our approach allows for reasonable estimates of the dependance of various remainders $O(\dots)$ on N and M, which leads to an estimate of allowed internuclear distance. (We note that the Lieb-Thiring upper bound in [24] on the interaction energy has a fairly good control of the allowed internuclear distance.)

Remark. Following arguments and ideas of this paper, the first author obtained a simple and well-behaved in N upper bound, not requiring Condition (E) and proved lower bounds with the reasonable $N-$dependence of the remainder estimates ([1]). We outline the proof of the upper bound from [1] in Appendix B.

Atomic number	Element	first Ionization energy (kcal/mol)	Electron affinity (kcal/mol)
1	H	313.5	17.3
2	He	566.9	-
3	Li	124.3	(14)
4	Be	214.9	-
5	B	191.3	(7)
6	C	259.6	29
7	N	335.1	-
8	O	314.0	34
9	F	401.8	79.5
10	Ne	497.2	-
11	Na	118.5	(19)
12	Mg	176.3	-
13	Al	138.0	(12)
14	Si	187.9	(32)
15	P	254	(17)
16	S	238.9	47
17	Cl	300.0	83.4
18	Ar	363.4	(16)
19	K	100.1	-
20	Ca	140.9	-
21	Sc	151.3	-
22	Ti	158	-
23	V	155	-
24	Cr	156	-
25	Mn	171.4	-

(Values in parentheses are estimated by quantum-mechanical calculation and have not been verified experimentally.)

Outline of the approach. Our approach is based on the Feshbach-Schur perturbation argument, with the small parameter being the reciprocal of in the distance between the nuclei. Recall that the Hamiltonian $H_N(y)$ acts on the space $\mathcal{H}_A = \bigwedge_1^N L^2(\mathbb{R}^3)$ of purely antisymmetric functions. In what follows we omit the argument y and write E and H for $E(y)$ and $H_N(y)$, respectively.

Feshbach-Schur method. This method originates in the works of H. Feshbach and I. Schur and was reformulated and generalized in [4]. We follow the textbook presentation of [12]. Let P be an orthogonal projection and $P^\perp = 1 - P$. Introduce the notation $H^\perp = P^\perp H P^\perp$. Assume

(a) $\mathrm{Ran}(P) \subset D(H)$ (domain of H) and therefore $\|HP\| < \infty$;

(b) The operator $(H^\perp - \lambda)$ is invertible.

The Feshbach-Schur method, as applied to the quantum Hamiltonian H, states that if Conditions (a) and (b) are satisfied, then the Feshbach-Schur map

$$F_P(\lambda) = (PHP - U(\lambda))|_{\mathrm{Ran}\, P}, \tag{8}$$

where

$$U(\lambda) := PHP^\perp (H^\perp - \lambda)^{-1} P^\perp HP, \tag{9}$$

is well defined and

$$\lambda \text{ eigenvalue of } H \iff \lambda \text{ eigenvalue of } F_P(\lambda). \tag{10}$$

Moreover, the eigenfunctions of H and $F_P(\lambda)$ corresponding to the eigenvalue λ are connected as

$$H\psi = \lambda\psi \quad \Leftrightarrow \quad F_P(\lambda)\phi = \lambda\phi, \tag{11}$$

where ϕ, ψ are related by the following equations $\phi = P\psi$, $\psi = Q(\lambda)\phi$. Here the family of operators $Q(\lambda)$ is defined as $Q(\lambda) = P - P^\perp(H^\perp - \lambda)^{-1} P^\perp HP$. (Remember that we do not display the $y-$dependence of various objects.)

Orthogonal projection P. To apply the Feshbach - Schur map, we have to choose the orthogonal projection P. We denote by $\mathcal{A}^{at}$ the collection of all decompositions $a = (A_1, ..., A_M)$, with $|A_j| = Z_j$ for all $j = 1, ..., M$. (Its elements correspond to decompositions of our system to neutral atoms.) Let $V_{a,R}$ be the ground state subspace of H_a (the corresponding energies define $E(\infty)$), cut-off at large distances, so that $V_{a,R}$ for various $a \in \mathcal{A}^{at}$ become mutually orthogonal. Let P be the orthogonal projection on span $\{V_{a,R},\ a \in \mathcal{A}^{at}\}$. (Here R signifies the scale on which we perform the cut-off.) As a result of cutting-off the ground state subspaces of H_a, we have the representation

$$P = \sum_{a \in \mathcal{A}^{at}} P_{a,R}, \tag{12}$$

where $P^{\alpha}_{a,R}$ are the orthogonal projections onto $V_{a,R}$.

To show that the Feshbach - Schur map exists, we note that $\mathrm{Ran}(P) \subset D(H)$ and therefore the condition (a) for the existence of the Feshbach map holds. We will prove in Section 4 that, under Property (E), there is a $\gamma > 0$, independent of y s.t. for R large enough, the following *stability bound* holds:

$$H^{\perp} \geq E(\infty) + 2\gamma, \tag{13}$$

where $H^{\perp} = P^{\perp} H P^{\perp}$ and $E(\infty)$ is the lowest possible energy for all possible break-ups of the system, defined in (1) and γ is a positive number independent of N and M. Thus the condition (b) is also satisfied and, for all $\lambda \leq E(\infty)+\gamma$, the Feshbach - Schur map $F_P(\lambda)$ is well defined. Now, to use (11) we have to estimate the two terms on the r.h.s of (8).

Estimate of PHP. First, we use the equations (12), the decomposition $H = H_a + I_a$, where I_a is the sum of the intercluster interactions in the decomposition a, and $H_a P_{a,R} \doteq E(\infty) P_{a,R}$, where the dot above the relations stands for exponentially small additive terms omitted, to obtain $PHP \doteq \sum_{a,b} P_{a,R} I_b P_{b,R}$. Since the supports of $V_{a,R}$, are mutually disjoint, this gives $PHP \doteq \sum_a P_{a,R} I_a P_{a,R}$. Clearly, the map $P_{a,R} I_a P_{a,R}\big|_{\mathrm{Ran}\, P_{a,R}}$ leaves the space $V_{a,R} = \mathrm{Ran}\, P_{a,R}$ invariant and commutes with T_π, $\forall \pi \in S(a)$. Since by Condition (D) $\mathrm{Ran}\, P_{a,R}$ is a space of an irreducible representation of $S(a)$, we conclude that it is a multiple of the identity,

$$P_{a,R} I_a P_{a,R}\big|_{\mathrm{Ran}\, P_{a,R}} = \frac{1}{\mathrm{rank}(P_{a,R})} \mathrm{Tr}(I_a P_{a,R}) P_{a,R},$$

where $\mathrm{rank}(P_{a,R})$ is the rank of $P_{a,R}$. Hence $P_{a,R} I_a P_{a,R}$ can be written as

$$\mathrm{Tr}(I_a P_{a,R}) = e^2 \sum_{i<j}^{1,M} Z_i Z_j \int\int I(z, z'; y_{ij}) \rho_{A_i,R}(z) \rho_{A_j,R}(z') dz dz', \tag{14}$$

where ρ_{A_j} are the one-electron densities of atoms A_j, defined in (21) below, $y_{ij} = y_i - y_j$, and, by the charge neutrality for each atom A_j,

$$I(z, z'; y) = -\frac{1}{|y+z|} - \frac{1}{|y-z'|} + \frac{1}{|y|} + \frac{1}{|y+z-z'|}. \tag{15}$$

Since as shown below, these densities are spherically symmetric, Newton's screening theorem gives

$$PHP = E(\infty) P + \text{ exponentially small terms.}$$

The above implies that the van der Waals decay law could only come from $U(\lambda)$.

Estimate of $U(\lambda)$. To estimate $U(\lambda)$, we use the equations (12), $H = H_a + I_a$ and $H_a P_{a,R} \doteq E(\infty) P_{a,R}$ again and use $P^\perp P_{a,R} = 0$, to obtain $P^\perp H P \doteq \sum_a P^\perp I_a P_{a,R}$. Now, using this in the definition of $U(\lambda)$, we obtain

$$U(E) \doteq \sum_{a,b} P_{a,R} I_a P^\perp R^\perp(E) P^\perp I_b P_{b,R}. \tag{16}$$

Next, we use the variables $z_{km} = x_k - y_m$, $\forall k \in A_m$, $m = 1, \ldots, M$, and expand the intercluster interaction I_a in $|y_{ij}|^{-1}$ and use properties of $P^\alpha_{a,R}$ and the charge neutrality for each atom A_j, to obtain

$$I_a P_{a,R} = \sum_{i<j} \left[\frac{e^2}{|y_{ij}|^3} f_{ij}(z, \widehat{y_{ij}}) P_{a,R} + O\left(\frac{e^2}{|y_{ij}|^4}\right)\right], \tag{17}$$

where $f_{ij}(z, \widehat{y_{ij}}) = \sum_{k \in A_i, l \in A_j} [z_{ki} \cdot z_{lj} - 3(z_{ki} \cdot \widehat{y_{ij}})(z_{lj} \cdot \widehat{y_{ij}})]$. This gives, in particular, $\|I_a P^\alpha_{a,R}\| \lesssim \sum_{i<j} \frac{1}{|y_{ij}|^3}$. Using this bound and elementary geometrical localization arguments, which allow us to pass from $(H^\perp - E)^{-1}$ to $(H^\perp_a - E)^{-1}$, where $H^\perp_a := H_a P^\perp_a$, with $P^\perp_a$ the projection onto the orthogonal complement of the ground state subspace of H_a, and to estimate the terms with $a \neq b$, and pass from E to $E(\infty)$, we find

$$U(E) = \sum_a Q U_{aa} Q + O\left(\frac{e^4}{R^7}\right), \tag{18}$$

where $U_{aa} := P_{a,R} I_a P^\perp_a R^\perp_a P^\perp_a I_a P_{a,R}$, with $R^\perp_a := (H^\perp_a - E(\infty))^{-1}$. Finally, since $P_{a,R}$ is a rank-one orthogonal projection, the above implies

$$U_{aa} - \frac{1}{\operatorname{rank} P_{a,R}} \operatorname{Tr}(U_{aa} P_{a,R}) P_{a,R}. \tag{19}$$

The terms $\operatorname{Tr}(U_{aa} P_{a,R})$ can be easily computed: using (17) and showing that the terms with $(ij) \neq (i'j')$ or $(ij) = (i'j')$, but $(kl) \neq (k'l')$, coming from different I_a's in the expression for $\operatorname{Tr}(U_{aa} P_{a,R})$ can be written as integrals of odd function and therefore vanish. This gives

$$U_{aa} = \frac{1}{\operatorname{rank} P_{a,R}} \sum_{i<j} \frac{e^4}{|y_{ij}|^6} \operatorname{Tr}(f_{ij} R^\perp_a f_{ij} P_{a,R}) + O\left(\frac{e^4}{R^7}\right),$$

which yields the van der Waals decay law. (The terms in the sum can be further simplified.)

Remark. Due to (17), for the charge neutral systems, the term PHP always starts, at least, with $\sum_{i<j} |y_i - y_j|^{-3}$.

Beyond Born-Oppenheimer approximation. Let $\psi_{\mathrm{BO}}(x, y)$ be the ground state of $H_N(y)$, normalized as $\int |\psi_{\mathrm{BO}}(x, y)|^2 dx = 1$. Using the Feshbach - Schur map with the orthogonal projection

$$(Pf)(x,y) = \psi_{\mathrm{BO}}(x,y) \int \overline{\psi_{\mathrm{BO}}(x,y)} f(x,y) dx$$

(i.e. integrating out the electronic degrees of freedom), as it is done e.g. in [12], we can show that

$$\lambda \in \sigma_d(H_{\mathrm{mol}}) \;\leftrightarrow\; \lambda \in \sigma_d(H_{\mathrm{nucl}}(\lambda)),$$

with the corresponding eigenfunctions related accordingly. Here $H_{\mathrm{nucl}}(\lambda)$ is the operator on $L^2(\mathbb{R}^{3M})$, defined by

$$H_{\mathrm{nucl}}(\lambda) := -\sum_{j=1}^{M} \frac{1}{2m_j} \Delta_{y_j} + E_\kappa(y,\lambda),$$

where $\kappa := m / \min_j m_j$ and

$$E_\kappa(y,\lambda) = E(y) + \sum_1^M \frac{1}{2m_j} \int |\nabla_{y_j} \psi_{BO}|^2 dx + O(\kappa^2),$$

with $O(\kappa^2)$ standing for a non-local operator of the indicated order. The energy $E_\kappa(y)$ (not potential anymore) can be used to define the interaction energy in all orders.

Organization of the paper. In Section 2 we discuss preliminaries of quantum many body systems. In Section 3 we prove Theorem 1.1 modulo the stability estimate (13) which is proven in Section 4. In Appendix A, we prove several statements about Property (E) formulated above.

Notation. We collect here general notation used in this paper. In the text C will denote a positive constant which might be different from one equation to the other, but which is independent of the nuclear co-ordinates $y_1, \ldots, y_M$. We will use the notation $\lesssim$ for inequalities that are true up to such a constant. We will write $A \doteq B$, and $A \dot{\leq} B$, $A \dot{\geq} B$ if $\|A - B\| \lesssim e^{-\frac{1}{6}\theta R}$, $A \leq B + Ce^{-\frac{1}{6}\theta R}$ and $A \geq B - Ce^{-\frac{1}{6}\theta R}$, for some $C > 0$, respectively, in an appropriate norm, where θ is the constant appearing in (29) and (30).

For any Banach space X, we denote $B(X) := \{f : X \to X : f \text{ linear and bounded}\}$. For an operator A, the symbols $\sigma(A)$ and $\sigma_{\mathrm{ess}}(A)$ stand for the spectrum and the essential spectrum, correspondingly. We also use the notation $\langle x \rangle = (1 + |x|^2)^{\frac{1}{2}}$. Finally, $\|\cdot\|$ will denote the L^2−norm of a function or the $B(L^2)$−norm of an operator, depending on the context, and the symbol $O(\delta)$ is understood in this norm. For an orthogonal projection P, we denote by $P^{\perp}$, the complementary projection $P^{\perp} := \mathbf{1} - P$ and for a vector Φ, we denote by P_{Φ} the rank-one orthogonal projection onto Φ.

Acknowledgements. The first author is grateful to Volker Bach, Alessandro Giuliani and Artem Dudko for useful discussions, and especially to Marcel Griesemer for numerous inspiring discussions. His research was supported in part by NSERC under Grant No. NA7901 and in part by the German Science Foundation under Grant No. 3213/1-1. The second author is grateful to Jérémy Faupin, Alessandro Pizzo and, especially Jürg Fröhlich and Elliott Lieb for very useful and inspiring discussions. His interest in the problem of the van der Waals forces derived from several talks by and conversations with Herbert Spohn on the Casimir- Polder effect. His research was supported in part by NSERC under Grant No. NA7901.

2. Preliminaries about many body systems

General properties of $H_N(y)$. The general information on the spectrum of the Hamiltonian (3) is given in the following theorem which is a special case of the HVZ Theorem (see e.g. [6, 12, 17]).

Theorem 2.1 (HVZ theorem). $\sigma_{ess}(H_N(y)) = [\Sigma, \infty)$, *where*

$$\Sigma = \inf \sigma(H_{N-1}(y)).$$

This theorem says that the essential (continuous) spectrum of $H_N(y)$ originates from the molecule shedding of an electron which moves freely at infinity and therefore whose energy spectrum changes continuously. The next results, due to G. Zhislin and J.-M. Combes and L. Thomas, respectively, show that $H_N(y)$, as well as each atom, has a well-localized ground state (see e.g. [6, 12, 17]):

Theorem 2.2 (Zhislin theorem). *The operator $H_N(y)$ has infinite number of eigenvalues, E_j, below its essential spectrum, $E_j < \Sigma$.*

Theorem 2.3 (Combes - Thomas bound). *The eigenfunctions, Φ_j, of $H_N(y)$, corresponding to the eigenvalues, $E_j < \Sigma$, are exponentially localized:*

$$|\Phi_j(x)| \leq Ce^{-\delta|x|}, \tag{20}$$

for any $\delta < \sqrt{\Sigma - E_j}$. Here $x = (x_1, \ldots, x_N)$.

The last two theorems show that the atoms and Born-Oppenheimer molecules are stable in the sense that at sufficiently low energy they are well localized in the space. However, it says nothing about stability of true molecules. The likely source of instability of molecules is not shedding of an electron but breaking up into atoms or ions which in total have the same energy as the molecule. So far the only molecule proven to be stable is the hydrogen molecule H_2 (see [10]).

Our next result concerns properties of the one-electron densities, ρ_A, mentioned in the introduction. Consider a Hamiltonian H_A of an atom or ion (i.e. H_A is of the form (25) below) and let H_A be this Hamiltonian restricted to the subspace of the symmetry type α. Let, furthermore, P_A be the orthogonal projection onto the ground state subspace of H_A. We define the one-electron density, ρ_A, for H_A, say, through the trace relation by the condition

$$\mathrm{Tr}(b\rho_A) = \mathrm{Tr}[(b \otimes \mathbf{1})P_A], \tag{21}$$

for any one-electron operator b. (ρ_A can be written explicitly in terms of any orthonormal basis in $\mathrm{Ran}\, P_A$, see Appendix B of [2]. Also, the one-electron density can be associated to any orthogonal projection on $L^2(\mathbb{R}^{3N})$.) We have

Proposition 2.1. *Let H_A be a Hamiltonian of an atom or ion. Then the one electron density of the ground state subspace of H_A is spherically symmetric. (If H_A is defined on the entire space (no symmetry restriction) then the ground state of H_A is spherically symmetric.)*

Proof. By the Riesz formula for eigen-projections, P_A and therefore $P_{A,R}$ commutes with any rotation, $P_{A,R}T_R = T_R P_{A,R}$, $\forall R \in O(3)$, where

$$T_R\Phi(z_1, ..., z_{|A|}) = \Phi(R^{-1}z_1, ..., R^{-1}z_{|A|}), \tag{22}$$

we have $\mathrm{Tr}[T_R^{-1}(b \otimes \mathbf{1})P_{A,R}T_R] = \mathrm{Tr}[(t_R^{-1}bt_R \otimes \mathbf{1})P_{A,R}]$, where t_R denotes the one-electron rotation. This, together with the definition $\mathrm{Tr}(b\rho_A) = \mathrm{Tr}[(b \otimes \mathbf{1})P_{A,R}]$ of ρ_A and the cyclic property of the trace, gives $\mathrm{Tr}(b\rho_A) = \mathrm{Tr}[(t_R^{-1}bt_R)\rho_A] = \mathrm{Tr}[b(t_R\rho_A t_R^{-1})]$ and therefore $\rho_A = t_R\rho_A t_R^{-1}$.

With no symmetry restriction, $P_{A,R}$ is a rank-one projection and therefore $P_{A,R}T_R = T_R P_{A,R}$, $\forall R \in O(3)$, implies that the ground state of H_A is spherically symmetric. □

We mention here that the definition of ρ_A implies that for any orthonormal basis, $\{\Psi_A^i, i = 1, ..., n_A\}$ in $\mathrm{Ran}\, P_{A,R}$, where $n_A := \dim \mathrm{Ran}\, P_{A,R}$, we have

$$\rho_A(z_1) := \sum_{i=1}^{n_A} \int |\Psi_A^i(z_1, ..., z_{|A|})|^2 dz_2...dz_{|A|}. \tag{23}$$

Decompositions. Recall that M and N are the numbers of the nuclei and electrons. Let $a = (A_1, ..., A_M)$ be a partition of $\{1, 2, ..., N\}$ into disjoint subsets some of which might be empty. With the set A_j we associate the j-th nucleus of charge eZ_j by assigning to it the electrons with labels in A_j. This gives a decomposition of the system into atoms/ions, $A_1, ..., A_M$, called also clusters. We denote the collection of all such decompositions by $\mathcal{A}$. The set of all $a \in \mathcal{A}$ with $|A_j| = Z_j$ for all $j = 1, ..., M$ will be denoted by $\mathcal{A}^{at}$. Its elements correspond to decompositions of our system to neutral atoms.

Permutations $\pi \in S_N$ act naturally on decompositions $a \in \mathcal{A}^{at}$. For any two non-equal decompositions, a and b, there is a unique permutation $\pi \in S_N/S(a)$ such that $b = \pi a$. Here, recall, $S(a)$ is the subgroup of S_N, which leaves a invariant.

For each decomposition $a = (A_1, ..., A_M) \in \mathcal{A}$ we define the Hamiltonian

$$H_a = \sum_{m=1}^{M} H_{A_m}, \tag{24}$$

where H_{A_m} is the Hamiltonian of the m-th atom or ion,

$$H_{A_m} := \sum_{i \in A_m} (-\Delta_{x_i} - \frac{e^2 Z_m}{|x_i - y_m|}) + \sum_{i,j \in A_m, i<j} \frac{e^2}{|x_i - x_j|}, \tag{25}$$

and the inter-cluster interaction $I_a := H_N(y) - H_a$, the sum of all interactions between the different atoms/ions in the decomposition a. We have that

$$H_N(y) = H_a + I_a. \tag{26}$$

Let E_{A_m} and ϕ_{A_m} and E_a and Φ_a be the ground state energy and ground state of H_{A_m} and H_a, respectively, $H_{A_m}\Phi_{A_m} = E_{A_m}\Phi_{A_m}$ and $H_a\Phi_a = E_a\Phi_a$. We have that $E_a = \sum_{m=1}^{M} E_{A_m}$ and

$$\Phi_a(x_1, ..., x_N) = \prod_{m=1}^{M} \phi_{A_m}(x_{A_m}), \tag{27}$$

where $x_{A_m} = (x_i : i \in A_m)$. Furthermore, if ϕ_m the ground state of the m-th atom with the nucleus fixed at the origin, then

$$\phi_{A_m}(x_{A_m}) = \phi_m(x_{A_m} - y_m), \tag{28}$$

with $x_{A_k} - y_m = (x_i - y_m : i \in A_m)$. Throughout the text we will always assume that $\|\phi_m\| = 1$ for all $m = 1, ..., M$. Standard estimates (see [17]) give that there exists $\theta > 0$ such that

$$\|e^{\theta \langle x_{A_m} \rangle} \partial^\alpha \phi_m\| \lesssim 1, \forall \alpha \text{ with } 0 \leq |\alpha| \leq 2, \tag{29}$$

where α is a multiindex with each index corresponding to differentiation in some body variable. For $a = \{A_1, ..., A_M\} \in \mathcal{A}$ we define $y_a = (y_{a,1}, ..., y_{a,N})$, where

$y_{a,i} = y_m$ for $i \in A_m$. In other words $y_{a,i}$ is the coordinate of the nucleus that x_i is assigned to. From (27), (28) and (29) we obtain that

$$\|e^{\theta\langle x-y_a\rangle}\partial^\alpha \Phi_a\| \lesssim 1, \forall \alpha \text{ with } 0 \le |\alpha| \le 2, \tag{30}$$

where $x = (x_1, ..., x_N)$.

By the definition of a's, $\min_{a\in\mathcal{A}} E_a = E(\infty)$, where $E(\infty)$ was defined in (1). If Property (E) holds, as we expect it always does, than $E_a = E(\infty)$, $\forall a \in \mathcal{A}^{at}$.

3. Proof of Theorem 1.1 assuming stability estimates

In this section we prove Theorem 1.1 stating van der Waals law in the highest spin case, under a technical assumption, which are then verified in later sections. In what follows we omit the subindex N and the argument y and write E and H for $E(y)$ and $H_N(y)$.

For the system to have the highest spin, H has to act on the subspace $\mathcal{H}_A = \bigwedge_1^N L^2(\mathbb{R}^3)$ corresponding to one-dimensional, anti-symmetric representation of S_N (with the one-column Young diagram). For each a, this representation induces the unique, the one-dimensional, anti-symmetric representation of $S(a)$. Hence we omit, without a danger of confusion, the label for this representation, having in mind that all operators below act on the subspace $\mathcal{H}_A = \bigwedge_1^N L^2(\mathbb{R}^3)$ and, as can be easily verified leave this subspace invariant.

Orthogonal projection *P*. We will now define the projection P, to be used in the Feshbach-Schur method described above. We cut off the ground state energy Φ_a defined in (27) as follows. Recall that ϕ_j denotes the ground state of the j-th atom centered at the origin. Let $\chi_R : \mathbb{R}^3 \to \mathbb{R}$ be a spherically symmetric, smoothed out characteristic function of the ball $B(0, \frac{R}{6})$ supported in the same ball. Let, as in (27) - (28),

$$\psi_{A_k}(x_{A_k}) := \psi_k(z_{A_k}), \quad \text{where} \quad \psi_k(z_{A_k}) := \frac{(\phi_k \chi_R^{\otimes Z_k})(z_{A_k})}{\|\phi_k \chi_R^{\otimes Z_k}\|}, \tag{31}$$

with $Z_k = |A_k|$. Here $z_{A_k} := (z_i := x_i - y_k : i \in A_k)$, where, recall, $x_{A_k} = (x_i : i \in A_k)$, and $\chi_R^{\otimes|A_k|}(z_{A_k}) := \prod_{i\in A_k} \chi_R(z_i)$. Furthermore, we define

$$\Psi_a(x_1, ..., x_N) := \prod_{k=1}^{M} \psi_{A_k}(x_{A_k}). \tag{32}$$

From (27), (28), (30), $H_a\Phi_a = E(\infty)\Phi_a$, where $E(\infty)$ was defined in (1), and from the construction of Ψ_a we obtain the estimates

$$\Phi_a \doteq \Psi_a, \qquad H_a\Psi_a \doteq E(\infty)\Psi_a, \tag{33}$$

$$\|e^{\theta\langle x-y_a\rangle}\partial^\alpha\Psi_a\| \lesssim 1, \forall\alpha,\ 0\le|\alpha|\le 2, \tag{34}$$

$$\Psi_a\Psi_b = 0, \forall a,b\in\mathcal{A}^{at},\ a\neq b. \tag{35}$$

We choose P as the orthogonal projection on $\mathrm{span}\{\Psi_a : a\in\mathcal{A}^{at}\}$. According to (8), we have to compute PHP and $U(\lambda)$. To this end we use (35) and as a consequence, we have

$$P = \sum_{a\in\mathcal{A}^{at}} P_{a,R} = \sum_{a\in\mathcal{A}^{at}} P_{\Psi_a} \tag{36}$$

(acting on the subspace $\mathcal{H}_A = \bigwedge_1^N L^2(\mathbb{R}^3)$). In the present context, $P_{a,R} = P_{\Psi_a}$ is the same the projection as in Introduction.

To show that the Feshbach map exists, we note that $\mathrm{Ran}(P)\subset D(H)$ since the range of P is spanned by $(\Psi_a)_{a\in\mathcal{A}^{at}}$ and by (34) each Ψ_a is in $H^2(\mathbb{R}^{3N})$. Hence the condition (a) for the existence of the Feshbach map holds. We will prove in Section 4 that, under Property (E), there is a $\gamma>0$, independent of y s.t. for R large enough, the following *stability bound* holds:

$$H^\perp \ge E(\infty)+2\gamma, \tag{37}$$

where $E(\infty)$ was defined in (1) and, recall, the notation $H^\perp = P^\perp HP^\perp$, acting on the subspace $\mathcal{H}_A = \bigwedge_1^N L^2(\mathbb{R}^3)$. We *assume* this for now. Using (37) we obtain that $H^\perp-\lambda$ is invertible for all $\lambda\le E(\infty)+\gamma$. Thus the condition (b) is also satisfied and, for all $\lambda\le E(\infty)+\gamma$, the Feshbach Schur map $F_P(\lambda)$ is well defined.

Our goal now is to prove the following

Theorem 3.1. *Assume the estimate* (37) *holds. Then so does the van der Waals law for the maximal spin (Theorem* 1.1*).*

To this end we have to estimate the two terms on the r.h.s of (8). We begin with the first one.

Estimate of PHP. We show that (on the subspace $\mathcal{H}_A = \bigwedge_1^N L^2(\mathbb{R}^3)$)

$$PHP \doteq E(\infty)P. \tag{38}$$

Indeed, using (36), that Ψ_a and Ψ_b have disjoint supports and that H is a local operator we obtain that $PHP = \sum_{a\in\mathcal{A}^{at}} P_{a,R}HP_{a,R} = \sum_{a\in\mathcal{A}^{at}}\langle\Psi_a, H\Psi_a\rangle P_{a,R}$.

Next using (36) and that by (33), we have $H\Psi_a = H_a\Psi_a + I_a\Psi_a \doteq E(\infty)\Psi_a + I_a\Psi_a$, $\forall a \in \mathcal{A}^{at}$, we obtain that

$$PHP \doteq E(\infty)P + \sum_{a\in\mathcal{A}^{at}} \langle\Psi_a, I_a\Psi_a\rangle P_{a,R}. \tag{39}$$

Note that since the group S_N acts transitively on partitions from $\mathcal{A}^{at}$, the r.h.s. of this expression commutes with the permutations T_π, $\pi \in S_N$, and therefore leaves the subspace $\mathcal{H}_A = \bigwedge_1^N L^2(\mathbb{R}^3)$ invariant. Eq. (38) follows from (39) and the following

Lemma 3.1. *For all $a \in \mathcal{A}^{at}$, the average intercluster interaction, $\langle\Psi_a, I_a\Psi_a\rangle = \mathrm{Tr}(I_a P_{a,R})$, vanishes*

$$\langle\Psi_a, I_a\Psi_a\rangle = \mathrm{Tr}(I_a P_{a,R}) = 0. \tag{40}$$

Proof. Let $a = \{A_1, ..., A_M\}$. Due to charge neutrality for each atom A_j, we have that

$$I_a = e^2 \sum_{i<j}^{1,M} \sum_{k\in A_i, l\in A_j} I_{ij}^{kl},\ I_{ij}^{kl} = [-\frac{1}{|y_j - x_k|} - \frac{1}{|y_i - x_l|} + \frac{1}{|y_i - y_j|} + \frac{1}{|x_k - x_l|}]. \tag{41}$$

Let $\rho_{A,R}$ be the one-electron density associated with the projection $P_{A,R}$ (by (31) and (32), $\rho_{A,R}(z) = \chi(z)\rho_A(z)$, where $\rho_A(z)$ is defined in (21)). Using (32), or $P_{a,R} = \prod_i P_{\psi_i}$, (41), and passing to the variables

$$z_{km} = x_k - y_m,\ \forall l \in A_m,\ \forall m = 1, ..., M, \tag{42}$$

we obtain (14) - (15). By Proposition 2.1 and the choice of the cut-off, the one-electron densities $\rho_{A_j,R}(z) = \chi(z)\rho_A(z)$ are spherically symmetric. This, together with Newton's theorem (see for example [22]) and support properties due to the cut-off (e.g. $\int \rho_{A_i,R}(z)\frac{1}{|y_{ij}+z|}dz = \frac{1}{|y_{ij}|}\int_{|z|\le|y_{ij}|}\rho_{A_i,R}(z)dz = \frac{1}{|y_{ij}|}$, $y_{ij} = y_i - y_j$), implies (40). □

Remark. By the charge neutrality of the clusters A_j, without spherical symmetry, the leading term in $\mathrm{Tr}(I_a P_{a,R})$ is of the order at least R^{-3} if at least two atoms have non-zero dipoles and of a higher order otherwise (see (53)).

Eq. (38) and the variational inequality $EP \le PHP$ imply

$$E \dot{\le} E(\infty). \tag{43}$$

By (37) and (43), the Feshbach map is well defined for E and by (8) - (10) and (38), we have that E is an eigenvalue of

$$F_P(E) \doteq E(\infty) + U(E)|_{\mathrm{Ran}\,P} \tag{44}$$

(acting on the subspace $\mathcal{H}_A = \bigwedge_1^N L^2(\mathbb{R}^3)$). Now we estimate $U(E)$.

Estimate of $U(E)$ We recall that H_A denotes the Hamiltonian of an atom A and E_A denotes its ground state energy. We let P_{ij} stand for the projection onto the ground state of $H_{A_i} + H_{A_j}$ and $P_{ij}^{\perp} := \mathbf{1} - P_{ij}$. Let $|y_{ij}| = |y_i - y_j|$. Our goal is to show

Lemma 3.2. *There are positive constants σ_{ij}, independent of y and depending only on the nature of the atoms i and j, s.t.*

$$U(E) = f(y)P + O(\frac{1}{R^7}), \quad \textit{where} \quad f(y) = \sum_{i<j}^{1,N} \frac{e^4 \sigma_{ij}}{|y_{ij}|^6}. \tag{45}$$

Moreover, σ_{ij} are given by the equations

$$\sigma_{ij} = \langle f_{ij}\phi_i\phi_j, R_{ij}^{\perp} f_{ij}\phi_i\phi_j \rangle, \tag{46}$$

where $R_{ij}^{\perp} := (P_{ij}^{\perp}(H_{A_i} + H_{A_j})P_{ij}^{\perp} - E_{A_i} - E_{A_j})^{-1}$ and

$$f_{ij} := \sum_{k \in A_i, l \in A_j} [z_{ki} \cdot z_{lj} - 3(z_{ki} \cdot \widehat{y_{ij}})(z_{lj} \cdot \widehat{y_{ij}})]. \tag{47}$$

Proof. By (37) and (43), the operator $H^{\perp} - E$, where, recall, $H^{\perp} = P^{\perp}HP^{\perp}$, has a bounded inverse, which we denote by $R^{\perp}(E) := (H^{\perp} - E)^{-1}$. It follows from (36), (40) and (35) and from (36), (35) and (33) that

$$P^{\perp} I_a \Psi_a = I_a \Psi_a \quad \text{and} \quad P^{\perp} H_a \Psi_a \doteq 0. \tag{48}$$

Using (36) and (48) we obtain that

$$P^{\perp}HP \doteq \sum_{a \in \mathcal{A}^{at}} I_a P_{a,R}. \tag{49}$$

The relations (9), (37), (49), $PHP^{\perp} = (P^{\perp}HP)^*$ and $P_{a,R} I_a P^{\perp} = P_{a,R} I_a$ give that

$$U(E) \doteq \sum_{a,b} P_{a,R} I_a R^{\perp}(E) I_b P_{b,R}. \tag{50}$$

By the same remark, as the one after (39), the r.h.s. of this expression commutes with the permutations T_π, $\pi \in S_N$, and therefore leaves the subspace $\mathcal{H}_A = \bigwedge_1^N L^2(\mathbb{R}^3)$ invariant.

Next, we estimate $I_a \Psi_a$, which is done in the next lemma.

Lemma 3.3.

$$\|I_a P_{a,R}\| \lesssim \sum_{i<j}^{1,M} \frac{1}{|y_i - y_j|^3}. \tag{51}$$

Proof. Let $a = \{A_1, ..., A_M\}$. Recall that, due to charge neutrality for each atom A_j, be have that by (41) and (15), the intercluster interaction I_a can be written as (41). Passing to the variables (42) and using that

$$|z \pm y_{ij}|^{-1} = |y_{ij}|^{-1}(1 \pm \frac{2z \cdot \widehat{y_{ij}}}{|y_{ij}|} + \frac{|z|^2}{|y_{ij}|^2})^{-\frac{1}{2}}, \tag{52}$$

where, recall, $y_{ij} = y_i - y_j$, we expand the function I_{ij}^{kl}, given in (41) (and expressed in the variables (42), cf. (15)), in z_{ki} and z_{lj}, we obtain

$$I_{ij}^{kl} = \frac{1}{|y_{ij}|}\left[-\frac{(z_{ki} - z_{lj}) \cdot \widehat{y_{ij}}}{|y_{ij}|} + \frac{z_{ki} \cdot \widehat{y_{ij}}}{|y_{ij}|} - \frac{z_{lj} \cdot \widehat{y_{ij}}}{|y_{ij}|} - \frac{|z_{ki} - z_{lj}|^2 - 3((z_{ki} - z_{lj}) \cdot \widehat{y_{ij}})^2}{2|y_{ij}|^2}\right]$$

$$+\frac{e^2}{|y_{ij}|}\left[\frac{|z_{ki}|^2 - 3(z_{ki} \cdot \widehat{y_{ij}})^2}{2|y_{ij}|^2} + \frac{|z_{lj}|^2 - 3(z_{lj} \cdot \widehat{y_{ij}})^2}{2|y_{ij}|^2}\right] + O(\frac{|z|^3}{|y_{ij}|^4}),$$

where $|z| := \max_{k,l}(|z_{ki}|, |z_{lj}|)$, provided $|z| \leq \frac{1}{3}|y_{ij}|$, together with the fact that $\operatorname{supp}\Psi_a \subset \{|z| \leq \frac{1}{3}|y_{ij}|\}$ gives that

$$\sum_{k\in A_i, l\in A_j} I_{ij}^{kl} = \frac{1}{|y_{ij}|^3} f_{ij}(z, \widehat{y_{ij}}) + O(\frac{|z|^3}{|y_{ij}|^4}) \quad \text{on } \operatorname{supp}\Psi_a, \tag{53}$$

where $f_{ij}(z, \widehat{y_{ij}})$ is given in (47). This estimate together with (34) gives (51). □

For each $a \in \mathcal{A}^{\text{at}}$, we introduce the sets

$$\hat{\Omega}_a^\nu := \{x \in \mathbb{R}^{3N} : |x_i - y_m| \leq \nu R,\ \forall i \in A_m,\ \forall A_m \in a\}, \tag{54}$$

and smoothed out characteristic functions, χ_a, $a \in \mathcal{A}^{\text{at}}$, symmetric under all permutations in $S(a)$ and such that

$$\chi_a(x) = 1 \ \text{ on } \ \hat{\Omega}_a^{1/6} \text{ and } \ = 0 \text{ outside } \ \hat{\Omega}_a^{1/5}, \tag{55}$$

$$|\partial^\alpha \chi_a| \lesssim R^{-|\alpha|}, \text{ for any multi-index } \alpha. \tag{56}$$

Eq. (55) and the definition of $P_{a,R}$ imply

$$I_a = O(\frac{1}{R}) \text{ on } \operatorname{supp}\chi_a. \tag{57}$$

$$\chi_a P_{b,R} = \delta_{ab} P_{a,R}, \quad \forall a, b \in \mathcal{A}^{at}. \tag{58}$$

Let $R_a^\perp(E) := (H_a^\perp - E)^{-1}$, where $H_a^\perp := H_a P_a^\perp$, with the notation $P_a = P_{\Phi_a} := |\Phi_a\rangle\langle\Phi_a|$ (the orthogonal projection onto the ground state of H_a). Our goal is to show the following estimate

$$\chi_a R^\perp(E) - R_a^\perp(E)\chi_a = O(\frac{1}{R}), \tag{59}$$

where, recall, $R^\perp(E) := (H^\perp - E)^{-1}$. Factoring out the inverse operators on the l.h.s. of (59), we obtain that

$$\chi_a R^\perp(E) - R_a^\perp(E)\chi_a = R_a^\perp(E) V R^\perp(E), \tag{60}$$

where $V := P_a^\perp H_a \chi_a - \chi_a P^\perp H P^\perp$. Due to the cut off of the ground states we have that Eq. (58) and the definition of $P^\perp$ give

$$\chi_a P^\perp = P_{a,R}^\perp \chi_a \text{ and } [\chi_a, P_{a,R}^\perp] = 0. \tag{61}$$

The latter relations, together with the fact that H_a is a local operator and the relation $H = H_a + I_a$, imply that $\chi_a P^\perp H P^\perp \doteq P_{a,R}^\perp \chi_a H_a + P_{a,R}^\perp \chi_a I_a P_{a,R}^\perp$, which, in turn, gives

$$V = P_{a,R}^\perp (H_a\chi_a - \chi_a H_a + \chi_a I_a P_{a,R}^\perp) + (P_{a,R} - P_a) H_a \chi_a. \tag{62}$$

Using (56), (57), $P_{a,R} \doteq P_a$ and the H_a boundedness of thegradient, ∇, we obtain that $\|V(-\Delta+1)^{-1/2}\| \lesssim \frac{1}{R}$, which together with (9) implies (59).

Since χ_a commutes with I_a, we have $P_{a,R} I_a = P_{a,R} I_a \chi_a$. Using this to insert χ_a into (50), and using (61), (59), (51) and (58) gives

$$\begin{aligned} U(E) &= \sum_{a,b} P_{a,R} I_a R_a^\perp(E) \chi_a I_b P_{b,R} + O(\frac{1}{R^7}) \\ &= \sum_a P_{a,R} I_a R_a^\perp(E) I_a P_{a,R} + O(\frac{1}{R^7}). \end{aligned} \tag{63}$$

Again, it is easy to see that the r.h.s. of this expression commutes with the permutations T_π, $\pi \in S_N$, and therefore leaves the subspace $\mathcal{H}_A = \bigwedge_1^N L^2(\mathbb{R}^3)$ invariant.

Now, by a standard estimate and (51), we have $0 \le \langle \psi, \sum_a P_{a,R} I_a R_a^\perp(E) I_a P_{a,R}\psi\rangle \lesssim \frac{1}{R^6}\langle \psi, \sum_a P_{a,R}\psi\rangle$, which by (36) and (63), gives $\langle \psi, U(E)\psi\rangle \lesssim \frac{1}{R^6}\|P\psi\|^2$, which in turn implies $\|U(E)|_{\text{Ran } P}\| \lesssim \frac{1}{R^6}$. Using, this estimate, together with (44) and the fact that E is an eigenvalue of $F_P(E)$, we estimate

$$|E - E(\infty)| \lesssim R^{-6}. \tag{64}$$

Using this and (51) in (63), and using the notation $R_a^\perp = R_a^\perp(E(\infty))$, we obtain

$$U(E) = \sum_a U_{aa} + O(\frac{1}{R^7}), \; U_{aa} = P_{a,R} I_a R_a^\perp I_a P_{a,R} \tag{65}$$

We insert (41) with (53) into the expression for U_{aa} in (65) and use that $O(\frac{|z|^3}{|y_{ij}|^4})P_{a,R} = O(\frac{1}{|y_{ij}|^4})$, to obtain

$$U_{aa} = \sum_{ij,\beta} \frac{e^4 \sigma_{ij,kl}^a}{|y_{ij}|^3 |y_{kl}|^3} P_{a,R} + O(\sum_{ij,kl} \frac{e^4}{|y_{ij}|^3 |y_{kl}|^4}), \tag{66}$$

where ij, kl run in pairs of nuclei in the decomposition a and $\sigma_{ij,kl}^a = \langle f_{ij}\Phi_a, R_a^\perp f_{kl}\Phi_a\rangle$.

Finally, we prove the following lemma

Lemma 3.4.

$$\sigma^a_{ij,kl} = \sigma_{ij}\delta_{ij,kl} + O(\frac{1}{R}), \tag{67}$$

where σ_{ij} are given in (46).

Proof. Consider an atomic decomposition $a = (A_1, ..., A_M)$. Recall the notation from Section 2 and write the orthogonal projection P_a on Φ_a as $P_a = P_{kl} \otimes P^{kl}$, where $P_{ij} = P_{A_i} \otimes P_{A_j}$ and $P^{ij} = \prod_{k \neq i,j} P_{A_k}$, with $P_{A_i} = P_{\phi_{A_i}}$, the orthogonal projection on ϕ_{A_i}. Then we have $P_a^{\perp} P^{kl} = P_{kl}^{\perp} \otimes P^{kl}$ and therefore, on the invariant subspace $\operatorname{Ran} P^{kl}$,

$$H_a P_a^{\perp} = P_{kl}^{\perp}(H_{A_k} + H_{A_l})P_{kl}^{\perp} + \sum_{i \neq k,l} E_{A_i}. \tag{68}$$

This, the notation $R_{kl}^{\perp} := (P_{kl}^{\perp}(H_{A_k} + H_{A_l})P_{kl}^{\perp} - E_{A_k} - E_{A_l})^{-1}$ and the fact that $f_{kl}\Phi_a = P^{kl} f_{kl}\Phi_a \in \operatorname{Ran} P^{kl}$ and $E(\infty) = \sum_{j=1}^N E_{A_j}$ give

$$R_a^{\perp}(E) f_{kl}\Phi_a = R_{kl}^{\perp} f_{kl}\Phi_a, \tag{69}$$

which implies

$$\langle f_{ij}\Phi_a, R_a^{\perp} f_{kl}\Phi_a\rangle = \langle f_{ij}\Phi_a, R_{kl}^{\perp} f_{kl}\Phi_a\rangle. \tag{70}$$

We now prove that

$$\langle f_{ij}\Phi_a, R_{kl}^{\perp} f_{kl}\Phi_a\rangle = \begin{cases} 0, \text{ if } ij \neq kl \\ \sigma_{ij}, \text{ if } ij = kl \end{cases}, \tag{71}$$

where σ_{ij} are given in (46). We denote the l.h.s. of (71) by $W_{ij;kl}$. If $ij \neq kl$, then we may assume without loss of generality that $i \neq k, l$. Observing that the inner product on the right hand side of (70) reduces to an integral that is odd in $z_{mi}, m \in A_i$ we obtain $W_{ij;kl} = 0$ for $ij \neq kl$. For the case $ij = kl$, $W_{ij;kl} = \sigma_{ij}$ follows immediately from the relations (27) - (28) and the fact that the ground states of the atoms are normalized. Hence $W_{ij;kl}$ follows.

This, together with the definition $\sigma^a_{ij,kl} = \langle f_{ij}\Phi_a, R_a^{\perp} f_{kl}\Phi_a\rangle$ and (70), proves the relation (67). □

The relations (65), (66) and (67), together with (36), imply (45) modulo the fact that σ_{ij} are positive and independent of y which we prove below.

First we note that the only dependence of σ_{ij} on $\widehat{y_{ij}} := \frac{y_i - y_j}{|y_i - y_j|}$ appears on f_{ij}. To display this dependence we will write $f_{ij}^{\widehat{y_{ij}}}$ and $\sigma_{ij}^{\widehat{y_{ij}}}$ for f_{ij} and σ_{ij}. For any rotation R in $\mathbb{R}^3$ we define T_R acting on the space $L^2(\mathbb{R}^{3(|A_i|+|A_j|)})$ (recall that $|A_k|$ is the number of electrons of the k-th atom) of functions of the variables

$z_{kl} = x_k - y_l$, $k = 1, \ldots, |A_i| + |A_j|$, $l = 1, 2$, by rotating them. Since T_R commutes with $H_{A_i} + H_{A_j}$ and since $\phi_i\phi_j$ is a ground state of $H_{A_i} + H_{A_j}$, we have that $T_R\phi_i\phi_j$ is also a ground state of $H_{A_i} + H_{A_j}$. Therefore, since the ground state is non degenerate and T_R is unitary we obtain that $T_R\phi_i\phi_j = c(R)\phi_i\phi_j$ where $|c(R)| = 1$. This implies

$$T_R P_{ij}^{\perp} T_R^{-1} = P_{ij}^{\perp}.$$

Using the last relation and the fact that $H_{A_i} + H_{A_j}$ commutes with T_R, together with (46) and the fact that T_R is unitary, we obtain that

$$\sigma_{ij}^{\widehat{y_{ij}}} = \langle T_{R^{-1}}(f_{ij}^{\widehat{y_{ij}}}\phi_i\phi_j), R_{kl}^{\perp} T_{R^{-1}}(f_{ij}^{\widehat{y_{ij}}}\phi_i\phi_j)\rangle. \tag{72}$$

On the other hand using (47) and the fact that R is unitary we obtain that

$$T_R^{-1} f_{ij}^{\widehat{y_{ij}}} = f_{ij}^{\widehat{R^{-1}y_{ij}}}. \tag{73}$$

But using that $T_R^{-1}\phi_i\phi_j = c(R^{-1})\phi_i\phi_j$, with $|c(R^{-1})| = 1$, together with (73), we obtain that

$$T_{R^{-1}}(f_{ij}^{\widehat{y_{ij}}}\phi_i\phi_j) = c(R^{-1}) f_{ij}^{\widehat{R^{-1}y_{ij}}}\phi_i\phi_j.$$

The last relation together with (72) and $|c(R^{-1})| = 1$ implies that

$$\sigma_{ij}^{\widehat{y_{ij}}} = \langle f_{ij}^{\widehat{R^{-1}y_{ij}}}\phi_i\phi_j, R_{kl}^{\perp}(f_{ij}^{\widehat{R^{-1}y_{ij}}}\phi_i\phi_j)\rangle = \sigma_{ij}^{\widehat{Ry_{ij}}},$$

which implies that σ_{ij} is independent of y_{ij}.

Since the operator $P_{ij}^{\perp}(H_{A_i} + H_{A_j})P_{ij}^{\perp} - E_i - E_j$ is a positive operator, then so is its inverse, $R_{kl}^{\perp} \geq 0$. Hence, by (46), we have that $\sigma_{ij} > 0$. □

Conclusion of the argument. From (44) and (45) we have that

$$F_P(E) = E(\infty) - f(y) + O(\frac{1}{R^7}). \tag{74}$$

Since the ground state energy E is an eigenvalue of H, we conclude from (10), (74) and the definition $W(y) = E - E(\infty)$ that (4) holds. This proves Theorem 3.1. □

Proof of the necessity of Property (E). In this paragraph we will show that if Property (E) fails to hold then so does the van der Waals - London law. We denote by $\mathcal{A}^{\min}$ the set of all $a \in \mathcal{A}$ for which $\inf\sigma(H_a) = \min_{b\in\mathcal{A}}\inf\sigma(H_b)$, and $\inf\sigma(H_a)$ is an isolated eigenvalue. Property (E) holds if and only if $\mathcal{A}^{\min} = \mathcal{A}^{at}$.

Now assume that Property (E) fails. To prove that the van der Waals London law also fails we use, as before, the Feshbach map but with P, the orthogonal projection on $\mathrm{span}\{\Psi_a : a \in \mathcal{A}^{\min}\}$. Proceeding as in the proof of (36) we can show that $P = \sum_{a\in\mathcal{A}^{\min}} P_{a,R}$. Furthermore, repeating the arguments of the proof of (37) (with $\mathcal{A}^{at}$ replacing $\mathcal{A}^{\min}$) we can prove that there exists $\gamma > 0$ such that

$H^{\perp} \geq E_{\min} + 2\gamma$, where $E_{\min} = \min_{a\in\mathcal{A}} \inf \sigma(H_a)$, which implies that the Feshbach Schur method can be used.

We will now estimate PHP. Using the relation $P = \sum_{a\in\mathcal{A}^{\min}} P_{a,R}$ and proceeding as in the proof of (39), we can show that

$$PHP \doteq E_{\min}P + \sum_{a\in\mathcal{A}^{\min}} \langle\Psi_a, I_a\Psi_a\rangle P_{a,R}.$$

Since the Property (E) is not satisfied it means that there exists $a \in \mathcal{A}^{\min}$ with $a \notin \mathcal{A}^{at}$. For such a decomposition $a = (I_1, \ldots, I_M)$, we define the charges $q_{ai} := (Z_i - |I_i|)e$. Next, we apply Newton's theorem to show that

$$\langle\Psi_a, I_a\Psi_a\rangle = \sum_{i\neq j} \frac{q_{ai}q_{aj}}{|y_i - y_j|}, \quad \forall a \notin \mathcal{A}^{at}.$$

Clearly, $\sum_{i\neq j} \frac{q_{ai}q_{aj}}{|y_{ai}-y_{aj}|} \asymp \frac{1}{R}$. The last three relations imply that

$$PHP \doteq E_{\min}P + \sum_{a\in\mathcal{A}^{\min}/\mathcal{A}^{at}} \sum_{i\neq j} \frac{q_{ai}q_{aj}}{|y_i - y_j|} P_{a,R}. \tag{75}$$

To estimate $U(\lambda)$, we proceeding as in the proof of (50), to obtain that

$$U(\lambda) = O(\frac{1}{R^2}), \qquad \forall\lambda \leq E_{\min} + \gamma. \tag{76}$$

The relations (8), (75) and (76) give give, $\forall\lambda \leq E_{\min} + \gamma$, that

$$F_P(\lambda) = E_{\min}P + \sum_{a\in\mathcal{A}^{\min}/\mathcal{A}^{at}} \sum_{i\neq j} \frac{q_{ai}q_{aj}}{|y_i - y_j|} P_{a,R} + O(\frac{1}{R^2}),$$

and therefore, by $\sum_{i\neq j} \frac{q_{ai}q_{aj}}{|y_i-y_j|} \asymp \frac{1}{R}$ and (10), the van der Waals law fails. As a consequence Property (E) is necessary for van der Waals law to hold.

4. Proof of stability bound (37) assuming Property (E)

Let $E_a^{'}$ be the first excited state energy of the Hamiltonian H_a and $E(\infty)$ as defined in (1). We define

$$\gamma_1 = \min_{a\in\mathcal{A}^{\rm at}} E_a^{'} - E(\infty), \quad \gamma_2 = \min_{a\in\mathcal{A}/\mathcal{A}^{\rm at}} E_a - E(\infty) \tag{77}$$

(recall that $E(\infty) = E_a$ for $a \in \mathcal{A}^{\rm at}$). By Property (E) we have that $\gamma_2 > 0$. Finally, we let

$$\gamma_0 := \min\{\gamma_1, \gamma_2\} > 0. \tag{78}$$

The stability bound (37) follows immediately from the following proposition:

Proposition 4.1. *There exists a constant $C > 0$ such that $H^{\perp} \geq E(\infty) + \gamma_0 - \frac{C}{R}$. In particular, (37) holds for R large enough.*

Proof. The idea here is to localize the Hamiltonian H to different domains of the configuration space $\mathbb{R}^{3N}$, each of which corresponding to a different break-up of the system into independent subsystems, and use geometry of these domains to estimate H. Since H is a differential operator, we have to pay a prize for this. A neat way way of doing such a localization is by constructing - following [32] - a partition of unity $\{J_a\}$, labeled by decompositions from $\mathcal{A}$ (see Subsection 2) and satisfying $\sum_{a\in\mathcal{A}} J_a^2 = 1$ and then using the IMS localization formula (see for example [6])

$$H = \sum_{a\in\mathcal{A}} \left(J_a H J_a - |\nabla J_a|^2\right). \tag{79}$$

We will split the configuration space $\mathbb{R}^{3N}$ into domains. For each $a = (A_1, ..., A_M) \in \mathcal{A}$, we define

$$\Omega_a^\nu = \{(x_1, ..., x_N) : |x_i - y_j| \geq \nu R,\ \forall j = 1, ..., M,\ \forall i \notin A_j\}. \tag{80}$$

Following [32] we will now construct a partition of unity $(J_a)_{a\in\mathcal{A}}$ having the properties:

$$0 \leq J_a \leq 1, \operatorname{supp} J_a \subset \Omega_a^{\frac{1}{5}},\ \|\partial^\alpha J_a\|_{L^\infty} \lesssim R^{-|\alpha|}, \tag{81}$$

$$J_a \text{ commute with all elements of } S(a), \tag{82}$$

where, recall, $S(a)$ is the group of permutations that keeps the clusters of a invariant, for all $a \in \mathcal{A}$, and

$$\sum_{a\in\mathcal{A}} J_a^2 = 1. \tag{83}$$

We consider the functions $F_a = \chi_{\Omega_a^{\frac{3}{10}}} * \phi$ where $\phi : \mathbb{R}^{3N} \to \mathbb{R}$ is a C_c^∞ function supported in $B_{\frac{1}{10}}(0)$ (ball of radius $\frac{1}{10}$ centered at the origin) symmetric with respect to particle coordinates, with $\phi \geq 0$ and $\int_{\mathbb{R}^{3N}} \phi = 1$ and χ_A denotes the characteristic function of the set A. Then $F_a \in C^\infty$ and $F_a \geq 0$. Furthermore using the triangle inequality and the fact that ϕ is supported in $B_{\frac{1}{10}}(\vec{0})$ we obtain that $\operatorname{supp}(F_a) \subset \Omega_a^{\frac{1}{5}}$ and that $F_a|_{\Omega_a^{\frac{2}{5}}} = 1$. The last relation together with the fact that $\cup_{a\in\mathcal{A}}\Omega_a^{\frac{2}{5}} = \mathbb{R}^{3N}$ gives that $\sum_{a\in\mathcal{A}} F_a \geq 1$. Therefore, if we define

$$J_a = \frac{F_a}{\sqrt{\sum_{a\in\mathcal{A}} F_a^2}},$$

then J_a is a partition of unity satisfying all the desired properties.

Now we use the IMS localization formula (79), together with (81), to obtain $H \geq \sum_{a\in\mathcal{A}} J_a H J_a - \frac{C}{R}$. The last relation together with the secomposition $H = H_a + I_a$ (see (26)) and $I_a \geq -\sum_{j=1}^M \sum_{i=1, i\notin A_j}^N \frac{e^2 Z_j}{|x_i - y_j|} \geq -\frac{C}{R}$ on $\operatorname{supp} J_a$ implies that

$$H \geq \sum_{a\in\mathcal{A}} J_a H_a J_a - \frac{C}{R}. \tag{84}$$

The last relation together with the fact that $P^{\perp}$ is self-adjoint and $P^{\perp} \leq 1$ implies that

$$P^{\perp}HP^{\perp} \geq \sum_{a\in\mathcal{A}} P^{\perp}J_aH_aJ_aP^{\perp} - \frac{C}{R}. \tag{85}$$

Proposition 4.1 now follows from the equations (83), (85) and the lemma below which provides estimates - (86) - of each of the terms on the right hand side of (85). □

Lemma 4.1. *Fix arbitrary $a \in \mathcal{A}$. With γ_0 defined in (78), we have, on functions that are antisymmetric with respect to permutations particles inside the clusters of a,*

$$P^{\perp}J_aH_aJ_aP^{\perp} \geq (E(\infty)+\gamma_0)P^{\perp}J_a^2P^{\perp} - \frac{C}{R}. \tag{86}$$

(Note that the restriction to $\mathcal{H}_A = \bigwedge_1^N L^2(\mathbb{R}^3)$, as any restriction, only lifts the lower bound.)

Proof. We will consider the following two cases

Case (1): $a \in \mathcal{A}/\mathcal{A}^{at}$ (i.e. some clusters of the decomposition a include ions). In this case, Condition (E) implies that $H_a \geq E(\infty)+\gamma_0$ and, as a consequence,

$$P^{\perp}J_aH_aJ_aP^{\perp} \geq (E(\infty)+\gamma_1)P^{\perp}J_a^2P^{\perp}. \tag{87}$$

Case (2): $a \in \mathcal{A}^{at}$ (i.e. a is a decomposition of the system to M (neutral) atoms). We recall the notation $P_{a,R} = P_{\Psi_a}$ and $P_a = P_{\Phi_a}$. Using the relation $H_aP_a^{\perp} \geq (E(\infty)+\gamma_0)P_a^{\perp}$, we find

$$\begin{aligned} J_aH_aJ_a &\geq J_a[(E(\infty)+\gamma_2)P_a^{\perp} + E(\infty)P_a]J_a \\ &= (E(\infty)+\gamma_2)J_a^2 - \gamma_2 J_aP_aJ_a. \end{aligned} \tag{88}$$

By (33), (80) and (81) and the support properties of Ψ_b, we have that

$$\|P_a - P_{a,R}\| \doteq 0 \quad \text{and} \quad J_aP_{b,R} = P_{b,R}J_a = \delta_{a,b}P_{a,R}. \tag{89}$$

Therefore, by (36), we have $P^{\perp}J_aP_aJ_aP^{\perp} \doteq P^{\perp}P_{a,R}P^{\perp} \doteq 0$. Hence

$$P^{\perp}J_aH_aJ_aP^{\perp} \dot{\geq} (E(\infty)+\gamma_2)J_a^2, \tag{90}$$

which, together with the previous case, implies (86) for all $a \in \mathcal{A}^{at}$. □

Appendix A. More about Property (E)

In this appendix we prove several statements about Property (E) formulated in the introduction. We begin with

Lemma A.1 (Property (E) for hydrogen atoms). *Property (E) holds for a system of several hydrogen atoms.*

Proof. Let $E^{(m)}$, $m \geq 0$, be the ground state energy of the hydrogen ion (or atom, if $m = 0$) with charge $-me$, i.e. the lowest eigenvalue the Hamiltonian

$$H^{(m)} = \sum_{j=1}^{m+1} H_j + \sum_{i<j}^{1,m} \frac{e^2}{|x_i - x_j|}, \tag{A.1}$$

where $H_j = -\Delta_{x_j} - \frac{e^2}{|x_j|}$ is the Hamiltonian for the hydrogen atom in the j-th coordinate. Property (E) is reduced to the property that for any $m \geq 1$, $E^{(m)}$ satisfies

$$E^{(m)} > (m+1)E^{(0)}. \tag{A.2}$$

Let $m_* := \max\{m' \leq m |\ H^{(m')}$ has a ground state$\}$ and let $\psi^{(m_*)}$ be the ground state of $H^{(m_*)}$ corresponding to $E^{(m_*)}$. (We know from [15] that $m_* \geq 1$ for $m \geq m_*$.) By the definition of m_* we have $E^{(m)} = E^{(m_*)}$ and therefore $E^{(m)} = \langle \psi^{(m_*)}, H^{(m_*)}\psi^{(m_*)}\rangle$. Since $H_j \geq E_0$ for any j, we have $E^{(m)} > (m+1)E^{(0)} + \delta$, with $\delta := \langle \psi^{(m_*)}, \sum_{i<j}^{1,m_*} \frac{e^2}{|x_i - x_j|}\psi^{(m_*)}\rangle$. □

Next, we show that

Lemma A.2. *Property (E′) implies Property (E).*

Proof. We prove (E) by induction in the number of the atoms k. For $k = 2$, Property (E) follows immediately from Property (E′). We assume that (E) holds for M, replaced by $k-1$ and show it holds for $M = k$. Indeed, let $n_1, ..., n_k$ be numbers satisfying the assumptions of the Property (E) for $M = k$. By relabelling the nuclei, if necessary, we can assume that $|n_1| \geq |n_k|$ and $n_1 n_k < 0$. By Property (E′), we have that $E_{1,n_1} + E_{k,n_k} > E_{1,n_1+n_k} + E_k$. Therefore, $E_{1,n_1} + ... + E_{k-1,n_{k-1}} + E_{k,n_k} > E_{1,n_1+n_k} + E_{2,n_2} + ... + E_{k-1,n_{k-1}} + E_k$, which together with the induction hypothesis implies Property (E) for $M = k$. □

This proves properties (a) and (b) of the introduction. For (c), it follows from the fact that $E^{(0)}$ remains the same whereas $E^{(m)}$ increases if the statistics is taken into account.

Appendix B. Constructive upper bound from [1]

In this appendix we sketch the proof of the following result

Theorem B.1. *Assume Condition (D). Then, there exist constants $C_1, C_2 > 0$ depending only on* $\max Z_j, j = 1, ..., M$ *such that if $R \geq C_1 M^{\frac{2}{3}}$, then*

$$W(y) \leq -\sum_{i<j}^{1,M} \frac{e^4 \sigma_{ij}}{|y_i - y_j|^6} + C_2 \left(\sum_{i<j}^{1,M} \frac{1}{|y_i - y_j|^7} + \frac{M^4}{R^9} \right), \tag{B.1}$$

where $\sigma_{ij} > 0$ are the same constants as in Theorem 1.1 *(see also* (4)*).*

The idea is to use as a test function an approximation of the ground state of H as given by the Feshbach map, namely,

$$\psi = \phi - (H^{\perp} - E)^{-1} P^{\perp} H \phi \tag{B.2}$$

for $\phi \in \operatorname{Ran} P$ (see the paragraph "Outline of the approach" of the introduction for the notation and e.g. [4], for the justification of (B.2)).

For simplicity we consider two atoms only, without taking into account the Fermionic statistics. We decompose the full Hamiltonian of the system as $H = H_{12} + I_{12}$ where H_{12} is the sum of the Hamiltonians of the two atoms and I_{12} has the interaction terms of the atoms. We denote by ϕ the ground state of H_{12}, so that $H_{12}\phi = E(\infty)\phi$. Let also P_ϕ be the orthogonal projection onto ϕ and $P_\phi^{\perp} = 1 - P_\phi$. The test function we consider is the normalization of the function $\tilde{\psi} = \phi - R_{12}^{\perp} I_{12} \phi$, where $R_{12}^{\perp} = (H_{12} P_\phi^{\perp} - E(\infty))^{-1}$. This approximation of the ground state of H, as given by the Feshbach map, namely, (B.2), mentioned above. We modified the resolvent by omitting the interaction between the atoms.

Since the interaction energy $W(y)$ is the ground state energy of $H - E(\infty)$, we have that

$$W(y) \leq \frac{1}{\|\tilde{\psi}\|^2} \langle \tilde{\psi}, (H - E(\infty)) \tilde{\psi} \rangle.$$

Expanding the inner product on the right hand side of the last equation and using the equality $(H - E(\infty))\phi = I_{12}\phi$, we obtain that

$$\begin{aligned} \langle \tilde{\psi}, (H - E(\infty)) \tilde{\psi} \rangle = \langle \phi, I_{12} \phi \rangle - 2 \langle I_{12} \phi, R_{12}^{\perp} I_{12} \phi \rangle + \langle R_{12}^{\perp} I_{12} \phi, I_{12} R_{12}^{\perp} I_{12} \phi \rangle \\ + \langle R_{12}^{\perp} I_{12} \phi, (H_{12} - E(\infty)) R_{12}^{\perp} I_{12} \phi \rangle, \end{aligned} \tag{B.3}$$

where the last two terms originated from the decomposition $(H - E(\infty)) = (H_{12} - E(\infty)) + I_{12}$. From Newton's theorem it follows that $\langle \phi, I_{12} \phi \rangle \approx 0$, as we explained in the sketch of the proof of Theorem 1.1, where $\approx$ means that the error decays exponentially in $|y_1 - y_2|$, where y_1, y_2 are the positions of the nuclei of the atoms. Therefore, we also have that $P_\phi^{\perp} I_{12} \phi \approx I_{12} \phi$, so that in the second line of equation (B.3) we have the simplification $(H_{12} - E(\infty)) R_{12}^{\perp} I_{12} \phi \approx I_{12} \phi$. As a consequence, the term on the second line of (B.3) is, up to an exponentially decaying in $|y_1 - y_2|$ error, equal to $\langle I_{12} \phi, R_{12}^{\perp} I_{12} \phi \rangle$. With these observations we arrive at

$$\langle \tilde{\psi}, (H - E(\infty)) \tilde{\psi} \rangle \approx -\langle I_{12} \phi, R_{12}^{\perp} I_{12} \phi \rangle + \langle R_{12}^{\perp} I_{12} \phi, I_{12} R_{12}^{\perp} I_{12} \phi \rangle. \tag{B.4}$$

As we explained in the sketch of proof of Theorem 1.1, if we take Taylor expansion of the interaction terms I_{12} between the atoms, it turns out that I_{12} is in leading term of the order $|y_1 - y_2|^{-3}$. If we drop the higher order terms of I_{12}, the term $-\langle I_{12}\phi, R_{12}^{\perp} I_{12}\phi\rangle$ gives exactly the van der Waals interaction term $-\frac{\sigma_{12}}{|y_1-y_2|^6}$. The remainder of this term can be easily proven to be of the order $|y_1 - y_2|^{-7}$. Due to the fact that I_{12} appears three times in the last term of (B.4), it turns out that this term is at most of the order $|y_1 - y_2|^{-9}$. Making the last statement precise is harder, because one of the three I_{12} terms is not multiplied with the exponentially decaying function ϕ. For this reason, we will push exponential weights through the resolvent $R_{12}^{\perp}$ using boosted Hamiltonians. Finally, observing that $\|\tilde{\psi}\| \leq 1 + O(|y_1 - y_2|^{-3})$, the theorem follows.

Bibliography

[1] I. Anapolitanos, Remainder estimates for the long range behavior of van der Waals force, arXiv: 1308.4808 [math-ph] (2013).

[2] I. Anapolitanos and I.M. Sigal, Long range behavior of van der Waals force, arXiv.

[3] E.A.G. Armour, J.-M. Richard and K.Varga: Stability of few-charge systems in quantum mechanics. *Physics Reports* **413** 1-90 (2005).

[4] V. Bach, J. Fröhlich and I.M. Sigal: Renormalization Group Analysis of Spectral Problems in Quantum field theory. *Adv. in Math.* **137**, 205-298 (1998).

[5] J.-M. Combes, P. Duclos and R. Seiler: The Born-Oppenheimer approximation, in: Rigorous Atomic and Molecular Physics (eds. G. Velo, A. Wightman), New York, Plenum, 185-212 (1981).

[6] H.L. Cycon, R.G. Froese, W. Kirsch and B. Simon : Schrödinger Operators with application to quantum mechanics and global geometry. *Texts and Monographs in Physics. Springer study edition. Springer-Verlag Berlin* (1987).

[7] G. Feinberg and J. Sucher: General theory of the van der Waals interaction: a model independent approach. *Phys. Rev. A* **9**, 2395-2415 (1970).

[8] C.L. Fefferman and L.A. Seco: Asymptotic neutrality of large ions. *Commun. Math. Phys.* **128**, 109-130 (1990).

[9] C.L. Fefferman and L.A. Seco: The spin of the ground state of an atom. *Rev. Mat. Iberoamericana* **12**, No. 1, 19-36 (1996).

[10] J. Fröhlich, G.M. Graf, J.-M. Richard and M. Seifert : Proof of stability of the hydrogen molecule. *Phys. Rev. Lett.*, **71**, No.9, 30 1332-1334 (1993).

[11] J. Fröhlich: Spin - or, actually: Spin and Quantum Statistics. *Prog. Math. Phys.*, **55**, Birkhäuser Verlag, Basel, 1-60 (2009).

[12] S. Gustafson and I.M. Sigal: Mathematical concepts of quantum mechanics. 2nd edition *Springer Berlin, Heidelberg, New York* (2011).

[13] G.Hagedorn: High order corrections to the time-dependent Born-Oppenheimer approximation. I. Smooth potentials. *Ann. of Math. (2)* **124** No. 3, 571-590 (1986). Erratum. *Ann. of Math. (2)* **126** No. 1, 219 (1987).

[14] M. Hamermesh: Group theory and its application to physical problems. *Addison-Wesley Pub. Co.* (1962).

[15] R. N. Hill: Proof that the H- ion has only one bound state. Details and extension to finite nuclear mass. *J. Math. Phys.* **18**, 2316-2330 (1977).

[16] W. Hunziker and C. Günther: Bound states in dipole fields and continuity properties of electronic spectra. *Helv. Phys. Acta, t.* **53**, 201-208 (1980).

[17] W. Hunziker and I.M. Sigal: The quantum $N-$body problem. *J. Math. Phys.* **41** No.6, 3448-3510 (2000).

[18] Th. Jecko, M. Klein, X.P. Wang: Existence and Born-Oppenheimer asymptotics of the total scattering cross-section in ion-atom collisions. in "Long time behaviour of classical and quantum systems", proceedings of the Bologna AP-TEX Int. Conf., 1999, A. Martinez and S. Graffi, eds.

[19] M. Klein, A. Martinez, R. Seiler and X. P. Wang.: On the Born-Oppenheimer expansion for polyatomic molecules. *Commun. Math. Phys.* **143** No. 3, 607-639 (1992).

[20] C. Le Bris and P.-L. Lions: From atoms to crystals: a mathematical journey. *Bulletin AMS* **42**, No. 3, 291 - 363 (2005).

[21] E.H. Lieb: Bound on the maximum negative ionization of atoms and molecules. *Phys. Rev. A* **29**, 3018-3028 (1984).

[22] E.H. Lieb and M. Loss: Analysis. *Graduate Studies in Mathematics* **14** AMS, Providence, RI, second edition (2001).

[23] E. H. Lieb, I. M. Sigal, B. Simon, and W. Thirring: Asymptotic neutrality of large-Z ions. *Commun. Math. Phys.* **116**, 635-644 (1988).

[24] E.H. Lieb and W. Thirring : Universal nature of van der Waals forces for Coulomb systems. *Phys. Rev. A* **34** No.1, 40-46 (1986).

[25] B. H. Mahan and R.J. Myers: University Chemistry. *Addison-Wesley Publishing company, inc* (1987).

[26] Phan Thành Nam. New bounds on the maximum ionization of atoms. *Commun. Math. Phys.* **312**, 427-445 (2012).

[27] G. Panati, H. Spohn and S. Teufel: The time-dependent Born-Oppenheimer Approximation. *ESIAM: Math. Model. and Num. Anal.* **41**, 297-314 (2007).

[28] M. Reed and B. Simon: Methods of modern Mathematical Physics IV: Analysis of Operators. *Academic Press Inc.* (1980).

[29] M.B. Ruskai: Absence of discrete spectrum in highly negative ions, II. Extension to Fermions. *Commun. Math. Phys.* **82** No.2, 325-327 (1982).

[30] L. A. Seco, I. M. Sigal, and J. P. Solovej: Bound on the ionization energy of large atoms. *Commun. Math. Phys.* **131** No.2, 307-315 (1990).

[31] F. Senese: General Chemistry Online! Revised 2010. Available under http://antoine.frostburg.edu/chem/senese/101/index.shtml.

[32] I.M. Sigal: Geometric methods in the quantum many-body problem. Non-existence of very negative ions. *Comm. Math. Phys.* **85**, 309-324 (1982).

[33] I.M. Sigal: How many electrons can a nucleus bind? *Ann. Phys.* **157** No.2, 307-320 (1984).

[34] J.P. Solovej: Proof of the ionization conjecture in a reduced Hartree-Fock model. *Invent. Math.* **104** No.1, 291-311 (1991).

[35] J.P. Solovej: The ionization conjecture in Hartree-Fock theory. *Ann. of Math.* **158**, 509-576 (2003).

Quantum spin correlations and random loops

D. Ueltschi

Department of Mathematics, University of Warwick,
Coventry, CV4 7AL, United Kingdom
E-mail: daniel@ueltschi.org
www.ueltschi.org

We review the random loop representations of Tóth and Aizenman-Nachtergaele for quantum Heisenberg models. They can be combined and extended so as to include the quantum XY model and certain SU(2)-invariant spin 1 systems. We explain the calculations of correlation functions.

Keywords: Quantum spin systems, Heisenberg model, random loop model.

1. Introduction

Random loop approaches to quantum spin systems offer an elegant and different perspective to quantum correlations. They find their origin in Feynman-Kac representations of quantum lattice systems. Motivated by earlier work of Conlon and Solovej [6], Tóth introduced a representation of the $S = \frac{1}{2}$ ferromagnetic Heisenberg model that is based on the random interchange model [16]. It allowed him to propose a bound for the free energy (it has been improved recently by Correggi, Giuliani, and Seiringer [7], who have reached the best possible constant). A similar representation was introduced by Aizenman and Nachtergaele for the $S = \frac{1}{2}$ antiferromagnet model [1] and certain models with higher spins. It allowed them to relate the one-dimensional quantum chain to two-dimensional Potts and random cluster models, yielding new insights on the quantum spin chain. This work was reviewed and extended in [14, 15]. See also [12] for a pedagogical introduction. Recently, Bachmann and Nachtergaele used the representation in their study of the classification of gapped ground states [3].

A synthesis of these two representations was proposed in [19]. In the case $S = \frac{1}{2}$, it applies to models that interpolate between the Heisenberg ferromagnetic and antiferromagnetic models such as the quantum XY model. It also applies to certain SU(2) invariant models of spin 1. Thanks to this representation, the existence of spin nematic long-range order was established in the model with $S = 1$ in dimension

$d \geq 3$ [19]. It also plays a rôle in the recent proof of emptyness formation of quantum spin chains of Crawford, Ng, and Starr [8].

This article reviews some of the material treated in [19], and also complements it. We consider the case of an external magnetic field, possibly disordered. We detail formulæ for the matrix elements of the operator $e^{-\beta H}$ and use them to calculate correlation functions. Since there are few loop correlation functions, and seemingly more quantum spin correlation functions, the relations provide useful identities; these identities do not seem otherwise immediate.

2. Quantum spin models

Let $(\Lambda, \mathcal{E})$ be a graph, with Λ the (finite) set of vertices and $\mathcal{E}$ the set of edges. Given $S \in \frac{1}{2}\mathbb{N}$, the Hilbert space is

$$\mathcal{H}_\Lambda = \bigotimes_{x \in \Lambda} \mathcal{H}_x, \tag{1}$$

where each $\mathcal{H}_x$ is a copy of $\mathbb{C}^{2S+1}$. The spin operators are denoted S_x^i, with $x \in \Lambda$ and $i = 1, 2, 3$. They satisfy the commutation relations $[S_x^1, S_y^2] = \mathrm{i}\delta_{x,y} S_x^3$, and further relations obtained by cyclic permutations of the indices $1, 2, 3$. Recall that "classical configurations" $\sigma \in \{-S, \dots, S-1, S\}^\Lambda$ form a basis of $\mathcal{H}_\Lambda$ where the operators S_x^3 are diagonal: Using Dirac's notation,

$$S_x^3 |\sigma\rangle = \sigma_x |\sigma\rangle. \tag{2}$$

We consider the three operators T_{xy}, P_{xy}, Q_{xy} on $\mathcal{H}_{\{x,y\}}$ (and their extensions on $\mathcal{H}_\Lambda$ by identifying T_{xy} with $T_{xy} \otimes \mathtt{Id}_{\Lambda \setminus \{x,y\}}$, etc...):

- T_{xy} is the transposition operator:

$$T_{xy}|a, b\rangle = |b, a\rangle. \tag{3}$$

- P_{xy} is the operator

$$P_{xy} = \sum_{a,b=-S}^{S} (-1)^{a-b} |a, -a\rangle\langle b, -b|. \tag{4}$$

  Equivalently, the matrix coefficients of P_{xy} are given by

$$\langle a, b|P_{xy}|c, d\rangle = (-1)^{a-c} \delta_{a,-b} \delta_{c,-d}. \tag{5}$$

  Notice that $\frac{1}{2S+1} P_{xy}$ is the projector onto the spin singlet.

- Q_{xy} is identical to P_{xy} except for the signs:

$$\langle a, b|Q_{xy}|c, d\rangle = \delta_{a,b} \delta_{c,d}. \tag{6}$$

These operators can be written in terms of spin operators. The form depends on the spin. In the case $S = \frac{1}{2}$, we have

$$\begin{aligned} T_{xy} &= 2\big(S^1_x S^1_y + S^2_x S^2_y + S^3_x S^3_y\big) + \tfrac{1}{2}, \\ Q_{xy} &= 2\big(S^1_x S^1_y - S^2_x S^2_y + S^3_x S^3_y\big) + \tfrac{1}{2}. \end{aligned} \tag{7}$$

In the case $S = 1$, we have

$$\begin{aligned} T_{xy} &= \vec{S}_x \cdot \vec{S}_y + (\vec{S}_x \cdot \vec{S}_y)^2 - 1, \\ P_{xy} &= (\vec{S}_x \cdot \vec{S}_y)^2 - 1. \end{aligned} \tag{8}$$

Here, we used the notation $\vec{S}_x \cdot \vec{S}_y = S^1_x S^1_y + S^2_x S^2_y + S^3_x S^3_y$.

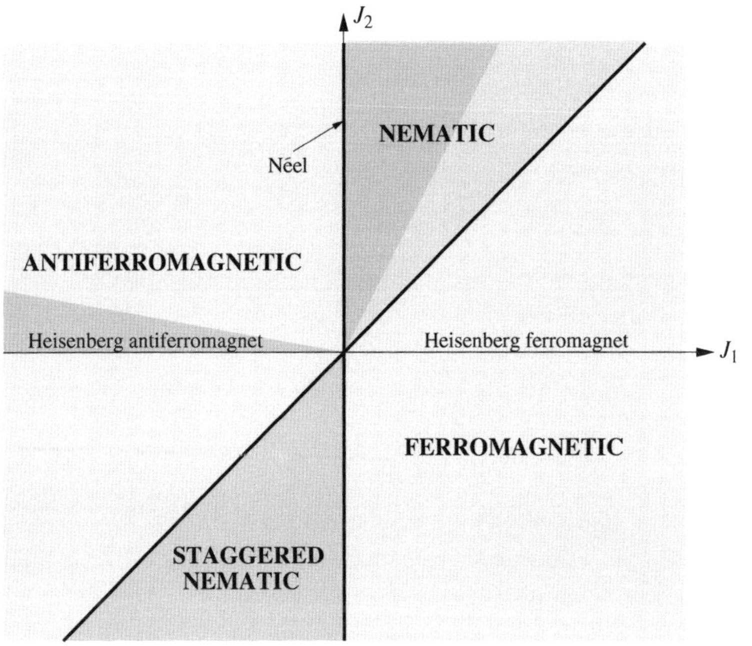

Figure 1. Phase diagram of the general spin 1 model with Hamiltonian $H = -\sum[J_1 \vec{S}_x \cdot \vec{S}_y + J_2(\vec{S}_x \cdot \vec{S}_y)^2]$, in dimension $d \geq 3$. The random loop representation applies to the half-quadrant $0 \leq J_1 \leq J_2$. The phase diagram is expected to show four phases (ferromagnetic, spin nematic, antiferromagnetic, staggered spin nematic). This is supported by rigorous results in the dark region around $J_1 < 0$ and small $J_2 > 0$ [9, 13], and in the dark region $0 \leq J_1 \leq \frac{1}{2} J_2$ [19].

Let $\boldsymbol{h} = (h_x)_{x \in \Lambda}$ denote external magnetic fields. We consider two distinct families of Hamiltonians, indexed by the parameter $u \in [0, 1]$:

$$H^{(u)}_{\Lambda,\boldsymbol{h}} = -\sum_{\{x,y\} \in \mathcal{E}} \Big(u T_{xy} + (1-u) Q_{xy} - 1 \Big) - \sum_{x \in \Lambda} h_x S^3_x, \tag{9}$$

$$\tilde{H}^{(u)}_{\Lambda,\boldsymbol{h}} = -\sum_{\{x,y\} \in \mathcal{E}} \Big(u T_{xy} + (1-u) P_{xy} - 1 \Big) - \sum_{x \in \Lambda} h_x S^3_x. \tag{10}$$

Let $Z^{(u)}(\beta, \Lambda, \boldsymbol{h})$ and $\tilde{Z}^{(u)}(\beta, \Lambda, \boldsymbol{h})$ denote the corresponding partition functions:

$$Z^{(u)}(\beta, \Lambda, \boldsymbol{h}) = \operatorname{Tr}_{\mathcal{H}_\Lambda} \mathrm{e}^{-\beta H^{(u)}_{\Lambda,\boldsymbol{h}}}, \tag{11}$$

$$\tilde{Z}^{(u)}(\beta, \Lambda, \boldsymbol{h}) = \operatorname{Tr}_{\mathcal{H}_\Lambda} \mathrm{e}^{-\beta \tilde{H}^{(u)}_{\Lambda,\boldsymbol{h}}}. \tag{12}$$

The Hamiltonians of Eqs (9) and (10) can also be expressed in terms of spin operators. In the case $S = \frac{1}{2}$, we have

$$H^{(u)}_{\Lambda,\boldsymbol{h}} = -2 \sum_{\{x,y\}\in\mathcal{E}} \left(S^1_x S^1_y + (2u-1) S^2_x S^2_y + S^3_x S^3_y - \tfrac{1}{4}\right) - \sum_{x\in\Lambda} h_x S^3_x. \tag{13}$$

The case $u = 1$ is the Heisenberg ferromagnet. The case $u = \frac{1}{2}$ is the quantum XY model. If the graph is bipartite, the case $u = 0$ is unitarily equivalent to the Heisenberg antiferromagnet.

In the case $S = 1$, we have

$$\tilde{H}^{(u)}_{\Lambda,\boldsymbol{h}} = -\sum_{\{x,y\}\in\mathcal{E}} \Big(u\vec{S}_x \cdot \vec{S}_y + (\vec{S}_x \cdot \vec{S}_y)^2 - 2\Big) - \sum_{x\in\Lambda} h_x S^3_x. \tag{14}$$

It is well-known that any two-body SU(2)-invariant interaction for $S = 1$ can be written as $J_1 \vec{S}_x \cdot \vec{S}_y + J_2 (\vec{S}_x \cdot \vec{S}_y)^2$. The phase diagram of this model is very interesting and it has been investigated by several authors [4, 10, 17, 18]. It is displayed in Fig. 1. The line $J_2 = 0$ corresponds to the usual Heisenberg models.

3. Random loop models

Let us first describe the models of random loops. The connection to quantum spin systems will be described in the next two sections.

At each edge $\{x, y\} \in \mathcal{E}$ is attached the interval $[0, \beta]$ and a Poisson point measure where "crosses" occur with intensity u and "double bars" occur with intensity $1-u$. Let ω denote a realization and $\rho(\mathrm{d}\omega)$ denote independent Poisson point measures on $\mathcal{E} \times [0, \beta]$.

To a given realization ω of the Poisson point measure corresponds a set of loops, denoted $\mathcal{L}(\omega)$. The loops consist of vertical lines connected by crosses or bars. This is best understood by looking at pictures, see Fig. 2. A mathematically precise definition can be found in [19].

The relevant probability distribution involves multiplicative weights with respect to loops. We consider a function $w(\gamma)$ that assigns a real number to each loop γ. We will consider explicit weights below; for now, we just assume that $w(\gamma)$ depends on the loop in a continuous fashion, so all integrals below are well defined. In the case

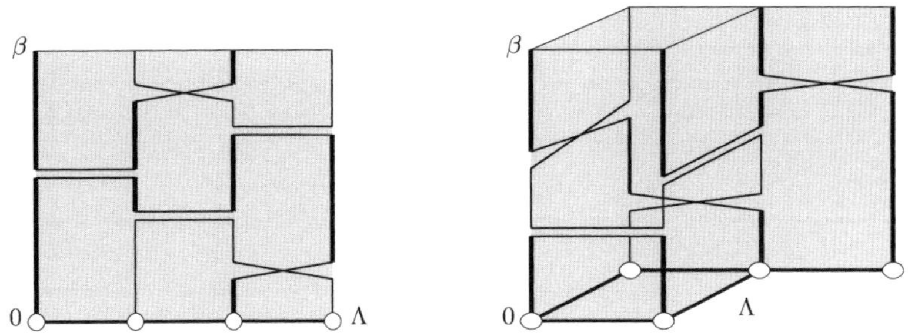

Figure 2. Graphs and realizations of Poisson point measures, and their loops. In both cases, the number of loops is $|\mathcal{L}(\omega)| = 2$.

where $w(\gamma)$ is nonnegative for all γ we have a probabilistic setting. But it is useful to include the possibility of negative weights as well.

We define the *partition function* as

$$Y(\beta, \Lambda) = \int \rho(\mathrm{d}\omega) \prod_{\gamma \in \mathcal{L}(\omega)} w(\gamma). \tag{15}$$

We will always consider cases where $Y(\beta, \Lambda) \neq 0$. The relevant measure for the model of random loops is given by

$$\frac{1}{Y(\beta, \Lambda)} \Big(\prod_{\gamma \in \mathcal{L}(\omega)} w(\gamma) \Big) \rho(\mathrm{d}\omega). \tag{16}$$

It is a probability measure when the weights are positive.

It is not hard to show that for β small, and under some conditions on $w(\gamma)$, the loops have small lengths and the probability that two sites belong to the same loop shows exponential decay with respect to the distance between the sites. See e.g. Theorem 6.1 in [12].

The special case of constant weights, $w(\gamma) = \theta$, is interesting, and actually relevant to quantum systems without external magnetic fields. Under some additional assumptions, namely that the graph $(\Lambda, \mathcal{E})$ be a d-dimensional cube with even side lengths L and $d \geq 3$, that $u \in [0, \frac{1}{2}]$, and that $\theta = 2, 3, \ldots$, one can prove the existence of *macroscopic loops* when β is sufficiently large. Let ℓ_0 denote the random variable for the length of the loop that contains the point $(0, 0) \in \Lambda \times [0, \beta]$. The length of the loop is defined as the sum of the length of all its vertical elements.

Theorem 3.1. *Under the assumptions listed in the paragraph above, there exists $c > 0$ such that for all L,*

$$\mathbb{E}\Big(\frac{\ell_0}{\beta L^d} \Big) \geq c.$$

See Section 5 in [19] for the statement with precise conditions. This theorem can be proved using the method of infrared bounds and reflection positivity introduced and developed in [2, 9, 11, 13]; see Biskup [5] for an excellent survey. This theorem implies the occurrence of long-range order in some quantum systems. The main novel result is the occurrence of spin nematic order for the spin 1 model with Hamiltonian $\tilde{H}^{(u)}_{\Lambda,\boldsymbol{h}}$ defined in Eq. (14). Indeed, it follows from Theorem 3.1 that

- $\frac{1}{|\Lambda|}\sum_x \langle S_0^3 S_x^3\rangle > c$ for the model $H = -\sum(\vec{S}_x \cdot \vec{S}_y)^2$, with $c > 0$ independent of Λ, $d \geq 5$, β large enough. (This is actually Néel order.)
- $\frac{1}{|\Lambda|}\sum_x \big(\langle (S_0^3)^2 (S_x^3)^2\rangle - \langle (S_0^3)^2\rangle\langle (S_x^3)^2\rangle\big) > c$ for the model $H = -\sum\big(J_1 \vec{S}_x \cdot \vec{S}_y + J_2(\vec{S}_x \cdot \vec{S}_y)^2\big)$, with $c > 0$ independent of Λ, $0 \leq J_1 \leq \frac12 J_2$, $d \geq 5$, β large enough. (When $J_1 \lesssim \frac12 J_2$ the result holds for $d \geq 3$.)

It does not seem possible to prove this using the method of infrared bounds and reflection positivity directly for quantum systems. The method in [19] consists in studying the model $H^{(u)}_\Lambda$, which is not related to $\tilde{H}^{(u)}_\Lambda$ in any obvious way when $u \neq 0, 1$. The Gibbs operator $\mathrm{e}^{-\beta H^{(u)}_\Lambda}$ can be expanded in random loops and "space-time spin configurations" (see next section), which gives a sort of classical model that is reflection positive. This allows to prove "Gaussian domination", leading to infrared bounds for the Duhamel two-point function. Combining with the Falk-Bruch inequality, as in [9, 13], one obtains Theorem 3.1. The results for $\tilde{H}^{(u)}_\Lambda$ are then consequences of the loop representation.

4. Gibbs operator and partition functions

The first result is a formula for the Gibbs operator $\mathrm{e}^{-\beta H}$ in terms of the Poisson point measure $\rho(\mathrm{d}\omega)$. To a realization ω corresponds a sequence $(A_1, t_1), \dots, (A_n, t_n)$ where $0 < t_1 < \dots < t_n < \beta$ are the times for the occurrence of events in ω, and A_j is the operator T_{xy} if the event of time t_j is a cross at $\{x, y\} \in \mathcal{E}$; A_j is the operator Q_{xy} if the event of time t_j is a double bar at $\{x, y\}$.

Theorem 4.1. *We have*

$$\mathrm{e}^{-\beta H^{(u)}_{\Lambda,\boldsymbol{h}}} = \int \rho(\mathrm{d}\omega)\, \mathrm{e}^{-(\beta - t_n)\sum h_x S_x^3} A_n\, \mathrm{e}^{-(t_n - t_{n-1})\sum h_x S_x^3} A_{n-1} \dots A_1\, \mathrm{e}^{-t_1 \sum h_x S_x^3}.$$

The same representation applies to the operator $\mathrm{e}^{-\beta \tilde{H}^{(u)}_{\Lambda,\boldsymbol{h}}}$, but with P_{xy} instead of Q_{xy} when double bars occur. The proof can be done by discretizing the time interval $[0, \beta]$, linearizing the Poisson point measure, grouping terms wisely and invoking the Trotter product formula.

Next, we consider partition functions. Given a loop γ, we denote by $\ell_x(\gamma)$ the length of the vertical element(s) of the loop at site $x \in \Lambda$. We have $0 \leq \ell_x(\gamma) \leq \beta$ and, for almost all realizations ω,

$$\sum_{\gamma \in \mathcal{L}(\omega)} \sum_{x \in \Lambda} \ell_x(\gamma) = \beta|\Lambda|. \tag{17}$$

Theorem 4.2. *Given $S \in \frac{1}{2}\mathbb{N}$, let*

$$w(\gamma) = \sum_{a=-S}^{S} \exp\Big\{a \sum_{x \in \Lambda} h_x \ell_x(\gamma)\Big\}.$$

Then for all $u \in [0,1]$, we have

$$Z^{(u)}(\beta, \Lambda, \boldsymbol{h}) = \int \rho(\mathrm{d}\omega) \prod_{\gamma \in \mathcal{L}(\omega)} w(\gamma).$$

In the case where $h_x \equiv 0$, we have $w(\gamma) = 2S+1$ for all loops, and the partition function is equal to $\int (2S+1)^{|\mathcal{L}(\omega)|} \rho(\mathrm{d}\omega)$.

The corresponding formula for the model with Hamiltonian $\tilde{H}^{(u)}_{\Lambda,\boldsymbol{h}}$ is more complicated, as it involves vertical directions of loops. Namely, let us choose an orientation for the loops, and let $\ell_x^+(\gamma)$ (resp. $\ell_x^-(\gamma)$) denote the vertical length of the elements of γ at x that move up (resp. that move down). We have $\ell_x^+(\gamma) + \ell_x^-(\gamma) = \ell_x(\gamma)$. We only state the theorem in the case of integer S, as there are inelegant signs when S is half integer.

Theorem 4.3. *Given $S \in \mathbb{N}$, let*

$$w(\gamma) = \sum_{a=-S}^{S} \exp\Big\{a \sum_{x \in \Lambda} h_x \big[\ell_x^+(\gamma) - \ell_x^-(\gamma)\big]\Big\}.$$

Then for all $u \in [0,1]$, we have

$$\tilde{Z}^{(u)}(\beta, \Lambda, \boldsymbol{h}) = \int \rho(\mathrm{d}\omega) \prod_{\gamma \in \mathcal{L}(\omega)} w(\gamma).$$

In order to prove Theorems 4.2 and 4.3, we need the concept of *space-time configurations*. This is also useful in the calculation of correlation functions. A space-time spin configuration is a function

$$\sigma : \Lambda \times [0, \beta] \longrightarrow \{-S, -S+1, \ldots, S\} \tag{18}$$

such that $\sigma_{x,t}$ is piecewise constant in t, for any x. Given a realization ω of the Poisson point measure, let $\Sigma_{\mathrm{per}}(\omega)$ denote the set of space-time spin configurations that take constant values along each loop. See Fig. 3 for an illustration. Let $\Sigma(\omega)$

denote the set of configurations that are compatible with ω, but without requiring that $\sigma_{x,0} = \sigma_{x,\beta}$. Notice that

$$|\Sigma_{\rm per}(\omega)| = (2S+1)^{|\mathcal{L}(\omega)|}. \tag{19}$$

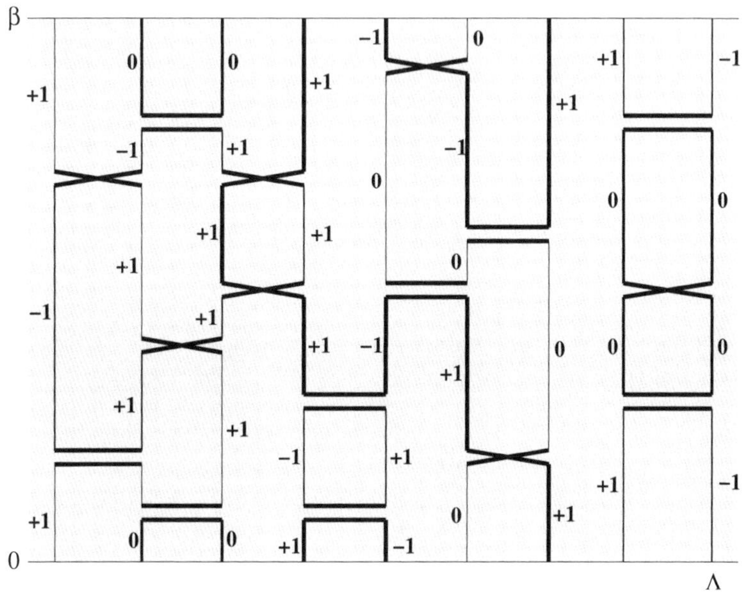

Figure 3. Illustration for a realization of the measure ρ_ι and a compatible space-time spin configuration.

We use Theorem 4.1 and we insert the resolution of the identity $\texttt{Id} = \sum_\sigma |\sigma\rangle\langle\sigma|$ on the left of each transition A_j. Because of the definitions of T_{xy} and Q_{xy}, we get

$$\begin{aligned} \operatorname{Tr} \mathrm{e}^{-\beta H^{(u)}_{\Lambda,\boldsymbol{h}}} &= \int \rho(\mathrm{d}\omega) \sum_{\sigma \in \Sigma_{\rm per}(\omega)} \exp\Big\{ -\sum_{x\in\Lambda} \int_0^\beta h_x \sigma_{x,t} \mathrm{d}t \Big\} \\ &= \int \rho(\mathrm{d}\omega) \prod_{\gamma\in\mathcal{L}(\omega)} \sum_{a=-S}^{S} \mathrm{e}^{-a\sum_x h_x \ell_x(\gamma)} . \end{aligned} \tag{20}$$

This gives the claim of Theorem 4.2. The proof of Theorem 4.3 is similar but with different sets of space-time spin configurations. Let $\tilde\Sigma(\omega)$, $\tilde\Sigma_{\rm per}(\omega)$ with the prescription that the sign of the spin changes when the vertical direction of the loop changes. The calculation is then the same, with additional signs due to the double bars. Namely, a double bar at $\{x,y\} \times t$ gives the sign $(-1)^{\sigma_{x,t-} + \sigma_{x,t+}}$. Fortunately each loop involves an even number of minus signs, so the weight is positive.

5. Correlation functions

We restrict ourselves to two-point correlation functions. We also make the important simplification $h_x \equiv 0$, although expressions can certainly be derived for nonzero external magnetic fields. The loop correlations are given by just three events:

- $E^+_{x,y}$ is the set of all realizations ω such that x and y belong to the same loop, and with identical vertical direction at these points.
- $E^-_{x,y}$ is the set of all ω such that x and y belong to the same loop, and with opposite vertical directions at these points.
- $E_{x,y} = E^+_{x,y} \cup E^-_{x,y}$ is the set of all ω such that x and y belong to the same loop.

These events are illustrated in Fig. 4.

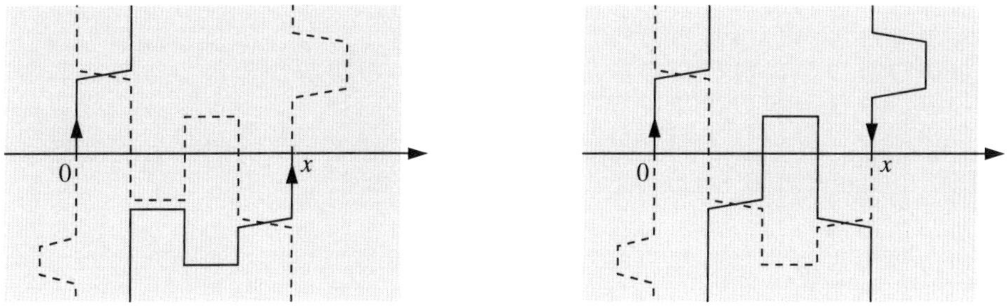

(a) (b)

Figure 4. Illustration for (a) the event $E^+_{0,x}$; (b) the event $E^-_{0,x}$.

Let A_x be an operator of the form $A \otimes \mathtt{Id}_{\Lambda\setminus\{x\}}$ and B_y be an operator of the form $B \otimes \mathtt{Id}_{\Lambda\setminus\{y\}}$, where A, B are operators on $\mathbb{C}^{2S+1}$. We consider the two-point function

$$\langle A_x B_y \rangle = \frac{1}{Z^{(u)}(\beta, \Lambda, 0)} \operatorname{Tr} A_x B_y \, \mathrm{e}^{-\beta H^{(u)}_{\Lambda,0}} . \tag{21}$$

We use the notation tr for the trace in $\mathbb{C}^{2S+1}$. Let $\mathbb{P}(\cdot)$ denote the probability with respect to the random loop measure $Z^{(u)}(\beta, \Lambda, 0)^{-1}(2S+1)^{|\mathcal{L}(\omega)|}\rho(\mathrm{d}\omega)$.

Theorem 5.1. *For $x \neq y$, the correlation function above is given by*

$$\begin{aligned}\langle A_x B_y \rangle = {} & \tfrac{1}{2S+1} (\operatorname{tr} AB)\, \mathbb{P}(E^+_{x,y}) + \tfrac{1}{2S+1} (\operatorname{tr} AB^{\mathrm{t}})\, \mathbb{P}(E^-_{x,y}) \\ & + \tfrac{1}{(2S+1)^2} (\operatorname{tr} A)(\operatorname{tr} B)\, \mathbb{P}(E^{\mathrm{c}}_{x,y}).\end{aligned}$$

Here, $B^{\rm t}$ denotes the transpose of the matrix B in the basis where $\{S_x^3\}$ is diagonal. Choosing $B = \mathtt{Id}$, we get the formula for the one-point function (it is relevant for truncated correlation functions):

$$\langle A_x \rangle = \tfrac{1}{2S+1} {\rm tr}\, A. \tag{22}$$

An interesting special case of correlation function is $S = \frac{1}{2}$ and $A = B = S^i$. We find that

$$\langle S_x^i S_y^i \rangle = \begin{cases} \frac{1}{4} \mathbb{P}(E_{x,y}) & \text{if } i = 1, 3, \\ \frac{1}{4} [\mathbb{P}(E_{x,y}^+) - \mathbb{P}(E_{x,y}^-)] & \text{if } i = 2. \end{cases} \tag{23}$$

Proof of Theorem 5.1. We use Theorem 4.1 and space-time spin configurations, and we get

$${\rm Tr}\, A_x B_y\, {\rm e}^{-\beta H_{\Lambda,0}^{(u)}} = \int \rho({\rm d}\omega) \sum_{\sigma \in \Sigma(\omega)} \langle \sigma_{\cdot,0} | A_x B_y | \sigma_{\cdot,\beta} \rangle. \tag{24}$$

Next we decompose

$$\int \cdot\, \rho({\rm d}\omega) = \int_{E_{x,y}^+} \cdot\, \rho({\rm d}\omega) + \int_{E_{x,y}^-} \cdot\, \rho({\rm d}\omega) + \int_{E_{x,y}^{\rm c}} \cdot\, \rho({\rm d}\omega) \tag{25}$$

and we treat each case separately. If $\omega \in E_{x,y}^+$, we find

$$\sum_{\sigma \in \Sigma(\omega)} \langle \sigma_{\cdot,0} | A_x B_y | \sigma_{\cdot,\beta} \rangle = (2S+1)^{|\mathcal{L}(\omega)|-1} \sum_{a,b=-S}^{S} \langle a, b | A_x B_y | b, a \rangle. \tag{26}$$

The term $(2S+1)^{|\mathcal{L}(\omega)|-1}$ is due to the sum of spin configurations on all the loops except the one that contains x and y. The sum over a, b represents the possible values of spins along this loop. Now we have

$$\langle a, b | A_x B_y | b, a \rangle = \langle a | A | b \rangle \langle b | B | a \rangle, \tag{27}$$

and the sum over a, b gives ${\rm tr}\, AB$. The case where $\omega \in E_{x,y}^-$ is similar, but the matrix elements involving A, B are

$$\langle a, a | A_x B_y | b, b \rangle = \langle a | A | b \rangle \langle a | B | b \rangle. \tag{28}$$

The sum over a, b gives ${\rm tr}\, AB^{\rm t}$. Finally, the case $\omega \in E_{x,y}^{\rm c}$ involves two special loops, those containing x and y, and we get

$$\sum_{\sigma \in \Sigma(\omega)} \langle \sigma_{\cdot,0} | A_x B_y | \sigma_{\cdot,\beta} \rangle = (2S+1)^{|\mathcal{L}(\omega)|-2} {\rm tr}\, A\, {\rm tr}\, B. \tag{29}$$

□

The case of the Hamiltonian $\tilde{H}_{\Lambda,0}^{(u)}$ is more complicated due to the signs. They lead to signed measures when S is half-integer but not integer (except when $u = 0$ on a bipartite lattice). We restrict here to integer S and we consider the two-point function

$$\langle A_x B_y \rangle^{\sim} = \frac{1}{\tilde{Z}^{(u)}(\beta,\Lambda,0)} \operatorname{Tr} A_x B_y \, \mathrm{e}^{-\beta \tilde{H}^{(u)}_{\Lambda,0}}. \tag{30}$$

We also write

$$\tilde{\mathbb{P}}(E_{x,y}) = \frac{1}{\tilde{Z}^{(u)}(\beta,\Lambda,0)} \int 1_{E_{x,y}}(\omega)(2S+1)^{|\mathcal{L}(\omega)|} \rho(\mathrm{d}\omega). \tag{31}$$

Theorem 5.2. *For $x \neq y$, the correlation above is given by*

$$\begin{aligned}\langle A_x B_y \rangle^{\sim} &= \tfrac{1}{2S+1} (\operatorname{tr} AB) \, \tilde{\mathbb{P}}(E^+_{x,y}) \\ &+ \tfrac{1}{2S+1} \Bigl(\sum_{a,b=-S}^{S} (-1)^{a+b} \langle a|A|b\rangle \langle -a|B|-b\rangle \Bigr) \, \tilde{\mathbb{P}}(E^-_{x,y}) \\ &+ \tfrac{1}{(2S+1)^2} (\operatorname{tr} A)(\operatorname{tr} B) \, \tilde{\mathbb{P}}(E^{\mathrm{c}}_{x,y}).\end{aligned}$$

The formula for one-point functions follow, $\langle A_x \rangle^{\sim} = \frac{1}{2S+1} \operatorname{tr} A$. The proof is similar to that of Theorem 5.1, but there are extra difficulties due to the minus signs. We do not write it explicitly. We find in particular (see [19] for more details)

$$\begin{aligned}&\langle S^i_x S^i_y \rangle = \tfrac{1}{3} S(S+1) \left[\mathbb{P}(E^+_{x,y}) - \mathbb{P}(E^-_{x,y}) \right] \\ &\langle (S^i_x)^2 (S^i_y)^2 \rangle - \langle (S^i_x)^2 \rangle \langle (S^i_y)^2 \rangle = \tfrac{1}{45} S(S+1)(2S-1)(2S+3) \, \mathbb{P}(E_{x,y}).\end{aligned} \tag{32}$$

It is remarkable that many spin correlation functions can be expressed with a handful of loop correlation functions.

Acknowledgments: I am grateful to Benjamin Lees for useful discussions. I would like to thank Pavel Exner, Wolfgang König, and Hagen Neidhardt, for organizing the conference QMATH 12 in Berlin, 10–13 September 2013, where this work was presented. This work is partially supported by EPSRC grant EP/G056390/1.

Bibliography

[1] M. Aizenman, B. Nachtergaele, *Geometric aspects of quantum spin states*, Comm. Math. Phys., 164, 17–63 (1994)

[2] C. Albert, L. Ferrari, J. Fröhlich, B. Schlein, *Magnetism and the Weiss exchange field — a theoretical analysis motivated by recent experiments*, J. Statist. Phys. 125, 77–124 (2006)

[3] S. Bachmann, B. Nachtergaele, *On gapped phases with a continuous symmetry and boundary operators*, J. Statist. Phys. 154, 91–112 (2014)

[4] C.D. Batista, G. Ortiz, *Algebraic approach to interacting quantum systems*, Adv. Phys. 53, 1–82 (2004)

[5] M. Biskup, *Reflection positivity and phase transitions in lattice spin models*, in Methods of contemporary mathematical statistical physics, Lect. Notes Math. 1970, 1–86 (2009)

[6] J. Conlon, J.P. Solovej, *Upper bound on the free energy of the spin 1/2 Heisenberg ferromagnet*, Lett. Math. Phys. 23, 223–231 (1991)

[7] M. Correggi, A. Giuliani, R. Seiringer, *Validity of the spin-wave approximation for the free energy of the Heisenberg ferromagnet*, arXiv:1312.7873 [math-ph] (2013)

[8] N. Crawford, S. Ng, S. Starr, in preparation

[9] F.J. Dyson, E.H. Lieb, B. Simon, *Phase transitions in quantum spin systems with isotropic and nonisotropic interactions*, J. Statist. Phys. 18, 335–383 (1978)

[10] Yu.A. Fridman, O.A. Kosmachev, Ph.N. Klevets, *Spin nematic and orthogonal nematic states in $S = 1$ non-Heisenberg magnet*, J. Magnetism and Magnetic Materials 325, 125–129 (2013)

[11] J. Fröhlich, B. Simon, T. Spencer, *Infrared bounds, phase transitions and continuous symmetry breaking*, Comm. Math. Phys. 50, 79–95 (1976)

[12] C. Goldschmidt, D. Ueltschi, P. Windridge, *Quantum Heisenberg models and their probabilistic representations*, in Entropy and the Quantum II, Contemp. Math. 552, 177–224 (2011); arXiv:1104.0983 [math-ph]

[13] T. Kennedy, E.H. Lieb, B.S. Shastry, *Existence of Néel order in some spin-$\frac{1}{2}$ Heisenberg antiferromagnets*, J. Statist. Phys. 53, 1019–1030 (1988)

[14] B. Nachtergaele, *Quasi-state decompositions for quantum spin systems*, in Probability Theory and Mathematical Statistics, B Grigelionis et al (eds), pp 565–590 (1994); arXiv:cond-mat/9312012

[15] B. Nachtergaele, *A stochastic geometric approach to quantum spin systems*, in Probability and Phase Transitions, G. Grimmett (ed.), Nato Science series C 420, pp 237–246 (1994)

[16] B. Tóth, *Improved lower bound on the thermodynamic pressure of the spin* 1/2 *Heisenberg ferromagnet*, Lett. Math. Phys. 28, 75–84 (1993)

[17] T.A. Tóth, A.M. Läuchli, F. Mila, K. Penc, *Competition between two- and three-sublattice ordering for $S = 1$ spins on the square lattice*, Phys. Rev. B 85, 140403 (2012)

[18] H.-H. Tu, G.-M. Zhang, T. Xiang, *Class of exactly solvable $SO(n)$ symmetric spin chains with matrix product ground states*, Phys. Rev. B 78, 094404 (2008)

[19] D. Ueltschi, *Random loop representations for quantum spin systems*, J. Math. Phys. 54, 083301 (2013)

Equidistribution estimates for eigenfunctions and eigenvalue bounds for random operators

D. Borisov

Institute of Mathematics USC RAS, Chernyshevskii str., 112, Ufa, 450000, Russia and Bashkir State Pedagogical University, October rev. st., 3a, Ufa, 450008, Russia
E-mail: borisovdi@yandex.ru
matem.anrb.ru and www.bspu.ru

M. Tautenhahn

Fakultät für Mathematik, Reichenhainer Str. 41, Chemnitz, D-09126, Germany
www.tu-chemnitz.de/~mtau

I. Veselić

Fakultät für Mathematik, Reichenhainer Str. 41, Chemnitz, D-09126, Germany
www.tu-chemnitz.de/mathematik/stochastik/

We discuss properties of L^2-eigenfunctions of Schrödinger operators and elliptic partial differential operators. The focus is set on unique continuation principles and equidistribution properties. We review recent results and announce new ones.

Keywords: Scale-free unique continuation property, equidistribution property, observability estimate, uncertainty relation, Carleman estimate, Schrödinger operator, elliptic differential equation.

1. Introduction

In this note we present recent results in Harmonic Analysis for solutions of (time-independent) Schrödinger equations and other partial differential equations. They are motivated by interest in techniques relevant for proving localization for random Schrödinger operators. The mentioned Harmonic Analysis results which we present are a quantitative unique continuation principle and an equidistribution property for eigenfunctions, which is scale-uniform. These results, and variants thereof, go under various names, depending on the particular field of mathematics: They are called observability estimate, uncertainty relation, scale-free unique continuation principle, or local positive definiteness. The latter term signifies that a self-adjoint operator is (strictly) positive definite when restricted to a relevant subspace, while it is not so on the whole Hilbert space. For the purpose of motivation we discuss this property in the next section.

The term *localization* refers to the phenomenon, that quantum Hamiltonians describing the movement of electrons in certain disordered media exhibit pure point spectrum in appropriately specified energy regions. The corresponding eigenfunctions decay exponentially in space. The (time-dependent) wavepackets describing electrons stay localized essentially in a compact region of space for all times. Nota bene, all mentioned properties hold *almost surely*. This is natural in the context of random operators.

An important partial result for deriving localization are Wegner estimates. These are bounds on the expected number of eigenvalues in a bounded energy interval of a random Schrödinger operator restricted to a box.

The localization problem has been studied for other classes of random operators beyond those of Schrödinger type. An example are random divergence type operators, see e.g. [1] and [2]. This are partial differential operators with randomness in coefficients of higher order terms. In particular, the second order term is no longer the Laplacian, but a variable coefficient operator. In this context one is again lead to consider the above mentioned questions of Harmonic Analysis for eigenfunctions of differential operators. In this note we present an exposition of recently published results, and an announcement of a quantitative unique continuation principle and an equidistribution estimate for eigenfunctions for a class of elliptic operators with variable coefficients.

1.1. *Motivation: Moving and lifting of eigenvalues*

Here we discuss some aspects of eigenvalue perturbation theory. It will provide an accessible explanation why one is interested in the results presented in Sections 2 and 3 below in the context of random Schrödinger operators and elliptic differential operators, respectively. In fact, to illustrate the main questions it will be for the moment completely sufficient to restrict our attention to the finite dimensional situation, i.e. to perturbation theory for finite symmetric matrices. The focus will be on how (local) positive definiteness of the perturbation relates to lifting of eigenvalues.

Let A and B be symmetric $n \times n$ matrices, with $B \geq b > 0$ positive definite. The variational min-max principle for eigenvalues shows that for any $k \in \{1, \ldots, n\}$ and $t \geq 0$

$$\lambda_k(A + tB) \geq \lambda_k(A) + bt \tag{1}$$

where $\lambda_k(M)$ denotes the kth lowest eigenvalue, counting multiplicities, of a symmetric matrix M. Note that the dimension n does not enter in the bound (1). Without the positive definiteness assumption on B this universal bound will fail, most blatantly if

$$A = \begin{pmatrix} A_1 & 0 \\ 0 & A_2 \end{pmatrix} \quad \text{and} \quad B = \begin{pmatrix} \mathrm{Id} & 0 \\ 0 & -\mathrm{Id} \end{pmatrix}.$$

In this case, all eigenvalue $\lambda_k(A + tB)$ *will move*, even with constant speed w.r.t. the variable t, albeit in different directions. If B is singular, some eigenvalues may not move at all. However, for appropriate classes of symmetric matrices A, and of positive semidefinite matrices B, one may still aim to prove

$$\forall t \geq 0, k \in \{1, \ldots, n\} \, \exists \kappa > 0 \text{ such that } \lambda_k(A + tB) \geq \lambda_k(A) + \kappa t \tag{2}$$

Note however, that κ is now not a uniform bound but depends on

- the class of symmetric matrices from which A is chosen,
- the class of semidefinite matrices from which B is chosen,
- the range from which the coupling t is chosen, and
- the range from which the index $k \in \{1, \ldots, n\}$ is chosen.

In the case of random operators or matrices one in is interested in the situation where

$$A(\omega) = A_0 + \sum_{j \in Q} \omega_j B_j = \Big(A_0 + \sum_{j \in Q, j \neq 0} \omega_j B_j\Big) + \omega_0 B_0 \tag{3}$$

is a multi-parameter pencil. Here Q is some subset of $\mathbb{Z}^d$ containing 0. The real variables ω_j model random coupling constants determining the strength of the perturbation B_j in each configuration $\omega = (\omega_j)_{j \in Q}$. Now, (3) already suggest to write $A(\omega)$ as

$$A(\omega_0^\perp) + tB \quad \text{where} \quad t = \omega_0, \; B = B_0, \text{ and } \omega_0^\perp = (\omega_j)_{j \in Q, j \neq 0}.$$

This highlights that if we consider $A(\omega)$ as a function of the single variable $t = \omega_0$, it is clearly a one-parameter family of operators, albeit the "unperturbed part" $A(\omega_0^\perp)$ of $A(\omega) = A(\omega_0^\perp) + tB$ is not a single operator, but varying over the ensemble $(A(\omega_0^\perp))_{\omega_0^\perp}$. To have a useful version of (2) in this situation, the constant κ needs to have a uniform lower bound $\inf_A \kappa$ where $A = A(\omega_0^\perp)$ varies over all matrices in the ensemble.

In what follows we present rigorous results of the type (2), but where A and B are not finite matrices, but differential and multiplication operators. The relevant operators have all compact resolvent, ensuring that the entire spectrum consists of eigenvalues.

2. Equidistribution property of Schrödinger eigenfunctions

The following result is taken from [3]. It is an equidistribution estimate for Schrödinger eigenfunctions, which is uniform w.r.t. the naturally arising length scales, and has strong implications for the spectral theory of random Schrödinger operators.

We fix some notation. For $L > 0$ we denote by $\Lambda_L = (-L/2, L/2)^d$ a cube in $\mathbb{R}^d$. For $\delta > 0$ the open ball centered at $x \in \mathbb{R}$ with radius δ is denoted by $B(x, \delta)$. For a sequence of points $(x_j)_j$ indexed by $j \in \mathbb{Z}^d$ we denote the collection of balls $\cup_{j\in\mathbb{Z}^d} B(x_j, \delta)$ by S and its intersection with Λ_L by S_L. We will be dealing with certain subspaces of the standard second order Sobolev space $W^{2,2}(\Lambda_L)$ on the cube. Let Δ be the d-dimensional Laplacian. Its restriction to the cube $\Lambda = \Lambda_L$ needs boundary conditions to be self-adjoint. The domain of the Dirichlet Laplacian will be denoted by $\mathcal{D}(\Delta_{\Lambda,0})$ and the domain of the Laplacian with periodic boundary conditions by $\mathcal{D}(\Delta_{\Lambda,\mathrm{per}})$. Let $V\colon \mathbb{R}^d \to \mathbb{R}$ be a bounded measurable function, and $H_L = (-\Delta + V)_{\Lambda_L}$ a Schrödinger operator on the cube Λ_L with Dirichlet or periodic boundary conditions. The corresponding domains are still $\mathcal{D}(\Delta_{\Lambda,0})$ and $\mathcal{D}(\Delta_{\Lambda,\mathrm{per}})$, respectively. Note that we denote a multiplication operator by the same symbol as the corresponding function.

The following theorem was proven in [3].

Theorem 2.1 (Scale-free unique continuation principle). *Let $\delta, K_V > 0$. Then there exists $C_{\mathrm{sfUC}} \in (0, \infty)$ such that for all $L \in 2\mathbb{N} + 1$, all measurable $V : \mathbb{R}^d \to [-K_V, K_V]$, all real-valued $\psi \in \mathcal{D}(\Delta_{\Lambda,0}) \cup \mathcal{D}(\Delta_{\Lambda,\mathrm{per}})$ with $(-\Delta + V)\psi = 0$ almost everywhere on Λ_L, and all sequences $(x_j)_{j\in\mathbb{Z}^d} \subset \mathbb{R}^d$, such that for all $j \in \mathbb{Z}^d$ the ball $B(x_j, \delta) \subset \Lambda_1 + j$, we have*

$$\int_{S_L} \psi^2 \geq C_{\mathrm{sfUC}} \int_{\Lambda_L} \psi^2. \tag{4}$$

The value of the result is not in the *existence* of the constant C_{sfUC}, but in the *quantitative control* of the dependence of C_{sfUC} on parameters entering the model. The very formulation of the theorem states that C_{sfUC} is independent of the position of the balls $B(x_j, \delta)$ within $\Lambda_1 + j$, and independent of the scale $L \in 2\mathbb{N} + 1$. From the estimates given in Section 2 of [3] one infers that C_{sfUC} depends on the potential V only through the norm $\|V\|_\infty$ (on an exponential scale), and it depends on the small radius $\delta > 0$ polynomially, i.e. $C \gtrsim \delta^N$, for some $N \in \mathbb{N}$ which depends on the dimension on d and $\|V\|_\infty$.

The theorem states a property of functions in the kernel of the operator. It is easily applied to eigenfunctions corresponding to other eigenvalues since

$$H_L\psi = E\psi \Leftrightarrow (H_L - E)\psi = 0.$$

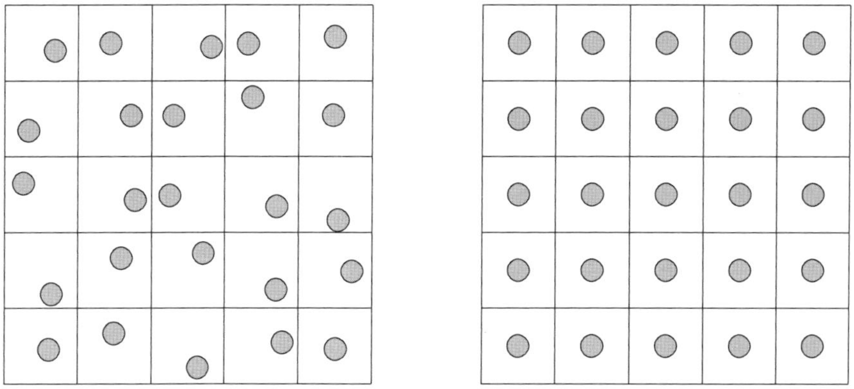

(a) non-periodic (b) periodic arrangement of balls

Figure 1. Examples of collections of balls S_L within region $\Lambda_L \subset \mathbb{R}^2$.

As a consequence of the energy shift the constant K_V has to be replaced with K_{V-E}, which may be larger than K_V. It may always be estimated by $K_{V-E} \leq K_V + |E|$.

There is a very natural question supported by earlier results, which was spelled out in [3], namely does the following generalisation of Theorem 2.1 hold: Given $\delta > 0$, $K \geq 0$ and $E \in \mathbb{R}$ there is a constant $C > 0$ such that for all measurable $V\colon \mathbb{R}^d \to [-K, K]$, all $L \in 2\mathbb{N}+1$, and all sequences $(x_j)_{j\in\mathbb{Z}^d} \subset \mathbb{R}^d$ with $B(x_j, \delta) \subset \Lambda_1 + j$ for all $j \in \mathbb{Z}^d$ we have

$$\chi_{(-\infty,E]}(H_L)\, W_L\, \chi_{(-\infty,E]}(H_L) \geq C\, \chi_{(-\infty,E]}(H_L), \tag{5}$$

where $W_L = \chi_{S_L}$ is the indicator function of S_L and $\chi_I(H_L)$ denotes the spectral projector of H_L onto the interval I. Here $C = C_{\delta,K,E}$ is determined by δ, K, E alone.

Klein obtained a positive answer to the question for sufficiently short subintervals of $(-\infty, E]$.

Theorem 2.2 ([4]). *Let $d \in \mathbb{N}$, $E \in \mathbb{R}$, $\delta \in (0, 1/2]$ and $V : \mathbb{R}^d \to \mathbb{R}$ be measurable and bounded. There is a constant $M_d > 0$ such that if we set*

$$\gamma = \frac{1}{2}\delta^{M_d\left(1+(2\|V\|_\infty+E)^{2/3}\right)},$$

then for all energy intervals $I \subset (-\infty, E]$ with length bounded by 2γ, all $L \in 2\mathbb{N}+1$, $L \geq 72\sqrt{d}$ and all sequences $(x_j)_{j\in\mathbb{Z}^d} \subset \mathbb{R}^d$ with $B(x_j, \delta) \subset \Lambda_1 + j$ for all $j \in \mathbb{Z}^d$

$$\chi_I(H_L)\, W_L\, \chi_I(H_L) \geq \gamma^2 \chi_I(H_L). \tag{6}$$

This does not answer the above posed question question completely due to the restriction $|I| \leq 2\gamma$. However, the result is sufficient for many questions in spectral theory of random Schrödinger operators. For a history of the questions discussed here and earlier results we refer to [3].

2.1. *Random Schrödinger operators*

Let Λ_L be a cube of side $L \in 2\mathbb{N}+1$, $(\Omega, \mathbb{P})$ a probability space, $V_0 \colon \Lambda_L \to \mathbb{R}$ a bounded, measurable deterministic potential, $V_\omega \colon \Lambda_L \to \mathbb{R}$ a bounded random potential and $H_{\omega,L} = (-\Delta + V_0 + V_\omega)_{\Lambda_L}$ a random Schrödinger operator on $L^2(\Lambda_L)$ with Dirichlet or periodic boundary conditions. We assume that the random potential is of Delone-Anderson form

$$V_\omega(x) := \sum_{j \in \mathbb{Z}^d} \omega_j u_j(x).$$

The random variables $\omega_j, j \in \mathbb{Z}^d$, are independent with probability distributions μ_j, such that for some $m > 0$ an all $j \in \mathbb{Z}^d$ we have $\operatorname{supp} \mu_j \subset [-m, m]$. Fix $0 < \delta_- < \delta_+ < \infty$ and $0 < C_- \leq C_+ < \infty$. The sequence of measurable functions $u_j \colon \mathbb{R}^d \to \mathbb{R}$, $j \in \mathbb{Z}^d$, is such that

$$\forall j \in \mathbb{Z}^d : \quad C_- \chi_{B(z_j,\delta_-)} \leq u_j \leq C_+ \chi_{B(z_j,\delta_+)}, \text{ and } B(z_j, \delta_-) \subset \Lambda_1 + j.$$

2.2. *Lifting of eigenvalues*

Let $\lambda_k^L(\omega)$ denote the eigenvalues of $H_{\omega,L}$ enumerated in non-decreasing order and counting multiplicities and $\psi_k = \psi_k^L(\omega)$ the normalised eigenvectors corresponding to $\lambda_k^L(\omega)$. While we suppress the dependence of ψ_k on L and ω in the notation, it should be kept in mind. Then

$$\lambda_k^L(\omega) = \langle \psi_k, H_{\omega,L} \psi_k \rangle = \int_{\Lambda_L} \overline{\psi_k} (H_{\omega,L} \psi_k).$$

Define the vector $e = (e_j)_{j \in \mathbb{Z}^d}$ by $e_j = 1$ for $j \in \mathbb{Z}^d$. Consider the monotone shift of V_ω

$$V_{\omega + t \cdot e} = \sum_{j \in \mathbb{Z}^d} (\omega_j + t) u_j$$

and set $Q = Q_L = \Lambda_L \cap \mathbb{Z}^d$. By first order perturbation theory we have

$$\frac{\mathrm{d}}{\mathrm{d}\tau} \lambda_k^L(\omega + \tau \cdot e)|_{\tau = t} = \langle \psi_k, \sum_{k \in Q} u_j \, \psi_k \rangle.$$

Note that the right hand side depends on t implicitly through the eigenfunction ψ_k. Let us fix some $E_0 \in \mathbb{R}$ and restrict our attention only to those eigenvalues satisfying $\lambda_n^L(\omega) \leq E_0$. By Theorem 2.1 there exists a constant C_{sfUC} depending on the energy E_0, δ_- and the overall supremum

$$\sup_{|s| \leq m} \sup_{|\omega_j| \leq m} \sup_{x \in \mathbb{R}^d} \Big| V_0(x) + V_\omega(x) + s \sum_{j \in Q} u_j \Big|$$

of the potential, such that

$$\sum_{k\in Q}\langle \psi_k, u_j\,\psi_k\rangle \geq C_- \sum_{k\in Q}\langle \psi_k, \chi_{B(z_k,\delta_-)}\psi_k\rangle \geq C_- \cdot C_{\mathrm{sfUC}} =: \kappa.$$

Here we used that $\|\psi\|_{L^2(\Lambda)} = 1$. (Note that the quantity κ depends a-priori on the model parameters.) Integrating the derivative gives

$$\begin{aligned}\lambda_k^L(\omega + t\cdot e) &= \lambda_k^L(\omega) + \int_0^t \frac{\mathrm{d}\lambda_k^L(\omega+\tau\cdot e)}{\mathrm{d}\tau}\Big|_{\tau=s}\,\mathrm{d}s\\ &\geq \lambda_k^L(\omega) + \int_0^t \kappa\,\mathrm{d}s = \lambda_k^L(\omega) + t\kappa. \end{aligned} \tag{7}$$

This is the lifting estimate for eigenvalues of random (Schrödinger) operators alluded to in §1.1. It should be compared with (2) there. Indeed, due to the uniform nature of the estimate in Theorem 2.1 we have

$$\inf_{L\in 2\mathbb{N}+1}\ \inf_{\omega \text{ s.t. } \forall k: |\omega_j|\leq m}\ \inf_{|t|\leq m}\ \inf_{n \text{ s.t. } \lambda_n^L(\omega)\leq E_0} \kappa > 0. \tag{8}$$

Thus eigenvalues lifting estimate is almost as uniform as (1). A parameter, with respect to which the lifting estimate is *not* uniform is the cut-off energy E_0. Indeed, if we add in (8) an infimum over $E_0 > 0$ on the left hand side, it becomes zero, unless $\sum_k \chi_{B(z_k,\delta_-)} \geq 1$ almost everywhere on $\mathbb{R}^d$.

2.3. *Wegner estimates*

Here we present a Wegner estimate. Such estimates play an important role in the proof of localization via the multiscale analysis. The latter is an induction argument over increasing length scales. The Wegner bound is used to prove the induction step.

Let $s\colon [0,\infty) \to [0,1]$ be the global modulus of continuity of the family $\{\mu_j\}_{j\in\mathbb{Z}^d}$, that is,

$$s(\epsilon) := \sup_{j\in\mathbb{Z}^d}\sup_{a\in\mathbb{R}} \mu_j\Big(\Big[a-\frac{\epsilon}{2}, a+\frac{\epsilon}{2}\Big]\Big)$$

The main result of [3] on the model described in the last paragraph is a Wegner estimate which is valid for all compact energy intervals.

Theorem 2.3 ([3]). *Let $H_{\omega,L}$ be a random Schrödinger operator as in §2.1. Then for each $E_0 \in \mathbb{R}$ there exists a constant C_W, such that for all $E \leq E_0$, $\epsilon \leq 1/3$, and all $L \in 2\mathbb{N}+1$ we have*

$$\mathbb{E}\{\mathrm{Tr}\,[\chi_{[E-\epsilon,E+\epsilon]}(H_{\omega,L})]\} \leq C_W\ s(\epsilon)\,|\ln\,\epsilon|^d\ |\Lambda_L|.$$

The Wegner constant C_W depends only on E_0, $\|V_0\|_\infty$, m, C_-, C_+, δ_-, and δ_+. Klein [4] obtains an improvement over this result based on his above quoted Theorem 2.2. There are many earlier, related Wegner estimates. For an overview we refer to [3].

2.4. *Comparison of local L^2-norms*

An important step in the proof of Theorem 2.1 is the following result which compares L^2-norms of the restrictions of a PDE-solution to two distinct subsets. In our applications the solution will be an eigenfunction of the Schrödinger operator. Various estimates of this type have been given in [5], [6] and [3]. We quote here the version from the last mentioned paper.

Theorem 2.4. *Let $K, R, \beta \in [0, \infty), \delta \in (0, 1]$. There exists a constant $C_{\mathrm{qUC}} = C_{\mathrm{qUC}}(d, K, R, \delta, \beta) > 0$ such that, for any $G \subset \mathbb{R}^d$ open, any $\Theta \subset G$ measurable, satisfying the geometric conditions*

$$\operatorname{diam}\Theta + \operatorname{dist}(0, \Theta) \leq 2R \leq 2\operatorname{dist}(0, \Theta), \quad \delta < 4R, \quad B(0, 14R) \subset G,$$

and any measurable $V : G \to [-K, K]$ and real-valued $\psi \in W^{2,2}(G)$ satisfying the differential inequality

$$|\Delta\psi| \leq |V\psi| \quad a.e. on\ G \quad as\ well\ as \quad \int_G |\psi|^2 \leq \beta \int_\Theta |\psi|^2,$$

we have

$$\int_{B(0,\delta)} |\psi|^2 \geq C_{\mathrm{qUC}} \int_\Theta |\psi|^2. \tag{9}$$

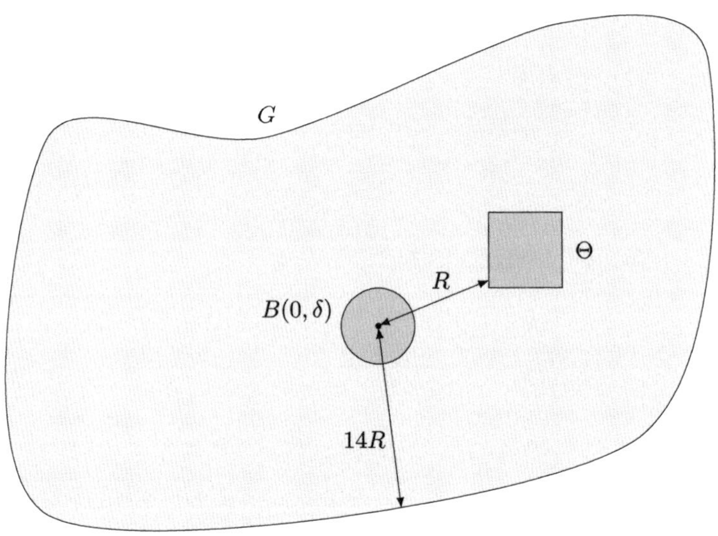

Figure 2. Assumptions in Theorem 2.4 on the geometric constellation of G, Θ, and $B(0, \delta)$

3. Equidistribution property eigenfunctions of second order elliptic operators

3.1. *Notation*

Let $\mathcal{L}$ be the second order partial differential operator

$$\mathcal{L}u = -\sum_{i,j=1}^{d} \partial_i \left(a^{ij}\partial_j u\right)$$

acting on functions u on $\mathbb{R}^d$. Here ∂_i denotes the ith weak derivative. Moreover, we introduce the following assumption on the coefficient functions a^{ij}.

Assumption 3.1. *Let $r, \vartheta_1, \vartheta_2 > 0$. The operator $\mathcal{L}$ satisfies $A(r, \vartheta_1, \vartheta_2)$, if and only if $a^{ij} = a^{ji}$ for all $i, j \in \{1, \ldots, d\}$ and for almost all $x, y \in B(0, r)$ and all $\xi \in \mathbb{R}^d$ we have*

$$\vartheta_1^{-1}|\xi|^2 \leq \sum_{i,j=1}^{d} a^{ij}(x)\xi_i\xi_j \leq \vartheta_1|\xi|^2 \quad \textit{and} \quad \sum_{i,j=1}^{d} |a^{ij}(x) - a^{ij}(y)| \leq \vartheta_2|x-y|.$$

3.2. *A quantitative unique continuation principle*

We first present an extension of the quantitative continuation principle, formulated for Schrödinger operators in Theorem 2.4, to elliptic operators with variable coefficients.

Theorem 3.1 ([7]). *Let $R \in (0, \infty)$, $K_V, \beta \in [0, \infty)$ and $\delta \in (0, 4R]$. There is an $\epsilon > 0$, such that if $A(14R, 1+\epsilon, \epsilon)$ holds then there is a constant $C_{\mathrm{qUC}} > 0$, such that for any open $G \subset \mathbb{R}^d$ containing the origin and $\Theta \subset G$ measurable satisfying*

$$\operatorname{diam}\Theta + \operatorname{dist}(0, \Theta) \leq 2R \leq 2\operatorname{dist}(0, \Theta) \quad \textit{and} \quad B(0, 14R) \subset G,$$

any measurable $V : G \to [-K_V, K_V]$ and real-valued $\psi \in W^{2,2}(G)$ satisfying the differential inequality

$$|\mathcal{L}\psi| \leq |V\psi| \quad \textit{a.e. on } G \quad \textit{as well as} \quad \frac{\|\psi\|_G^2}{\|\psi\|_\Theta^2} \leq \beta,$$

we have

$$\|\psi\|_{B(x,\delta)}^2 \geq C_{\mathrm{qUC}}\|\psi\|_\Theta^2. \tag{10}$$

3.3. Scale-free unique continuation principle

We move on to discuss the equidistribution property or scale-free unique continuation principle for eigenfunctions. The aim is to formulate an analog of Theorem 2.1 for variable coefficient elliptic operators. As presented below, for the moment we have solved only the situation where the second order term is sufficiently close to the Laplacian.

As before, we denote by Λ_L a box of side $L \in \mathbb{N}$. By V we indicate a bounded measurable potential on $\mathbb{R}^d$ taking values in $[-K_V, K_V]$, where K_V is a positive constant. We restrict the operator $\mathcal{L}$ on $\Lambda_L(0)$ and add either periodic or Dirichlet boundary conditions. In the former case we denote such an operator by $\mathcal{L}_{L,0}$, and its domain $\mathcal{D}(\mathcal{L}_{L,0})$ is the subspace of $W^{2,2}(\Lambda_L)$ consisting of functions vanishing on $\partial\Lambda_L$. The notation for the operator with periodic boundary condition is $\mathcal{L}_{L,\text{per}}$ and its domains $\mathcal{D}(\mathcal{L}_{L,\text{per}})$ consists of the functions in $W^{2,2}(\Lambda_L)$ satisfying periodic boundary conditions.

Assumption 3.2. *For each pair i, j the function $a^{ij} : \mathbb{R}^d \to \mathbb{R}$ is $\mathbb{Z}^d$-periodic.*

Assume that in the case of operator $\mathcal{L}_{L,0}$ its coefficients a^{ij}, $i \neq j$ vanish on the sides of box Λ_L, while the coefficients a^{ii} satisfy periodic boundary conditions on the sides of box Λ_L. In the case of operator $\mathcal{L}_{L,\text{per}}$ suppose that all its coefficients satisfy periodic boundary conditions on the sides of box Λ_L.

Theorem 3.2. *Fix $K_V \in [0, +\infty)$, $\delta \in (0, 1]$. Assume $A(\sqrt{d}, 1+\epsilon, \epsilon)$ with $\epsilon > 0$ as in Theorem 3.1. Assume 3.2.*

Then there exists a constant $C_{sfUC} > 0$ such that for any $L \in 2\mathbb{N}+1$, any sequence

$$Z := \{z_k\}_{k\in\mathbb{Z}^d} \quad \text{in} \quad \mathbb{R}^d \quad \text{such that } B(z_k, \delta) \subset \Lambda_1(k) \text{ for each } k \in \mathbb{Z}^d,$$

any measurable $V : \Lambda_L \mapsto [-K_V, K_V]$ and any real-valued $\psi \in \mathcal{D}(\mathcal{L}_{L,0})$, respectively $\psi \in \mathcal{D}(\mathcal{L}_{L,\text{per}})$ satisfying

$$|\mathcal{L}\psi| \leqslant |V\psi| \quad \text{a.e.} \quad \Lambda_L$$

we have

$$\int_{S_L} |\psi(x)|^2 dx = \sum_{k\in Q_L} \|\psi\|^2_{L_2(B(z_k,\delta))} \geqslant C_{sfUC}\|\psi\|^2_{L_2(\Lambda_L)}, \tag{11}$$

where $S_L := S \cap \Lambda_L = \cup_{k\in Q_L} B(z_k, \delta)$, $Q_L = \Lambda_L \cap \mathbb{Z}^d$, and $S := \cup_{k\in\mathbb{Z}^d} B(z_k, \delta)$.

As a *Corollary* we obtain immediately an eigenvalue lifting estimate analogous to (7), where κ is again uniform w.r.t. many parameters, as spelled out in subsection 2.2 explicitly.

The proof of Theorem 3.2 is based on the strategy implemented in [3]. First one uses the conditions on the coefficients a^{ij} described in Assumption 3.2 to extend ψ as well as the differential expression $\mathcal{L}$ to the whole of $\mathbb{R}^d$ while keeping the $W^{2,2}$-regularity and the differential inequality originally satisfied by ψ. Then one uses the comparison Theorem 3.1 for local L^2-norms. Note that now the condition concerning the minimal distance to the boundary of G plays no role, since ψ has been extended to the whole of $\mathbb{R}^d$. From this point the combinatorial and geometric arguments of [3] take over. In fact, one can prove a abstract meta-theorem: Once the comparison of local L^2-norms of ψ holds up to the boundary, an equidistribution property for ψ follows. Interestingly, such an argument no longer uses the fact that ψ is a solution of an differential equation or inequality.

Acknowledgments

D.B. was partially supported by RFBR, the grant of the President of Russia for young scientists - doctors of science (MD-183.2014.1), and the fellowship of Dynasty foundation for young mathematicians.

M.T. and I.V. have been partially supported by the DAAD and the Croatian Ministry of Science, Education and Sports through the PPP-grant "Scale-uniform controllability of partial differential equations". M.T. and I.V. have been partially supported by the DFG.

Bibliography

[1] A. Figotin and A. Klein, *Commun. Math. Phys.* **180**, 265 (1996).
[2] P. Stollmann, *Isr. J. Math.* **107**, 125 (1998).
[3] C. Rojas-Molina and I. Veselić, *Commun. Math. Phys.* **320**, 245 (2013).
[4] A. Klein, *Commun. Math. Phys.* **323**, 1229 (2013).
[5] F. Germinet and A. Klein, *J. Eur. Math. Soc.* **15**, 53 (2013).
[6] J. Bourgain and A. Klein, *Invent. Math.* **194**, 41 (2013).
[7] D. I. Borisov, M. Tautenhahn and I. Veselić, Equidistribution properties of eigenfunctions of divergence form operators, in preparation.

PART B

Invited section talks

Vector quantum fields

J. Dereziński

Department of Mathematical Methods in Physics,
Faculty of Physics, University of Warsaw,
Hoza 74, 00-682 Warszawa, Poland
E-mail: Jan.Derezinski@fuw.edu.pl

Various approaches to quantization of vector fields, both massive (Proca) and massless (Maxwell), are described. The massless limit of the Proca fields is discussed.

1. Introduction

In these notes I would like to discuss quantization of free vector fields on a flat Minkowski space – both massless Maxwell fields and massive Proca fields. This subject, especially in the massless case, is covered in many standard textbooks [2, 8–11, 13]. Therefore the reader may wonder why I consider such a topic.

In my opinion, this subject has several interesting and confusing aspects, which are not sufficiently discussed in the literature. Therefore, I would like to discuss it in this short note, based on my more comprehensive manuscript [4].

First, there is more than one way to formulate its classical theory. One option is to initially reduce the constraints and then formulate the theory. Or one can start with the unreduced theory and then reduce.

Second, there are several ways to formulate the quantized theory of vector fields. If one starts from the unreduced classical formalism, one can use either the formalism with a (positive definite) Hilbert space or with an indefinite scalar product. One could argue that the former formalism is more physical. Nowadays it is rarely presented in the literature, however it is implicit in older works such as [9], and in a more mathematical form in [12]. The latter approach, known under the name of the Gupta-Bleuler formalism, dates back only to the early 50's [6, 7]. It is nowadays much more popular in the literature. However, it is based on an obviously nonphysical trick. At the end the physical theory does not depend on the approach that we chose. However the route to the physical theory is quite different depending on the approach.

Third, the limit $m \searrow 0$ is for vector fields rather subtle. Indeed, for $m > 0$ the fields have 3 spin degrees of freedom, whereas for $m = 0$ only 2. In practice, physical quantities are continuous wrt m, including the continuity at zero even though the number of spin degree of freedom is discontinuous.

Note that both the massive and massless theories are Poincaré covariant. I will argue, however, that to describe the limit $m^2 \searrow 0$ one has to choose a frame of reference. Then the longitudinal spin decouples from the theory and the transversal degrees of freedom converge to the potential in the Coulomb gauge, which is not manifestly covariant. Thus the Poincaré covariance is broken twice, and the resulting massless theory is covariant!

My note will start with a discussion of scalar quantum fields. Of course, this subject is quite straightforward and covered in all textbooks on quantum field theory. I do this in order to fix the notation and describe the general philosophy of quantizing free fields [1, 5]. According to this philosophy one starts with a classical Hamiltonian systems, and then describes its quantization. Classical Hamiltonians for free fields are always quadratic – therefore quantization is rather straightforward and the main problem is the infinite number of degrees of freedom.

One can sometimes hear the statement that the topic of free quantum fields is very simple, because it is just the 2nd quantization of a unitary representation of the Poincaré group – the spin 0 representation in the case of scalar fields. This view does not capture the most important feature of the fields – their locality. Actually, even the simple topic of scalar quantum fields can be formulated in a few equivalent but seemingly different ways. Some of them we describe, because we need them in the slightly more complicated case of vector fields.

Acknowledgments. The research of the author was supported in part by the National Science Center (NCN) grant No. 2011/01/B/ST1/04929.

2. Scalar fields

2.1. *Notation*

We equip the Minkowski space $\mathbb{R}^{1,3}$ with the pseudo-Euclidean form of signature $(-+++)$

$$x_\mu x^\mu = g_{\mu\nu} x^\mu x^\nu = -(x^0)^2 + \sum_{i=1}^{3} (x^i)^2.$$

$\mathbb{R}^3$ we will typically denote the *spatial part* of the Minkowski space obtained by setting $x^0 = 0$. If $x \in \mathbb{R}^{1,3}$, then $\vec{x}$ will denote the projection of x onto its spatial $\mathbb{R}^3$.

A function on $\mathbb{R}^{1,3}$ is called *space-compact* iff there exists a compact $K \subset \mathbb{R}^{1,3}$ such that supp$f$ is contined in the causal shadow of K.

2.2. *Classical theory*

Let $\mathcal{Y}_{\mathrm{KG}}$ denote the *space of real smooth space-compact solutions of the Klein-Gordon equation*

$$(-\Box + m^2)\zeta = 0.$$

Let $\dot{\zeta}$ denote the time derivative of ζ.

The space $\mathcal{Y}_{\mathrm{KG}}$ has a natural *symplectic form*

$$\zeta_1 \omega \zeta_2 = \int \Big(-\dot{\zeta}_1(t,\vec{x})\zeta_2(t,\vec{x}) + \zeta_1(t,\vec{x})\dot{\zeta}_2(t,\vec{x})\Big)\,\mathrm{d}\vec{x}.$$

The Poincaré group $\mathbb{R}^{1,3} \rtimes O(1,3)$ acts on $\mathcal{Y}_{\mathrm{KG}}$ and $\mathbb{C}\mathcal{Y}_{\mathrm{KG}}$ by

$$r_{(a,\Lambda)}\zeta(x) := \zeta\left((a,\Lambda)^{-1}x\right).$$

For $x \in \mathbb{R}^{1,3}$, the *classical fields* $\phi(x)$, $\pi(x)$ are defined as the functionals on $\mathcal{Y}_{\mathrm{KG}}$ given by

$$\langle \phi(x)|\zeta\rangle := \zeta(x), \qquad \langle \pi(x)|\zeta\rangle := \dot{\zeta}(x).$$

Clearly,

$$(-\Box + m^2)\phi(x) = 0, \quad \dot{\phi}(x) = \pi(x).$$

The symplectic structure on the space $\mathcal{Y}_{\mathrm{KG}}$ leads to a *Poisson bracket* on functions on $\mathcal{Y}_{\mathrm{KG}}$:

$$\begin{aligned}\{\phi(t,\vec{x}), \phi(t,\vec{y})\} = \{\pi(t,\vec{x}), \pi(t,\vec{y})\} &= 0,\\ \{\phi(t,\vec{x}), \pi(t,\vec{y})\} &= \delta(\vec{x}-\vec{y}).\end{aligned}$$

We can smear the fields with help of the *pairing given by the symplectic form.*

$$\phi(\!(\zeta)\!) = \int \Big(-\overline{\dot{\zeta}(t,\vec{x})}\phi(t,\vec{x}) + \overline{\zeta(t,\vec{x})}\pi(t,\vec{x})\Big)\,\mathrm{d}\vec{x},$$

where $\zeta \in \mathbb{C}\mathcal{Y}_{\mathrm{KG}}$. Note that

$$\{\phi(\!(\zeta_1)\!), \phi(\!(\zeta_2)\!)\} = \overline{\zeta}_1 \omega \overline{\zeta}_2.$$

We introduce the *(total) Hamiltonian*, which is a quadratic function on $\mathcal{Y}_{\mathrm{KG}}$:

$$H := \int \frac{1}{2}\left(\pi(t,\vec{x})^2 + \left(\vec{\partial}\phi(t,\vec{x})\right)^2 + m^2\phi(t,\vec{x})^2\right)\mathrm{d}\vec{x}.$$

H generates the dynamics:

$$\dot{\phi}(x) = \{H, \phi(x)\}, \quad \dot{\pi}(x) = \{H, \pi(x)\}.$$

Set $\varepsilon(\vec{k}) := \sqrt{\vec{k}^2 + m^2}$. For $\big(\pm\varepsilon(\vec{k}), \vec{k}\big)$, the *plane wave* $|k)$ is defined as

$$(x|k) = \frac{1}{(2\pi)^{3/2}\sqrt{2\varepsilon(\vec{k})}}\mathrm{e}^{\mathrm{i}kx}. \tag{1}$$

The *plane wave functionals* act on $\mathcal{Y}_{\mathrm{KG}}$ and are defined by

$$\begin{aligned} a(k) &= \mathrm{i}\phi((|k))), \\ a^*(k) &= -\mathrm{i}\phi((|-k))). \end{aligned}$$

The fields can be written in terms of $a^*(k)$, $a(k)$:

$$\phi(x) = (2\pi)^{-\frac{3}{2}}\int\frac{\mathrm{d}\vec{k}}{\sqrt{2\varepsilon(\vec{k})}}\left(\mathrm{e}^{\mathrm{i}kx}a(k) + \mathrm{e}^{-\mathrm{i}kx}a^*(k)\right).$$

$a(k)$, $a^*(k)$ diagonalize simultaneously the Hamiltonian and symplectic form:

$$\begin{aligned} H &= \int \mathrm{d}\vec{k}\varepsilon(\vec{k})a^*(k)a(k), \\ \mathrm{i}\omega &= \int \mathrm{d}\vec{k}a^*(k)\wedge a(k). \end{aligned}$$

$\mathcal{W}_{\mathrm{KG}}^{(+)}$ will denote the subspace of $\mathbb{C}\mathcal{Y}_{\mathrm{KG}}$ consisting of *positive frequency solutions*, that is,

$$\mathcal{W}_{\mathrm{KG}}^{(+)} := \{g \in \mathbb{C}\mathcal{Y}_{\mathrm{KG}} \ : \ \langle a^*(k)|g\rangle = 0\}.$$

For $g_1, g_2 \in \mathcal{W}_{\mathrm{KG}}^{(+)}$ we define the scalar product

$$(g_1|g_2) := \mathrm{i}\overline{g}_1\omega g_2 = \int\overline{\langle a(k)|g_1\rangle}\langle a(k)|g_2\rangle\mathrm{d}\vec{k}$$

The *Hilbert space of positive energy solutions* is denoted $\mathcal{Z}_{\mathrm{KG}}$, and is the completion of $\mathcal{W}_{\mathrm{KG}}^{(+)}$ in this scalar product.

We have a natural identification $\mathcal{Y}_{\mathrm{KG}} \ni \zeta \mapsto \zeta^{(+)} \in \mathcal{W}_{\mathrm{KG}}^{(+)}$ given by

$$\zeta = \zeta^{(+)} + \overline{\zeta^{(+)}}.$$

This identification allows us to define a real scalar product on $\mathcal{Y}_{\mathrm{KG}}$:

$$\begin{aligned}\langle\zeta_1|\zeta_2\rangle_{\mathcal{Y}} &:= \mathrm{Re}(\zeta_1^{(+)}|\zeta_2^{(+)})\\ &= \int\int \dot{\zeta}_1(0,\vec{x})(-\mathrm{i})D^{(+)}(0,\vec{x}-\vec{y})\dot{\zeta}_2(0,\vec{y})\mathrm{d}\vec{x}\mathrm{d}\vec{y}\\ &+\int\int \zeta_1(0,\vec{x})(-\Delta_{\vec{x}}+m^2)(-\mathrm{i})D^{(+)}(0,\vec{x}-\vec{y})\zeta_2(0,\vec{y})\mathrm{d}\vec{x}\mathrm{d}\vec{y},\end{aligned}$$

where $D^{(+)}$ is the standard positive frequency solution of the Klein-Gordon equation, sometimes called the *Wightman function.*

2.3. *Quantization*

Let us describe the quantization of the Klein-Gordon equation. We will use the "hat" to denote the quantized objects. It can be formulated in an axiomatized way. There are several equivalent formulations – we will present three of them.

We start with an axiomatization which is the closest to the presentation given in most standard textbooks on QFT. We want to construct a Hilbert space, a self-adjoint operator and normalized vector $(\mathcal{H},\hat{H},\Omega)$ such that Ω is the unique ground state of $\hat{H}$. We will assume that $H\Omega=0$. We also want to have a self-adjoint operator valued distribution

$$\mathbb{R}^{1,3}\ni x\mapsto\hat{\phi}(x), \tag{2}$$

such that, with $\hat{\pi}(x):=\dot{\hat{\phi}}(x)$,

(1) $(-\Box+m^2)\hat{\phi}(x)=0$,

(2) $[\hat{\phi}(0,\vec{x}),\hat{\phi}(0,\vec{y})]=[\hat{\pi}(0,\vec{x}),\hat{\pi}(0,\vec{y})]=0$,
$[\hat{\phi}(0,\vec{x}),\hat{\pi}(0,\vec{y})]=\mathrm{i}\delta(\vec{x}-\vec{y})$.

(3) $\mathrm{e}^{\mathrm{i}t\hat{H}}\hat{\phi}(x^0,\vec{x})\mathrm{e}^{-\mathrm{i}t\hat{H}}=\hat{\phi}(x^0+t,\vec{x})$.

(4) Ω is cyclic for $\hat{\phi}(x)$.

The above problem has a solution, which is unique up to a unitary equivalence. Let us describe this solution.

For the Hilbert space we choose the bosonic Fock space $\mathcal{H}=\Gamma_{\mathrm{s}}(\mathcal{Z}_{\mathrm{KG}})$ and for Ω the Fock vacuum. Suppose the annihilation operator is denoted by $\hat{a}(k)$. Then

$$\hat{\phi}(x):=(2\pi)^{-\frac{3}{2}}\int\frac{\mathrm{d}\vec{k}}{\sqrt{2\varepsilon(\vec{k})}}\left(\mathrm{e}^{\mathrm{i}kx}\hat{a}(k)+\mathrm{e}^{-\mathrm{i}kx}\hat{a}^*(k)\right)$$

and

$$\hat{H}:=\int\hat{a}^*(k)\hat{a}(k)\varepsilon(\vec{k})\mathrm{d}\vec{k}$$

satisfy our requirements.

The orthochronous Poincaré group $\mathbb{R}^{1,3} \rtimes O^{\uparrow}(1,3)$ is automatically unitarily implemented on $\mathcal{H}$, even though we did not demand it beforehand.

There exists an alternative equivalent formulation of the quantization program, which uses *smeared fields* instead of point fields. It avoids the use of distributions on the Minkowski space and uses more directly the natural degrees of freedom of the fields.

Again, we want to construct $(\mathcal{H}, \hat{H}, \Omega)$ as above and a linear function

$$\mathcal{Y}_{\mathrm{KG}} \ni \zeta \mapsto \hat{\phi}((\zeta))$$

with values in self-adjoint operators such that

(1) $[\hat{\phi}((\zeta_1)), \hat{\phi}((\zeta_2))] = \mathrm{i}\zeta_1\omega\zeta_2$.

(2) $\hat{\phi}((r_{(t,\vec{0})}\zeta)) = \mathrm{e}^{\mathrm{i}t\hat{H}}\hat{\phi}((\zeta))\mathrm{e}^{-\mathrm{i}t\hat{H}}$.

(3) Ω is cyclic for the algebra generated by $\hat{\phi}((\zeta))$.

One can pass between these two versions of the quantization by

$$\hat{\phi}((\zeta)) = \int \Big(-\dot{\zeta}(t,\vec{x})\hat{\phi}(t,\vec{x}) + \zeta(t,\vec{x})\hat{\pi}(t,\vec{x})\Big)\,\mathrm{d}\vec{x}. \tag{3}$$

Let us mention yet another equivalent approach to quantization, using the language of *C^*-algebras*. This approach has its own conceptual advantages and is popular in a part of the mathematical physics community.

Let $\mathrm{CCR}(\mathcal{Y}_{\mathrm{KG}})$ denote the *(Weyl) C^*-algebra of the CCR over* $\mathcal{Y}_{\mathrm{KG}}$. By definition, it is generated by $W(\zeta)$, $\zeta \in \mathcal{Y}_{\mathrm{KG}}$, such that

$$W(\zeta_1)W(\zeta_2) = \mathrm{e}^{-\mathrm{i}\frac{\zeta_1\omega\zeta_2}{2}}W(\zeta_1+\zeta_2), \quad W(\zeta)^* = W(-\zeta).$$

$\mathbb{R}^{1,3} \rtimes O^{\uparrow}(1,3)$ acts on $\mathrm{CCR}(\mathcal{Y}_{\mathrm{KG}})$ by $*$-automorphisms defined by

$$\hat{r}_{(a,\Lambda)}\left(W(\zeta)\right) := W\left(r_{(a,\Lambda)}(\zeta)\right).$$

We are looking for a cyclic representation of this algebra with the time evolution generated by a positive Hamiltonian.

The solution is provided by the state on on $\mathrm{CCR}(\mathcal{Y}_{\mathrm{KG}})$ defined by

$$\psi\big(W(\zeta)\big) = \exp\Big(-\frac{1}{2}\langle y|y\rangle_{\mathcal{Y}}\Big).$$

Let $(\mathcal{H}_\psi, \pi_\psi, \Omega_\psi)$ be the GNS representation generated by the state ψ. Then this representation has the required properties. $\mathcal{H}_\psi$ can be identified with $\Gamma_{\mathrm{s}}(\mathcal{Z}_{\mathrm{KG}})$ and the fields are related to the Weyl operators by

$$\pi_\psi(W(\zeta)) = \mathrm{e}^{\mathrm{i}\hat{\phi}((\zeta))}.$$

Note that there are other formulations of quantization of free fields. In particular, instead of starting with the Hamiltonian formalism with fields satisfying the equations of motion (the *on shell approach*) one can start with the Lagrangian formalism where at the beginning fields are not subject to the equations of motion (the *off-shell* approach) see eg. [3]. In our notes, however, we restrict ourselves to the on-shell approach.

3. Massive photons

3.1. *Classical theory based on the Proca equation*

Let $\mathcal{Y}_{\rm Pr}$ denote the set of real smooth space-compact solutions of the Proca equation

$$-\partial^\mu(\partial_\mu\zeta_\nu - \partial_\nu\zeta_\mu) + m^2\zeta_\nu(x) = 0. \tag{1}$$

Note that (1) is equivalent to the Klein-Gordon equation together with the Lorentz condition

$$\begin{aligned}(-\Box + m^2)\zeta_\nu &= 0,\\ \partial^\nu\zeta_\nu &= 0.\end{aligned}$$

$\mathcal{Y}_{\rm Pr}$ is a symplectic space with the *symplectic form*

$$\begin{aligned}&\zeta_1\omega_{\rm Pr}\zeta_2\\ &= \int \left(-\left(\dot{\vec{\zeta}}_1(t,\vec{x}) - \vec{\partial}\zeta_{10}(t,\vec{x})\right)\vec{\zeta}_2(t,\vec{x}) + \vec{\zeta}_1(t,\vec{x})\left(\dot{\vec{\zeta}}_2(t,\vec{x}) - \vec{\partial}\zeta_{10}(t,\vec{x})\right)\right)\mathrm{d}\vec{x}\end{aligned}$$

The Poincaré group $\mathbb{R}^{1,3} \rtimes O(1,3)$ acts on $\mathcal{Y}_{\rm Pr}$ by

$$r_{(a,\Lambda)}\zeta_\mu(x) := \Lambda_\mu^{\ \nu}\zeta_\nu\left((a,\Lambda)^{-1}x\right).$$

Introduce the functionals $A_\mu(x)$, called *potentials*, acting on $\zeta \in \mathcal{Y}_{\rm Pr}$ as

$$\langle A_\mu(x)|\zeta\rangle := \zeta_\mu(x).$$

We also introduce the *field tensor* and the *electric field vector*:

$$\begin{aligned}F_{\mu\nu}(x) &:= \partial_\mu A_\nu(x) - \partial_\nu A_\mu(x),\\ E_i(x) &:= F_{0i}(x) = \dot{A}_i - \partial_i A_0.\end{aligned}$$

The symplectic form leads to a *Poisson bracket* on functions on $\mathcal{Y}_{\rm Pr}$:

$$\begin{aligned}\{A_i(t,\vec{x}), A_j(t,\vec{y})\} = \{E_i(t,\vec{x}), E_j(t,\vec{y})\} &= 0,\\ \{A_i(t,\vec{x}), E_j(t,\vec{y})\} &= \delta_{ij}\delta(\vec{x}-\vec{y}).\end{aligned} \tag{2}$$

For $\zeta \in \mathcal{Y}_{\rm Pr}$, the corresonding *spatially smeared potential* is the functional on $\mathcal{Y}_{\rm Pr}$ given by

$$A(\!(\zeta)\!) = \int \left(-\overline{(\dot{\vec{\zeta}}(t,\vec{x}) - \partial\zeta^0(t,\vec{x}))}\vec{A}(t,\vec{x}) + \overline{\vec{\zeta}(t,\vec{x})}\vec{E}(t,\vec{x}) \right) \mathrm{d}\vec{x}. \tag{3}$$

Note that

$$\{A(\!(\zeta_1)\!), A(\!(\zeta_2)\!)\} = \overline{\zeta}_1 \omega_{\rm Pr} \overline{\zeta}_2.$$

The Hamiltonian generates the equations of motion:

$$\begin{aligned} H := \int \Big(& \frac{1}{2}\vec{E}^2(t,\vec{x}) + \frac{1}{2m^2}(\mathrm{div}\vec{E})^2(t,\vec{x}) + (\vec{\mathrm{rot}}\vec{A})^2(t,\vec{x}) \\ & + \frac{m^2}{2}\vec{A}^2(t,\vec{x})\Big)\mathrm{d}\vec{x}. \end{aligned} \tag{4}$$

For $\vec{k} \in \mathbb{R}^3$, $\vec{k} \neq \vec{0}$ fix two spatial vectors $\vec{e}_1(\vec{k}), \vec{e}_2(\vec{k})$ that form an oriented orthonormal basis of the plane orthogonal to $\vec{k}$. Define

$$\vec{e}(\vec{k}, \pm 1) := \frac{1}{\sqrt{2}} \left(\vec{e}_1(\vec{k}) \pm \mathrm{i}\vec{e}_2(\vec{k}) \right).$$

Introduce

$$u(k,0) := \Big(\frac{|\vec{k}|}{m}, \frac{\varepsilon(\vec{k})\vec{k}}{m|\vec{k}|} \Big), \tag{5}$$

$$u(k,\pm 1) := \left(0, \vec{e}(\vec{k}, \pm 1) \right). \tag{6}$$

A *plane wave* $|k,\sigma)$ is defined as

$$(x|k,\sigma) = \frac{1}{(2\pi)^{3/2}\sqrt{2\varepsilon(\vec{k})}} u_\mu(k,\sigma)\mathrm{e}^{\mathrm{i}kx}, \tag{7}$$

We also introduce the plane wave functionals

$$\begin{aligned} a(k,\sigma) &= -\mathrm{i}A(\!(|k,\sigma))\!) \\ a^*(k,\sigma) &= \mathrm{i}A(\!(|-k,\sigma))\!). \end{aligned}$$

The potentials can be written in terms of $a^*(k,\sigma)$, $a(k,\sigma)$ as

$$A_\mu(x) = (2\pi)^{-\frac{3}{2}} \sum_{\sigma=0,\pm1} \int \frac{\mathrm{d}\vec{k}}{\sqrt{2\varepsilon(\vec{k})}} \left(u_\mu(k,\sigma)\mathrm{e}^{\mathrm{i}kx} a(k,\sigma) + \overline{u_\mu(k,\sigma)}\mathrm{e}^{-\mathrm{i}kx} a^*(k,\sigma) \right).$$

We have accomplished the diagonalization of the Hamiltonian:

$$\begin{aligned} H &= \sum_{\sigma=0,\pm1} \int \mathrm{d}\vec{k}\varepsilon(\vec{k}) a^*(k,\sigma) a(k,\sigma), \\ \mathrm{i}\omega_{\rm Pr} &= \sum_{\sigma=0,\pm1} \int a^*(k,\sigma) \wedge a(k,\sigma) \mathrm{d}\vec{k}. \end{aligned}$$

$\mathcal{W}^{(+)}_{\rm Pr}$ will denote the subspace of $\mathbb{C}\mathcal{Y}_{\rm Pr}$ consisting of *positive frequency solutions*:

$$\mathcal{W}^{(+)}_{\rm Pr} := \{g \in \mathbb{C}\mathcal{Y}_{\rm Pr} \; : \; \langle a^*(k,\sigma)|g\rangle = 0, \quad \sigma = \pm, 0\}.$$

For $g_1, g_2 \in \mathcal{W}^{(+)}_{\rm Pr}$ we define the scalar product

$$(g_1|g_2) := -\mathrm{i}\overline{g}_1\omega_{\rm Pr}g_2 = \sum_{\sigma=0,\pm1} \int \overline{\langle a(k,\sigma)|g_1\rangle}\langle a(k,\sigma)|g_2\rangle \mathrm{d}\vec{k}$$

We set $\mathcal{Z}_{\rm Pr}$ to be the completion of $\mathcal{W}^{(+)}_{\rm Pr}$ in this scalar product.

3.2. *Quantization of the Proca equation*

We want to construct a Hilbert space $\mathcal{H}$, a self-adjoint operator $\hat{H}$ having a ground state Ω with $H\Omega = 0$ and a self-adjoint operator-valued distribution $\mathbb{R}^{1,3} \ni x \mapsto \hat{A}_\mu(x)$ such that, setting $\vec{\hat{E}} = \dot{\vec{\hat{A}}} - \vec{\partial}\hat{A}_0$, we have

(1) $-\partial^\mu(\partial_\mu\hat{A}_\nu - \partial_\nu\hat{A}_\mu) + m^2\hat{A}_\nu(x) = 0,$

(2) $[\hat{A}_i(0,\vec{x}), \hat{A}_j(0,\vec{y})] = [\hat{E}_i(0,\vec{x}), \hat{E}_j(0,\vec{y})] = 0,$
$[\hat{A}_i(0,\vec{x}), \hat{E}_j(0,\vec{y})] = \mathrm{i}\delta_{ij}\delta(\vec{x}-\vec{y}),$

(3) $\mathrm{e}^{\mathrm{i}t\hat{H}}\hat{A}_\mu(x^0,\vec{x})\mathrm{e}^{-\mathrm{i}t\hat{H}} = \hat{A}_\mu(x^0+t,\vec{x}),$

(4) Ω is cyclic for $\hat{A}_\mu(x)$.

The above problem has a solution, which is unique up to unitary equivalence, which we describe below.

For the Hilbert space we should take the bosonic Fock space $\mathcal{H} = \Gamma_{\rm s}(\mathcal{Z}_{\rm Pr})$ and for Ω the Fock vacuum. The annihilation operator is denoted by $\hat{a}(k,\sigma)$. The *quantized potentials* are

$$\hat{A}_\mu(x) = (2\pi)^{-\frac{3}{2}} \int \frac{\mathrm{d}\vec{k}}{\sqrt{2\varepsilon(\vec{k})}} \sum_{\sigma=0,\pm1} \left(u_\mu(k,\sigma)\mathrm{e}^{\mathrm{i}kx}\hat{a}(k,\sigma) + \overline{u_\mu(k,\sigma)}\mathrm{e}^{-\mathrm{i}kx}\hat{a}^*(k,\sigma)\right)$$

The *quantum Hamiltonian* is

$$\hat{H} = \sum_{\sigma=0,\pm1} \int \varepsilon(\vec{k})\hat{a}^*(k,\sigma)\hat{a}(k,\sigma)\mathrm{d}\vec{k}.$$

3.3. *Classical theory based on the Klein-Gordon equation*

In an alternative approach to vector fields one considers first the Klein-Gordon equation on functions with values in $\mathbb{R}^{1,3}$:

$$(-\Box + m^2)\zeta_\mu(x) = 0 \tag{8}$$

(without the Lorentz condition). The space of smooth real space-compact solutions of (8) will be denoted by $\mathcal{Y}_{\text{vec}}$. We have a symplectic form on $\mathcal{Y}_{\text{vec}}$

$$\zeta_1\omega_{\text{vec}}\zeta_2 = \int \Big(-\dot{\zeta}_{1\nu}(t,\vec{x})\zeta_2^{\nu}(t,\vec{x}) + \zeta_{1\nu}(t,\vec{x})\dot{\zeta}_2^{\nu}(t,\vec{x})\Big)\,\mathrm{d}\vec{x}.$$

$A^\mu(x)$ is the linear functional on $\mathcal{Y}_{\text{vec}}$ given by

$$\langle A^\mu(x)|\zeta\rangle := \zeta^\mu(x).$$

We clearly have

$$(-\Box + m^2)A_\mu(x) = 0. \tag{9}$$

The conjugate variable is $\Pi_\mu(x) := \dot{A}_\mu(x)$. The Poisson structure is given by the equal time brackets

$$\begin{aligned}\{A_\mu(t,\vec{x}), A_\nu(t,\vec{y})\} &= \{\Pi_\mu(t,\vec{x}), \Pi_\nu(t,\vec{y})\} = 0,\\ &\quad\ \{A_\mu(t,\vec{x}), \Pi_\nu(t,\vec{y})\} = g_{\mu\nu}\delta(\vec{x}-\vec{y}).\end{aligned}$$

The Hamiltonian is

$$H = \int \Big(\frac{1}{2}\Pi_\mu(t,\vec{x})\Pi^\mu(t,\vec{x}) + \frac{1}{2}A_{\mu,i}(t,\vec{x})A^{\mu,i}(t,\vec{x}) + \frac{m^2}{2}A_\mu(t,\vec{x})A^\mu(t,\vec{x})\Big)\mathrm{d}\vec{x}.$$

The Hamiltonian is unbounded from below.

Introduce two subspaces of $\mathcal{Y}_{\text{vec}}$

$$\begin{aligned}\mathcal{Y}_{\text{Lor}} &:= \{\zeta \in \mathcal{Y}_{\text{vec}} \ : \ \partial_\mu\zeta^\mu = 0\},\\ \mathcal{Y}_{\text{sc}} &:= \{\zeta \in \mathcal{Y}_{\text{vec}} \ : \ \zeta^\mu = \partial^\mu\chi, \ \ \chi \in \mathcal{Y}_{\text{KG}}\}.\end{aligned}$$

Note that $\mathcal{Y}_{\text{vec}} = \mathcal{Y}_{\text{Lor}} \oplus \mathcal{Y}_{\text{sc}}$ is a decomposition into symplectically orthogonal subspaces each preserved by the Poincaré group. If $\zeta \in \mathcal{Y}_{\text{vec}}$, then its projection onto $\mathcal{Y}_{\text{sc}}$ is

$$\zeta_{\text{sc}}^\mu := \frac{1}{m^2}\partial^\mu\partial_\nu\zeta^\nu.$$

Elements of $\mathcal{Y}_{\text{Lor}}$ satisfy the Proca equation, so that we can make the identification

$$\mathcal{Y}_{\text{Lor}} = \mathcal{Y}_{\text{Pr}}.$$

On $\mathcal{Y}_{\text{Lor}}$ the forms ω_{vec} and ω_{Pr} coincide.

Clearly, we are back with the theory introduced at the beginning. In particular, the Hamiltonian restricted to $\mathcal{Y}_{\text{Lor}}$ is now positive.

In order to diagonalize the Hamiltonian, besides the polarization vectors $u(k,\sigma)$ with $\sigma = 0, \pm 1$, we will need the scalar polarization vectors:

$$u(k, \text{sc}) := \frac{1}{m}(\varepsilon(\vec{k}), \vec{k}).$$

The potentials can be decomposed as

$$A_\mu(x) = (2\pi)^{-\frac{3}{2}} \sum_\sigma \int \frac{d\vec{k}}{\sqrt{2\varepsilon(\vec{k})}} \left(u_\mu(k,\sigma) e^{ikx} a(k,\sigma) + \overline{u_\mu(k,\sigma)} e^{-ikx} a^*(k,\sigma) \right).$$

Clearly, the restriction to $\mathcal{Y}_{\rm Lor}$ amounts to dropping all scalar components.

We diagonalize the Hamiltonian

$$H = \sum_{\sigma=0,\pm1} \int d\vec{k} \varepsilon(\vec{k}) a^*(k,\sigma) a(k,\sigma) - \int d\vec{k} \varepsilon(\vec{k}) a^*(k,{\rm sc}) a(k,{\rm sc}).$$

$\mathcal{W}_{\rm vec}^{(+)}$ will denote the subspace of $\mathbb{C}\mathcal{Y}_{\rm vec}$ consisting of *positive frequency solutions*:

$$\mathcal{W}_{\rm vec}^{(+)} := \{ g \in \mathbb{C}\mathcal{Y}_{\rm Pr} \ : \ a^*(k,\sigma) g = 0, \ \sigma = \pm, 0, {\rm sc} \}.$$

For $g_1, g_2 \in \mathcal{W}_{\rm vec}^{(+)}$ we have a natural scalar product

$$\begin{aligned}
(g_1|g_2) &:= {\rm i}\overline{g}_1 \omega_{\rm vec} g_2 \\
&= \sum_{\sigma=0,\pm1} \int \overline{\langle a(k,\sigma)|g_1\rangle} \langle a(k,\sigma)|g_2\rangle d\vec{k} - \int \overline{\langle a(k,{\rm sc})|g_1\rangle} \langle a(k,{\rm sc})|g_2\rangle d\vec{k} \\
&= \int g^{\mu\nu} \overline{\langle a_\mu(k)|g_1\rangle} \langle a_\nu(k)|g_2\rangle d\vec{k}.
\end{aligned}$$

Unfortunately, the above definition gives an indefinite scalar product. We can also introduce a positive definite scalar product, which unfortunately is not covariant:

$$(g_1|g_2)_+ := \sum_\mu \int \overline{\langle a_\mu(k)|g_1\rangle} \langle a_\mu(k)|g_2\rangle d\vec{k}.$$

The positive frequency space $\mathcal{W}_{\rm vec}^{(+)}$ completed in the norm given by $(\cdot|\cdot)_+$ will be called $\mathcal{Z}_{\rm vec}$.

$\mathcal{W}_{\rm vec}^{(+)}$ can be in the obvious way decomposed into the direct sum of orthogonal subspaces $\mathcal{W}_{\rm Lor}^{(+)}$ and $\mathcal{W}_{\rm sc}^{(+)}$. On $\mathcal{W}_{\rm Lor}^{(+)}$ the scalar product (10) is positive definite, on $\mathcal{W}_{\rm sc}^{(+)}$ it is negative definite. Their completions will be denoted $\mathcal{Z}_{\rm Lor}$ and $\mathcal{Z}_{\rm sc}$.

As usual, any $\zeta \in \mathcal{Y}_{\rm vec}$ can be projected on $\mathcal{W}_{\rm vec}^{(+)}$ which allows us to define a real scalar product on $\mathcal{Y}_{\rm vec}$:

$$\begin{aligned}
\langle \zeta_1|\zeta_2\rangle_{\mathcal{Y}} &:= {\rm Re}(\zeta_1^{(+)}|\zeta_2^{(+)}) \\
&= \int\int \dot{\zeta}_{1\mu}(0,\vec{x})(-{\rm i}) D^{(+)}(0,\vec{x}-\vec{y}) \dot{\zeta}_2^\mu(0,\vec{y}) d\vec{x} d\vec{y} \\
&\quad + \int\int \zeta_{1\mu}(0,\vec{x})(-\Delta_{\vec{x}} + m^2)(-{\rm i}) D^{(+)}(0,\vec{x}-\vec{y}) \zeta_2^\mu(0,\vec{y}) d\vec{x} d\vec{y}.
\end{aligned}$$

Again, this scalar product is positive definite on $\mathcal{Y}_{\rm Lor}$ and negative definite on $\mathcal{Y}_{\rm sc}$.

3.4. *Quantizations based on the Klein-Gordon equation*

There exist at least two methods of quantization, which use the symplectic space $\mathcal{Y}_{\mathrm{vec}}$ introduced in (8) as the starting point.

(1) The first insists on using only *positive definite Hilbert spaces.* Unfortunately, the Hamiltonian turns out to be unbounded from below.

(2) In the *Gupta-Bleuler approach* the potentials $\hat{A}^{\mu}(x)$ evolve with positive frequencies. Unfortunately, it uses an indefinite scalar product.

In the quantization on a positive definite Hilbert space we use

$$\Gamma_{\mathrm{s}}(\mathcal{Z}_{\mathrm{Lor}} \oplus \overline{\mathcal{Z}}_{\mathrm{sc}}) \tag{10}$$

equipped with a positive definite scalar product. More explicitly, we replace $a(k,\sigma)$ with $\hat{a}(k,\sigma)$ for $\sigma = 0, \pm 1$. We replace $a(k, \mathrm{sc})$ with $\hat{b}^*(k, \mathrm{sc})$. They satisfy the standard commutation relations

$$[\hat{a}(k,\sigma), \hat{a}^*(k',\sigma')] = \delta_{\sigma,\sigma'}\delta(\vec{k}-\vec{k}'),$$
$$[\hat{b}(k,\mathrm{sc}), \hat{b}^*(k',\mathrm{sc})] = \delta(\vec{k}-\vec{k}').$$

$\hat{a}(k,\sigma)$, $\hat{b}(k,\sigma)$ kill the vacuum:

$$\hat{a}(k,\sigma)\Omega = \hat{b}(k,\mathrm{sc})\Omega = 0.$$

The quantized potentials and Hamiltonian become

$$\hat{A}_{\mu}(x) = (2\pi)^{-\frac{3}{2}} \sum_{\sigma=0,\pm1} \int \frac{\mathrm{d}\vec{k}}{\sqrt{2\varepsilon(\vec{k})}} \left(u_{\mu}(k,\sigma)\mathrm{e}^{\mathrm{i}kx}\hat{a}(k,\sigma) + \overline{u_{\mu}(k,\sigma)}\mathrm{e}^{-\mathrm{i}kx}\hat{a}^*(k,\sigma)\right)$$
$$+(2\pi)^{-\frac{3}{2}} \int \frac{\mathrm{d}\vec{k}}{\sqrt{2\varepsilon(\vec{k})}} \left(u_{\mu}(k,\mathrm{sc})\mathrm{e}^{\mathrm{i}kx}\hat{b}^*(k,\mathrm{sc}) + \overline{u_{\mu}(k,\mathrm{sc})}\mathrm{e}^{-\mathrm{i}kx}\hat{b}(k,\mathrm{sc})\right),$$

$$\hat{H} = \sum_{\sigma=0,\pm1} \int \mathrm{d}\vec{k}\varepsilon(\vec{k})\hat{a}^*(k,\sigma)\hat{a}(k,\sigma) - \int \mathrm{d}\vec{k}\varepsilon(\vec{k})\hat{b}^*(k,\mathrm{sc})\hat{b}(k,\mathrm{sc}).$$

Vectors built by applying fields satisfying the Lorentz condition to the vacuum will be called *physical.* Equivalently, physical vectors are elements of the Fock space built on $\mathcal{W}_{\mathrm{Lor}}^{(+)}$. After the completion the physical space coincides with $\Gamma_{\mathrm{s}}(\mathcal{Z}_{\mathrm{Lor}})$. Thus we obtain the same space as in the method "first reduce, then quantize".

It will be convenient to reformulate this method in the C^*-algebraic language. Let $\mathrm{CCR}(\mathcal{Y}_{\mathrm{vec}})$ denote the *(Weyl) C^*-algebra of the CCR over* $\mathcal{Y}_{\mathrm{vec}}$, that is, the C^*-algebra generated by $W(\zeta)$, $\zeta \in \mathcal{Y}_{\mathrm{vec}}$, such that

$$W(\zeta_1)W(\zeta_2) = \mathrm{e}^{-\mathrm{i}\frac{\zeta_1\omega_{\mathrm{vec}}\zeta_2}{2}} W(\zeta_1+\zeta_2), \quad W(\zeta)^* = W(-\zeta).$$

Choose the state on $\mathrm{CCR}(\mathcal{Y}_{\mathrm{vec}})$ defined by

$$\psi(W(\zeta)) \\ = \exp\Big(-\frac{1}{2}\langle\zeta_1|\zeta_2\rangle_{\mathcal{Y}} - \frac{1}{m^2}\langle\partial_\mu\zeta^\mu|\partial_\nu\zeta^\nu\rangle_{\mathcal{Y}}\Big) \tag{11}$$

Let $(\mathcal{H}_\psi, \pi_\psi, \Omega_\psi)$ be the GNS representation generated by the state ψ. $\mathcal{H}_\psi$ can be identified with $\Gamma_{\mathrm{s}}(\mathcal{Z}_{\mathrm{Lor}} \oplus \overline{\mathcal{Z}}_{\mathrm{sc}})$ and the fields are related to the Weyl operators by

$$\pi_\psi(W(\zeta)) = \mathrm{e}^{\mathrm{i}\hat{A}(\!(\zeta)\!)}.$$

The Gupta-Bleuler approach also uses the symplectic space $\mathcal{Y}_{\mathrm{vec}}$ as the basic input. It follows almost verbatim the usual steps of quantization of the Klein-Gordon equation. We introduce the bosonic Fock space $\Gamma_{\mathrm{s}}(\mathcal{Z}_{\mathrm{vec}})$, which has an indefinite scalar product and can be viewed as a Krein space.

We replace $a(k,\sigma)$ by $\hat{a}(k,\sigma)$. The commutation relations have a wrong sign for the scalar component:

$$[\hat{a}(k,\sigma), \hat{a}^*(k',\sigma')] = \kappa_{\sigma,\sigma'}\delta(\vec{k}-\vec{k}').$$

The annihillation operators kill the vacuum:

$$\hat{a}(k,\sigma)\Omega = 0.$$

The expressions for the Hamiltonian and potentials are the same as in the classical case:

$$\hat{H} = \sum_{\sigma=0,\pm1}\int \mathrm{d}\vec{k}\varepsilon(\vec{k})\hat{a}^*(k,\sigma)\hat{a}(k,\sigma) - \int \mathrm{d}\vec{k}\varepsilon(\vec{k})\hat{a}^*(k,\mathrm{sc})\hat{a}(k,\mathrm{sc}).$$

$$\hat{A}_\mu(x) = (2\pi)^{-\frac{3}{2}}\sum_\sigma\int\frac{\mathrm{d}\vec{k}}{\sqrt{2\varepsilon(\vec{k})}}\Big(u_\mu(k,\sigma)\mathrm{e}^{\mathrm{i}kx}\hat{a}(k,\sigma) + \overline{u_\mu(k,\sigma)}\mathrm{e}^{-\mathrm{i}kx}\hat{a}^*(k,\sigma)\Big).$$

Similarly as in the previous method, vectors created by applying fields satisfying the Lorentz condition to the vacuum will be called *physical*. On the space of physical vectors the scalar product is positive definite and after the completion coincides with $\Gamma_{\mathrm{s}}(\mathcal{Z}_{\mathrm{Lor}})$.

4. Massless photons

4.1. *Classical theory based on the Maxwell equation*

If we set $m = 0$ in the Proca equation we obtain

$$-\partial_\mu\left(\partial^\mu\zeta^\nu(x) - \partial^\nu\zeta^\mu(x)\right) = 0, \tag{1}$$

which we will call the *Maxwell equation.* It has to be treated separately from the Proca equation, because of the *gauge invariance.* It is invariant w.r.t. the replacement of ζ_μ with $\zeta_\mu + \partial_\mu \chi$, where χ is an arbitrary smooth function on the space-time. This property poses problems both for the classical and quantum theory.

The space of smooth real space compact solutions of the Maxwell equation is denoted $\mathcal{Y}_{\widetilde{\mathrm{Max}}}$. It is equipped with a gauge-invariant presymplectic form:

$$\begin{aligned}&\zeta_1 \omega_{\widetilde{\mathrm{Max}}} \zeta_2 \\ &= \int \Big(- \big(\dot{\vec{\zeta}}_1(t,\vec{x}) - \vec{\partial}\zeta_{10}(t,\vec{x}) \big) \vec{\zeta}_2(t,\vec{x}) + \vec{\zeta}_1(t,\vec{x}) \big(\dot{\vec{\zeta}}_2(t,\vec{x}) - \vec{\partial}\zeta_{10}(t,\vec{x}) \big) \Big) \mathrm{d}\vec{x}.\end{aligned}$$

We say that a solution ζ of the Maxwell equation is in the *Coulomb gauge* if

$$\zeta_0 = 0, \quad \mathrm{div}\vec{\zeta} = 0.$$

Note that every $\zeta \in \mathcal{Y}_{\widetilde{\mathrm{Max}}}$ is gauge-equivalent to a unique solution of the Maxwell equation in the Coulomb gauge, denoted by ζ^{Coul}.

We have three equivalent characterizations of solutions that physically describe the zero configuration, as described in the following easy proposition:

Proposition 4.1. *Let $\zeta \in \mathcal{Y}_{\widetilde{\mathrm{Max}}}$. We have the following equivalence:*

$$\zeta \in \mathrm{Ker}\omega_{\widetilde{\mathrm{Max}}} \Leftrightarrow \zeta^{\mathrm{Coul}} = 0 \Leftrightarrow \zeta = \partial\chi.$$

Define $\mathcal{Y}_{\mathrm{Max}}$ to be $\mathcal{Y}_{\widetilde{\mathrm{Max}}}$ divided by the gauge equivalence. By Proposition 4.1 $\mathcal{Y}_{\mathrm{Max}}$ coincides with the *symplectic reduction* of $\mathcal{Y}_{\widetilde{\mathrm{Max}}}$:

$$\mathcal{Y}_{\mathrm{Max}} := \mathcal{Y}_{\widetilde{\mathrm{Max}}} / \mathrm{Ker}\omega_{\widetilde{\mathrm{Max}}}.$$

Obviously, it is equipped with a natural *symplectic form* ω_{Max}.

Let $A^\mu(x)$ denote the functional on $\mathcal{Y}_{\widetilde{\mathrm{Max}}}$ given by

$$\langle A^\mu(x)|\zeta\rangle := \zeta^\mu(x). \tag{2}$$

Obviously, $A^\mu(x)$ is not defined on $\mathcal{Y}_{\mathrm{Max}}$.

We introduce also the functional $A_\mu^{\mathrm{Coul}}(x)$ on $\mathcal{Y}_{\widetilde{\mathrm{Max}}}$, called the *classical potential in the Coulomb gauge,*

$$A_0^{\mathrm{Coul}}(x) := 0, \quad \vec{A}^{\mathrm{Coul}}(x) := \vec{A}(x) - \vec{\partial}\Delta^{-1}\mathrm{div}\vec{A}(x).$$

Note that

$$\langle A_\mu^{\mathrm{Coul}}(x)|\zeta\rangle = \langle A_\mu(x)|\zeta^{\mathrm{Coul}}\rangle = \zeta_\mu^{\mathrm{Coul}}(x).$$

$A^{\mathrm{Coul}}(x)$ does not depend on the gauge, hence can be interpreted as a functional on $\mathcal{Y}_{\mathrm{Max}}$. It is not, however, Lorentz covariant.

We also introduce Lorentz covariant functionals $F_{\mu\nu}(x)$ on $\mathcal{Y}_{\widetilde{\mathrm{Max}}}$, called the *fields*:

$$\langle F_{\mu\nu}(x)|\zeta\rangle := \partial_\mu\zeta_\nu(x) - \partial_\nu\zeta_\mu(x).$$

They also do not depend on the gauge, hence can be interpreted as functionals on $\mathcal{Y}_{\mathrm{Max}}$.

We will write $E_i(x) = F_{0i}(x)$. Clearly, $\vec{E} = \partial_t\vec{A}^{\mathrm{Coul}}$ and

$$\mathrm{div}\vec{A}^{\mathrm{Coul}}(x) = 0, \quad \mathrm{div}\vec{E}(x) = 0. \tag{3}$$

In what follows we will usually drop the subscript Coul from $A^{\mathrm{Coul}}(x)$, even though this introduces a possible ambiguity with $A(x)$ acting on $\mathcal{Y}_{\widetilde{\mathrm{Max}}}$.

The symplectic structure on the space $\mathcal{Y}_{\mathrm{Max}}$ leads to a *Poisson bracket* on the level of functions on $\mathcal{Y}_{\mathrm{Max}}$:

$$\begin{aligned}\{A_i(t,\vec{x}), A_j(t,\vec{y})\} = \{E_i(t,\vec{x}), E_j(t,\vec{y})\} = 0,&\\ \{A_i(t,\vec{x}), E_j(t,\vec{y})\} = \left(\delta_{ij} - \frac{\partial_i\partial_j}{\Delta}\right)\delta(\vec{x}-\vec{y}).&\end{aligned}$$

For $\zeta \in \mathcal{Y}_{\mathrm{Max}}$ the corresponding *spatially smeared potential* is a functional on $\mathcal{Y}_{\mathrm{Max}}$ given by

$$A(\!(\zeta)\!) = \int \left(-\overline{\dot{\zeta}_\mu(t,\vec{x})}A^\mu(t,\vec{x}) + \overline{\zeta_\mu(t,\vec{x})}E^\mu(t,\vec{x})\right)\mathrm{d}\vec{x}.$$

Note that

$$\{A(\!(\zeta_1)\!), A(\!(\zeta_2)\!)\} = \overline{\zeta}_1\omega\overline{\zeta}_2.$$

$A(\!(\zeta)\!)$ depends on ζ only modulo gauge transformations and is Lorentz covariant. In the mass zero case the plane waves $|k,0)$ and the plane wave functionals $a^*(k,0)$ and $a(k,0)$ are ill defined. However the plane waves $k,\pm1)$ and the corresponding plane wave functionals $a^*(k,\pm1)$ and $a(k,\pm1)$ are well defined and can be used to express the potentials as

$$\begin{aligned}&A_\mu(x)\\ &= (2\pi)^{-\frac{3}{2}}\sum_{\sigma=\pm1}\int\frac{\mathrm{d}\vec{k}}{\sqrt{2\varepsilon(\vec{k})}}\left(u_\mu(x,\sigma)\mathrm{e}^{\mathrm{i}kx}a(k,\sigma) + \overline{u_\mu(x,\sigma)}\mathrm{e}^{-\mathrm{i}kx}a^*(k,\sigma)\right).\end{aligned}$$

Plane wave functionals diagonalize the Hamiltonian and the symplectic form:

$$\begin{aligned}H &= \sum_{\sigma=\pm1}\int\mathrm{d}\vec{k}\varepsilon(\vec{k})a^*(k,\sigma)a(k,\sigma),\\ \mathrm{i}\omega_{\mathrm{Max}} &= \sum_{\sigma=\pm1}\int a^*(k,\sigma)\wedge a(k,\sigma)\mathrm{d}\vec{k}.\end{aligned}$$

$\mathcal{W}_{\rm Max}^{(+)}$ will denote the subspace of $\mathbb{C}\mathcal{Y}_{\rm Max}$ consisting of classes of solutions that in the Coulomb gauge have positive frequencies.

For $g_1, g_2 \in \mathcal{W}_{\rm Max}^{(+)}$ we define the scalar product

$$(g_1|g_2) := \mathrm{i}\overline{g}_1\omega_{\rm Max}g_2 = \sum_{\sigma=\pm1}\int \overline{\langle a(k,\sigma)|g_1\rangle}\langle a(k,\sigma)|g_2\rangle \mathrm{d}\vec{k}.$$

The definition of $\mathcal{W}_{\rm Max}^{(+)}$ depends on the choice of coordinates. It is however easy to see that the space $\mathcal{W}_{\rm Max}^{(+)}$ is invariant w.r.t. $\mathbb{R}^{1,3} \rtimes O^{\uparrow}(1,3)$.

We set $\mathcal{Z}_{\rm Max}$ to be the completion of $\mathcal{W}_{\rm Max}^{(+)}$ in this scalar product.

We can identify $\mathcal{Y}_{\rm Max}$ with $\mathcal{W}_{\rm Max}^{(+)}$ and transport the scalar product onto $\mathcal{Y}_{\rm Max}$, which for ζ_1, ζ_2 is given by

$$\begin{aligned}\langle\zeta_1|\zeta_2\rangle_{\mathcal{Y}} &:= \mathrm{Re}(\zeta_1^{(+)}|\zeta_2^{(+)})\\ &= \int\int (\dot{\zeta}_{1i}^{\rm Coul}(0,\vec{x})(-\mathrm{i})D^{(+)}(0,\vec{x}-\vec{y})\dot{\zeta}_{2i}^{\rm Coul}(0,\vec{y})\mathrm{d}\vec{x}\mathrm{d}\vec{y}\\ &+\int\int \zeta_{1i}^{\rm Coul}(0,\vec{x})(-\Delta_{\vec{x}})(-\mathrm{i})D^{(+)}(0,\vec{x}-\vec{y})\zeta_{2i}^{\rm Coul}(0,\vec{y})\mathrm{d}\vec{x}\mathrm{d}\vec{y}.\end{aligned}$$

4.2. *Quantization based on the Maxwell equation*

The quantization of the Maxwell equation is similar to that of the Proca equation described in Subsect. 3.2. Condition (1) is replaced with

$$-\Box\hat{A}_i(x) = 0,\ \partial_i\hat{A}_i(x) = 0, \quad \hat{A}_0(x) = 0.$$

Condition (2) is replaced by

$$\begin{aligned}[\hat{A}_i(0,\vec{x}), A_j(0,\vec{y})] = [\hat{E}_i(0,\vec{x}), \hat{E}_j(0,\vec{y})] = 0,&\\ [\hat{A}_i(0,\vec{x}), \hat{E}_j(0,\vec{y})] = \mathrm{i}\left(\delta_{ij} - \frac{\partial_i\partial_j}{\Delta}\right)\delta(\vec{x}-\vec{y}).&\end{aligned}$$

The above problem has a solution unique up to a unitary equivalence. We set $\mathcal{H} := \Gamma_{\rm s}(\mathcal{Z}_{\rm Max})$. The annihilation operators are denoted by $\hat{a}(k,\sigma)$. Ω will be the Fock vacuum. We set

$$\hat{A}_{,i}(x) := (2\pi)^{-\frac{3}{2}}\int \frac{\mathrm{d}\vec{k}}{\sqrt{2\varepsilon}}\sum_{\sigma=\pm1}\left(u_i(k,\sigma)\mathrm{e}^{\mathrm{i}kx}\hat{a}(k,\sigma) + \overline{u_i(k,\sigma)}\mathrm{e}^{-\mathrm{i}kx}\hat{a}^*(k,\sigma)\right).$$

The quantum Hamiltonian is

$$\hat{H} := \sum_{\sigma=\pm1}\int \hat{a}^*(k,\sigma)\hat{a}(k,\sigma)\varepsilon(\vec{k})\mathrm{d}\vec{k}.$$

The group $\mathbb{R}^{1,3} \rtimes O^{\uparrow}(1,3)$ acts on $\mathcal{H}$.

Here is the C^*-algebraic version of the above construction. Let $\mathrm{CCR}(\mathcal{Y}_{\widetilde{\mathrm{Max}}})$ denote the *(Weyl) C^*-algebra of canonical commutation relations* over $\mathcal{Y}_{\widetilde{\mathrm{Max}}}$. By definition, it is generated by $W(\zeta)$, $\zeta \in \mathcal{Y}_{\widetilde{\mathrm{Max}}}$, such that

$$W(\zeta_1)W(\zeta_2) = \mathrm{e}^{-\mathrm{i}\frac{\zeta_1 \omega_{\widetilde{\mathrm{Max}}} \zeta_2}{2}} W(\zeta_1 + \zeta_2), \quad W(\zeta)^* = W(-\zeta).$$

$\mathbb{R}^{1,3} \rtimes O^{\uparrow}(1,3)$ acts on $\mathrm{CCR}(\mathcal{Y}_{\widetilde{\mathrm{Max}}})$ by $*$-automorphisms defined by

$$\hat{r}_{(a,\Lambda)}\left(W(\zeta)\right) := W\left(r_{(a,\Lambda)}(\zeta)\right).$$

We are looking for a cyclic representation of this algebra with the time evolution generated by a positive Hamiltonian.

Consider the state on $\mathrm{CCR}(\mathcal{Y}_{\widetilde{\mathrm{Max}}})$ defined for $\zeta \in \mathcal{Y}_{\widetilde{\mathrm{Max}}}$ by

$$\psi\big(W(\zeta)\big) = \exp\big(-\frac{1}{2}\langle\zeta|\zeta\rangle_{\mathcal{Y}}\big).$$

Note that the state is gauge and Poincare invariant. Let $(\mathcal{H}_\psi, \pi_\psi, \Omega_\psi)$ be the GNS representation. $\mathcal{H}_\psi$ is naturally isomorphic to $\Gamma_{\mathrm{s}}(\mathcal{Z}_{\mathrm{Max}})$. Ω_ψ can be identified with the vector Ω. $\pi_\psi(W(\zeta))$ can be identified with $\mathrm{e}^{\mathrm{i}\hat{A}(\!(\zeta)\!)}$. In particular, if ζ_1 and ζ_2 are gauge equivalent, then $\hat{A}(\!(\zeta_1)\!) = \hat{A}(\!(\zeta_2)\!)$. However, $\hat{A}(x)$ (in the original sense, not in the Coulomb gauge) is not well defined.

4.3. *Classical theory based on the d'Alembert equation*

So far, our treatment of the massless vector fields was based on the Coulomb gauge, which depends on the choice of the temporal coordinate. One can ask whether massless vector fields can be studied in a manifestly covariant fashion.

Let Ξ be an arbitrary space-time function. The Maxwell equation allow us to impose a *generalized Lorentz condition*

$$\partial_\mu A^\mu = \Xi. \tag{4}$$

The Maxwell equation together with (4) imply

$$-\Box A^\mu = \partial^\mu \Xi. \tag{5}$$

The function Ξ has no physical meaning. Therefore it is natural to adopt the simplest choice $\Xi = 0$, that is the usual *Lorentz condition*, for which (5) reduces to the d'Alembert's equation for vector valued functions

$$-\Box A^\mu = 0.$$

We will discuss this approach in what follows.

Recall that the Proca equation is equivalent to the Klein-Gordon equation for vector fields together with the Lorentz condition. Therefore, one can first develop its theory on the symplectic space $\mathcal{Y}_{\text{vec}}$, and then reduce it to the subpace $\mathcal{Y}_{\text{Lor}}$, as described before.

One can follow a similar route for the Maxwell equation. However, there is a difference: the reduction by the Lorentz condition is insufficient, one has to make an additional reduction.

Anyway, let us start by introducing the space $\mathcal{Y}_{\text{vec}}$, the form ω_{vec}, the subspace $\mathcal{Y}_{\text{Lor}}$ and the potentials $A_\mu(x)$, as for the Proca equation except that now $m = 0$.

In the massive case $\mathcal{Y}_{\text{Lor}}$ was symplectic (that means, the form ω_{vec} restricted to $\mathcal{Y}_{\text{Lor}}$ was nondegenerate). This is no longer true in the massless case. Instead, $\mathcal{Y}_{\text{Lor}}$ is coisotropic. (That means, if ζ is symplectically orthogonal to $\mathcal{Y}_{\text{Lor}}$, then $\zeta \in \mathcal{Y}_{\text{Lor}}$).

$\mathcal{Y}_{\text{Lor}}$ is a subspace of $\mathcal{Y}_{\widetilde{\text{Max}}}$ and on $\mathcal{Y}_{\text{Lor}}$ the forms $\omega_{\widetilde{\text{Max}}}$ and ω_{vec} coincide. The following fact is easy to show:

Proposition 4.2. *Any $\zeta \in \mathcal{Y}_{\widetilde{\text{Max}}}$ is gauge equivalent to an element of $\mathcal{Y}_{\text{Lor}}$.*

Therefore, the symplectically reduced $\mathcal{Y}_{\text{Lor}}$ coincides with the symplectically reduced $\mathcal{Y}_{\widetilde{\text{Max}}}$, that is, with $\mathcal{Y}_{\text{Max}}$. This shows that both approaches to the Maxwell equation are equivalent.

$\mathcal{W}_{\text{Lor}}^{(+)}$ will denote the subspace of $\mathbb{C}\mathcal{Y}_{\text{Lor}}$ consisting of solutions that have positive frequencies.

For $g_1, g_2 \in \mathcal{W}_{\text{Lor}}^{(+)}$ we define the scalar product

$$\begin{aligned}(g_1|g_2) &:= \mathrm{i}\overline{g}_1\omega_{\text{vec}}g_2 \\ &= \mathrm{i}\overline{g}_1^{\text{Coul}}\omega_{\text{vec}}g_2^{\text{Coul}}.\end{aligned} \tag{6}$$

Note that the definition of (6) does not depend on the choice of coordinates and is invariant wrt. the group $\mathbb{R}^{1,3} \rtimes O^{\uparrow}(1,3)$.

The scalar product is positive semidefinite, but not strictly positive definite. Let $\mathcal{W}_{\text{Lor},0}^{(+)}$ be the subspace of elements $\mathcal{W}_{\text{Lor}}^{(+)}$ with a zero norm. $\mathcal{W}_{\text{Lor},0}^{(+)}$ consists of pure gauges. The space $\mathcal{W}_{\text{Lor}}^{(+)}/\mathcal{W}_{\text{Lor},0}^{(+)}$ has a nondegenerate scalar product. Its completion is naturally isomorphic to the space $\mathcal{Z}_{\text{Max}}$.

We have a natural identification of $\mathcal{Y}_{\text{Lor}}$ with $\mathcal{W}_{\text{Lor}}^{(+)}$ given by the obvious projection. For $\zeta \in \mathcal{Y}_{\text{Lor}}$ we will denote by $\zeta^{(+)}$ the corresponding element of $\mathcal{W}_{\text{Lor}}^{(+)}$. This identification allows us to define a positive semidefinite scalar product on $\mathcal{Y}_{\mathrm{Lor}}$:

$$\begin{aligned}\langle\zeta_1|\zeta_2\rangle_{\mathcal{Y}} &:= \mathrm{Re}(\zeta_1^{(+)}|\zeta_2^{(+)}) \\ &= \int\int \dot{\zeta}_{1i}^{\mathrm{Coul}}(0,\vec{x})(-\mathrm{i})D^{(+)}(0,\vec{x}-\vec{y})\dot{\zeta}_{2i}^{\mathrm{Coul}}(0,\vec{y})\mathrm{d}\vec{x}\mathrm{d}\vec{y} \\ &+ \int\int \zeta_{1i}^{\mathrm{Coul}}(0,\vec{x})(-\Delta_{\vec{x}})(-\mathrm{i})D^{(+)}(0,\vec{x}-\vec{y})\zeta_{2i}^{\mathrm{Coul}}(0,\vec{y})\mathrm{d}\vec{x}\mathrm{d}\vec{y}.\end{aligned}$$

4.4. *Quantizations based on the d'Alembert equation*

One can try to use the symplectic space $\mathcal{Y}_{\mathrm{vec}}$ of real vector valued solutions of the Klein-Gordon equation as the basis for quantization. In the literature, this starting point is employed by two approaches.

(1) The first, which we call the *approach with a subsidiary condition* has the advantage that it uses only positive definite Hilbert spaces. Unfortunately, in this approach there are problems with the potential $\hat{A}^\mu(x)$. Besides, the full Hilbert space turns out to be non-separable.

(2) In the *Gupta-Bleuler approach* the potentials $\hat{A}^\mu(x)$ are well defined and covariant. Unfortunately it uses indefinite scalar product spaces.

Let us start to describe the first approach. It is convenient to use the C^*-algebraic formulation described in Subsect. 3.4. In particular, $\mathrm{CCR}(\mathcal{Y}_{\mathrm{vec}})$, the *(Weyl) C^*-algebra of canonical commutation relations* over $\mathcal{Y}_{\mathrm{vec}}$, is well defined also for $m = 0$ and is invariant wrt the Poincaré group.

Strictly speaking, the spaces $\mathcal{Y}_{\mathrm{vec}}$ and hence the algebras $\mathrm{CCR}(\mathcal{Y}_{\mathrm{vec}})$ are different for various m. We can identify them by using the initial conditions at $t = 0$.

Recall that in the massive case

$$(\Omega|\hat{A}(\langle\zeta\rangle)^2\Omega) = \langle\zeta|\zeta\rangle_{\mathcal{Y}} + \frac{2}{m^2}\langle\partial_\mu\zeta^\mu|\partial_\nu\zeta^\nu\rangle_{\mathcal{Y}}. \tag{7}$$

Clearly, if $m \searrow 0$, (7) is divergent to $+\infty$ for $\zeta \notin \mathcal{Y}_{\mathrm{Lor}}$ and equals $\langle\zeta|\zeta\rangle_{\mathcal{Y}}$ for $\zeta \in \mathcal{Y}_{\mathrm{Lor}}$. So, the following state on $\mathrm{CCR}(\mathcal{Y}_{\mathrm{vec}})$ is the limit of the state that was considered for the Proca equation:

$$\psi(W(\zeta)) = \begin{cases} \exp\left(-\frac{1}{2}\langle\zeta|\zeta\rangle_{\mathcal{Y}}\right), & \zeta \in \mathcal{Y}_{\mathrm{Lor}}, \\ 0, & \zeta \notin \mathcal{Y}_{\mathrm{Lor}}. \end{cases}$$

Let $(\mathcal{H}_\psi, \pi_\psi, \Omega_\psi)$ denote the *GNS representation* for this state. We have an identification

$$J : \mathcal{H}_\psi \to l^2\left(\mathcal{Y}_{\mathrm{vec}}/\mathcal{Y}_{\mathrm{Lor}}, \Gamma_{\mathrm{s}}(\mathcal{Z}_{\mathrm{Max}})\right). \tag{8}$$

To describe this identification, first note that $\mathcal{Y}_{\rm vec}/\mathcal{Y}_{\rm Lor}$ can be parametrized by smooth space-compact functions

$$\Xi = \partial_\mu \zeta^\mu,$$

which can be called the *values of the Lorentz condition.* For each Ξ choose $\zeta_\Xi \in \mathcal{Y}_{\rm vec}$ such that $\partial_\mu \zeta_\Xi^\mu = \Xi$. We demand that

$$\Big(J\pi_\psi\big(W(\zeta_\Xi)\big)\Omega_\psi\Big)(\Xi) = \begin{cases} \Omega, & \partial_\mu\zeta^\mu = \Xi, \\ 0, & \partial_\mu\zeta^\mu \neq \Xi. \end{cases}$$

Then J is given by

$$\left(J\pi_\psi\big(W(\zeta)\big)\Omega_\psi\right)(\Xi) = \begin{cases} \mathrm{e}^{\frac{\mathrm{i}}{2}\zeta\omega_{\rm vec}\zeta_\Xi}\mathrm{e}^{\mathrm{i}\hat{A}((\zeta-\zeta_\Xi))}\Omega, & \partial_\mu\zeta^\mu = \Xi, \\ 0, & \partial_\mu\zeta^\mu \neq \Xi. \end{cases}$$

Note that $\mathcal{H}_\psi$ is *non-separable* – it is a direct sum of *superselection sectors* corresponding to various values of the *Lorentz condition.*

Special role is played by the (separable) subspace corresponding to the Lorentz condition $\partial_\mu A^\mu(x) = 0$. On this subspace, the fields are equivalent to the usual quantization obtained by the method "first reduce, then quantize".

Note that $\pi_\psi(W(\zeta))$ maps between various sectors of (8) if $\partial_\mu\zeta^\mu \neq 0$. The unitary group $\mathbb{R} \ni t \mapsto \pi_\psi\left(W(t\zeta)\right)$ is strongly continuous if and only if $\partial_\mu\zeta^\mu = 0$. If this is the case, we can write $\pi_\psi(W(\zeta)) = \mathrm{e}^{\mathrm{i}\hat{A}((\zeta))}$. We have $\hat{A}((\zeta_1)) = \hat{A}((\zeta_2))$ if in addition ζ_1 differs from ζ_2 by a pure gauge. $\hat{A}((\zeta))$ is ill defined if $\partial_\mu\zeta^\mu \neq 0$.

The approach that we described above, restricted to the 0th sector was typical for older presentations, eg. Jauch and Rohrlich [9]. However, without the language of C^*-algebras it is somewhat awkward to describe. One usually says that the Lorentz condition $\partial_\mu\hat{A}^\mu(x) = 0$ is enforced on the Hilbert space of states and constitutes a *subsidiary condition.*

The *Gupta-Bleuler* approach follows the same lines as in the massive case until we arrive at the algebraic Fock space built on $\mathcal{W}_{\rm Lor}^{(+)}$. As we know, the scalar product on $\mathcal{W}_{\rm Lor}^{(+)}$ is only semidefinite. We factor $\mathcal{W}_{\rm Lor}^{(+)}$ by the null space of its scalar product, obtaining $\mathcal{W}_{\rm Max}^{(+)}$. We complete it, obtaining $\mathcal{Z}_{\rm Max}$ and we take the corresponding Fock space $\Gamma_{\rm s}(\mathcal{Z}_{\rm Max})$ – thus the resulting physical theory is the same as with the other quantizations.

5. The $m \to 0$ limit

In order to understand the $m \to 0$ limit let us consider quantum vector field interacting with an external *conserved current*, that is a function $\mathbb{R}^{1,3} \ni x \mapsto J^\mu(x)$ satisfying

$$\partial_\mu J^\mu(x) = 0.$$

We assume that J^μ is a *Schwarz function on the space-time.* (Massive and massless) quantum fields interacing with J^μ satisfy the equation

$$-\partial^\mu\big(\partial_\mu \hat{A}_\nu(x) - \partial_\nu \hat{A}_\mu(x)\big) + m^2 \hat{A}_\nu(x) = J_\nu(x).$$

The corresponding *scattering operator* can be computed exactly. In particular, for a positive mass it is

$$\begin{aligned}
\hat{S} = \exp\Bigg(&\frac{\mathrm{i}}{2}\int \mathrm{d}k\overline{J^i(k)}\frac{1}{m^2+k^2-\mathrm{i}0}\left(g_{ij}-\frac{k_ik_j}{m^2+\vec{k}^2}\right)J^i(k)\\
&-\frac{\mathrm{i}}{2}\int \mathrm{d}k\frac{1}{\vec{k}^2+m^2}|J^0(k)|^2\Bigg)\\
&\times\exp\left(-\mathrm{i}\sum_{\sigma=0,\pm1}\int \mathrm{d}\vec{k}a^*(k,\sigma)\frac{\overline{u_\mu(k,\sigma)}}{\sqrt{2\varepsilon(\vec{k})}}J^\mu(k)\right)\\
&\times\exp\left(-\mathrm{i}\sum_{\sigma=0,\pm1}\int \mathrm{d}\vec{k}a(k,\sigma)\frac{u_\mu(k,\sigma)}{\sqrt{2\varepsilon(\vec{k})}}\overline{J^\mu(k)}\right)\\
=\ &\hat{S}_{\mathrm{tr}}\otimes\hat{S}_{\mathrm{lg}},
\end{aligned}$$

($J^\mu(k)$ denotes the Fourier transform of $J^\mu(x)$). Here the *transversal scattering operator* is

$$\begin{aligned}
\hat{S}_{\mathrm{tr}} = \exp\Bigg(&\frac{\mathrm{i}}{2}\int \mathrm{d}k\overline{J^i(k)}\frac{1}{m^2+k^2-\mathrm{i}0}\left(g_{ij}-\frac{k_ik_j}{k^2}\right)J^j(k)\\
&-\frac{\mathrm{i}}{2}\int \mathrm{d}k\frac{1}{\vec{k}^2+m^2}|J^0(k)|^2\Bigg)\\
&\times\exp\left(-\mathrm{i}\sum_{\sigma=\pm1}\int \mathrm{d}\vec{k}a^*(k,\sigma)\frac{\overline{u_\mu(k,\sigma)}}{\sqrt{2\varepsilon(\vec{k})}}J^\mu(k)\right)\\
&\times\exp\left(-\mathrm{i}\sum_{\sigma=\pm1}\int \mathrm{d}\vec{k}a(k,\sigma)\frac{u_\mu(k,\sigma)}{\sqrt{2\varepsilon(\vec{k})}}\overline{J^\mu(k)}\right)
\end{aligned}$$

and converges as $m \searrow 0$ to the scattering operator for the massless theory in the Coulomb gauge. The *longitudinal scattering operator* is

$$\hat{S}_{\rm lg} = \exp\left(\frac{\rm i}{2}m^2\int {\rm d}k\overline{J^i(k)}\frac{1}{m^2+k^2-{\rm i}0}\frac{k_ik_j}{(m^2+\vec{k}^2)\vec{k}^2}J^j(k)\right)$$
$$\times\exp\left(-{\rm i}\int {\rm d}\vec{k}a^*(k,0)\frac{\overline{u_\mu(k,0)}}{\sqrt{2\varepsilon(\vec{k})}}J^\mu(k)\right)$$
$$\times\exp\left(-{\rm i}\int {\rm d}\vec{k}a(k,0)\frac{u_\mu(k,0)}{\sqrt{2\varepsilon(\vec{k})}}\overline{J^\mu(k)}\right).$$

This can be rewritten as

$$\hat{S}_{\rm lg} = \exp\left(\frac{\rm i}{2}m^2\int {\rm d}k\frac{|\vec{J}\cdot\vec{k}|^2}{(m^2+k^2)(m^2+\vec{k}^2)\vec{k}^2}\right)$$
$$\times\exp\left(-\frac{1}{2}\int {\rm d}\vec{k}\frac{m^2|J^0(k)|^2}{2\varepsilon(\vec{k})\vec{k}^2}\right)$$
$$\times\exp\left({\rm i}\int {\rm d}\vec{k}a^*(k,0)\frac{mJ^0(k)}{|\vec{k}|\sqrt{2\varepsilon(\vec{k})}}\right)$$
$$\times\exp\left({\rm i}\int {\rm d}\vec{k}a(k,0)\frac{m\overline{J^0(k)}}{|\vec{k}|\sqrt{2\varepsilon(\vec{k})}}\right),$$

where the integral on the first line should be understood as the principal value. Thus $\hat{S}_{\rm lg}$, under rather general circumstances, converges to $1\!\!1$ as $m \searrow 0$.

Bibliography

[1] Baez, J.C., Segal, I.E., Zhou, Z., 1991: *Introduction to algebraic and constructive quantum field theory*, Princeton University Press.

[2] Bialynicki-Birula, I., Bialynicka-Birula, Z. 1975: Quantum Electrodynamics, Pergamon, Oxford 1975,

[3] Brunetti, R., Fredenhagen, K., 2009: Quantum field theory on curved backgrounds, in Bär, C. and Fredenhagen, K., editors, *Quantum Field Theory on Curved Spacetimes*, Lecture Notes in Physics 786, Springer

[4] Dereziński, J. 2014: Quantum fields with classical perturbations, Preprint arXiv:1307.1162 [math-ph]

[5] Dereziński, J., Gérard, G. 2013: *Mathematics of quantization and quantum fields*, Cambridge University Press

[6] Gupta, S. 1950: Proc. Phys. Soc. A63, 681691

[7] Bleuler, K. 1950: Helv.Phys.Acta 23, 567586

[8] Itzykson, C., Zuber, J.-B. 1980: Quantum Field Theory, McGraw Hill
[9] Jauch, J. M., Röhrlich, F., 1976: *The theory of photons and electrons*, 2nd edition, Springer.
[10] Schweber, S. S., 1962: *Introduction to non-relativistic quantum field theory*, Harper and Row.
[11] Srednicki, M., 2007: *Quantum Field Theory*, Cambridge University Press.
[12] Thirring, W., Narnhofer, H., 1992: Covariant QED without indefinite metric, Rev. Math. Phys. SI1 197-211
[13] Weinberg, S., 1995: *The quantum theory of fields, vol. I, Foundations*, Cambridge.

On the BCS gap equation for superfluid fermionic gases

G. Bräunlich* and C. Hainzl†

Mathematical Institute, University of Tübingen,
Auf der Morgenstelle 10, 72076 Tübingen, Germany
**E-mail: gerhard.braeunlich@uni-tuebingen.de*
†E-mail: christian.hainzl@uni-tuebingen.de

R. Seiringer

Institute of Science and Technology Austria,
Am Campus 1, 3400 Klosterneuburg, Austria
Email: robert.seiringer@ist.ac.at

We present a rigorous derivation of the BCS gap equation for superfluid fermionic gases with point interactions. Our starting point is the BCS energy functional, whose minimizer we investigate in the limit when the range of the interaction potential goes to zero.

Keywords: BCS theory, scattering length, superfluidity.

1. Introduction

In BCS theory [1–3] of superfluid fermionic gases the interaction between the spin-1/2 fermions is usually modeled by a contact potential. This approximation is justified for the low density atomic gases usually observed in the lab, since the range of the effective interaction is much smaller than the mean particle distance. In the theoretical physics literature [2–4] the states of superfluidity are usually characterized via the simplified BCS gap equation

$$-\frac{1}{4\pi a} = \frac{1}{(2\pi)^3} \int_{\mathbb{R}^3} \left(\frac{\tanh\left(\frac{\sqrt{(p^2-\mu)^2+|\Delta|^2}}{2T}\right)}{\sqrt{(p^2-\mu)^2+|\Delta|^2}} - \frac{1}{p^2} \right) \mathrm{d}^3 p, \tag{1}$$

where μ is a fixed chemical potential and Δ is the corresponding order parameter of the system. This order parameter does not vanish below the critical temperature T_c, uniquely defined by

$$\frac{1}{4\pi a} = \frac{1}{(2\pi)^3} \int_{\mathbb{R}^3} \left(\frac{\tanh\left(\frac{p^2-\mu}{2T_c}\right)}{p^2-\mu} - \frac{1}{p^2} \right) \mathrm{d}^3 p$$

for $a < 0$. The parameter a is the scattering length of the corresponding interaction and the usual argument for the derivation of equation (1) involves an ad-hoc renormalization scheme. The main goal of this paper is to give a rigorous derivation of equation (1) starting from the BCS functional of superfluidity [2, 3, 5] for a sequence of interaction potentials V_ℓ with range tending to zero, and scattering length $a(V_\ell)$ converging to a negative a. The present result is a consequence of our previous work [6] where we treated the more general case of BCS-Hartree-Fock theory and where we allowed for potentials with strong repulsive core. Dropping the direct and exchange term, however, as we do here, allows us to give a shorter, more transparent derivation of (1), and also to work with simpler assumptions on the interaction potentials.

2. The Model

We consider a gas of spin 1/2 fermions in the thermodynamic limit at temperature $T \geq 0$ and chemical potential $\mu \in \mathbb{R}$. The particles interact via a local two-body potential which we denote by V. The state of the system is described by two functions $\hat{\gamma} : \mathbb{R}^3 \to \mathbb{R}_+$ and $\hat{\alpha} : \mathbb{R}^3 \to \mathbb{C}$, which are conveniently combined into a 2×2 matrix

$$\Gamma(p) = \begin{pmatrix} \hat{\gamma}(p) & \hat{\alpha}(p) \\ \overline{\hat{\alpha}(p)} & 1 - \hat{\gamma}(-p) \end{pmatrix}, \tag{2}$$

required to satisfy $0 \leq \Gamma \leq \mathbb{1}_{\mathbb{C}^2}$ at every point $p \in \mathbb{R}^3$. The function $\hat{\gamma}$ is interpreted as the momentum distribution of the gas, while α (the inverse Fourier transform of $\hat{\alpha}$) is the Cooper pair wave function. Note that there are no spin variables in Γ; the full, spin dependent Cooper pair wave function is the product of $\alpha(x - y)$ with an antisymmetric spin singlet.

The *BCS functional* $\mathcal{F}_T^V$, whose infimum over all states Γ describes the negative of the pressure of the system, is given as

$$\mathcal{F}_T^V(\Gamma) = \int_{\mathbb{R}^3} (p^2 - \mu)\hat{\gamma}(p)\, \mathrm{d}^3p + \int_{\mathbb{R}^3} |\alpha(x)|^2 V(x)\, \mathrm{d}^3x - TS(\Gamma), \tag{3}$$

where

$$S(\Gamma) = -\int_{\mathbb{R}^3} \mathrm{Tr}_{\mathbb{C}^2}\big(\Gamma(p) \ln \Gamma(p)\big)\, \mathrm{d}^3p$$

is the entropy of the state Γ. The functional (3) can be obtained by restricting the many-body problem on Fock space to translation-invariant and spin-rotation invariant quasi-free states, and dropping the direct and exchange term in the interaction energy, see [5, Appendix A] and [7].

The *normal state* Γ_0 is the minimizer of the functional (3) restricted to states with $\alpha = 0$. It is given by

$$\hat{\gamma}_0(p) = \frac{1}{1 + e^{\frac{p^2-\mu}{T}}}.$$

The system is said to be in a superfluid phase if and only if the minimum of $\mathcal{F}_T^V$ is not attained at a normal state, and we call a normal state Γ_0 *unstable* in this case.

In a previous work [5] we thoroughly studied the functional (3). It is not difficult to see that this functional has a (not necessarily unique) minimizer (γ, α). More difficult is the question under which circumstances it is possible to guarantee that α does not vanish. Such a non-vanishing α in fact describes a macroscopic coherence of pairs such that the system displays a superfluid behavior. The corresponding Euler-Lagrange equations for γ and α can be equivalently expressed via $\Delta = 2(2\pi)^{-3/2}\hat{V} * \hat{\alpha}$ in the form of the BCS *gap equation*

$$\Delta(p) = -\frac{1}{(2\pi)^{3/2}} \int_{\mathbb{R}^3} \hat{V}(p-q) \frac{\Delta(q)}{E(q)} \tanh \frac{E(q)}{2T}\, dq \tag{4}$$

with $E(p) = \sqrt{(p^2-\mu)^2 + |\Delta(p)|^2}$; here, $\hat{V}$ denotes the Fourier transform of V. The function $\Delta(p)$ is the order parameter and is related to the wavefunction of the *Cooper pairs.* The equation (4) is highly non-linear; nonetheless, it is possible to show [5] that the existence of a non-trivial solution to (4) at some temperature T is equivalent to the fact that a certain *linear operator* has a negative eigenvalue. For $T = 0$ this operator is given by the Schrödinger-type operator $|-\Delta-\mu| + V$. This rather astonishing fact that one can reduce a non-linear to a linear problem, allowed for a more thorough mathematical study. Using spectral-theoretic methods, the class of potentials leading to a non-trivial solution for (4) has been precisely characterized. For instance, in [8] it was shown that if $\int V(x)dx < 0$, then there exists a critical temperature $T_c(V) > 0$ such that (4) attains a non-trivial solution for all $T < T_c(V)$, whereas there is no solution for $T \geq T_c(V)$. Additionally, in [8] the precise asymptotic behavior of $T_c(\lambda V)$ in the small coupling limit $\lambda \to 0$ was determined; the resulting expression generalizes well-known formulas in the physics literature [3, 9] valid only at low density. The low density limit $\mu \to 0$ of the critical temperature was studied in [10].

3. Main Results

We study the case of short-range interaction potentials V_ℓ, with range ℓ tending to zero in such a way that V_ℓ converges to a contact interaction. Such contact interactions are thoroughly studied in the literature [11, chap I.1.2-4] and are known to arise as a one parameter family of self-adjoint extensions of the Laplacian on $\mathbb{R}^3 \setminus \{0\}$. The relevant parameter uniquely determining the extension is, in fact, the scattering length, which we assume to be negative, in which case the resulting operator is non-negative, i.e., there are no bound states. In other words, we require that the scattering length $a(V_\ell)$ converges to a negative value as $\ell \to 0$, i.e.,

$$\lim_{\ell \to 0} a(V_\ell) = a < 0.$$

It was pointed out in [10, Equ. (3)] that the scattering length of any potential $V \in L^1 \cap L^{3/2}$ can be written as

$$a(V) = \frac{1}{4\pi} \left\langle |V|^{1/2} \middle| \frac{1}{1 + V^{1/2} \frac{1}{p^2} |V|^{1/2}} V^{1/2} \right\rangle \tag{5}$$

where $V^{1/2}$ is defined by $V^{1/2}|V|^{-1/2} = V$. Note that in case of a two-body interaction that does allow bound states the system would display features of a Bose-Einstein condensate of fermion pairs in the low density limit, see [4, 12–15].

In order to obtain a non-vanishing limit of the sequence of scattering lengths $a(V_\ell)$, we have to adjust the sequence of potentials so that the corresponding Schrödinger operator just barely fails to have a bound state. We shall follow the method of [11, chap I.1.2-4] and first choose a potential V such that $p^2 + V$ is non-negative and has a simple zero-energy resonance. Equivalently this means that the corresponding Birman-Schwinger operator

$$V^{1/2} \frac{1}{p^2} |V|^{1/2}$$

has -1 as lowest, simple, eigenvalue. Next we scale this potential and multiply it by a factor $\lambda(\ell) < 1$, such that the corresponding V_ℓ does no longer have a zero resonance, but a negative scattering length. Recall that a potential with zero resonance has an infinite scattering length.

To be precise, we define V_ℓ according to

$$V_\ell(x) = \lambda(\ell)\ell^{-2} V(\tfrac{x}{\ell}), \tag{6}$$

where $\lambda(0) = 1$, $\lambda < 1$ for all $\ell > 0$ and $1 - \lambda(\ell) = O(\ell)$. We are interested in the limit $\ell \to 0$ meaning that the range of the potential converges to zero. This scaling essentially leaves the $L^{3/2}$ norm of V invariant, but the L^1 norm vanishes linearly in ℓ. This is a major difference to the work [6] where we allowed the point interaction to be approximated by a sequence V_ℓ, whose L^1 norm converges to a positive number, with its $L^{3/2}$-norm even diverging. This required a new approach in the proof and led to a more general statement about contact interactions [16]. In the case considered here it suffices to rely on results of [11, chap I.1.2-4].

Our main objective now is to consider the solution Δ_ℓ of the BCS gap-equation (4), coming from a minimizer of the functional $\mathcal{F}_T^{V_\ell}$, and to show that in the limit where the range ℓ goes to zero, i.e., the potentials V_ℓ tend to a contact interaction, the order parameter Δ_ℓ converges to a constant function Δ, which satisfies the simplified equation (1). This equation appears throughout the physics literature as the one describing superfluid systems.

It is obvious that the critical temperature T_c is defined as the temperature where $\Delta = 0$ satisfies the equation (1). More precisely:

Definition 3.1 (Critical temperature). *Let $\mu > 0$. The* critical temperature *T_c corresponding to the scattering length $a < 0$ is given by the equation*

$$-\frac{1}{4\pi a} = \frac{1}{(2\pi)^3} \int_{\mathbb{R}^3} \left(\frac{\tanh\left(\frac{p^2-\mu}{2T_c}\right)}{p^2-\mu} - \frac{1}{p^2} \right) \mathrm{d}^3 p. \tag{7}$$

Since the function

$$T \mapsto \int_{\mathbb{R}^3} \left(\frac{\tanh\left(\frac{p^2-\mu}{2T}\right)}{p^2-\mu} - \frac{1}{p^2} \right) \mathrm{d}^3 p$$

is strictly monotone in T the critical temperature T_c is unique.

As our main theorem we reproduce the BCS gap equation for contact interactions, see, e.g., [2, Eq. (10)], [4, Eq. (7)].

Theorem 3.1 (Effective Gap equation). *Let $T \geq 0$, $\mu \in \mathbb{R}$ and assume $V \in L^{3/2}(\mathbb{R}^3) \cap L^1(\mathbb{R}^3)$ and $|x|V(x) \in L^2(\mathbb{R}^3)$. Let further $(\hat{\gamma}_\ell, \hat{\alpha}_\ell)$ be a minimizer of $\mathcal{F}_T^{V_\ell}$ with corresponding $\Delta_\ell = 2(2\pi)^{-3/2}\hat{V}_\ell * \hat{\alpha}_\ell$. Then there exist $\Delta \geq 0$ such that $|\Delta_\ell(p)| \to \Delta$ pointwise as $\ell \to 0$. If $\Delta \neq 0$ then it satisfies the equation*

$$\boxed{-\frac{1}{4\pi a} = \frac{1}{(2\pi)^3} \int_{\mathbb{R}^3} \left(\frac{1}{K_{T,\mu}^{\Delta}} - \frac{1}{p^2} \right) \mathrm{d}^3 p,} \tag{8}$$

where we use the abbreviations

$$K_{T,\mu}^{\Delta}(p) = \frac{E_\mu^\Delta(p)}{\tanh\left(\frac{E_\mu^\Delta(p)}{2T}\right)}, \quad E_\mu^\Delta(p) = \sqrt{(p^2-\mu)^2 + |\Delta|^2}\,.$$

Furthermore, the limiting Δ does not vanish if and only if $T < T_c$.

4. Proofs

Recall that we chose V so that $V^{1/2}\frac{1}{p^2}|V|^{1/2}$ has -1 as lowest simple eigenvalue, i.e., there is a unique ϕ, with

$$\left(V^{1/2}\tfrac{1}{p^2}|V|^{1/2} + 1\right)\phi = 0.$$

With U_ℓ denoting the unitary operator $(U_\ell\varphi)(x) = \ell^{-3/2}\varphi(\frac{x}{\ell})$, we can rewrite

$$V_\ell(x) = \lambda(\ell)\ell^{-2}V(\tfrac{x}{\ell}) = \lambda(\ell)\ell^{-2}U_\ell V U_\ell^{-1}.$$

Since $U_\ell p^2 U_\ell^{-1} = \ell^2 p^2$, it is easy to see that

$$U_\ell V^{1/2}\frac{1}{p^2}|V|^{1/2}U_\ell^{-1} = \frac{1}{\lambda(\ell)}V_\ell^{1/2}\frac{1}{p^2}|V_\ell|^{1/2}.$$

Denoting $\phi_\ell = U_\ell \phi$, this implies

$$V_\ell^{1/2} \frac{1}{p^2} |V_\ell|^{1/2} \phi_\ell = \lambda(\ell) U_\ell V^{1/2} \frac{1}{p^2} |V|^{1/2} \phi = -\lambda(\ell) \phi_\ell. \tag{9}$$

showing that the lowest eigenvalue of $1 + V_\ell^{1/2} \frac{1}{p^2} |V_\ell|^{1/2}$ is $1 - \lambda(\ell) = O(\ell)$. Moreover, note that

$$\|V_\ell\|_p = \lambda(\ell) \ell^{3/p-2} \|V\|_p \tag{10}$$

for $p \geq 1$.

Remark 4.1. In [6, Appendix A.1], we show that the scattering length corresponding to the potential V_ℓ, indeed, converges to the negative value

$$\lim_{\ell \to 0} a(V_\ell) = -\frac{1}{\lambda'(0)} \frac{|\langle |V|^{1/2} |\phi\rangle|^2}{\langle \mathrm{sgn}(V)\phi|\phi\rangle} < 0.$$

In the next Lemma, we derive a lower bound for the BCS functional which is uniform in ℓ. This will allow us to obtain limits for the order parameter Δ_ℓ.

Lemma 4.1. *There exists $C_1 > 0$, independent of ℓ, such that*

$$\mathcal{F}_T^{V_\ell}(\Gamma) \geq -C_1 + \frac{1}{2} \int_{\mathbb{R}^3} (1+p^2)(\hat{\gamma} - \hat{\gamma}_0)^2 \, \mathrm{d}^3 p + \frac{1}{2} \int_{\mathbb{R}^3} |p|^b |\hat{\alpha}|^2 \, \mathrm{d}^3 p, \tag{11}$$

for $0 \leq b < 1$, where we denote $\hat{\gamma}_0(p) = \frac{1}{1+\mathrm{e}^{(p^2-\mu)/T}}$.

Proof. For details of the proof of this Lemma we refer to [6, Lemma 3]. The main observation in the proof is that we may express the difference $\mathcal{F}_T^{V_\ell}(\Gamma) - \mathcal{F}_T^{V_\ell}(\Gamma_0)$ as

$$\mathcal{F}_T^{V_\ell}(\Gamma) - \mathcal{F}_T^{V_\ell}(\Gamma_0) = \frac{T}{2} \mathcal{H}(\Gamma, \Gamma_0) + \int_{\mathbb{R}^3} V_\ell(x) |\alpha(x)|^2 \, \mathrm{d}^3 x,$$

where $\mathcal{H}(\Gamma, \Gamma_0)$ is the relative entropy of Γ and Γ_0. By means of [17, Lemma 3], which is an extension of [18, Theorem 1], giving a bound on the relative entropy one obtains

$$\begin{aligned} \mathcal{F}_T^{V_\ell}(\Gamma) - \mathcal{F}_T^{V_\ell}(\Gamma_0) &\geq \langle \alpha \, |p^2 + V_\ell - \mu| \, \alpha \rangle \\ &\quad + \frac{1}{2} \int_{\mathbb{R}^3} (1+p^2)(\hat{\gamma} - \hat{\gamma}_0)^2 \, \mathrm{d}^3 p - C \end{aligned}$$

for an appropriate constant C. By means of a Birman-Schwinger type argument one can further show that

$$p^2 + V_\ell \geq |p|^b - C_1,$$

uniformly in ℓ for $0 \leq b < 1$ and an appropriate C_1, which, together with the constraint $|\hat{\alpha}|^2 \leq 1$, then implies the statement.

□

It was shown in [5, Theorem 1] that the functional $\mathcal{F}_T^{V_\ell}$ attains a minimizer $(\gamma_\ell, \alpha_\ell)$ for each V_ℓ. Lemma 4.1, with $\hat{\alpha}_\ell(p) \leq 1$, immediately tells us that the terms

$$\int_{\mathbb{R}^3} (1 + |p|^b)|\hat{\alpha}_\ell(p)| \, \mathrm{d}^3p \qquad \text{and} \qquad \int_{\mathbb{R}^3} (1 + |p|^2)(\hat{\gamma}_\ell - \hat{\gamma}_0)^2 \, \mathrm{d}^3p$$

are uniformly bounded in ℓ. Let us further mention the following useful relations between Δ_ℓ and the minimizer $(\gamma_\ell, \alpha_\ell)$,

$$\hat{\gamma}_\ell = \frac{1}{2} - \frac{1}{2}\frac{p^2 - \mu}{K_{T,\mu}^{\Delta_\ell}}, \qquad \Delta_\ell = 2K_{T,\mu}^{\Delta_\ell}\hat{\alpha}_\ell, \tag{12}$$

which follow from the corresponding Euler-Lagrange equations. One immediate consequence of these relations and Lemma 4.1 is the uniform boundedness of $\int_{\mathbb{R}^3} \hat{\gamma}_\ell(p)|p|^b \, \mathrm{d}^3p$, which implies

$$\lim_{R\to\infty}\lim_{\ell\to 0}\int_{|p|^2 \geq R} \hat{\gamma}_\ell(p) \, \mathrm{d}^3p = 0. \tag{13}$$

In the following lemma we show that, as $\ell \to 0$, pointwise limits for the main quantities exist. To this aim we introduce the notation

$$m_\mu^{\Delta_\ell}(T) = \frac{1}{(2\pi)^3}\int_{\mathbb{R}^3}\left(\frac{1}{K_{T,\mu}^{\Delta_\ell}} - \frac{1}{p^2}\right)\mathrm{d}^3p\,.$$

Lemma 4.2. *Let $(\gamma_\ell, \alpha_\ell)$ be a sequence of minimizers of $\mathcal{F}_T^{V_\ell}$ and $\Delta_\ell = 2(2\pi)^{-3/2}\hat{V}_\ell * \hat{\alpha}_\ell$. Then there is a subsequence of Δ_ℓ, which we continue to denote by Δ_ℓ, and a $\Delta \in \mathbb{R}_+$ such that*

(i) $|\Delta_\ell(p)|$ converges pointwise to the constant function Δ as $\ell \to 0$,

(ii) $\lim_{\ell\to 0} m_\mu^{\Delta_\ell}(T) = m_\mu^{\Delta}(T)$.

We shall see later that it is not necessary to restrict to a subsequence, the result holds in fact for the whole sequence.

Proof. i Set $c_\ell = \frac{1}{(2\pi)^{3/2}}\int_{\mathbb{R}^3} V_\ell(x)\alpha_\ell(x)\,\mathrm{d}^3x$. Then

$$\begin{aligned}|\Delta_\ell(p) - c_\ell| &\leq \frac{1}{(2\pi)^{3/2}}\int_{\mathbb{R}^3}\left|(e^{-ip\cdot x} - 1)V_\ell(x)\alpha_\ell(x)\right|\mathrm{d}^3x \\ &\leq (2\pi)^{-3/2}\|\alpha_\ell\|_2\left(\int_{\mathbb{R}^3}\left|(e^{-ip\cdot x} - 1)V_\ell(x)\right|^2\mathrm{d}^3x\right)^{1/2}.\end{aligned}$$

Now $\|\alpha_\ell\|_2$ is uniformly bounded in ℓ and $|\cdot|V \in L^2(\mathbb{R}^3)$ by assumption, so

$$\begin{aligned}\int_{\mathbb{R}^3}\left|(e^{-ip\cdot x} - 1)V_\ell(x)\right|^2\mathrm{d}^3x &= \ell^{-1}\lambda(\ell)^2\int_{\mathbb{R}^3}\left|(e^{-i\ell p\cdot x} - 1)V(x)\right|^2\mathrm{d}^3x \\ &\leq \ell\lambda(\ell)^2|p|^2\|\,|\cdot|V\|_2^2.\end{aligned} \tag{14}$$

Hence, $|\Delta_\ell(p) - c_\ell|$ converges to zero pointwise. Since $\hat{\alpha}_\ell = -2(K_{T,\mu}^{\Delta_\ell})^{-1}\Delta_\ell$ is uniformly bounded in L^2, it is straightforward to see, using $E_\mu^{\Delta_\ell} \geq |\Delta_\ell|$, that the same holds for the sequence $\Delta_\ell/E_\mu^{\Delta_\ell}$. This fact can now be used to show that the sequence $|c_\ell|$ is a uniformly bounded. Assume on the contrary that $\bar{c} = \limsup_{\ell\to 0}|c_\ell| = \infty$. Then by dominated convergence

$$\begin{aligned} &\limsup_{\ell\to 0}\int_{|p|\leq R}\frac{|\Delta_\ell|^2}{(p^2-\mu)^2+|\Delta_\ell|^2}\,\mathrm{d}^3p \\ &\qquad = \int_{|p|\leq R}\limsup_{\ell\to 0}\frac{1}{(p^2-\mu)^2/|c_\ell|^2+1}\,\mathrm{d}^3p = \frac{4}{3}\pi R^3. \end{aligned} \tag{15}$$

However, the divergence of right side of Eq. (15) as $R\to\infty$ contradicts the uniform boundedness of $\Delta_\ell/E_\mu^{\Delta_\ell}$ in $L^2(\mathbb{R}^3)$. Hence $\bar{c} < \infty$ and $\lim_{\ell\to 0}|\Delta_\ell(p)| = \bar{c}$ for a suitable subsequence.

ii Obviously by i the integrand of $m_\mu^{\Delta_\ell}(T)$ converges pointwise to the integrand of $m_\mu^{\Delta}(T)$. By (12) we are able to rewrite the integrand as

$$\frac{1}{K_{T,\mu}^{\Delta_\ell}(p)} - \frac{1}{p^2} = \frac{1}{p^2-\mu} - \frac{1}{p^2} - \frac{2\hat{\gamma}_\ell(p)}{p^2-\mu}. \tag{16}$$

Using (13) we now conclude that

$$\lim_{R\to\infty}\lim_{\ell\to 0}\int_{|p|^2\geq R}\left(\frac{1}{K_{T,\mu}^{\Delta_\ell}(p)} - \frac{1}{p^2}\right)\mathrm{d}^3p = 0,$$

which together with the dominated convergence inside $|p|^2 \leq R$, implies the statement of ii.

□

Proposition 4.1. *Let* $a = \lim_{\ell\to 0} a(V_\ell)$, *then*

$$\lim_{\ell\to 0}\frac{1}{(2\pi)^3}\int_{\mathbb{R}^3}\left(\frac{1}{K_{T,\mu}^{\Delta_\ell}} - \frac{1}{p^2}\right)\mathrm{d}^3p = -\frac{1}{4\pi a}. \tag{17}$$

Proof. We again follow the proof of [6, Theorem 2]. Observe that with help of the second relation in (12) the BCS gap equation (4) for α_ℓ can be conveniently written in the form

$$(K_{T,\mu}^{\Delta_\ell} + V_\ell)\alpha_\ell = 0, \qquad \text{with } \alpha_\ell \in H^1(\mathbb{R}^3)\,.$$

By means of the Birman–Schwinger principle one concludes that $K_{T,\mu}^{\Delta_\ell} + V_\ell$ having 0 as an eigenvalue is equivalent to $V_\ell^{1/2}\frac{1}{K_{T,\mu}^{\Delta_\ell}}|V_\ell|^{1/2}$ having -1 as eigenvalue.

We now rewrite $V_\ell^{1/2}\frac{1}{K_{T,\mu}^{\Delta_\ell}}|V_\ell|^{1/2}$ as

$$V_\ell^{1/2}\frac{1}{K_{T,\mu}^{\Delta_\ell}}|V_\ell|^{1/2} = V_\ell^{1/2}\frac{1}{p^2}|V_\ell|^{1/2} + m_\mu^{\Delta_\ell}(T)|V_\ell^{1/2}\rangle\langle|V_\ell|^{1/2}| + A_{\mu,T,\ell}, \tag{18}$$

where $A_{\mu,T,\ell}$ is given in terms of the integral kernel

$$A_{\mu,T,\ell}(x,y) = \frac{V_\ell(x)^{\frac{1}{2}}|V_\ell(y)|^{\frac{1}{2}}}{(2\pi)^3}\int_{\mathbb{R}^3}\left(\frac{1}{K_{T,\mu}^{\Delta_\ell}} - \frac{1}{p^2}\right)(\mathrm{e}^{-i(x-y)\cdot p} - 1)\,\mathrm{d}^3p, \tag{19}$$

which can be estimated, e.g., by

$$|A_{\mu,T,\ell}(x,y)| \leq \frac{|V_\ell(x)|^{\frac{1}{2}}|V_\ell(y)|^{\frac{1}{2}}}{(2\pi)^3}\int_{\mathbb{R}^3}\left|\frac{1}{K_{T,\mu}^{\Delta_\ell}} - \frac{1}{p^2}\right|(|x-y|\;|p|)^{1/2}\,\mathrm{d}^3p. \tag{20}$$

Using the relation (16) as well as the uniform boundedness of $\int_{\mathbb{R}^3}\hat{\gamma}_\ell(p)|p|^{1/2}\,\mathrm{d}^3p$ we are able to conclude that the integral

$$\int_{\mathbb{R}^3}\left|\frac{1}{K_{T,\mu}^{\Delta_\ell}} - \frac{1}{p^2}\right||p|^{1/2}\,\mathrm{d}^3p = \int_{\mathbb{R}^3}\left|\frac{1}{p^2-\mu} - \frac{1}{p^2} - \frac{2\hat{\gamma}_\ell(p)}{p^2-\mu}\right||p|^{1/2}\,\mathrm{d}^3p$$

is uniformly bounded in ℓ. We can thus bound the Hilbert-Schmidt norm of $A_{\mu,T,\ell}$ by

$$\|A_{\mu,T,\ell}\|_2 \leq \text{const}\,\|V_\ell|\cdot|\|_1^{1/2}\|V_\ell\|_1^{1/2} \leq O(\ell^{3/2})\,. \tag{21}$$

We further proceed with equation (18). By construction, $1 + V_\ell^{1/2}\frac{1}{p^2}|V_\ell|^{1/2}$ is invertible and thus can be factored out, i.e.,

$$1 + V_\ell^{1/2}\frac{1}{K_{T,\mu}^{\Delta_\ell}}|V_\ell|^{1/2} = \left(1 + V_\ell^{1/2}\frac{1}{p^2}|V_\ell|^{1/2}\right)\times$$
$$\times\left[1 + \frac{1}{1+V_\ell^{1/2}\frac{1}{p^2}|V_\ell|^{1/2}}\left(m_\mu^{\Delta_\ell}(T)|V_\ell^{1/2}\rangle\langle|V_\ell|^{1/2}| + A_{T,\mu,\ell}\right)\right],$$

with the second term on the right hand side necessarily having an eigenvalue 0. With $J = V_\ell(x)/|V_\ell(x)|$ and $X = |V_\ell|^{1/2}\frac{1}{p^2}|V_\ell|^{1/2}$, we are able to rewrite

$$\frac{1}{1+V_\ell^{1/2}\frac{1}{p^2}|V_\ell|^{1/2}} = \frac{1}{1+JX} = 1 - JX^{1/2}\frac{1}{1+X^{1/2}JX^{1/2}}X^{1/2}\,.$$

This allows us to bound

$$\left\|\frac{1}{1+V_\ell^{1/2}\frac{1}{p^2}|V_\ell|^{1/2}}\right\| \leq 1 + \|X\|\left\|\frac{1}{1+X^{1/2}JX^{1/2}}\right\| \leq O(\ell^{-1}),$$

where we have used that, due to the HLS-inequality, $\|X\| \leq C\|V_\ell\|_{3/2}$, as well as the fact that $1 + X^{1/2}JX^{1/2}$ is self-adjoint with its lowest eigenvalue of order $O(\ell)$. Indeed, $X^{1/2}JX^{1/2}$ has the same spectrum as JX. Hence,

$$\left\|\frac{1}{1+V_\ell^{1/2}\frac{1}{p^2}|V_\ell|^{1/2}}A_{\mu,T,\ell}\right\| < \left\|\frac{1}{1+V_\ell^{1/2}\frac{1}{p^2}|V_\ell|^{1/2}}\right\|\|A_{\mu,T,\ell}\| \leq O(\ell^{1/2}).$$

Since

$$\frac{1}{1+V_\ell^{1/2}\frac{1}{p^2}|V_\ell|^{1/2}}\left(m_\mu^{\Delta_\ell}(T)|V_\ell^{1/2}\rangle\langle|V_\ell|^{1/2}|+A_{T,\mu,\ell}\right)$$

has an eigenvalue -1 and $1+(1+V_\ell^{1/2}\frac{1}{p^2}|V_\ell|^{1/2})^{-1}A_{\mu,T,\ell}$ is invertible for small enough ℓ, we can argue by factoring out the term $1+(1+V_\ell^{1/2}\frac{1}{p^2}|V_\ell|^{1/2})^{-1}A_{\mu,T,\ell}$ that the rank one operator

$$m_\mu^{\Delta_\ell}(T)\left(1+\frac{1}{1+V_\ell^{1/2}\frac{1}{p^2}|V_\ell|^{1/2}}A_{\mu,T,\ell}\right)^{-1}\frac{1}{1+V_\ell^{1/2}\frac{1}{p^2}|V_\ell|^{1/2}}|V_\ell^{1/2}\rangle\langle|V_\ell|^{1/2}|$$

has an eigenvalue -1, which, by taking the trace, implies

$$-1=m_\mu^{\Delta_\ell}(T)\left\langle|V_\ell|^{1/2}\middle|\left[1+\frac{1}{1+V_\ell^{1/2}\frac{1}{p^2}|V_\ell|^{1/2}}A_{\mu,T,\ell}\right]^{-1}\frac{1}{1+V_\ell^{1/2}\frac{1}{p^2}|V_\ell|^{1/2}}\middle|V_\ell^{1/2}\right\rangle. \tag{22}$$

With the aid of Eq. (5) and the resolvent identity, we can rewrite Eq. (22) as

$$\begin{aligned}&4\pi a(V_\ell)+\frac{1}{m_\mu^{\Delta_\ell}(T)}\\&=\left\langle|V_\ell|^{1/2}\right|\frac{1}{1+V_\ell^{1/2}\frac{1}{p^2}|V_\ell|^{1/2}}A_{\mu,T,\ell}\left[1+\frac{1}{1+V_\ell^{1/2}\frac{1}{p^2}|V_\ell|^{1/2}}A_{\mu,T,\ell}\right]^{-1}\times\\&\quad\times\frac{1}{1+V_\ell^{1/2}\frac{1}{p^2}|V_\ell|^{1/2}}\left|V_\ell^{1/2}\right\rangle,\end{aligned} \tag{23}$$

where the right-hand side is bounded by

$$\begin{aligned}\|V_\ell\|_1\left\|\frac{1}{1+V_\ell^{1/2}\frac{1}{p^2}|V_\ell|^{1/2}}\right\|\left\|\frac{1}{1+V_\ell^{1/2}\frac{1}{p^2}|V_\ell|^{1/2}}A_{\mu,T,\ell}\right\|\times&\\ \times\left\|\left[1+\frac{1}{1+V_\ell^{1/2}\frac{1}{p^2}|V_\ell|^{1/2}}A_{\mu,T,\ell}\right]^{-1}\right\|\le O(\ell^{1/2})\,.&\end{aligned}$$

This implies Eq. (17) and completes the proof.

□

With the aid of Lemma 4.2 and Proposition 4.1, we can now finish the proof of Theorem 3.1.

Proof of Theorem 3.1. We know from Lemma 4.2 and Proposition 4.1 that $|\Delta_\ell(p)|$ has a subsequence that converges to a constant function Δ, which satisfies the equation

$$-\frac{1}{4\pi a}=\frac{1}{(2\pi)^3}\int_{\mathbb{R}^3}\left(\frac{1}{K_{T,\mu}^{\Delta}}-\frac{1}{p^2}\right)\mathrm{d}^3p.$$

Since the solution $|\Delta|$ of (8) is unique we obtain that the sequence $|\Delta_\ell(p)|$ converges to the unique solution of (8). Furthermore, this shows that the limit of $|\Delta_\ell|$ does not vanish in the case that $T<T_c$, and that the limit vanishes for $T\ge T_c$. □

Bibliography

[1] J. Bardeen, L. Cooper, J. Schrieffer, *Phys. Rev.* **108**, 1175 (1957).

[2] A. J. Leggett, Diatomic molecules and cooper pairs, in *Modern Trends in the Theory of Condensed Matter*, eds. A. Pekalski and J. Przystawa, Lecture Notes in Physics, Vol. 115 (Springer Berlin / Heidelberg, 1980) pp. 13–27.

[3] P. Noziéres and S. Schmitt-Rink, *Journal of Low Temperature Physics* **59**, 195 (1985).

[4] M. Randeria, Crossover from BCS theory to Bose-Einstein condensation, in *Bose-Einstein Condensation*, eds. A. Griffin, D. W. Snoke and S. Stringari (Cambridge University Press, July 1996), pp. 355–392.

[5] C. Hainzl, E. Hamza, R. Seiringer, J. P. Solovej, *Comm. Math. Phys.* **281**, 349 (2008).

[6] G. Bräunlich, C. Hainzl, R. Seiringer, *preprint* arXiv/1305.5135.

[7] V. Bach, E. H. Lieb, J. P. Solovej, *J. Statist. Phys.* **76**, 3 (1994).

[8] R. L. Frank, C. Hainzl, S. Naboko, R. Seiringer, *J. Geom. Anal.* **17**, 559 (2007).

[9] L. P. Gor'kov, *ZH. Eksp. Teor. Fiz.* **36**, 1918 (1959).

[10] C. Hainzl and R. Seiringer, *Lett. Math. Phys.* **84**, 99 (2008).

[11] S. Albeverio, F. Gesztesy, R. Høegh-Krohn, H. Holden, *Solvable models in quantum mechanics*, Texts and monographs in physics (Springer-Verlag New York, 1988).

[12] M. Drechsler and W. Zwerger, *Ann. Phys.* **504**, 15 (1992).

[13] P. Pieri and G. C. Strinati, *Phys. Rev. Lett.* **91**, 030401 (2003).

[14] C. Hainzl and R. Seiringer, *Lett. Math. Phys.* **100**, 119 (2012).

[15] C. Hainzl and B. Schlein, *Journal of Functional Analysis* **265**, 399 (2013).

[16] G. Bräunlich, C. Hainzl, R. Seiringer, *Meth. Funct. Anal. Top.* **19**, 364 (2013).

[17] R. L. Frank, C. Hainzl, R. Seiringer, J. P. Solovej, *J. Amer. Math. Soc.* **25**, 667 (2012).

[18] C. Hainzl, M. Lewin, R. Seiringer, *Rev. Math. Phys.* **20**, 1283 (2008).

Improved Hardy inequality in twisted tubes

H. Kovařík

DICATAM, Sezione di Matematica, Università di Brescia,
Via Branze, 38 - 25123 Brescia, Italy
E-mail: hynek.kovarik@unibs.it

In this note we consider the Dirichlet Laplacian in twisted three-dimensional tubes. In particular we study a related Hardy type inequality, which has been establish in previous works, and show how such an inequality can be improved. We also discuss the optimality of the corresponding integral weight.

Keywords: Hardy inequality, twisted tubes.

1. Introduction

Let $\omega \subset \mathbb{R}^2$ be an open bounded set and let $\Omega_0 = \omega \times \mathbb{R}$ be a straight tube in $\mathbb{R}^3$. Let us now define the twisted tube Ω by

$$\Omega = \big\{ r_\theta(x_3)\, x \mid x = (x_1, x_2, x_3) \in \Omega_0 \big\},$$

where

$$r_\theta(x_3) = \begin{pmatrix} \cos\theta(x_3) & \sin\theta(x_3) & 0 \\ -\sin\theta(x_3) & \cos\theta(x_3) & 0 \\ 0 & 0 & 1 \end{pmatrix}$$

and $\theta : \mathbb{R} \to \mathbb{R}$ is the angle of rotation. Here and in the sequel we will denote by x the variable in the straight tube Ω_0 and by $\mathbf{x}$ the variable in the twisted tube Ω. Denote moreover by E_1 and ψ_1 the principal eigenvalue and the associated positive normalized eigenfunction of $-\Delta_\omega^D$, the Dirichlet Laplacian on ω;

$$-\Delta_\omega \, \psi_1 = E_1 \, \psi_1, \qquad \|\psi_1\|_{L^2(\omega)} = 1. \tag{1}$$

Recall that $E_1 > 0$.

We will work under the following hypotheses on ω:

Assumption 1.1. *$\omega \subset \mathbb{R}^2$ is an open bounded connected set with Lipschitz boundary. Assume moreover that ω contains the origin and that ω is not a disc centred at the origin.*

In [6, Thm. 3] it is proved that if ω satisfies assumption 1.1 and if $\theta \in C^{1,1}(\mathbb{R})$ is such that $\dot{\theta}$ has compact support and $\dot{\theta} \neq 0$, then there exists a positive constant $C = C(\theta, \omega)$ for which the inequality

$$Q[u] := \int_\Omega \left(|\nabla u(\mathbf{x})|^2 - E_1 \, |u(\mathbf{x})|^2 \right) d\mathbf{x} \, \geq \, C \int_\Omega \frac{|u(\mathbf{x})|^2}{1 + \mathbf{x}_3^2} \, d\mathbf{x} \qquad \forall \, u \in H_0^1(\Omega) \quad (2)$$

holds true. A slightly different proof of (2), still under the assumption that $\dot{\theta}$ has compact support, was given later in [13, Thm. 3.1], see also [12]. Inequality (2) shows, among other things, that a sufficiently weak attractive perturbation of $-\Delta_\Omega^D$ in twisted tubes, contrary to the straight ones, will not produce a discrete spectrum. The repulsive effect of twisting has been observed, without appeal to a Hardy type inequality, also in [4, 9, 10].

Apart from removing the weakly coupled discrete eigenvalues, the twisting has also consequences on scattering properties of $-\Delta_\Omega^D$, see [3] , and on the large time behaviour of the heat semigroup [8, 13]. Similar effects for two-dimensional strips with "twisted" Dirichlet-Neumann boundary conditions were studied in [11, 14].

The aim of this note is to improve inequality (2) in two directions; first *we remove the assumption on the compactness of the support of* $\dot{\theta}$, see Proposition 2.1, and then we add to the integral weight on the right hand side a *singular factor proportional to the inverse square of the distance to* $\partial\Omega$, see Theorem 2.1 for details. Finally, in Proposition 3.1 we briefly discuss the link between a long range twisting and the integral weight in (2).

2. Main result

2.1. *Notation*

Let us fix some notation. We denote by $x_t = (x_1, x_2) \in \omega$ a generic point of ω and by $B(\omega)$ the largest disc contained in ω and centered in the origin. Accordingly, let $\nabla_t := (\partial_1, \partial_2)$ denote the gradient operator on ω. Given a set $M \subset \mathbb{R}^n$ $(n = 2, 3)$ we denote by $\mathbb{1}_M$ its characteristic function. We also introduce the distance function on ω by

$$\delta(\mathbf{x}) = \mathrm{dist}\,(\mathbf{x}, \partial\Omega),$$

and the numerical constant

$$\gamma(\omega) := \int_{B(\omega)} \psi_1^2(x_t) \, dx_t. \qquad (3)$$

Throughout the text we assume, without loss of generality, that the all considered test functions are real-valued. The main result of this note is the following.

Theorem 2.1. *Assume that ω satisfies assumption* 1.1. *Suppose moreover that $\theta \in C^{1,1}(\mathbb{R})$ and that $\dot{\theta} \neq 0$. Then there exists a positive constant $K = K(\theta, \omega)$ such that*

$$\int_\Omega \Big(|\nabla u(\mathbf{x})|^2 \, d\mathbf{x} - E_1 \, u^2(\mathbf{x}) \Big) \, d\mathbf{x} \, \geq \, K \int_\Omega \frac{u^2(\mathbf{x})}{\delta^2(\mathbf{x}) \, (1 + \mathbf{x}_3^2)} \, d\mathbf{x} \tag{4}$$

holds for all $u \in H_0^1(\Omega)$.

Remark 2.1. The proofs of inequality (2) given in [6] and [13] hold under slightly weaker regularity conditions on θ. In particular [13] assumes $\theta \in C^1(\mathbb{R})$ while Theorem 2.1 assumes $\theta \in C^{1,1}(\mathbb{R})$.

On the other hand, both [6] and [13] require $\dot{\theta}$ to have a compact support in order for (2) to hold. In fact, a local version of (2), which is equivalent to inequality (15), holds independently of the compactness of $\operatorname{supp} \dot{\theta}$, see the proof of [6, Thm. 4] or [13, Lem.3.1]. The assumption that $\operatorname{supp} \dot{\theta}$ is compact is needed in [6] and [13] in order to extend the local Hardy inequality into the global form (2) with the integral weight decaying as $|\mathbf{x}|^{-2}$.

However, this assumption is unnatural since more twisting should not be an obstacle for the validity of (2) (but rather vice versa). In fact Theorem 2.1 shows that the compactness of $\operatorname{supp} \dot{\theta}$ is not only unnatural, but also unnecessary for (2) to hold.

Remark 2.2. In contrast to (2) inequality (4) reflects the fact that the test functions satisfy Dirichlet boundary conditions at $\partial\Omega$ through the presence of the singular factor $\delta(\mathbf{x})^{-2}$ on the right hand side. The inclusion of this factor is also the reason why we suppose that ω has Lipschitz boundary.

Indeed, it is well-known that the classical Hardy inequality on bounded domains with the integral weight δ^{-2}, cannot hold without certain regularity assumption on the boundary, see e.g. [17]. Therefore some regularity of ω is necessary for (4) to hold, contrary to inequality (2) which holds for any ω.

2.2. *Preliminaries*

Before proving the above claim we establish several preliminary results. We start by the following auxiliary Lemma in which E_2 denotes the second eigenvalue of $-\Delta_\omega^D$. It is well-known that $E_2 > E_1$.

Lemma 2.1. *Let $\varepsilon > 0$. Then for all $f \in H_0^1(\omega)$ we have*

$$\int_\omega \Big(|\nabla f(x_t)|^2 - E_1 \, f(x_t)^2 + \varepsilon \, \mathbb{1}_{B(\omega)} \, f(x_t)^2 \Big) \, dx_t \, \geq \, a(\varepsilon) \int_\omega f^2(x_t) \, dx_t, \tag{5}$$

where

$$a(\varepsilon) = \frac{\varepsilon\,(E_2 - E_1)\,\gamma(\omega)}{4\varepsilon + 2(E_2 - E_1)}. \tag{6}$$

Proof. We use the decomposition

$$f(x_t) = \alpha\,\psi_1(x_t) + g(x_t), \tag{7}$$

where $\alpha = (f, \psi_1)_{L^2(\omega)}$. Obviously we have

$$\int_\omega \psi_1(x_t)\,g(x_t)\,dx_t = 0. \tag{8}$$

Moreover, since $-\Delta_\omega^D \psi_1 = E_1 \psi_1$, the last equation in combination with integration by parts implies that

$$\int_\omega \nabla\psi_1(x_t) \cdot \nabla g(x_t)\,dx_t = 0. \tag{9}$$

With the help of (8) and (9) we thus obtain the lower bound

$$\begin{aligned}\int_\omega \Big(|\nabla f(x_t)|^2 - E_1\, f^2(x_t)\Big)\,dx_t &= \int_\omega \Big(|\nabla g(x_t)|^2 - E_1\, g^2(x_t)\Big)dx_t \\ &\geq (E_2 - E_1) \int_\omega g^2(x_t)\,dx_t, \end{aligned} \tag{10}$$

where we have used the fact that $\|\nabla g\|^2_{L^2(\omega)} \geq E_2\, \|g\|^2_{L^2(\omega)}$, see (8). Next we recall that for any $A, B \in \mathbb{R}$ it holds

$$2\,|A\,B| \leq s\,A^2 + s^{-1} B^2, \qquad s > 0. \tag{11}$$

Using the above estimate with $s = \frac{\varepsilon}{\varepsilon + E_2 - E_1}$ we find

$$\frac{1}{2}(E_2 - E_1)\,g^2(x_t) + \varepsilon\, f^2(x_t) \geq b(\varepsilon)\,(E_2 - E_1)\,\alpha^2 \psi_1^2(x_t),$$

with

$$b(\varepsilon) = \frac{\varepsilon}{2\varepsilon + E_2 - E_1}.$$

The last estimate together with (10) and (3) yields

$$\begin{aligned}\int_\omega \Big(|\nabla f|^2 - E_1\,|f|^2 + \varepsilon\,\mathbb{1}_{B(\omega)}\, f^2\Big)\,dx_t &\geq (E_2 - E_1) \int_\omega \Big(\frac{g^2}{2} + b(\varepsilon)\,\mathbb{1}_{B(\omega)}\,\alpha^2 \psi_1^2\Big)\,dx_t \\ &= (E_2 - E_1) \int_\omega \Big(\frac{g^2}{2} + b(\varepsilon)\,\gamma(\omega)\,\alpha^2 \psi_1^2\Big)\,dx_t. \end{aligned}$$

Note also that $0 < \gamma(\omega) < 1$ on view of the normalisation $\|\psi_1\|_{L^2(\omega)} = 1$. Hence $b(\varepsilon)\,\gamma(\omega) < 1/2$ and

$$\begin{aligned}\int_\omega \left(|\nabla f|^2 - E_1\, f^2 + \varepsilon\, \mathbb{1}_{B(\omega)}\, f^2\right) dx_t &\geq (E_2 - E_1)\, b(\varepsilon)\,\gamma(\omega) \int_\omega (g^2 + \alpha^2 \psi_1^2)\, dx_t \\ &\geq \frac{(E_2 - E_1)\, b(\varepsilon)\,\gamma(\omega)}{2} \int_\omega (g + \alpha\,\psi_1)^2\, dx_t \\ &= a(\varepsilon) \int_\omega f^2(x_t)\, dx_t.\end{aligned}$$

□

Following [6] we now introduce the transform

$$(U_\theta\, u)(x) := u\,(r_\theta(x_3)\, x)\,, \quad x \in \Omega_0, \quad u \in L^2(\Omega), \tag{12}$$

which maps $L^2(\Omega)$ unitarily onto $L^2(\Omega_0)$. Writing $U_\theta\, u = v$, straightforward calculation then shows that the quadratic form Q transforms as follows

$$\widetilde{Q}[v] := Q[U_\theta^{-1} v] = \int_{\Omega_0} \left(|\nabla_t v|^2 + |\partial_3 v + \dot\theta\, \partial_\tau v|^2\right) dx, \qquad v \in H_0^1(\Omega_0), \tag{13}$$

where $\partial_\tau := x_1 \partial_2 - x_2 \partial_1$. We have

Proposition 2.1. *Under the assumptions of Theorem* 2.1 *there exists* $c_1 = c_1(\theta, \omega) > 0$ *such that the inequality*

$$\int_{\Omega_0} \left(|\nabla_t v(x)|^2 + |\partial_3 v(x) + \dot\theta\, \partial_\tau v(x)|^2 - E_1\, v^2(x)\right) dx \geq c_1 \int_{\Omega_0} \frac{v^2(x)}{1 + x_3^2}\, dx \tag{14}$$

holds true for all $v \in H_0^1(\Omega_0)$.

Proof. Let $v \in H_0^1(\Omega_0)$. By assumption there exists a closed interval I such that $\inf_I |\dot\theta| > 0$. Without loss of generality we may assume that $I = [-a, a]$ for some $a > 0$. Since $\ddot\theta \in L^\infty_{\mathrm{loc}}(\mathbb{R})$, which is a consequence of the hypothesis on θ, it follows from the proof of [6, Thm. 4] that

$$\begin{aligned}&\int_{\omega \times I} \left(|\nabla_t v|^2 + |\partial_3 v + \dot\theta\, \partial_\tau v|^2 - E_1\, v^2\right) dx \\ &\quad \geq c_0 \int_{\omega \times I} v^2\, dx \geq c_0 \int_{B(\omega) \times I} v^2\, dx\end{aligned} \tag{15}$$

for some $c_0 > 0$. Let $u = U_\theta^{-1} v \in H_0^1(\Omega)$. By applying the transform (12) to inequality (15) we obtain

$$\int_\Omega \left(|\nabla u(\mathbf{x})|^2 - E_1\, u^2(\mathbf{x})\right) d\mathbf{x} \geq c_0 \int_{B(\omega) \times I} u^2(\mathbf{x})\, d\mathbf{x}. \tag{16}$$

We now extend the integral weight on the right hand side of the last inequality to the domain $B(\omega) \times \mathbb{R}$. This extension procedure is well-known, see [5, sec. 3.1.3] or [6, sec. 3.3], but we prefer to present its main steps for the convenience of the reader. To this end we define the function $\varphi : \mathbb{R} \to \mathbb{R}$ by

$$\varphi(s) := \begin{cases} 1 & \text{for } |s| \geq a\,, \\ |s|/a & \text{for } |s| < a\,, \end{cases}$$

and recall the one-dimensional Hardy inequality

$$\int_{\mathbb{R}} |h'(s)|^2 \, ds \geq \frac{1}{4} \int_{\mathbb{R}} \frac{h^2(s)}{s^2} \, ds \qquad \forall\, h \in H^1(\mathbb{R}) : \; h(0) = 0. \tag{17}$$

We now write $u(\mathbf{x}) = \varphi(\mathbf{x}_3)\, u(\mathbf{x}) + (1 - \varphi(\mathbf{x}_3))\, u(\mathbf{x})$ and keeping in mind that $\|\varphi\|_\infty = 1$ and $\|\varphi'\|_\infty = a^{-1}$, we obtain the upper bound

$$\begin{aligned} \int_{B(\omega)\times\mathbb{R}} \frac{u^2}{1+\mathbf{x}_3^2}\, d\mathbf{x} &\leq 2 \int_{B(\omega)\times\mathbb{R}} \frac{(\varphi\, u)^2}{\mathbf{x}_3^2}\, d\mathbf{x} + 2 \int_{B(\omega)\times I} (1-\varphi)^2\, u^2\, d\mathbf{x} \\ &\leq 8 \int_{B(\omega)\times\mathbb{R}} (\partial_3(\varphi\, u))^2\, d\mathbf{x} + 2 \int_{B(\omega)\times I} u^2\, d\mathbf{x} \\ &\leq 16 \int_{B(\omega)\times\mathbb{R}} |\partial_3 u|^2\, d\mathbf{x} + \left(\frac{16}{a^2} + 2\right) \int_{B(\omega)\times I} u^2\, d\mathbf{x}, \end{aligned}$$

where we applied inequality (17) to the function $\varphi(\cdot)\, u(\mathbf{x}_1, \mathbf{x}_2, \cdot)$ on the first line. Next we use inequality (16) in order to bound the last term on the right hand side. This gives

$$\int_{B(\omega)\times\mathbb{R}} \frac{u^2}{1+\mathbf{x}_3^2}\, d\mathbf{x} \leq 16 \int_{\Omega} |\partial_3 u|^2\, d\mathbf{x} + \left(\frac{16}{a^2} + 2\right) c_0^{-1} \int_{\Omega} \left(|\nabla u|^2 - E_1\, u^2\right) d\mathbf{x}.$$

The last estimate together with the Poincaré inequality

$$\int_{\Omega} \left(|\partial_1 u|^2 + |\partial_2 u|^2 - E_1\, u^2\right) d\mathbf{x} \geq 0$$

thus implies the bound

$$\varepsilon \int_{B(\omega)\times\mathbb{R}} \frac{u^2}{1+\mathbf{x}_3^2}\, d\mathbf{x} \leq \int_{\Omega} \left(|\nabla u|^2 - E_1\, u^2\right) d\mathbf{x}, \tag{18}$$

where

$$\varepsilon = \left(16 + \left(\frac{16}{a^2} + 2\right) c_0^{-1}\right)^{-1} > 0. \tag{19}$$

Passing back to the straight tube, via the transform (12), we deduce from inequality

(18) and Lemma 2.1 that

$$
\begin{aligned}
\int_{\Omega_0} \Big(|\nabla_t v|^2 + |\partial_3 v + \dot\theta\, \partial_\tau v|^2 - E_1\, v^2 \Big)\, dx \\
&\geq \frac{1}{2} \int_{\Omega_0} \left(|\nabla_t v|^2 - E_1\, v^2 + \frac{\varepsilon}{2}\, \mathbb{1}_{B(\omega)} \frac{v^2}{1+x_3^2} \right) dx \\
&\geq \frac{1}{2} \int_{\mathbb{R}} \frac{1}{1+x_3^2} \left(\int_\omega \Big(|\nabla_t v|^2 - E_1\, v^2 + \frac{\varepsilon}{2}\, \mathbb{1}_{B(\omega)}\, v^2 \Big)\, dx_t \right) dx_3 \\
&\geq c_1 \int_{\Omega_0} \frac{v^2}{1+x_3^2}\, dx,
\end{aligned}
$$

where $c_1 = \frac{1}{2}\, a(\varepsilon/2)$ with ε and $a(\varepsilon/2)$ given by (19) and (6) respectively. In the second step we have used the fact that

$$
\int_\omega (|\nabla_t v(\cdot, x_3)|^2 - E_1\, v^2(\cdot, x_3))\, dx_t \geq 0 \qquad \text{a. e. } x_3 \in \mathbb{R}. \tag{20}
$$

This completes the proof. □

2.3. *Proof of the main result*

For the proof of Theorem 2.1 we will need a well-known version of the Hardy inequality on bounded domains in $\mathbb{R}^2$, see for example [17].

Theorem 2.2 (Opic, Kufner). *Let*

$$
k(\omega) := \inf_{f \in H_0^1(\omega)} \frac{\int_\omega |\nabla f(x_t)|^2\, dx_t}{\int_\omega f^2(x_t)/D^2(x_t)\, dx_t}, \tag{21}
$$

where

$$
D(x_t) = \operatorname{dist}(x_t, \partial\omega).
$$

If ω satisfies assumption 1.1, *then* $k(\omega) > 0$.

Now we are in position to prove the main result.

Proof of Theorem 2.1. Fix $u \in H_0^1(\Omega)$ and let $v = U_\theta\, u$. Note that in (14) we may assume without loss of generality that $c_1 \leq E_1$. From Proposition 2.1 and inequality (20) we thus deduce that

$$\begin{aligned}
&\int_{\Omega_0} \left(|\nabla_t v|^2 + |\partial_3 v + \dot\theta\, \partial_\tau v|^2 - E_1\, v^2\right) dx \\
&\geq \frac{1}{2} \int_{\mathbb{R}} \int_\omega \left(|\nabla_t v|^2 - E_1\, v^2 + \frac{c_1\, v^2}{1+x_3^2}\right) dx_t\, dx_3 \\
&= \frac{1}{2} \int_{\mathbb{R}} \int_\omega \left(\frac{c_1\, |\nabla_t v|^2}{E_1(1+x_3^2)} + \frac{1}{E_1}\left(E_1 - \frac{c_1}{1+x_3^2}\right) |\nabla_t v|^2 - E_1\, v^2 + \frac{c_1\, v^2}{1+x_3^2}\right) dx_t dx_3 \\
&\geq \frac{1}{2} \int_{\mathbb{R}} \int_\omega \left(\frac{c_1}{E_1(1+x_3^2)}\, |\nabla_t v|^2\right) dx_t\, dx_3 .
\end{aligned}$$

Next, for a fixed $x_3 \in \mathbb{R}$ we apply Theorem 2.2 to the function $v(\cdot, x_3)$. We thus obtain

$$\begin{aligned}
&\int_{\Omega_0} \left(|\nabla_t v|^2 + |\partial_3 v + \dot\theta\, \partial_\tau v|^2 - E_1\, v^2\right) dx \\
&\geq \frac{1}{2} \int_{\mathbb{R}} \int_\omega \left(\frac{k(\omega)\, c_1\, v^2}{E_1\, D^2(x_t)\,(1+x_3^2)}\right) dx_t\, dx_3 \qquad (22) \\
&= \frac{k(\omega)\, c_1}{2\, E_1} \int_{\Omega_0} \frac{v^2}{\operatorname{dist}^2(x, \partial\Omega_0)\,(1+x_3^2)}\, dx.
\end{aligned}$$

It remains to note that under the hypothesis on θ there exists a constant $c_2 > 0$ such that

$$c_2^{-1} \operatorname{dist}(x, \partial\Omega_0) \leq \delta(r_\theta(x_3)\, x) \leq c_2 \operatorname{dist}(x, \partial\Omega_0) \qquad \forall\, x \in \Omega_0.$$

Hence applying the transform (12) to the inequality (22) we get

$$\int_\Omega \left(|\nabla u|^2 - E_1\, u^2\right) dx \geq K \int_\Omega \frac{u^2}{\delta^2(\mathbf{x})(1+\mathbf{x}_3^2)}\, d\mathbf{x}.$$

with

$$K = \frac{k(\omega)\, c_1\, c_2^2}{2\, E_1}$$

□

Remark 2.3. The value of the constant $k(\omega)$ in (21) depends in general on ω. However, it is known that if ω is convex, then $k(\omega) = \frac{1}{4}$, see [15, 16], and if ω is simply connected, then $k(\omega) \geq \frac{1}{16}$ [1] .

3. Long range twisting

In [6] is it also shown that the quadratic decay in $\mathbf{x}_3$ of the weight function in right hand side of inequality (2) is the best possible if $\dot\theta$ has compact support. It is not difficult to see that the same is true for inequality (4). The natural question is whether the decay of the weight in $\mathbf{x}_3$ might be improved if $\dot\theta$ is sufficiently long range, i.e. decays slowly enough at infinity. More precisely, we are interested in the validity of the inequality

$$\int_\Omega \Big(|\nabla u(\mathbf{x})|^2 \, d\mathbf{x} - E_1 u^2(\mathbf{x}) \Big) \, d\mathbf{x} \ \geq \ C_\beta \int_\Omega \frac{u^2(\mathbf{x})}{\delta^2(\mathbf{x}) \, (1+\mathbf{x}_3^2)^\beta} \, d\mathbf{x} \tag{23}$$

for all $u \in H_0^1(\Omega)$, and some $\beta < 1$, $C_\beta > 0$. We are going to give a *necessary* condition on $\dot\theta$ for (23) to hold.

Proposition 3.1. *Assume that the hypothesis of Theorem* 2.1 *are satisfied. Let* $\beta < 1$. *If* (23) *holds true, then*

$$\limsup_{x_3 \to \infty} |\dot\theta(x_3)| \, x_3^\beta > 0 \quad \wedge \quad \limsup_{x_3 \to -\infty} |\dot\theta(x_3)| \, |x_3|^\beta > 0. \tag{24}$$

Proof. From (23) it follows that

$$\int_{\Omega_0} \Big(|\nabla_t v|^2 + |\partial_3 v + \dot\theta \, \partial_\tau v|^2 - E_1 \, v^2 \Big) \, dx \geq c \int_{\Omega_0} \frac{v^2}{d^2(x_t)(1+x_3^2)^\beta} \, dx \tag{25}$$

holds true for all $v \in H_0^1(\Omega_0)$ and some $c > 0$. Inserting $v(x) = \psi_1(x_t) \, \varphi(x_3)$ into (25) and using

$$\int_\omega \psi_1 \, \partial_\tau \psi_1 \, dx_t = 0,$$

which follows by integration by parts, we find that

$$\int_{\mathbb{R}} \Big(|\varphi'(x_3)|^2 + \|\partial_\tau \psi_1\|^2_{L^2(\omega)} \, \dot\theta^2(x_3) \, \varphi^2(x_3) \Big) \, dx_3 \geq \tilde c \int_{\mathbb{R}} \frac{\varphi^2(x_3)}{(1+x_3^2)^\beta} \, dx_3 \tag{26}$$

holds for all $\varphi \in H^1(\mathbb{R})$ and some $\tilde c > 0$. Let $f \in H^1(\mathbb{R})$ be such that $f \neq 0$ and $f(s) = 0$ if $|s| \leq 1$. Define

$$\varphi_\lambda(x_3) = f(\lambda \, x_3), \qquad \lambda > 0.$$

Applying (26) to φ_λ and writing $s = \lambda \, x_3$ then implies

$$\begin{aligned} \lambda^{2-2\beta} \int_{\mathbb{R}} |f'|^2 \, ds + \|\partial_\tau \psi_1\|^2_{L^2(\omega)} \int_{\mathbb{R}} \left(\frac{s}{\lambda}\right)^{2\beta} \dot\theta^2(s/\lambda) \, \frac{f^2}{s^{2\beta}} \, ds \\ \geq \tilde c \int_{\mathbb{R}} \frac{f^2}{(\lambda^2+s^2)^\beta} \, ds. \end{aligned} \tag{27}$$

Note that by the monotone converge theorem

$$\lim_{\lambda \to 0} \int_{\mathbb{R}} \frac{f^2}{(\lambda^2+s^2)^\beta} \, ds = \int_{\mathbb{R}} \frac{f^2}{s^{2\beta}} \, ds > 0. \tag{28}$$

Assume now, for example, that

$$\limsup_{x_3 \to \infty} |\dot\theta(x_3)| \, x_3^\beta = 0,$$

and take $f = 0$ on $(-\infty, 1]$. Then the left hand side of (27) converges to zero as $\lambda \to 0$ in view of the dominated convergence theorem and the fact that $\beta < 1$.

This is in contradiction with (28). To prove that $\limsup_{x_3\to-\infty} |\dot\theta(x_3)|\,|x_3|^\beta > 0$ we choose $f = 0$ on $[-1, \infty)$ and follow the same reasoning. □

Remark 3.1. Inequality (23) in the case $\beta = 0$ and $\dot\theta$ =constant, which corresponds to periodically twisted waveguides, was proved in [7, Thm. 1]. Note also that under a stronger regularity assumption on ω, namely $\omega \in C^2$, we have $\|\partial_\tau \psi_1\|_{L^2(\omega)} > 0$, see [2, Prop. 3.4]. Proposition 3.1 says that if (23) should hold with $\beta < 1$, then $|\dot\theta|$ must not decay faster than $x_3^{-\beta}$ at infinity. Sufficiency of the condition (24) for $0 < \beta < 1$ remains an open question.

Acknowledgements

The supported by the MIUR-PRIN 2010-11 grant for the project "Calcolo delle Variazioni" is gratefully acknowledged.

Bibliography

[1] A. Ancona : On strong barriers and inequality of Hardy for domains in $\mathbb{R}^n$. *J. London Math. Soc.* **34** (1986) 274–290.

[2] Ph. Briet, H. Kovařík, G. Raikov and E. Soccorsi: Eigenvalue asymptotics in a twisted waveguide. *Comm. Partial Differential Equations* **34** (2009) 818–836.

[3] Ph. Briet, H. Kovařík, G. Raikov: Scattering in twisted waveguides. *J. Funct. Anal.* **266** (2014) 1–35.

[4] G. Bouchittè, M. L. Mascarenhas, and L. Trabucho: On the curvature and torsion effects in one dimensional waveguides. *Control, Optimisation and Calculus of Variations* **13** (2007) 793–808.

[5] T. Ekholm and H. Kovařík: Stability of the magnetic Schrödinger operator in a waveguide. *Comm. Partial Differential Equations* **30** (2005) 539-565.

[6] T. Ekholm, H. Kovařík, D. Krejčiřík: A Hardy inequality in twisted waveguides. *Arch. Rational Mech. Anal.* **188** (2) (2008), 245–264.

[7] P. Exner, H. Kovařík: Spectrum of the Schrödinger operator in a perturbed periodically twisted tube, *Lett. Math. Phys.* **73** (2005), 183-192.

[8] G. Grillo, H. Kovařík and Y. Pinchover: Sharp two-sided heat kernel estimates of twisted tubes and applications. *Arch. Rational Mech. Anal.* **213** (2014) 215-243.

[9] V.V. Grushin: On the eigenvalues of finitely perturbed Laplace operators in infinite cylindrical domains, *Math. Notes* **75** (2004), no. 3, 331-340; translation from *Mat. Zametki* **75** (2004), 360-371.

[10] V.V. Grushin: Asymptotic behavior of the eigenvalues of the Schrödinger operator with transversal potential in a weakly curved infinite cylinder,

Math. Notes **77** (2005), no. 5, 606-613; translation from *Mat. Zametki* **77** (2005) 656-664.

[11] H. Kovařík, D. Krejčiřík: A Hardy inequality in a twisted Dirichlet-Neumann waveguide, *Math. Nachr.* **281** (2008) 1159-1168.

[12] D. Krejčiřík: Twisting versus bending in quantum waveguides. arXiv:0712.3371.

[13] D. Krejčiřík, E. Zuazua: The Hardy inequality and the heat equation in twisted tubes. *J. Math. Pures Appl.* (9) **94** (2010) 277–303.

[14] Krejčiřík, D., Zuazua, E.: The asymptotic behaviour of the heat equation in a twisted Dirichlet-Neumann waveguide. J. Diff. Eq. **250** 2334–2346 (2011)

[15] M. Markus, V.J. Mizel, and Y. Pinchover: On the best constant for Hardy's inequality in $\mathbb{R}^n$, *Trans. Amer. Math. Soc.* **350** (1998), 3237–3255.

[16] T. Matskewich and P. E. Sobolevskii: The best possible constant in a generalized Hardy's inequality for convex domains in $\mathbb{R}^n$, *Nonlinear Analysis TMA* **28** (1997), 1601–1610.

[17] B. Opic and A. Kufner, *Hardy inequalities*, Pitman Research Notes in Mathematics **219**. Longman Scientific & Technical, Harlow, 1990.

Microscopic foundations of Ohm and Joule's laws – The relevance of thermodynamics

J.-B. Bru* and W. de Siqueira Pedra†

*Universidad del País Vasco, Apartado 644, 48080 Bilbao, Spain
*BCAM - Basque Center for Applied Mathematics, Mazarredo, 14. 48009 Bilbao, Spain
*IKERBASQUE, Basque Foundation for Science, 48011, Bilbao, Spain
†Departamento de Física Matemática, Instituto de Física - Universidade de São Paulo, 05314-970 Caixa Postal 66318, Brazil

We give a brief historical account on microscopic explanations of electrical conduction. One aim of this short review is to show that Thermodynamics is fundamental to the theoretical understanding of the phenomenon. We discuss how the 2nd law, implemented in the scope of Quantum Statistical Mechanics, can be naturally used to give mathematical sense to conductivity of very general quantum many–body models. This is reminiscent of original ideas of J.P. Joule. We start with Ohm and Joule's discoveries and proceed by describing the Drude model of conductivity. The impact of Quantum Mechanics and the Anderson model are also discussed. The exposition is closed with the presentation of our approach to electrical conductivity based on the 2nd law of Thermodynamics as passivity of systems at thermal equilibrium. It led to new rigorous results on linear conductivity of interacting fermions. One example is the existence of so–called AC–conductivity measures for such a physical system. These measures are, moreover, Fourier transforms of time correlations of current fluctuations in the system. I.e., the conductivity satisfies, for a large class of quantum mechanical microscopic models, Green–Kubo relations.

Keywords: Ohm's law, Joule's law, conductivity measure, 2nd law, fermions.

1. Electrical Conductivity and Classical Physics

1.1. *The genesis of Ohm and Joule's laws*

G.S. Ohm was born in 1789 in Erlangen and came from a modest background (son of a master locksmith). Nevertheless, he succeeded in learning basic mathematics and sciences and became for about a decade teacher of mathematics and physics in Cologne. During this time, he had been able to elaborate his own experiments on electrical resistivity. He was originally inspired by J. Fourier's work, published in 1822, about heat theory. Indeed, G.S. Ohm drew a comparison of heat conduction with the electrical one in a metallic bar. Based on this intuition, he published a few papers on experimental outcomes about electrical resistivity in metals. He concluded his work on electrical conduction by his famous theory [1], which was a theoretical deduction of his law from "*first principles*". Indeed, he states that the current in the steady regime is proportional to the voltage applied to the conducting material. The proportionality coefficient is the conductivity (or inverse resistivity) of the physical system. It is an empirical law which looks almost obvious nowadays.

At that time, however, his writings were almost unknown. His book [1] was at best completely ignored and at worst treated really negatively. Rather than scientific, some critics were more ethical as they were based on a priori conceptions on what science and nature are, probably on what L. Daston and P. Galison have called *truth-to-nature* [2]. Quoting [3, p. 243]:

...Ohm's theory, to quote one critic, was "a web of naked fancies", which could never find the semblance of support from even the most superficial observation of facts; "he who looks on the world", proceeds the writer, "with the eye of reverence must turn aside from this book as the result of an incurable delusion, whose sole effort is to detract from the dignity of nature". ... where he had looked for commendation he found at best complete indifference, and at worst open abuse and derision. ... The influence of this opposition (some school official) *reached the Minister of Education himself, and he, speaking officially, definitely pronounced it as his opinion that "a physicist who professed such heresies was unworthy to teach science".*

Retrospectively, such comments in "a country so well represented in the world of science by men of eminence and knowledge" [3] are marks of revolutionary ideas, but it was a real bitter blow for G.S. Ohm: He gave up his teacher position at Cologne and started six years of hard times. His work was nevertheless occasionally cited and rumors about Ohm's theory started to appear in different places. This includes America where the famous physicist J. Henry asked in 1833 his colleagues: "Could you give me any information about the theory of Ohm? Where is it to be found?" J. Henry succeeded in having this information by going to England in 1837 at a time when Ohm's work had already become famous, particularly outside his own country.

Although at the origin of Ohm's intuition, the relation between heat and electrical conduction has not been established by himself, but J.P. Joule, who was born in 1818 in England. The pivotal ingredient was the wide concept of energy. Joule's intuition seems to have been that the different physical properties appearing in nature can be tracked by the concept of energy. He thus studied different forms of energy in order to relate them. The conversion of mechanical work into heat is a famous topic of such studies. His works, although also very controversial at the beginning, were seminal and yielded the *1st law of Thermodynamics*, see, e.g., [4]. Recall also that all mechanical work can be converted to heat but the converse is not true, in general. This observation refers to the *2nd law of Thermodynamics* and the concept of *entropy* invented by R.J.E. Clausius in 1865.

Applied to electricity theory, Joule's intuition allowed to establish a relation between heat and electrical conduction. Indeed, more than one decade after Ohm's discovery [1] on linear electrical conduction, the physicist J. P. Joule observed [5] in 1840 that the heat (per second) produced within an electrical circuit is proportional to the electrical resistance and the square of the current:

...the calorific effects of equal quantities of transmitted electricity are proportional to the resistances opposed to its passage, whatever may be the length, thickness, shape, or kind of metal which closes the circuit: and also that, coeteris paribus, these effects are in the duplicate ratio of the quantities of transmitted electricity; and consequently also in the duplicate ratio of the velocity of transmission.

[Joule, 1840]

Nowadays, electrical conductivity usually refers to Ohm and Joule's laws. They are indeed among the most resilient laws of (classical) electricity theory. Materials are called ohmic or nonohmic, depending on whether they obey Ohm's law. Both assertions are empirical laws and, as usual, they generated at least as many theoretical problems as they solved. From a mathematically rigorous point of view, the microscopic origin of the phenomenological behavior of macroscopic conductors described by these laws is still not completely understood, specially in the DC regime. Moreover, as recent experiments show, Ohm's law is not only valid at macroscopic scales. Indeed, the validity of Ohm's law at the atomic scale for a purely quantum system has experimentally been verified [6] in 2012. Such a behavior was unexpected [7]:

...In the 1920s and 1930s, it was expected that classical behavior would operate at macroscopic scales but would break down at the microscopic scale, where it would be replaced by the new quantum mechanics. The pointlike electron motion of the classical world would be replaced by the spread out quantum waves. These quantum waves would lead to very different behavior. ... Ohm's law remains valid, even at very low temperatures, a surprising result that reveals classical behavior in the quantum regime.

[D.K. Ferry, 2012]

1.2. *Towards a microscopic theory of electrical conduction*

In the end of the nineteenth century, the so–called classical physics reached a very high level of completeness with Classical Mechanics, Electrodynamics, and Thermodynamics. However, borderline problems became manifestly more and more important and eventually led to the scientific revolution of the beginning of the twentieth century. For instance, the study of the link between Classical Mechanics and Thermodynamics yielded the so–called Statistical Physics via Gibbs and Boltzmann's genius intuitions.

Classical Mechanics is indeed a causal theory based on elementary physical objects satisfying Newton's laws of motion. The exceptional success of this theory together with new technologies like photography had propagated a new view point on science during the last part of the nineteenth century with the so–called *mechanical objectivity* [2]. By contrast, Thermodynamics emphasizes the concepts of energy and entropy of macroscopic systems. It speaks about reversible and irreversible processes, but it does not care about the concrete system under consideration. Classical Mechanics is in some sense a bottom–up or "local" approach, whereas Thermodynamics is a top–down or global one.

In order to bridge the gap between both theories, L. Boltzmann successfully tried to go from Classical Mechanics towards Thermodynamics via statistical or probabilist methods. For more details on Boltzmann's legacy, see [8]. His H–theorem (1872) was undoubtedly an important achievement [9] as it has provided a mechanical explanation of the 2nd law of Thermodynamics from the dynamics of rarefied gases. Boltzmann's vision of "atoms" as the physical objects satisfying Newton's laws was however again very controversial for a long time [10]: "have you seen any?" might have said the famous physicist and philosopher E. Mach as a reply to the issue of atoms (cf. [10, p. 20]). E. Mach had indeed a philosophical approach centered on the world of sensations in a similar spirit of mechanical objectivity, whereas L. Boltzmann also focussed on mathematical structures. See later the development of *structural objectivity* [2] (M. Planck (1906), B. Russell, H. Poincaré, C.S. Peirce, etc.). Similar ethical oppositions appeared in other sciences: S. Ramón y Cajal and C. Goldi were together Nobel laureates in 1906, but C. Goldi violently opposed S. Ramón y Cajal's theory of neurons (similar to Boltzmann's theory of atoms) to explain the global system which is the brain. For more details, see [11]. As explained in [2], the opposite conceptions of science were in this case truth–to–nature (Goldi) and mechanical objectivity (Ramón y Cajal) as well as continuous (Goldi) versus discontinuous (Ramón y Cajal) visions.

In the same spirit as Boltzmann, it was natural to rise the question of the microscopic origin of Ohm and Joule's laws. In 1846, W. Weber conjectured that currents were a flow of charged fluids and in 1881, H. von Helmholtz argued the existence of positive and negative charges as "atoms of electricity". The discovery of the electron took place in the last years of the nineteenth century by J.J. Thomson (Nobel Prize in Physics 1906) and others. It is the first discovered elementary particle. Based on the vision that current is a flow of electrons, the celebrated Drude model was next proposed [12] in 1900 to give a mechanical explanation of the phenomenon of conductivity. This model or its extension, the Drude–Lorentz model (1905), are still used as microscopic explanations of conductivity in textbooks. Indeed, although the motion of electrons and ions is treated classically and the interaction between these two species is modeled by perfectly elastic random collisions, this quite elementary model provides a qualitatively good description of DC– and AC–conductivities in metals.

2. Electrical Conductivity and Quantum Mechanics

2.1. *Emergence of Quantum Mechanics*

The main principles of physics were considered as well–founded by the end of the nineteenth century, even with, for instance, no satisfactory explanation of the phenomenon of thermal radiation, first discovered in 1860 by G. Kirchhoff. In contrast to classical physics, which deals with continuous quantities, Planck's intuition was to introduce an intrinsic discontinuity of energy to explain thermal radiation. Assuming the existence of a quantum of action h, the celebrated Planck's constant, he derived the well–known Planck's law of thermal radiation. Inspired by Planck's ideas, Einstein presented his famous discrete (corpuscular) theory of light to explain the photoelectric effect.

Emission spectra of chemical elements had also been known since the nineteenth century and no theoretical explanation was available at that time. It became clear that electrons play a key role in this phenomenon. However, the classical solar system model of the atom failed to explain the emitted or absorbed radiation. Following again Planck's ideas, N. Bohr proposed in 1913 an atomic model based on discrete energies that characterize electron orbits. It became clear that the main principles of classical physics are unable to describe atomic physics.

Planck's quantum of action, Einstein's quanta of light (photons), and Bohr's atomic model could not be a simple extension of classical physics, which, in turn, could also not be questioned in its field of validity. N. Bohr tried during almost a decade to conciliate the paradoxical looking microscopic phenomena by defining a radically different kind of logic. Bohr's concept of complementarity gave in 1928 a conceptual solution to that problem and revolutionized the usual vision of nature. See, e.g., [13]. Classical logic should be replaced by quantum logic as claimed [14] by G. Birkhoff and J. von Neumann in 1936.

On the level of theoretical physics, until 1925, quantum corrections were systematically included, in a rather *ad hoc* manner, into classical theories to allow explicit discontinuous properties. Then, as explained for instance in [15], two apparently complementary directions were taken by W.K. Heisenberg and E. Schrödinger, respectively, to establish basic principles of the new quantum physics, in contrast with the "old quantum theory" starting in 1900. Indeed, even with the so–called correspondence principle of N. Bohr, "many problems, even quite central ones like the spectrum of helium atom, proved inaccessible to any solution, no matter how elaborate the conversion", see [15, p. 18].

2.2. *Quantum Fermi liquids*

Electric current is carried by electrons, purely quantum objects (W.E. Pauli, 1925; E. Fermi, 1925; P.A.M. Dirac, 1929), whereas the Drude model describes *non–interacting classical* particles interacting with impurities via perfectly elastic collisions. Quantum Mechanics, which governs the microscopic world, represents a radical transformation of usual principles of classical physics and it is therefore not at all satisfactory to see the Drude (or the Drude–Lorentz) model as a proper microscopic explanation of conductivity, even with good agreements with experimental data. As one can see from the existence of superconducting phases first discovered in 1911, electrons can show collective behaviors while satisfying the celebrated Pauli exclusion principle.

A. Sommerfeld and H. Bethe modified in 1933 the Drude model to take into account quantum effects. Essentially, they replaced the classical point–like particles of Drude, carriers of electrical current, with fermions. In particular, the carriers present quantum coherences and obey the Fermi–Dirac statistics. However, the Drude–Sommerfeld model describes a system of non–interacting fermions although electrons strongly interact with each other via the Coulomb repulsion. A formal explanation of the success of this model is given [16] by L.D. Landau in the fifties. His theory is based on the concept of *Landau Fermi liquids* (or Fermi liquids).

Landau's idea is, in a caricatured view, that the low–energy exited states of a Fermi system with interparticle interactions can be mapped onto the states of an effective non–interacting (or ideal) Fermi system. The theoretical justification of such a behavior, i.e., the fact that the electron–electron scattering remains negligible to change the momentum distribution, results from the Pauli exclusion principle for energies near the Fermi level. More precisely, if the system is at initial time in a state closed to an ideal system (weakly excited), then its time–dependent state can be uniquely described by occupation numbers of *quasiparticles* (as approximated quantum numbers). Moreover, L.D. Landau postulates the existence of a function $\mathrm{f}_{k,k'}$, the so–called Landau interaction function, which quantifies the energy change of a quasiparticle of quasimomentum k in the presence of a quasiparticle of quasimomentum k'. The effective mass, another parameter of Fermi liquids, determines the dispersion relation of quasiparticles, i.e., the energy of quasiparticles as a function of their quasimomenta. This effective (or phenomenological) theory has been very successful in explaining the physics of some electron systems, called Fermi liquids. Fermion systems are called non–Fermi liquids if their behavior does not correspond to Landau's predictions. Non–Fermi liquid behaviors usually appear in low dimensions. For instance, in one dimension, the celebrated Luttinger liquid replaces the (Landau) Fermi liquid. For more details, see [17].

2.3. *From theoretical physics to mathematics: The Anderson model*

Resistivity of metals is believed to be due to interparticle interactions but also to inhomogeneities of the conducting crystal. Disordered electron liquids are therefore an important issue in this context. The theory of Fermi liquids can be extended to disordered systems but major differences appear as compared to the (space) homogeneous systems. New properties like the so–called *Anderson localization* are consequence of strong space inhomogeneities, even in the absence of interparticle interactions.

Anderson localization corresponds to the absence of electron transport at strong disorder and has been predicted [18] by the physicist P.W. Anderson in 1958. This allows to guess a metal–insulator transition in three dimensions. This theory has experimentally been investigated and P.W. Anderson, together with N.F. Mott and J.H. van Vleck, won the 1977 Nobel price in physics for "their fundamental theoretical investigations of the electronic structure of magnetic and disordered systems".

The Anderson model corresponds to a single quantum particle within a random potential. It is one of the most important types of random Schrödinger operators, which constitute nowadays an advanced and relatively mature branch of mathematics. In fact, random Schrödinger operators start to be studied in the seventies. The Anderson localization for an one–dimensional model was first proved by I. Goldsheid, S. Molchanov and L. Pastur in 1977, while a similar result for the one–dimensional Anderson model was obtained in 1981 by H. Kunz and B. Souillard. It is known that, in general, the one–dimensional Anderson model only has purely point spectrum with a complete set of localized eigenstates (Anderson localization) and it is thus believed that no steady current can exist in this case. For more detailed introduction to the Anderson model and more general random Schrödinger operators, see for instance the lecture notes of [19].

Nevertheless, mathematical studies usually focus on the existence of (dynamical or spectral) Anderson localizations and, even in absence of interactions, there are only few mathematical results on transport properties of such random models that yield Ohm's law in some form.

In 2007, A. Klein, O. Lenoble and P. Müller introduced [20] for the first time the concept of a "conductivity measure" for a system of non–interacting fermions subjected to a random potential. More precisely, the authors considered the Anderson tight–binding model in presence of a time–dependent spatially homogeneous electric field that is adiabatically switched on. The fermionic nature of charge carriers – electrons or holes in crystals – as well as thermodynamics of such systems were implemented by choosing the Fermi–Dirac distribution as the initial density matrix of particles. In [20] only systems at zero temperature with Fermi energy lying in the localization regime are considered, but it is shown in [21] that a conductivity measure can also be defined without the localization assumption and at any positive temperature. Their study can thus be seen as a mathematical derivation of Ohm's law for space–homogeneous electric fields having a specific time behavior.

[22] is another mathematical result on free fermions proving Ohm's law for graphene–like materials subjected to space–homogeneous and time–periodic electric fields. Joule's law and heat production are not considered, at least not explicitly, in these mathematical studies.

3. Electrical Conductivity and 2nd Law of Thermodynamics

Via [20, 21] one sees that measures (instead of functions or other types of distributions) are the natural mathematical objects to be used to describe conductivity starting form microscopic quantum dynamics. We claim that this is so because of the 2nd law of Thermodynamics. Indeed, such a principle guarantees the positivity of certain quadratic forms on external electric fields, which naturally appears by considering linear response. By Bochner's theorem, in a convenient form, such quadratic forms define measures. In the case of current response to external electric fields, one gets AC–conductivity measures. This approach permits to tackle the mathematical problem of a rigorous microscopic description of the phenomenon of linear conductivity starting from first principles. Moreover, it is general enough to be applied to interacting systems. We implement the 2nd law of Thermodynamics in the scope of algebraic Quantum Mechanics, by using the remarkable results [23] of W. Pusz and S. L. Woronowicz: We consider the 2nd law as a first principle of Physics which supplements Quantum Mechanics in the sense that it singles out special states of the considered systems. Indeed, states of infinite systems that are compatible with the 2nd law exist for a huge class of dynamics.

In fact, the 2nd law is "*one of the most perfect laws in physics*" [24, Section 1] and it has never been faulted by reproducible experiments. Its history starts with works of S. Carnot in 1824. Different popular formulations of the same principle have been stated by R.J.E. Clausius, W. Thomson or Lord Kelvin (and M. Planck), and C. Carathéodory. Our study is based on Kelvin–Planck statement while avoiding the concept of "cooling" [24, p. 49]:

No process is possible, the sole result of which is a change in the energy of a simple system (without changing the work coordinates) and the raising of a weight.

Using this formulation of the 2nd law, we define the concept of *thermal equilibrium* states by using algebraic Quantum Mechanics as mathematical framework. It is a well–known approach – originally due to J. von Neumann (cf. von Neumann algebras, C^*–algebras) – that extends the Hilbert space formulation of Quantum Mechanics. One important result of the theory of C^*–algebras, obtained in the forties, is the celebrated GNS (Gel'fand–Naimark–Segal) representation of states, which permits a natural relation between the new algebraic formulation and the usual Hilbert space based formulation of Quantum Mechanics to be established. Indeed, I.E. Segal proposed to leave the Hilbert space approach to consider quantum observables as elements of certain involutive Banach algebras, now known as C^*–algebra. The GNS representation has also led to very important applications of the Tomita–Takesaki theory, developed in seventies, to Quantum Field Theory and Statistical Mechanics. These developments mark the beginning of the algebraic approach to Quantum Mechanics and Quantum Field Theory. For more details, see, e.g., [25].

The algebraic formulation turned out to be extremely important and fruitful for the mathematical foundations of Quantum Statistical Mechanics and have been an important branch of research during decades with lots of works on quantum spin and Fermi systems. See, e.g., [26, 27] (spin) and [28–30] (Fermi). Basically, it uses some C^*–algebra $\mathcal{X}$, the self–adjoint elements of which are the so–called observables of the physical system. States on the C^*–algebra $\mathcal{X}$ are, by definition, continuous linear functionals $\rho \in \mathcal{X}^*$ which are normalized and positive, i.e., $\rho(\mathbf{1}) = 1$ and $\rho(A^*A) \geq 0$ for all $A \in \mathcal{X}$. They represent the state of the physical system.

To conveniently define equilibrium states in our case, [23] is pivotal because it gives a definition of equilibrium by using the Kelvin–Planck statement via the notion of *passive* states: The internal dynamics of the system is a strongly continuous one–parameter group $\tau \equiv \{\tau_t\}_{t\in\mathbb{R}}$ of $*$–automorphisms of $\mathcal{X}$ with (generally unbounded) generator δ. Usually, δ is a dissipative and closed derivation of $\mathcal{X}$. On this system, one applies a *cyclic* process of length $T \geq 0$, that is, a continuously differentiable family $\{A_t\}_{t\in\mathbb{R}} \subset \mathcal{X}$ of self–adjoint elements of $\mathcal{X}$ such that $A_t = 0$ for all $t \leq s$ and $t \geq T+s$. The perturbed dynamics is the solution $\{\tau_{t,s}\}_{t\geq s}$ of the non–autonomous evolution equation defined, for any $B \in \mathrm{Dom}(\delta)$, by

$$\forall s,t \in \mathbb{R},\ t \geq s: \quad \partial_t \tau_{t,s}(B) = \tau_{t,s}\left(\delta(B) + i\left[A_t, B\right]\right), \quad \tau_{s,s}(B) := B \ .$$

The state of the system evolves as $\rho_t = \rho \circ \tau_{t,s}$ for any $t \geq s$ at fixed initial state $\rho \in \mathcal{X}^*$. Then, as explained in [23, p. 276], a state $\rho \in \mathcal{X}^*$ is *passive* iff the full work performed by the external device is non–negative for all cyclic process $\{A_t\}_{t\geq s} \subset \mathcal{X}$ of any time–length $T \geq 0$, i.e.,

$$L_\rho^A := \int_s^T \rho \circ \tau_{t,s}\left(\partial_t A_t\right) \mathrm{d}t \geq 0 \ . \tag{1}$$

In this way the Kelvin–Planck statement of the 2nd law can be formulated in precise mathematical terms. If the product state $\otimes_{j=1}^{n}\rho$ is passive for any $n \in \mathbb{N}$ copies $(\mathcal{X}_1, \tau_1, \rho_1), \ldots, (\mathcal{X}_n, \tau_n, \rho_n)$ of the original system defined by $(\mathcal{X}, \tau, \rho)$, then ρ is called *completly passive* [23]. Such states are the *thermal equilibrium states* of our setting. [23, Theorem 1.4] shows that thermal equilibrium states in this sense are exactly the *KMS* [26] (Kubo–Martin–Schwinger) states of the corresponding C^*–dynamical system.

In our approach to electrical conduction, the C^*–algebra $\mathcal{X}$ is the CAR algebra associated to the d–dimensional cubic lattice $\mathfrak{L} := \mathbb{Z}^d$ $(d \in \mathbb{N})$ and particles of finite spin. The initial state is a thermal equilibrium state and cyclic processes are induced by electromagnetic potentials $\{\eta A_t\}_{t \geq s}$ with constant strength $\eta \geq 0$ within some finite region Λ of the lattice $\mathfrak{L}$. This yields to perturbed dynamics with discrete magnetic Laplacians in disordered media (like in the Anderson model). The quadratic response with respect to η of the full heat production or electromagnetic work per unit volume turns out to equal $Q_\rho^A = \varphi(A * \tilde{A})$, where φ is a distribution, $\tilde{A}(t) := A(-t)$ and $A \in C_0^\infty(\mathbb{R}, \mathbb{R})$. Indeed, we have shown [32, 36] that $|\Lambda|^{-1} L_\rho^{\eta A} - \eta^2 Q_\rho^A$ is of order $\mathcal{O}(\eta^3)$, uniformly w.r.t. A and the size $|\Lambda|$ of the region Λ where the electromagnetic field is applied. The 2nd law, that is, (1), implies then that $Q_\rho^A = \varphi(A * \tilde{A}) \geq 0$, i.e., φ is a *distribution of positive type*. By the Bochner–Schwarz theorem, there is a positive measure $\tilde{\mu}$ on $\mathbb{R}$ such that

$$Q_\rho^A = \int_{\mathbb{R}} \mathrm{d}\tilde{\mu}(\nu) |\hat{A}(\nu)|^2 = \int_{\mathbb{R}\backslash\{0\}} \mathrm{d}\tilde{\mu}(\nu) \nu^{-2} |\hat{E}(\nu)|^2$$

for all $A \in C_0^\infty(\mathbb{R}, \mathbb{R})$. Here, $E = -\partial_t A$ is the electric field in the Weyl gauge and $\hat{A}, \hat{E}$ are the Fourier transforms of A, E with support outside some neighborhood of $\nu = 0$.

The measure $\mathrm{d}\mu(\nu) := \nu^{-2} \mathrm{d}\tilde{\mu}(\nu)$ on $\mathbb{R}\backslash\{0\}$ turns out to be the AC–conductivity measure we are looking for and the quantity $\mathrm{d}\mu(\nu)|\hat{E}(\nu)|^2$ is the heat production due to the component $\hat{E}(\nu)$ of frequency ν of the electric field E, in accordance with Joule's law. It is directly related to the passivity property of thermal equilibrium states on the CAR algebra. In other words, the existence of AC–conductivity measures results from the 2nd law of Thermodynamics in connection with the full quantum microscopic dynamics of the considered system. Moreover, the approach to linear (or quadratic in the energetic view point) response we propose has also the following technical and conceptual advantages, even in the non–interacting case:

- The conductivity measure naturally appears as the Fourier transform of current–current time correlations, that is, four–point correlation functions, in this framework. This means that Green–Kubo relations are generally valid, from first principles.

- The algebraic formulation allows a clear link between macroscopic transport properties of fermion systems and the CCR algebra of current fluctuations associated to that systems. The latter is related to non–commutative central limit theorems.

- Moreover, this approach can be naturally used to define and analyze conductivity measures for *interacting* fermions as well.

In [31, 32] we study free lattice fermions subjected to a static bounded potential and a time– and space–dependent electromagnetic field. [31, 32] establish a form of Ohm and Joule's laws valid at microscopic scales, uniformly with respect to the size of the region on which the electric field is applied. It is in accordance with the validity of Ohm's law in the quantum world, i.e., at microscopic scales, see [6]. [33, 34] extend the results of [31, 32] to macroscopic scales and are reminiscent of [20, 21]. For more details, see the discussions in [33]. Part of the phenomenology of the Drude model can be derived from our more detailed study [34] of macroscopic conductivity measures of free–fermions in disordered media.

Therefore, [31–34] give a complete, mathematically rigorous, microscopic derivation of the phenomenon of linear conductivity from first principles of Thermodynamics and Quantum Mechanics, only. These studies are restricted to non–interacting fermion systems. However, it is believed in theoretical physics that electric resistance of conductors should also result from the interactions between charge carriers, and not only from the presence of inhomogeneities (impurities). In [35–37], we succeed to extend our previous results to fermion systems with interactions.

To conclude, we think that such an approach can be useful in other contexts since it gives appropriate tools to tackle mathematically what is known in Physics as *excitation spectrum*. Indeed, the concept of excitation spectrum is usually associated with the spectrum of a self-adjoint operator describing the energy of the system. In condensed mater physics, this notion mainly comes from superfluid helium 4, a quantum boson liquid which can be described by the spectrum of collective excitations via the celebrated Bogoliubov theory [38]. However, there is a plethora [39, 40] of other types of elementary excitations not covered by the Bogoliubov theory. We show [32, 33, 36] that the notion of conductivity measure we defined is nothing else than a spectral measure of the generator of dynamics in an appropriate representation.

Bibliography

[1] G.S. Ohm, *Die galvanische Kette, mathematisch bearbeitet* (Riemann, Berlin, 1827).

[2] L.J. Daston and P. L. Galison, *Objectivity* (Zone Books, 2007).

[3] I.B. Hart, *Makers of Science* (Oxford University Press, London, 1923).

[4] G. Emch, C. Liu, *The Logic of Thermostatistical Physics* (Springer-Verlag, New York, 2002).

[5] J. P. Joule, *Philosophical Transactions of the Royal Society of London* **4**, 280 (1840).
[6] B. Weber et al., *Science* **335**(6064), 64 (2012).
[7] D.K. Ferry, *Science* **335**(6064), 45 (2012).
[8] G. Gallavotti et al., *Boltzmann's Legacy* (ESI Lectures in Math. & Phys., EMS, 2007)
[9] C. Villani, *H-theorem and beyond: Boltzmann's entropy in today's mathematics.* In ESI Lectures in Math. & Phys., EMS, Ed. G. Gallavotti et al., 2007, p. 129–144.
[10] J. Renn, *Boltzmann and the end of the mechanistic worldview.* In ESI Lectures in Math. & Phys., EMS, Ed. G. Gallavotti et al., 2007, p. 7-26.
[11] J.A. De Carlos and J. Borrell, *Brain Research Reviews* **55**, 8 (2007).
[12] P. Drude, Annalen der Physik **306** (3), 566 (1900); ibid. **308**(11), 369 (1900).
[13] N. Bohr, *Physique atomique et connaissance humaine*, translation by E. Bauer and R. Omnès, Ed. C. Chevalley (Paris, Editions Gallimard, 1991).
[14] G. Birkhoff and J. Von Neumann, *Annals of Mathematics* **37**(4), 823 (1936).
[15] J. Renn, *Schrödinger and the genesis of wave mechanics.* In ESI Lectures in Math. & Phys., EMS, Ed. W.L. Reiter, J. Yngvason, 2013, p 9–36.
[16] L.D. Landau, *Zh. Eksp. Teor. Fiz.* **30**, 1058 (1956) [*Sov. Phys. JETP* **3**, 920 (1957)]; *Zh. Eksp. Teor. Fiz.* **32**, 59 (1957) [*Sov. Phys. JETP* **5**, 101 (1957)]
[17] G.F. Giuliani and G. Vignale, *Quantum Theory of the Electron Liquid* (Cambrigde Univ. Press., 2005).
[18] P.W. Anderson, *Phys. Rev.* **109**(5), 1492 (1958).
[19] Werner Kirsch, *An invitation to random Schrödinger operators* (with an appendix by F. Klopp). In Panoramas et Synthèses, Société Mathématique de France **25**, 2008.
[20] A. Klein, O. Lenoble, and P. Müller, *Annals of Mathematics* **166**, 549 (2007).
[21] A. Klein and P. Müller, *J. Math. Phys., Analysis, Geometry* **4**, 128 (2008).
[22] M.H. Brynildsen, H.D. Cornean, *Preprint* arXiv:1112.2613 (2011).
[23] W. Pusz and S. L. Woronowicz, *Commun. math. Phys.* **58**, 273 (1978).
[24] E. H. Lieb and J. Yngvason, *Phys. Rep.* **310**, 1 (1999).
[25] G.G. Emch, *Algebraic Methods in Statistical Mechanics and Quantum Field Theory* (Willey–Interscience, New York, 1972).
[26] O. Bratteli and D.W. Robinson, *Operator Algebras and Quantum Statistical Mechanics, Vol. II*, 2nd ed. (Springer-Verlag, New York, 1996).
[27] R.B. Israel, *Convexity in the theory of lattice gases* (Princeton Univ. Press, 1979).
[28] H. Araki and H. Moriya, *Rev. Math. Phys.* **15**, 93 (2003).
[29] J.-B. Bru and W. de Siqueira Pedra, *Memoirs of the AMS* **224**(1052) (2013).
[30] J.-B. Bru and W. de Siqueira Pedra, *J. Math. Phys.* **53**, 123301 (2012).
[31] J.-B. Bru, W. de Siqueira Pedra and C. Kurig (Hertling), Comm. Pure Appl. Math. (2014), mp_arc 13-29.

[32] J.-B. Bru, W. de Siqueira Pedra and C. Kurig (Hertling), mp_arc 13-68 (2013).

[33] J.-B. Bru, W. de Siqueira Pedra and C. Kurig (Hertling), mp_arc 13-88 (2013).

[34] J.-B. Bru, W. de Siqueira Pedra and C. Hertling (Kurig), to appear in Rev. Math. Phys (2014), mp_arc 13-93.

[35] J.-B. Bru and W. de Siqueira Pedra, mp_arc 14-27 (2014).

[36] J.-B. Bru and W. de Siqueira Pedra, soon in mp_arc (2014).

[37] J.-B. Bru and W. de Siqueira Pedra, in preparation (2014).

[38] V.A. Zagrebnov and J.-B. Bru, *Phys. Rep.* **350**, 291 (2001).

[39] N.M. Blagoveshchenskii et al., *Phys. Rev. B* **50**, 16550 (1994).

[40] A. Griffin, *Excitations in a Bose-Condensated Liquid* (Cambridge Univ. Press, 1993).

The quantum marginal problem

C. Schilling

Clarendon Laboratory, University of Oxford,
Oxford OX1 3PU, United Kingdom (new)
Institute for Theoretical Physics, ETH Zurich,
8093 Zurich, Switzerland (old)
E-mail: christian.schilling@physics.ox.ac.uk

The question of whether given density operators for subsystems of a multipartite quantum system are compatible to one common total density operator is known as the quantum marginal problem. We briefly review the solution of a subclass of such problems found just recently. In particular, this provides the solution of the 1-body N-representability problem. Its solution, the so-called generalized Pauli constraints, restrict the set of mathematically possible fermionic occupation numbers significantly, and strengthens Pauli's exclusion principle. Moreover, we review the study of a concrete physical model of interacting fermions confined to a harmonic trap. There, we found occupation numbers close, but not exactly on the boundary of the allowed region. This new effect of quasipinning is physically relevant since it corresponds to a simplified structure of the corresponding N-fermion quantum state.

Keywords: Quantum marginal problem, generalized Pauli constraints, quasipinning.

1. Introduction

Since most quantum effects emerge from the interaction between two or more quantum systems as e.g. the interaction of a system with an environment or the interaction of macroscopically many electrons the concept of a multipartite quantum system is fundamental. We describe the quantum state of such a system $\mathcal{J}$ built up from subsystems $A, B, C, \ldots$ by a density operator $\rho_{\mathcal{J}}$ acting on the Hilbert space $\mathcal{H}_{\mathcal{J}} = \mathcal{H}_{\mathcal{A}} \otimes \mathcal{H}_{\mathcal{B}} \otimes \ldots$, the tensor product of the separable Hilbert spaces of the subsystems $A, B, \ldots$. To describe properties of some subsystem $\mathcal{I}$ of $\mathcal{J}$, as e.g. B or AC (containing system A and C), it suffices to deal with the corresponding reduced density operator (marginal) of system $\mathcal{I}$

$$\rho_{\mathcal{I}} = \mathrm{Tr}_{\mathcal{J}\setminus\mathcal{I}}[\rho_{\mathcal{J}}], \tag{1}$$

obtained by tracing out the complementary system $\mathcal{J}\setminus\mathcal{I}$ of $\mathcal{I}$. According to Eq. (1) it is clear that marginals arising from the same total quantum state $\rho_{\mathcal{J}}$ need to fulfil certain compatibility conditions. This gives rise to the quantum marginal problem (QMP)

Definition 1.1 (quantum marginal problem). *For a given family $\mathcal{K}$ of subsystems $\mathcal{I}$ of $\mathcal{J}$ the quantum marginal problem $\mathcal{M}_{\mathcal{K}}$ is the problem of determining and describing the set $\Sigma_{\mathcal{K}}$ of tuples $(\rho_{\mathcal{I}})_{\mathcal{I}\in\mathcal{K}}$ of compatible marginals. Compatible here means that there exists a density operator $\rho_{\mathcal{J}}$ for the total system such that $\forall \mathcal{I} \in \mathcal{K}$*

$$\rho_{\mathcal{I}} = \mathrm{Tr}_{\mathcal{J}\setminus\mathcal{I}}[\rho_{\mathcal{J}}]\,. \tag{2}$$

In this work, due to mathematical reasons, we assume $\mathcal{H}_{\mathcal{J}}$ to be finite dimensional. Moreover, we may think of $\mathcal{I} \subset \mathcal{J}$ as one particle, a few particles, a system of macroscopically many particles or as just the spin degree of freedom of a single electron. The QMP is also illustrated in Fig. 1. There, every physical system

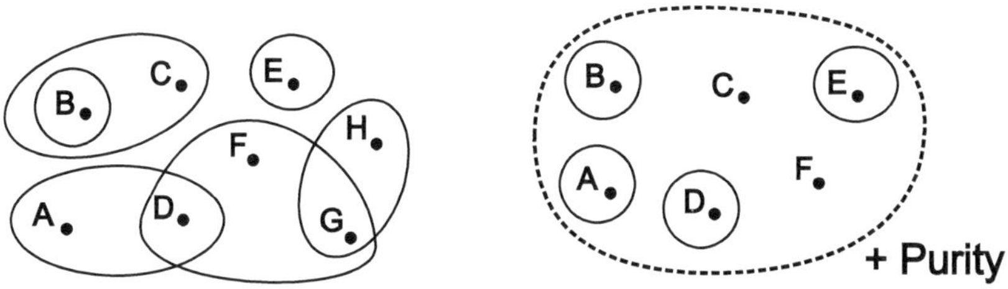

Figure 1. Illustration of the quantum marginal problem in its general form (left) and its pure univariant form (right). See also text.

$I = A, B, C, \ldots$ is symbolically described by a black dot. The family $\mathcal{K}$ of systems of interest is illustrated by 'black islands', where every island describes one subset $\mathcal{I} \in \mathcal{K}$. In addition, one may modify the QMP by exposing restrictions on the total state as e.g. purity or if all the subsystems are identical a fermionic or bosonic exchange symmetry.

One of the most important QMP is given by (see [1])

Definition 1.2 (r-body N-representability problem). *Given a system of N identical fermions, with corresponding Hilbert space $\mathcal{H}_N^{(f)} = \wedge^N[\mathcal{H}_1^{(d)}]$ of antisymmetric states, where the dimension d of the 1-particle Hilbert $\mathcal{H}_1^{(d)}$ may be infinite. For fixed $r \in \{1, 2, \ldots, N-1\}$ the problem of determining the set $D_r^{(p/e)}$ of possible r-reduced density operators (r-RDO) ρ_r arising via partial trace from a corresponding pure/ensemble N-fermion density operator ρ_N on $\mathcal{H}_N^{(f)}$ is the* r-body pure/ensemble-N-representability problem.

The paper is arranged as follows. In the next section we present the solution of the subclass of pure univariant QMP and give the reader an idea how to derive it. In Sec. 3 we focus on the 1-body pure N-representability problem and explain that the antisymmetry of the N-fermion state implies so-called generalized Pauli constraints, restrictions of fermionic occupation numbers stronger than Pauli's exclusion principle. Their role for ground states is studied in Sec. 4 by solving a model of few fermions confined by a harmonic trap. We find the effect of quasipinning. In the last section, Sec. 5, we briefly discuss the physical relevance of pinning and explain why this may lead to a generalized Hartree-Fock method.

2. Pure Univariant Quantum Marginal Problem and Solution

Solving the QMP in its general form is practically impossible. This was shown in [2] for the 2-body N-representability problem, by proving that it belongs to the quantum Mermin-Arthur complexity class (a generalization of the NP class). As a consequence one was tempted to focus on a subclass of the QMP, the so-called pure univariant QMP. There, the family $\mathcal{K}$ (recall Definition 1.1) is given by disjoint subsystems $\mathcal{I} \subset \mathcal{J}$ and the total state is required to be pure. This is illustrated on the right side of Fig. 1. There, all black islands are non-overlapping (univariant) and for each island we collect all its black dots by one. This restriction of non-overlapping 'black islands' leads to a significant simplification of the corresponding QMP:

Remark 2.1. Due to the unitary equivalence

$$(\rho_{\mathcal{I}})_{\mathcal{I}\in\mathcal{K}}\,\text{compatible} \Rightarrow (U_{\mathcal{I}}\rho_{\mathcal{I}}U_{\mathcal{I}}^{\dagger})_{\mathcal{I}\in\mathcal{K}}\,\text{compatible}\,, \quad \forall\,\text{unitaries}\,U_{\mathcal{I}}\,\text{on}\,\mathcal{H}_{\mathcal{I}} \tag{3}$$

the set $\Sigma_{\mathcal{K}}$ of compatible marginals for univariant QMP $\mathcal{M}_{\mathcal{K}}$ is described by conditions on their spectra $\vec{\lambda}_{\mathcal{I}}$, only.

In 2004, Klyachko [3] has solved the pure univariant QMP. Further important contributions came from Daftuar and Hayden [4] and Christandl and Mitchison [5]. We present the abstract solution and review briefly the main idea for deriving it.

Theorem 2.1. *Let $\mathcal{K}$ define a pure univariant QMP $\mathcal{M}_{\mathcal{K}}$ (recall Definition 1.1). According to Remark 2.1 the set $\Sigma_{\mathcal{K}}$ of compatible marginals is described by their decreasingly-ordered spectra $\vec{\lambda}_{\mathcal{I}}$. The set of possible spectra $\vec{\lambda} \equiv (\vec{\lambda}_{\mathcal{I}})_{\mathcal{I}\in\mathcal{K}}$ forms a polytope $\mathcal{P} \subset \mathbb{R}^d$. It highly depends on the dimensions of all local Hilbert spaces $\mathcal{H}_A$, $\mathcal{H}_B, \ldots$ and d is the corresponding dimension of those $\vec{\lambda}$-vectors.*

Since a polytope is nothing else but an intersection of finitely many Euclidean half spaces the set of possible vectors $\vec{\lambda}$ is described by a finite family of so-called marginal constraints, linear conditions on the eigenvalues $\vec{\lambda}$. Klyachko [3] provides an (quite abstract) algorithm for calculation those constraints.

2.1. *Derivation of marginal constraints*

In this section we give the reader an idea how to derive marginal constraints and follow quite closely [4]. For this it is instructive to study an elementary prototype of a QMP, the spectral version of $\mathcal{M}_{A,AB}$. It asks when spectra $\vec{\lambda}_A \in \mathbb{R}^{d_A}$, $\vec{\lambda}_{AB} \in \mathbb{R}^{d_A d_B}$ are compatible in the sense that there exists a total state ρ_{AB} with spectrum $\vec{\lambda}_{AB}$ such that its marginal ρ_A has spectrum $\vec{\lambda}_A$.

First, we need to introduce the concept of a flag induced by a hermitian operator, the complex Grassmanian, its Schubert cells and a variational principle due to Hersch and Zwahlen.

Definition 2.1. Let $\mathcal{H}$ be a d-dimensional complex Hilbert space. A *complete flag* $F_\bullet$ is a maximal sequence of nested linear subspaces, i.e.

$$F_\bullet := [0 = F_0 \lneq F_1 \lneq \ldots \lneq F_{d-1} \lneq F_d = \mathcal{H}] \tag{4}$$

In particular, complete flags can be induced by non-degenerate hermitian operators according

Definition 2.2. Given a hermitian operator A on a d-dimensional Hilbert space with non-degenerate spectrum $a = (a_1, \ldots, a_d)$ arranged in decreasing order. A then induces a complete flag $F_\bullet(A)$ according

$$F_i(A) = \langle v_1, \ldots, v_i \rangle, \quad \forall i = 0, 1, \ldots, d\,, \tag{5}$$

where v_j is the eigenvector corresponding to the eigenvalue a_j and $\langle \cdot \rangle$ denotes the span of vectors.

Definition 2.3. Let $\mathcal{H}$ be a d-dimensional Hilbert space and $F_\bullet$ a complete flag. Then for every binary sequence $\pi \in \{0,1\}^d$ we define the Grassmannian Schubert cell $S_\pi^\circ(F_\bullet)$ by

$$S_\pi^\circ(F_\bullet) := \{V \leq \mathcal{H} \,|\, \forall i = 1, \ldots, d : \dim((V \cap F_i)/(V \cap F_{i-1})) = \pi_i\} \tag{6}$$

These cells are subsets of the Grassmannian $\mathrm{Gr}_{\|\pi\|_1,d}$, which are defined as

$$\mathrm{Gr}_{\|\pi\|_1,d} := \{V \leq \mathcal{H} \,|\, \dim(V) = \|\pi\|_1\} \tag{7}$$

and $\|\pi\|_1 \equiv \sum_{k=1}^d \pi_k$.

Remark 2.2. The binary sequence π defines the indices at which the components (vector spaces) of the sequence $V \cap F_0 \leq V \cap F_1 \leq \ldots \leq V \cap F_d$ increase their dimension. The label $^\circ$ indicates that the Schubert cells are open w.r.t. the natural topology. The closures of these Schubert cells are called Schubert varieties. Moreover, it is well known (see e.g. [6]) that $\mathrm{Gr}_{k,d}$ is a projective algebraic variety and the Schubert varieties form subvarieties.

Now, we can express sums of *arbitrary* eigenvalues of a hermitian operator by a variational principle:

Lemma 2.1 (Hersch-Zwahlen). *Let ρ be a hermitian operator with non-degenerate spectrum $\vec{\lambda}$ arranged in decreasing order, $\pi \in \{0,1\}^d$ a binary sequence of length d. Then*

$$\sum_{j=1}^{d} \pi_j \lambda_j = \min_{V \in S_\pi^\circ(\rho)} (\mathrm{Tr}[P_V \rho]) \,, \tag{8}$$

where P_V is the orthogonal projection operator onto the subspace V.

The proof is elementary and can e.g. be found in [4, 6]. Lemma 2.1 can be used to derive necessary conditions on compatible spectra $\vec{\lambda}_A, \vec{\lambda}_{AB}$. For this, we denote their total state by ρ_{AB} with $\rho_A = \mathrm{Tr}_B[\rho_{AB}]$ and choose arbitrary binary sequences $\pi \in \{0,1\}^{d_A}$ and $\sigma \in \{0,1\}^{d_A d_B}$. Then, whenever the intersection property (the *dual binary sequence* $\hat{\sigma}$ of a sequence σ of length d is defined by $\hat{\sigma}_k := \sigma_{d-k+1}$)

$$\left(S_\pi^\circ(\rho_A) \otimes \mathcal{H}^{(B)}\right) \cap S_{\hat{\sigma}}^\circ(\rho_{AB}) \neq \emptyset \tag{9}$$

holds, we obtain

$$\begin{aligned}
&\sum_{j=1}^{d_A} \pi_j \lambda_j^{(A)} - \sum_{i=1}^{d_A d_B} \sigma_i \lambda_i^{(AB)} \\
&= \sum_{j=1}^{d_A} \pi_j \lambda_j^{(A)} + \sum_{i=1}^{d_A d_B} \sigma_i (-\lambda_i^{(AB)}) \\
&= \min_{V \in S_\pi^\circ(\rho_A)} (\mathrm{Tr}_A[P_V \rho_A]) + \min_{W \in S_{\hat{\sigma}}^\circ(-\rho_{AB})} (\mathrm{Tr}_{AB}[P_W(-\rho_{AB})]) \\
&= \min_{V \otimes \mathcal{H}^{(B)} \in S_\pi^\circ(\rho_A) \otimes \mathcal{H}^{(B)}} (\mathrm{Tr}_{AB}[P_{V \otimes \mathcal{H}^{(B)}} \rho_{AB}]) + \min_{W \in S_{\hat{\sigma}}^\circ(-\rho_{AB})} (\mathrm{Tr}_{AB}[P_W(-\rho_{AB})]) \\
&\leq \mathrm{Tr}_{AB}[P_{W_0} \rho_{AB}] + \mathrm{Tr}_{AB}[P_{W_0}(-\rho_{AB})] \\
&= 0 \,,
\end{aligned} \tag{10}$$

where we applied $S_\sigma^\circ(-\rho) = S_{\hat{\sigma}}^\circ(\rho)$ in the third line and (9) was used in the second last line, with an element $W_0 \in \left(S_\pi^\circ(\rho_A) \otimes \mathcal{H}^{(B)}\right) \cap S_{\hat{\sigma}}^\circ(\rho_{AB})$. Hence, we obtain a spectral inequality

$$\boxed{\sum_{j=1}^{d_A} \pi_j \lambda_j^{(A)} \leq \sum_{i=1}^{d_A d_B} \sigma_i \lambda_i^{(AB)}} \,. \tag{11}$$

Most of the work by Klyachko, Daftuar and Hayden concerns the intersection property Eq. (9). Using cohomology theory allows to map it to an algebraic level and study it there systematically. The structure of Schubert varieties as projective algebraic varieties is essential for that. It gives them an algebraic meaning, since they turn out to stand in a one-to-one correspondence with the generators of the cohomology ring of the complex Grassmannian (see e.g. [4]).

In addition, Klyachko did not only provide an algorithm for calculating these necessary marginal constraints, but also proved that they are sufficient [3] for the compatibility of spectra/marginals.

3. Generalized Pauli Constraints

In this section we emphasize that the solution of the pure univariant QMP implies a generalized (and stronger) Pauli exclusion principle. For this we introduce the

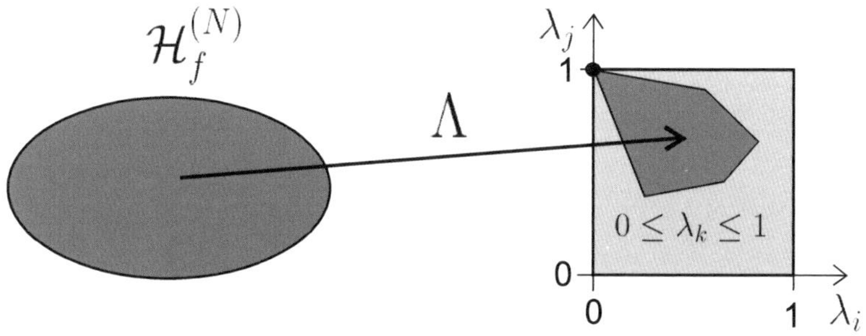

Figure 2. The family of antisymmetric $N-$particle states maps to the family of possible natural occupation numbers $\vec{\lambda}$, which turns out to be a proper subset of the Pauli hyper cube. The Hartree-Fock point is shown as black dot.

map

$$\begin{aligned} \Lambda : \wedge^N[\mathcal{H}_1^{(d)}] \quad &\rightarrow \quad \mathbb{R}^d \\ |\Psi_N\rangle \quad &\mapsto \quad \vec{\lambda} \equiv \mathrm{spec}\left(N\mathrm{Tr}_{N-1}[|\Psi_N\rangle\langle\Psi_N|]\right) . \end{aligned} \tag{12}$$

Λ maps pure antisymmetric quantum states via their 1-particle density operator $\rho_1 \equiv N\mathrm{Tr}_{N-1}[|\Psi\rangle\langle\Psi_N|]$ (trace-normalized to the particle number N) to its decreasingly-ordered *natural occupation numbers* (NON) $\vec{\lambda} \equiv (\lambda_1, \ldots, \lambda_d)$. Pauli's exclusion principle can then be formulated as

$$0 \leq \lambda_i \leq 1 , \tag{13}$$

which is a strong restriction of the range of Λ. This well-known implication of Pauli's exclusion principle from the antisymmetry is also illustrated in Fig. 2. A natural question arises: Are there further restrictions on fermionic NON $\vec{\lambda}$? This question is equivalent to the 1-body pure N-representability problem (see Definition 1.2 and Remark 2.1). Moreover, since this is a pure univariant QMP the answer to that question is 'yes'. Klyachko [7, 8] provides an algorithm that allows to calculated these so-called generalized Pauli constraints for each fixed N and d. They all take the form of linear inequalities,

$$D_i^{(N,d)}(\vec{\lambda}) = \kappa_i^{(0)} + \kappa_i^{(1)}\lambda_1 + \ldots + \kappa_i^{(d)}\lambda_d \geq 0 , \tag{14}$$

with affine coefficients $\kappa_i^{(j)} \in \mathbb{Z}$, $j = 0, 1, \ldots, d$ and $i = 1, 2, \ldots, r^{(N,d)}$. The number $r^{(N,d)}$ of such constraints (14) increases drastically with d. To give the reader an idea how non-trivial they are we consider the setting $\wedge^3[\mathcal{H}_1^{(6)}]$. Already in 1972, Borland and Dennis found the necessary and sufficient conditions on $\vec{\lambda}$. They are given by [9]

$$\lambda_1 + \lambda_6 = \lambda_2 + \lambda_5 = \lambda_3 + \lambda_4 = 1\,, \tag{15}$$

$$D^{(3,6)}(\vec{\lambda}) \equiv 2 - (\lambda_1 + \lambda_2 + \lambda_4) \geq 0\,, \tag{16}$$

where the NON are always ordered decreasingly, $\lambda_1 \geq \lambda_2 \geq \ldots \geq \lambda_6 \geq 0$. Notice that the inequality $D^{(3,6)}(\vec{\lambda}) \geq 0$ is manifestly stronger than Pauli's exclusion principle, which just states that $2 - (\lambda_1 + \lambda_2) \geq 0$. That some constraints take the form of equalities (instead of inequalities) is specific and happens only for this small setting of three fermions and a 6-dimensional 1-particle Hilbert space.

It is important to notice that the existence of that polytope $\mathcal{P}_{N,d}$ and the corresponding restriction of fermionic occupation numbers is purely kinematic and not related to any Hamiltonian. To use this beautiful new mathematical structure revealed by Klyachko for physics the first task is to understand where the occupation numbers of relevant fermionic quantum states do lie. If we consider e.g. ground states of non-interacting fermions confined by some external potential the position of the corresponding $\vec{\lambda}$-vector is obvious. The ground state $|\Psi_0\rangle$ is given by a single Slater determinant $|1, 2, \ldots, N\rangle$, the antisymmetrized tensor product of the 1-particle states $|i\rangle$ corresponding to the lowest N 1-particle energy levels of the external trap. This ground state yields the NON $\vec{\lambda} = (1, \ldots, 1, 0, \ldots)$ (black dot in Fig. 2). If we turn on some interaction with coupling strength κ the NON $\vec{\lambda}(\kappa)$ will move away from the so-called Hartree-Fock point $(1, \ldots, 1, 0, \ldots)$. The central question is then whether it moves towards the middle of the polytope or whether it still lies on the boundary of the polytope. In the latter case we say that the NON are *pinned to the boundary* of the polytope. Before investigating this question we explain why possible pinning may be interesting. The first point was suggested by Klyachko in [10], where he also introduced the effect of pinning.

(1) Given an N-fermion Hamiltonian H_N. Its ground state $|\Psi_N\rangle$ can be obtained via a minimization of the energy expectation value $E[\Psi_N] \equiv \langle\Psi_N|H_N|\Psi_N\rangle$. If the ground state turns out to be pinned to the boundary of the polytope a generalized Pauli constraint is active for the minimization in the sense that any further minimization of the energy would violate it. In that case the corresponding constraint would have a strong influence on the ground state and its energy.

(2) Consider an N-fermion system initially prepared in a quantum state $|\Psi_N\rangle$ with NON $\vec{\lambda}$ pinned to the boundary $\partial\mathcal{P}_{N,d}$ of the polytope. By switching on a unitary time evolution for that system the NON $\vec{\lambda}(t)$ will begin to move. For some very specific time evolutions $\vec{\lambda}(t)$ may like to leave the

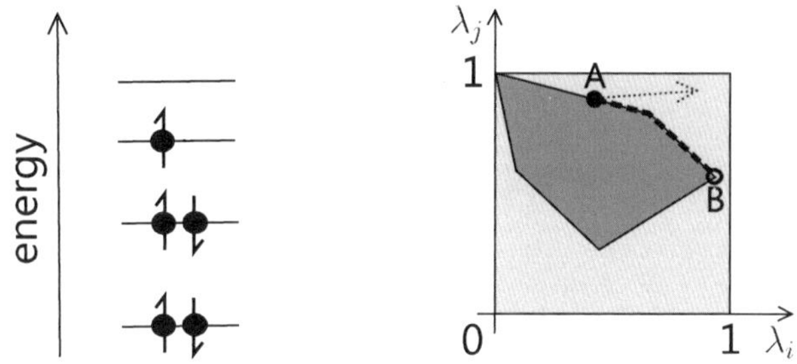

Figure 3. Right: Initial NONs on the boundary (A) of the polytope. Any time evolution that would like to drive them out of the polytope (dashed arrow) is dominated by the geometry of the polytope rather than by the Hamiltonian. This kinematical influence by the generalized Pauli constraints generalizes that by the Pauli exclusion principle (on the left). There e.g. the electron in the highest shell cannot decay to a lower one.

polytope. Such a scenario is illustrated on the right side of Fig. 3. Since $\vec{\lambda}(t)$ cannot leave the polytope, it will move along the boundary from the initial point A to the final point B. In that case the time evolution is dominated by the geometry of the polytope rather than by the Hamiltonian. This kinematical effect on time evolutions is a generalization of a similar more elementary effect shown on the left side of Fig. 3. There, we can see some (non-interacting) electrons occupying low-lying energy shells. Coupling this systems to photons may in principle lead to a decaying of the electrons in the higher energy shells to the lowest one. However, this is impossible due to Pauli's exclusion principle. In the same way the physics of solid bodies at low temperatures is dominated by the electrons close to the Fermi level.

4. Quasipinning for N-Harmonium

To understand how ground states of interacting fermions look like from the new viewpoint of generalized Pauli constraints we studied a model of few harmonically coupled spinless fermions in one dimension confined by a harmonic trap. For details of this study we refer to [6, 11, 12]. The Hamiltonian reads

$$H = \sum_{i=1}^{N} \left(\frac{p_i^2}{2m} + \frac{1}{2} m\omega^2 x_i^2 \right) + \frac{1}{2} K \sum_{i,j=1}^{N} (x_i - x_j)^2 \tag{17}$$

and acts on the fermionic Hilbert space $\mathcal{H}_N^{(f)} \equiv \wedge^N[\mathcal{H}_1]$, where the 1-particle Hilbert space is given by $\mathcal{H}_1 = L^2(\mathbb{R})$. Its ground state can easily be found (see e.g. [13]),

$$\Psi_N(\vec{x}) = c_0 \times \prod_{1 \le i < j \le N} (x_i - x_j) \times \exp\left[-c_1(x_1 + \ldots + x_N)^2 - c_2 \vec{x}^2\right] \tag{18}$$

and has some similarity to the famous Laughlin wave functions [14]. We determined analytically the corresponding 1−RDO depending on the relative interaction strength $\kappa \equiv \frac{NK}{m\omega^2}$ for arbitrary N. For $N = 3$ we calculated analytically with high effort by applying degenerate Rayleigh-Schrödinger perturbation theory its eigenvalues $\lambda_i(\kappa)$ for the regime of not too strong interaction κ. For the ground state we found for the smallest distance $D(\kappa)$ of the spectrum $\vec{\lambda}(\kappa)$ to the polytope boundary[a]

$$D(\kappa) \sim const \times \kappa^8 \,. \tag{19}$$

We called this surprising behavior, to be not on the boundary but very close to it, *quasipinning*. This effect is non-trivial in the sense that the distance (19) to the polytope boundary ∂P is by four orders in κ smaller than the distance to the Hartree-Fock point (lies on the boundary, see Figure 2), which behaves as κ^4. Moreover, quasipinning is not only present in the regime of week interaction ($|\kappa|$ small), but also for medium interaction strengths. E.g. for $\kappa \equiv \frac{3K}{m\omega^2} = \frac{1}{3}$ we found $D = 5.8 \cdot 10^{-8}$.

It is one of the open problems to explore whether few-fermion ground states exhibit generic quasipinning for not too strong interactions in the sense that it is independent of the concrete interaction form. Moreover, the mechanism behind it is not clear yet. Since quasipinning is weaker for the first few excited N-fermion states [6] and vanishes for higher excitations a promising candidate for this mechanism would be the conflict of the antisymmetry (implying the generalized Pauli constraints) and the energy minimization: If one skipped the antisymmetry for the N-particle quantum state one would find much lower ground state energies.

5. Physical Relevance of Pinning

In this section we explain that pinning as an effect in the 1-particle picture allows to reconstruct the structure of the corresponding N-fermion quantum state.

First, we introduce some notation. Given an N-fermion pure state $|\Psi_N\rangle \in \wedge^N[\mathcal{H}_1^{(d)}]$. Its 1-RDO $\rho_1 = N\mathrm{Tr}_{N-1}[|\Psi_N\rangle\langle\Psi_N|]$ can be diagonalized according to

$$\rho_1 \equiv \sum_{k=1}^{d} \lambda_k \, |k\rangle\langle k| \tag{20}$$

with decreasingly-ordered λ_k, the occupation numbers w.r.t. the *natural orbitals* (NO) $|k\rangle$. The NO define an orthonormal basis $\mathcal{B}_1$ for $\mathcal{H}_1^{(d)}$, which induces an orthonormal basis for $\mathcal{H}_N^{(f)}$, the Slater determinants $|\mathbf{k}\rangle$, $\mathbf{k} \equiv (k_1, \ldots, k_N)$ with $1 \leq k_1 < \cdots < k_N \leq d$. Moreover, we denote the fermionic creation and annihilation operators w.r.t. $\mathcal{B}_1$ by $a_k^\dagger$ and a_k, respectively.

[a]The analysis of the distance to the boundary is quite subtle. The underlying 1-particle Hilbert space is infinite-dimensional, but the polytopes are known only for dimensions $d \leq 10$. However, it turns out that all except the largest seven NON are very small and a mathematical result presented in [15] allows to neglect them and truncate the spectrum to the first few NON.

Now, we consider a generalized Pauli constraint (14) and define (for a fixed $|\Psi_N\rangle$ with NON $\vec{\lambda}$)

$$\hat{D} := \kappa^{(0)}\mathbf{1} + \kappa^{(1)} a_1^\dagger a_1 + \ldots + \kappa^{(d)} a_d^\dagger a_d \,. \tag{21}$$

Since $\kappa_i \in \mathbb{Z}$ we have $\mathrm{spec}(\hat{D}) \subset \mathbb{Z}$. One can prove (see e.g. [16]) the important result

Lemma 5.1. *For given generalized Pauli constraint $D(\cdot) \geq 0$ (see also Eq. (14)) and fixed state $|\Psi_N\rangle$ with NON $\vec{\lambda}$ define $\hat{D}$ according to Eq. (21). Then, when $\vec{\lambda}$ is pinned by D, $D(\vec{\lambda}) = 0$, it follows that*

$$\hat{D}|\Psi_N\rangle = 0 \,. \tag{22}$$

From Lemma 5.1 we can immediately conclude that whenever NON are pinned to some facet of the polytope the corresponding $|\Psi_N\rangle$ has weight only in the 0-eigenspace of the corresponding $\hat{D}$-operator (21). By expanding $|\Psi_N\rangle$ w.r.t. $\mathcal{B}_N$,

$$|\Psi_N\rangle = \sum_{\mathbf{i}} c_{\mathbf{i}} \, |\mathbf{i}\rangle, \tag{23}$$

we find a selection rule (recall (21)) due to Klyachko [10],

$$\hat{D}|\mathbf{i}\rangle \neq 0 \qquad \Rightarrow \qquad c_{\mathbf{i}} = 0 \,. \tag{24}$$

To emphasize the importance of (24) we apply the selection rule to an example. Consider a state $|\Psi_3\rangle \in \wedge^3[\mathcal{H}_1^{(6)}]$ with NON $\vec{\lambda}$. The generalized Pauli constraints are given by (15) and (16). The first three constraints take independent of $\vec{\lambda}$ the form of equalities. According to Eq. (24) this leads to universal structural implications for any arbitrary $|\Psi_3\rangle \in \wedge^3[\mathcal{H}_1^{(6)}]$. In the expansion

$$|\Psi_3\rangle = \sum_{1 \leq i_1 < i_2 < i_3 \leq 6} c_{i_1,i_2,i_3} \, |i_1, i_2, i_3\rangle \tag{25}$$

only those Slater determinants $|i_1, i_2, i_3\rangle$ can show up which have exactly one index i_k in each of the three sets $\{1,6\}$, $\{2,5\}$ and $\{3,4\}$. There are only $2^3 = 8$ such Slater determinants: $|1,2,3\rangle$, $|1,2,4\rangle$, $|1,3,5\rangle$, $|1,4,5\rangle$, $|2,3,6\rangle$, $|2,4,6\rangle$, $|3,5,6\rangle$ and $|4,5,6\rangle$. This universal statement is not in contradiction to the dimension $\binom{6}{3} = 20$ of the 3-fermion Hilbert space $\wedge^3[\mathcal{H}_1^{(6)}]$ since the 1-particle states $|k\rangle$ depend on $|\Psi_3\rangle$. If in addition $\vec{\lambda}$ is pinned to the facet described by saturation of (16) Eq. (24) implies that

$$|\Psi_3\rangle = \alpha|1,2,3\rangle + \beta|1,4,5\rangle + \gamma|2,4,6\rangle \,. \tag{26}$$

The three coefficients are free but should be chosen such that $\lambda_1 \geq \lambda_2 \geq \ldots \geq \lambda_6$.

We summarize these insights by

Remark 5.1. Pinning corresponds to specific and simplified structures of the corresponding N-fermion quantum state $|\Psi_N\rangle$. In that sense pinning is highly physically relevant. It is also remarkable that pinning as phenomenon in the elementary 1-particle picture allows to reconstruct the structure of $|\Psi_N\rangle$ as object in the important N-particle picture.

In [15] strong evidence is provided that the structural implications of exact pinning also hold (approximately) for quasipinning. Due to this stability quasipinning is highly physical relevant.

A first physical application was suggested in [11] in form of a variational method. If quasipinning for few-fermion ground states $|\Psi_N\rangle$ turns out to be generic for not too strong interactions one can use the structural insights and choose an ansatz for the ground state based on (24). By minimizing the energy expectation value w.r.t the corresponding non-zero coefficients $c_{\mathbf{i}}$ and NO $|i\rangle$ one would obtain a good approximation to the unknown exact ground state. Such an ansatz, a linear combination of several (up to 10'000) Slater determinants, is well-known in quantum chemistry as multi-configurational self-consistent field (MCSCF) method. However, for our variational optimization we would choose just a few, but very carefully chosen Slater determinants.

Acknowledgments

We thank M.Christandl for helpful discussions and acknowledge financial support from the Swiss State Secretariat for Education and Research supporting COST action MP1006.

Bibliography

[1] A. J. Coleman and V. I. Yukalov, *Reduced Density Matrices: Coulsons Challenge* (Springer, New York, 2000).
[2] Y.-K. Liu, M. Christandl and F. Verstraete, *Phys. Rev. Lett.* **98**, p. 110503 (2007).
[3] A. Klyachko, Quantum marginal problem and representations of the symmetric group (2004).
[4] S. Daftuar and P. Hayden, *Annals Phys.* **315**, 80 (2005).
[5] M. Christandl and G. Mitchison, *Commun. Math. Phys.* **261**, 789 (2006).
[6] C. Schilling, Quantum marginal problem and its physical relevance (2014), PhD thesis at ETH Zurich.
[7] A. Klyachko, *Journal of Physics: Conference Series* **36**, 72 (2006).
[8] M. Altunbulak and A. Klyachko, *Commu. Math. Phys.* **282**, 287 (2008).
[9] R. Borland and K. Dennis, *J. Phys. B* **5**, p. 7 (1972).

[10] A. Klyachko, *Preprint* **quant-ph/0904.2009** (2009).
[11] C. Schilling, D. Gross and M. Christandl, *Phys. Rev. Lett.* **110**, p. 040404 (2013).
[12] C. Schilling, *Phys. Rev. A* **88**, p. 042105 (2013).
[13] Z.-L. Wang, A. Wang, Y. Yang and X. Li, *Comm. Theor. Phys.* **58**, p. 639 (2012).
[14] R. B. Laughlin, *Phys. Rev. Lett.* **50**, 1395 (1983).
[15] C. Schilling, D. Gross and M. Christandl, (2013), Supplemental Material of [11].
[16] A. Lopes, PhD thesis at Freiburg University, in preparation.

Hartree-Fock dynamics for weakly interacting fermions

N. Benedikter

Institute of Applied Mathematics, University of Bonn,
53115 Bonn, Germany
E-mail: niels.benedikter@hcm.uni-bonn.de

M. Porta* and B. Schlein†

Institute for Mathematics, University of Zurich,
8057 Zurich, Switzerland
**E-mail: marcello.porta@gmail.com*
†E-mail: benjamin.schlein@math.uzh.ch

We review recent results [3] concerning the evolution of fermionic systems. We are interested in the mean field regime, where particles experience many weak collisions. For fermions, the mean field regime is naturally linked with a semiclassical limit. Assuming some regularity of the interaction potential we show that the many body evolution of initial states close to Slater determinants exhibiting the appropriate semiclassical structure can be approximated by the Hartree-Fock equation. Our method provides precise bounds on the rate of the convergence.

Keywords: Many body quantum dynamics, mean field regime, semiclassical limit.

1. Introduction

We consider systems of N fermions, described by wave functions $\psi_N \in L^2(\mathbb{R}^{3N}, dx_1 \dots dx_N)$, antisymmetric with respect to permutations. Since it does not play any role in our analysis, we neglect the spin of the particles.

Initially, particles are confined by a trapping potential in a volume of order one (we will consider the system at or close to zero temperature; hence the initial data will be chosen close to the ground state of the trapped Hamiltonian). We are interested in understanding the evolution of the system resulting from a change of the external fields. For example, we study the dynamics generated by the translation invariant Hamiltonian

$$H_N = \sum_{j=1}^{N} -\Delta_{x_j} + \lambda \sum_{i<j}^{N} V(x_i - x_j).$$

In this case, the external fields have been completely switched off. Because of the Pauli principle, the kinetic energy of the N fermions at time $t = 0$ is of the order $N^{5/3}$. The potential energy, on the other hand, is of the order λN^2. Hence, to observe a non-trivial effective dynamics, we have to choose the coupling constant λ to be of the order $N^{-1/3}$. Also, since the typical velocity of particles is large, of the order $N^{1/3}$, we can only hope to follow the evolution for times of the order $N^{-1/3}$. Rescaling time, we are led to the Schrödinger equation

$$iN^{1/3}\partial_t\psi_{N,t} = \left[\sum_{j=1}^{N} -\Delta_{x_j} + \frac{1}{N^{1/3}}\sum_{i<j}^{N} V(x_i - x_j)\right]\psi_{N,t}.$$

To write this equation in a more familiar form, we set $\varepsilon = N^{-1/3}$ and we multiply it by ε^2. We find

$$i\varepsilon\partial_t\psi_{N,t} = \left[\sum_{j=1}^{N} -\varepsilon^2\Delta_{x_j} + \frac{1}{N}\sum_{i<j}^{N} V(x_i - x_j)\right]\psi_{N,t}. \tag{1}$$

In Eq. (1) we recover the coupling constant of the order N^{-1} characterizing the mean field limit of bosonic systems. We observe, however, that in the fermionic case, the mean field regime is naturally linked with a semiclassical limit, with $\varepsilon = N^{-1/3}$ playing the role of Planck's constant and converging to zero as $N \to \infty$.

As explained above, we would like to study the solution of the Schrödinger equation (1) for initial data $\psi_{N,0}$ describing particles trapped in a volume of order one. In particular, at (or close to) zero temperature, we are interested in initial data close to the ground state of an Hamiltonian of the form

$$H_N^{\text{trap}} = \sum_{j=1}^{N}\left(-\varepsilon^2\Delta_{x_j} + V_{\text{ext}}(x_j)\right) + \frac{1}{N}\sum_{i<j}^{N} V(x_i - x_j) \tag{2}$$

where V_{ext} is a confining external potential.

The ground state of (2) is expected to be approximated by a Slater determinant

$$\psi_N^{\text{slater}}(x_1, \dots, x_N) = \frac{1}{\sqrt{N!}}\det\left(f_i(x_j)\right)_{1\le i,j\le N} \tag{3}$$

where $\{f_j\}_{j=1}^N$ is an orthonormal system of N functions in $L^2(\mathbb{R}^3)$. For an arbitrary $\psi_N \in L^2(\mathbb{R}^{3N})$, antisymmetric with respect to permutations, we define the reduced one-particle density by taking the partial trace of the orthogonal projection $|\psi_N\rangle\langle\psi_N|$ onto ψ_N over $(N-1)$ particles (the result is a non-negative trace class operator on $L^2(\mathbb{R}^3)$; we normalize reduced densities so that their trace is N). For the Slater determinant (3), the reduced one-particle density ω_N can be easily computed to be the orthogonal projection onto the N dimensional space spanned by $f_1, \dots, f_N$, i.e.

$$\omega_N = \sum_{j=1}^{N} |f_j\rangle\langle f_j|.$$

Slater determinants are quasi-free states; the expectation of any observable in the state ψ_N^{slater} can be expressed (through Wick's theorem) as a function of the reduced one-particle density ω_N. In particular, if we restrict our attention to Slater determinants, the expectation of the Hamiltonian (2) can be expressed as a function of ω_N. We obtain the Hartree-Fock energy functional

$$\begin{aligned}\mathcal{E}_{\text{HF}}(\omega_N) &= \operatorname{tr}\left(-\varepsilon^2\Delta + V_{\text{ext}}\right)\omega_N \\ &\quad + \frac{1}{2N}\int dxdy\, V(x-y)\left[\omega_N(x,x)\omega_N(y,y) - |\omega_N(x,y)|^2\right].\end{aligned} \tag{4}$$

Hence, coming back to the dynamics, we are interested in understanding the solution of the Schrödinger equation (1) for initial data close to Slater determinants whose reduced one-particle density ω_N minimizes the Hartree-Fock energy (4) among all orthogonal projections with trace equal to N. Because of the mean field character of the evolution, it is natural to expect that the evolution of such an initial data remains close to a Slater determinant, with an evolved reduced density $\omega_{N,t}$. If we assume for a moment that this is indeed the case, it is easy to show that the evolution of the reduced density $\omega_{N,t}$ must be governed by the time-dependent Hartree-Fock equation

$$i\varepsilon\partial_t\omega_{N,t} = \left[-\varepsilon^2\Delta + (V * \rho_t) - X_t, \omega_{N,t}\right] \tag{5}$$

with the initial data $\omega_{N,0} = \omega_N$. Here we defined the density $\rho_t(x) = N^{-1}\omega_{N,t}(x,x)$ and the exchange operator X_t, having the integral kernel $X_t(x,y) = N^{-1}V(x-y)\omega_{N,t}(x,y)$. In order to prove that the evolution of the initial Slater determinant, minimizing (4), remains close to a Slater determinant, we need to have a closer look at its reduced density ω_N. It turns out that ω_N is characterized by a special semiclassical structure which plays a crucial role in the analysis of the evolution.

Acknowledgements. M. Porta and B. Schlein gratefully acknowledge support by the ERC starting grant MAQD-240518.

2. Semiclassical structure

Let us begin this section by considering a system of N free fermions moving in the box $\Lambda = [0; 2\pi]^3 \subset \mathbb{R}^3$ (for example, with periodic boundary conditions). The one-particle Hamiltonian (i.e. the Laplace operator on Λ, with periodic boundary conditions) has the eigenmodes $e^{ip\cdot x}$, for $p \in \mathbb{Z}^3$, with energy p^2. The ground state of the N-particle system is therefore the Slater determinant (Fermi sea) obtained by filling the N eigenmodes of the one-particle Hamiltonian having the N smallest energies (because of the Pauli principle, we cannot occupy the same mode with more than one particle). Thus, the reduced one-particle density ω_N of the free ground state has the integral kernel

$$\omega_N(x,y) = \sum_{p\in\mathbb{Z}^3:|p|\leq cN^{1/3}} e^{ip\cdot(x-y)}$$

for an appropriate constant $c > 0$ of order one. Changing variable $p \to \varepsilon p = N^{-1/3}p$, we find

$$\begin{aligned}\omega_N(x,y) &= \sum_{p\in\varepsilon\mathbb{Z}^3:|p|\leq c} e^{ip\cdot(x-y)/\varepsilon} \\ &= N\sum_{p\in\varepsilon\mathbb{Z}^3:|p|<c} \varepsilon^3 e^{ip\cdot(x-y)/\varepsilon} \simeq N\int_{|p|\leq c} e^{ip\cdot(x-y)/\varepsilon}dp.\end{aligned}$$

We find that $\omega_N(x,y) \simeq N\varphi((x-y)/\varepsilon)$ for a function φ decaying to zero at infinity (it is possible to compute φ explicitly, but the result is not very important for our purposes). The important observation is that the kernel $\omega_N(x,y)$ is concentrated close to the diagonal $x = y$, and that it decays to zero, for $|x-y| \gg \varepsilon$.

If instead of imposing periodic boundary conditions we trap the particles with an external confining potential, we still expect the reduced one-particle density of the Slater determinant minimizing the energy to be concentrated close to the diagonal; in this case, however, it will also depend on the variable $(x+y)$, so that the density of the particles can vary locally, to better adapt to the profile of the external potential. In other words, we expect

$$\omega_N(x,y) \simeq N\varphi((x-y)/\varepsilon)\rho(x+y) \tag{6}$$

for appropriate functions φ, ρ (normalizing φ so that $\varphi(0) = 1$, $\rho(x) \simeq N^{-1}\omega_N(x,x)$ is the density of particles at point x). We expect therefore a clear separation of scales. The kernel $\omega_N(x,y)$ should vary on the microscopic scale ε in the $(x-y)$ direction; on the other hand, it should only vary on scales of order one in the $(x+y)$ direction.

In order to characterize the structure (6), it is useful to consider the commutators $[x,\omega_N]$ and $[\varepsilon\nabla,\omega_N]$, having the integral kernels

$$\begin{aligned}[x,\omega_N](x,y) &= (x-y)\omega_N(x,y) \\ [\varepsilon\nabla,\omega_N](x,y) &= \varepsilon(\nabla_x+\nabla_y)\omega_N(x,y).\end{aligned}$$

Assuming the decomposition (6), the factors $(x-y)$ and $\varepsilon(\nabla_x+\nabla_y)$ are both of size ε (at least if ρ has some regularity). Hence, reduced densities with the semiclassical structure (6) satisfy the bounds

$$\text{tr } |[x,\omega_N]| \leq CN\varepsilon \quad \text{and} \quad \text{tr } |[\varepsilon\nabla,\omega_N]| \leq CN\varepsilon. \tag{7}$$

So far, we argued that ground states of non-interacting systems are Slater determinants satisfying (7). What happens now if we turn on an interaction? Semiclassical analysis suggests that the general picture remains essentially unchanged. The minimizer of the Hartree-Fock functional (4), now with non-vanishing interaction, is expected to be close to the Weyl quantization

$$\omega_N(x,y) \simeq \mathrm{Op}^w_M(x,y) = \frac{1}{(2\pi\varepsilon)^3} \int dp\, M\left(\frac{x+y}{2}, p\right) e^{ip\cdot(x-y)/\varepsilon} \tag{8}$$

of the phase-space density $M(x,p) = \chi(|p| \leq (6\pi^2 \rho_{TF}(x))^{1/3})$, where ρ_{TF} minimizes the Thomas-Fermi energy functional

$$\mathcal{E}_{\mathrm{TF}}(\rho) = \frac{3}{5}(3\pi^2)^{2/3} \int \rho^{5/3}(x)dx + \int V_{\mathrm{ext}}(x)\rho(x)dx + \frac{1}{2}\int V(x-y)\rho(x)\rho(y)$$

under the conditions $\rho \geq 0$ and $\int \rho\, dx = 1$. One can interpret (8) as stating that, like in the case of free fermions, the minimizer of (4) can be constructed by filling the one-particle modes with the smallest momenta. Here, however, we fill the Fermi sea locally, depending on x, according to the value of the Thomas-Fermi density. Taking (8) for granted, we find that

$$[x, \omega_N] = -i\varepsilon \mathrm{Op}^w_{\nabla_p M}, \quad \text{and } [\varepsilon\nabla, \omega_N] = \varepsilon \mathrm{Op}^w_{\nabla_x M}.$$

Semiclassical analysis gives

$$\mathrm{tr}\,|[x,\omega_N]| \simeq \frac{\varepsilon}{(2\pi\varepsilon)^3} \int dxdp |\nabla_p M(x,p)| = CN\varepsilon \int dx\, \rho_{\mathrm{TF}}^{2/3}(x) \leq CN\varepsilon$$

and

$$\mathrm{tr}\,|[\varepsilon\nabla,\omega_N]| \simeq \frac{\varepsilon}{(2\pi\varepsilon)^3} \int dxdp |\nabla_x M(x,p)| = CN\varepsilon \int dx |\nabla \rho_{\mathrm{TF}}(x)| \leq CN\varepsilon,$$

in accordance with (7).

The heuristic argument we just presented motivates the expectation that the initial data we are interested in, namely data close to the ground state of a Hamiltonian of the form (2), are approximate Slater determinants with one particle reduced density ω_N satisfying the semiclassical bounds (7). From now on, this will be our assumption; in other words, our main theorem will describe the time evolution of initial data with these properties. For such initial data, we will prove that the evolution stays close to the Slater determinant with reduced density $\omega_{N,t}$ satisfying the time dependent Hartree-Fock equation (5).

3. Fock space representation

To state our theorem more precisely, we switch to a Fock space representation, so that the number of particles is allowed to fluctuate. We denote by

$$\mathcal{F} = \bigoplus_{n\geq 0} L^2_a(\mathbb{R}^{3n}, dx_1 \ldots dx_n)$$

the fermionic Fock space over $L^2(\mathbb{R}^3)$ ($L^2_a(\mathbb{R}^{3n})$ denotes the subspace of $L^2(\mathbb{R}^{3n})$ consisting of antisymmetric wave functions).

On $\mathcal{F}$, we introduce as usual creation and annihilation operators satisfying the canonical anticommutation relations

$$\{a(f), a^*(g)\} = \langle f, g\rangle, \quad \{a(f), a(g)\} = \{a^*(f), a^*(g)\} = 0$$

for all $f, g \in L^2(\mathbb{R}^3)$. We will also use the operator valued distributions a_x^*, a_x, which are formally creating and, respectively, annihilating a particle at the point x. In terms of these distributions, we define the number of particles operator

$$\mathcal{N} = \int dx\, a_x^* a_x.$$

More generally, given a self-adjoint operator A on $L^2(\mathbb{R}^3)$, we define its second quantization $d\Gamma(A)$ by

$$(d\Gamma(A)\psi)^{(n)} = \sum_{j=1}^{n} A^{(j)}\psi^{(n)}$$

where $A^{(j)} = 1 \otimes \cdots \otimes A \otimes \cdots \otimes 1$ denotes the operator A acting only on the j-th particle. It is easy to check that, if A has the integral kernel $A(x, y)$, its second quantization can be expressed in terms of the operator valued distributions as

$$d\Gamma(A) = \int dxdy\, A(x, y) a_x^* a_y.$$

With this notation, we have $\mathcal{N} = d\Gamma(1)$.

Next, we introduce an Hamilton operator on $\mathcal{F}$, by setting $(\mathcal{H}_N\psi)^{(n)} = \mathcal{H}_N^{(n)}\psi^{(n)}$, where

$$\mathcal{H}_N^{(n)} = \sum_{j=1}^{n} -\varepsilon^2\Delta_{x_j} + \frac{1}{N}\sum_{i<j}^{n} V(x_i - x_j).$$

By definition, the Hamiltonian $\mathcal{H}_N$ leaves the number of particles invariant. In particular, on the N-particle sector, $\mathcal{H}_N$ coincides exactly with the Hamiltonian generating the evolution (1). In terms of the operator valued distributions a_x^*, a_x, $\mathcal{H}_N$ can be expressed as

$$\mathcal{H}_N = \varepsilon^2 \int dx\, \nabla_x a_x^* \nabla_x a_x + \frac{1}{2N}\int dxdy\, V(x - y) a_x^* a_y^* a_y a_x.$$

On the Fock space $\mathcal{F}$, Slater determinants can be very conveniently generated by Bogoliubov transformations. Let $\omega_N = \sum_{j=1}^{N} |f_j\rangle\langle f_j|$ be the reduced density of an N-particle Slater determinant. The orthonormal family $\{f_j\}_{j=1}^{N}$ can be completed to an orthonormal basis $\{f_j\}_{j\in\mathbb{N}}$ of $L^2(\mathbb{R}^3)$. Then there exists a unitary operator $R_{\omega_N} : \mathcal{F} \to \mathcal{F}$ such that

$$R_{\omega_N}\Omega = a^*(f_1)\dots a^*(f_N)\Omega$$

is the Slater determinant with reduced density ω_N (here $\Omega = \{1, 0, 0, \ldots\}$ denotes the Fock space vacuum), and

$$R^*_{\omega_N} a^*(f_j) R_{\omega_N} = \begin{cases} a^*(f_j), & \text{if } j > N \\ a(f_j), & \text{if } j \leq N \end{cases}. \tag{9}$$

Taking the adjoint, we obtain a similar formula also for the action of R_{ω_N} on annihilation operators. The idea here is that the Bogoliubov transformation R_{ω_N} allows us to switch to a new representation of the canonical anticommutation relations. The new vacuum $R_{\omega_N}\Omega$ is the Slater determinant with reduced density ω_N. The new creation operators $R^*_{\omega_N} a(f_j) R_{\omega_N}$ create a particle with wave function f_j if $j > N$, while they create a hole in the Slater determinant if $j \leq N$. The new number of particles operator $R^*_{\omega_N} \mathcal{N} R_{\omega_N}$ measures the number of particles outside the Slater determinant combined with the number of holes in the Slater determinant. In other words, it measures the number of excitations w.r.t. the Slater determinant; since our goal is exactly to prove closeness to a Slater determinant, this explains why Bogoliubov transformations are so useful for us, and play such an important role in our analysis.

From (9), we conclude that, for arbitrary $f \in L^2(\mathbb{R}^3)$,

$$R^*_{\omega_N} a^*(f) R_{\omega_N} = a^*(u_N f) + a(\overline{v}_N \overline{f}) \tag{10}$$

where $u_N = 1 - \omega_N$ and $v_N = \sum_{j=1}^N |\overline{f}_j\rangle\langle f_j|$ (recall that creation operators are linear and annihilation operators are antilinear in their arguments; this explains the emergence of complex conjugation).

4. Main results

We are now ready to present our main theorem, which describes the many-body (Fock space) evolution of approximate initial Slater determinants in terms of the Hartree-Fock equation.

Theorem 4.1. *Let $V \in L^1(\mathbb{R}^3)$ with Fourier transform $\widehat{V}$ satisfying*

$$\int |\widehat{V}(p)|(1 + p^2) dp < \infty. \tag{11}$$

Let ω_N be a sequence of orthogonal projections on $L^2(\mathbb{R}^3)$ with $\operatorname{tr} \omega_N = N$ and such that

$$\operatorname{tr} |[x, \omega_N]| \leq CN\varepsilon, \quad \text{and} \quad \operatorname{tr} |[\varepsilon\nabla, \omega_N]| \leq CN\varepsilon. \tag{12}$$

Let $\xi_N \in \mathcal{F}$ be a sequence with $\|\xi_N\| = 1$ and $\langle \xi_N, \mathcal{N}\xi_N\rangle \leq C$, uniformly in N. We consider the evolution

$$\psi_{N,t} = e^{-i\mathcal{H}_N t/\varepsilon} R_{\omega_N} \xi_N$$

and we denote by $\gamma^{(1)}_{N,t}$ *the reduced one-particle density associated with* $\psi_{N,t}$. *Then there exist constants* $C, c > 0$ *such that*

$$\left\|\gamma^{(1)}_{N,t} - \omega_{N,t}\right\|_{HS} \leq C \exp\left(c \exp(c|t|)\right) \tag{13}$$

where $\omega_{N,t}$ *is the solution of the Hartree-Fock equation*

$$i\varepsilon\partial_t \omega_{N,t} = \left[-\varepsilon^2\Delta + (V * \rho_t) - X_t, \omega_{N,t}\right] \tag{14}$$

with the initial data $\omega_{N,0} = \omega_N$. *Assuming additionally that* $\langle \xi_N, \mathcal{N}^2 \xi_N\rangle \leq C$ *and the orthogonality condition* $d\Gamma(\omega_N)\xi_N = 0$, *we find constants* $C, c > 0$ *such that*

$$\operatorname{tr}\left|\gamma^{(1)}_{N,t} - \omega_{N,t}\right| \leq C N^{1/6} \exp\left(c \exp(c|t|)\right). \tag{15}$$

Remark 4.1.

i) Taking $\xi_N = \Omega$, the theorem describes the time-evolution of the initial Slater determinant $R_{\omega_N}\Omega$. Even if $\xi_N \neq \Omega$, the assumption that $\langle \xi_N, \mathcal{N}\xi_N\rangle \leq C$, uniformly in N, guarantees that the initial data is close to the Slater determinant with reduced density ω_N (for example, in the sense of (13), which holds, in particular, at $t = 0$). It is easy to extend the bound (13) to the evolution of initial data of the form $R_{\omega_N}\xi_N$, where $\langle \xi_N, \mathcal{N}\xi_N\rangle \leq CN^\alpha$ for $0 \leq \alpha < 1$. In this case, we have to replace (13) by

$$\left\|\gamma^{(1)}_{N,t} - \omega_{N,t}\right\|_{\mathrm{HS}} \leq CN^{\alpha/2} \exp(c\exp(c|t|)). \tag{16}$$

ii) The bound (13) is optimal in its N dependence; it should be compared with $\|\gamma^{(1)}_{N,t}\|_{\mathrm{HS}} \simeq N^{1/2}$ and $\|\omega_{N,t}\|_{\mathrm{HS}} = N^{1/2}$. The bound (15) is not expected to be optimal (the optimal estimate should probably be of the order one in N); still it gives more precise information on the many-body evolution (since it should be compared with the normalization $\operatorname{tr}\gamma^{(1)}_{N,t} = \operatorname{tr}\omega_{N,t} = N$).

iii) Results similar to Theorem 4.1 have been recently [4] obtained also for fermions with a relativistic dispersion; in this case, the many body Schrödinger evolution is approximated by a semirelativistic Hartree-Fock equation.

iv) It is easy to show that the exchange term appearing in the Hartree-Fock equation (5) is of smaller order, compared with the other terms. For example, using the formula

$$[X_t, \omega_{N,t}](x,y) = N^{-1}\int dz (V(x-z) - V(y-z))\omega_{N,t}(x,z)\omega_{N,t}(z,y)$$

we can estimate the Hilbert-Schmidt norm of the commutator $[X_t, \omega_{N,t}]$ (assuming the potential to be bounded, as follows from (11)) by

$$\begin{aligned}\|[X_t,&\omega_{N,t}]\|_{\rm HS}^2 \\ &= \frac{1}{N^2}\int dxdydz_1dz_2\omega_{N,t}(x,z_1)\omega_{N,t}(z_1,y)\omega_{N,t}(x,z_2)\omega_{N,t}(z_2,y) \\ &\qquad\qquad \times (V(x-z_1)-V(y-z_1))\,(V(x-z_2)-V(y-z_2)) \\ &\leq \frac{C}{N^2}\|\omega_{N,t}\|_{\rm HS}^4 \leq C.\end{aligned}$$

For this reason, it is possible to absorb the contribution of the exchange term in the error on the r.h.s. of (13) (and, similarly, on the r.h.s. of (15)). As a consequence, the bounds (13) and (15) remain valid if we replace the solution $\omega_{N,t}$ of the Hartree-Fock equation (5) with the solution $\widetilde{\omega}_{N,t}$ of the fermionic Hartree equation

$$i\varepsilon\partial_t\widetilde{\omega}_{N,t} = \left[-\varepsilon^2\Delta + (V*\widetilde{\rho}_t), \widetilde{\omega}_{N,t}\right] \qquad (17)$$

of course with the initial data $\widetilde{\omega}_{N,0} = \omega_N$ and with $\widetilde{\rho}_t(x) = N^{-1}\widetilde{\omega}_{N,t}(x,x)$.

v) The Hartree-Fock equation (14) and the Hartree equation (17) still depend on N through the semiclassical parameter $\varepsilon = N^{-1/3}$. As $N\to\infty$, the Hartree-Fock and the Hartree dynamics can be approximated by the Vlasov equation. We define the Wigner transform $W_{N,t}$ associated to the solution $\omega_{N,t}$ of the Hartree-Fock equation by setting

$$W_{N,t}(x,v) = \frac{1}{(2\pi)^3}\int dy\,\omega_{N,t}\left(x+\frac{\varepsilon y}{2}, x-\frac{\varepsilon y}{2}\right)e^{-iv\cdot y}.$$

As $N\to\infty$ we have, in an appropriate sense, $W_{N,t}\to W_{\infty,t}$, where $W_{\infty,t}$ solves the Vlasov equation

$$\partial_t W_{\infty,t} + 2v\cdot\nabla_x W_{\infty,t} - \nabla\left(V*\rho_{\infty,t}\right)\cdot\nabla_v W_{\infty,t} = 0$$

with the density $\rho_{\infty,t}(x) = \int W_{\infty,t}(x,v)dv$. It should be observed, however, that the Hartree-Fock and the Hartree equation give a better approximation to many body quantum mechanics, compared with the classical Vlasov equation. While the relative size of the corrections to the Vlasov equation is of the order $\varepsilon = N^{-1/3}$, (15) shows that the Hartree-Fock equation is correct up to errors of relative size $N^{-5/6}$.

Using Theorem 4.1, it is also possible to study the evolution of approximated Slater determinants with fixed number of particles N. This is the content of the next corollary.

Corollary 4.1. *Let $V\in L^1(\mathbb{R}^3)$ with Fourier transform $\widehat{V}$ satisfying (11). Let ω_N be a sequence of orthogonal projections on $L^2(\mathbb{R}^3)$ with $\operatorname{tr}\omega_N = N$ and satisfying*

(12). Let $\psi_N \in L^2_a(\mathbb{R}^{3N})$ be a sequence with $\|\psi_N\| = 1$ and with one-particle reduced density $\gamma_N^{(1)}$ satisfying

$$\operatorname{tr} \left|\gamma_N^{(1)} - \omega_N\right| \leq CN^\alpha \tag{18}$$

for some $\alpha \in [0;1)$. Let $\psi_{N,t} = e^{-iH_N t/\varepsilon}\psi_N$ and denote by $\gamma_{N,t}^{(1)}$ the reduced one-particle density associated with $\psi_{N,t}$. Then there exist constants $C, c > 0$ such that

$$\left\|\gamma_{N,t}^{(1)} - \omega_{N,t}\right\|_{HS} \leq CN^{\alpha/2} \exp(c \exp(c|t|))$$

where $\omega_{N,t}$ is the solution of the Hartree-Fock equation (14) or of the Hartree equation (17) with initial data $\omega_{N,0} = \omega_N$.

Proof. We identify ψ_N with the Fock space vector $\{0, \dots, 0, \psi_N, 0 \dots\}$ and we set

$$\xi_N = R^*_{\omega_N} \psi_N.$$

We compute

$$\begin{aligned}
\langle \xi_N, \mathcal{N} \xi_N \rangle &= \langle \psi_N, R_{\omega_N} \mathcal{N} R^*_{\omega_N} \psi_N \rangle \\
&= \langle \psi_N, (\mathcal{N} - 2d\Gamma(\omega_N) + N)\, \psi_N \rangle \\
&= 2\langle \psi_N, d\Gamma(1 - \omega_N)\psi_N \rangle = 2 \operatorname{tr} \gamma_N^{(1)}(1 - \omega_N) \\
&= 2 \operatorname{tr} (\gamma_N^{(1)} - \omega_N)\,(1 - \omega_N) \leq 2 \operatorname{tr} |\gamma_N^{(1)} - \omega_N|.
\end{aligned}$$

The assumption (18) implies $\langle \xi_N, \mathcal{N} \xi_N \rangle \leq CN^\alpha$. Therefore, the corollary follows from the observation (16). □

Let us compare Theorem 4.1 and Corollary 4.1 with previous results available in the literature. The first rigorous result on mean field evolution of fermions has been obtained by Narnhofer and Sewell [8], who showed the convergence of the solution of the many body Schrödinger equation towards the Vlasov equation, for analytic potentials. Spohn [9] extended the previous result proving convergence towards Vlasov for potentials $V \in C^2(\mathbb{R}^3)$. The last two results do not give bounds on the rate of the convergence. More recently [5], convergence towards Hartree dynamics has been established for analytic interaction potentials and for short times. Theorem 4.1 is comparable with this last result, but it improves it because it holds for all times (of order one) and for a larger class of interaction potentials. Convergence of the many body dynamics towards the Hartree-Fock equation has also been established for different scalings by Bardos-Golse-Gottlieb-Mauser [2] (for bounded potentials) e by Knowles-Fröhlich [6] (for a Coulomb interaction); in the regime considered in these papers there is no semiclassical limit involved. Related results proving that Hartree-Fock theory approximates the many body ground state energy, up to errors of the (relative) size $N^{-2/3-\delta}$, for a $\delta > 0$, have been obtained for models of atoms and molecules (in which electrons interact through a Coulomb potential) by Bach [1] and by Graf and Solovej [7].

5. Strategy of the proof

We define the fluctuation vector $\xi_{N,t} \in \mathcal{F}$ by requiring that

$$\psi_{N,t} = e^{-i\mathcal{H}_N t/\varepsilon} R_{\omega_N} \xi_N = R_{\omega_{N,t}} \xi_{N,t}$$

where $\omega_{N,t}$ is the solution of the Hartree-Fock equation (5). Equivalently, $\xi_{N,t} = \mathcal{U}_N(t)\xi_N$, where we defined the fluctuation dynamics

$$\mathcal{U}_N(t) = R^*_{\omega_{N,t}} e^{-i\mathcal{H}_N t/\varepsilon} R_{\omega_N} .$$

Having $\xi_{N,t} = \Omega$ would imply that $\psi_{N,t}$ is exactly the Slater determinant with reduced density $\omega_{N,t}$. Of course, for $t \neq 0$, this will never be the case, even if initially $\xi_N = \Omega$. Still, this remark suggests that, in order to prove that $\psi_{N,t}$ is close to a Slater determinant, it is enough to show that, in an appropriate sense, $\xi_{N,t}$ is close to the vacuum Ω. In fact, using (10), it is easy to show that

$$\begin{aligned} \|\gamma^{(1)}_{N,t} - \omega_{N,t}\|^2_{\text{HS}} &\leq 2 \operatorname{tr} \gamma^{(1)}_{N,t}(1-\omega_{N,t}) \\ &= 2\langle \psi_{N,t}, d\Gamma(1-\omega_{N,t})\psi_{N,t}\rangle = \langle \xi_{N,t}, \mathcal{N}\xi_{N,t}\rangle. \end{aligned} \tag{19}$$

Hence, (13) follows from bounds on the expectation of $\mathcal{N}$ in the state $\xi_{N,t}$ (the estomate (15) requires substantially more work [3]; we will not discuss it here).

To obtain these bounds, we intend to use Gronwall's lemma. Hence, we compute the derivative

$$\begin{aligned} &i\varepsilon\, \partial_t \langle \xi_{N,t}, \mathcal{N}\, \xi_{N,t}\rangle \\ &= i\varepsilon\, \partial_t \left\langle R_{\omega_N}\xi_N, e^{i\mathcal{H}_N t/\varepsilon} \left(\mathcal{N} - 2d\Gamma(\omega_{N,t}) + N\right) e^{-i\mathcal{H}_N t/\varepsilon} R_{\omega_N}\xi_N \right\rangle \\ &= 2\left\langle e^{-i\mathcal{H}_N t/\varepsilon} R_{\omega_N}\xi_N, \{[\mathcal{H}_N, d\Gamma(\omega_{N,t})] - d\Gamma(i\varepsilon\,\partial_t\omega_{N,t})\}\, e^{-i\mathcal{H}_N t/\varepsilon} R_{\omega_N}\xi_N \right\rangle. \end{aligned}$$

There are many cancellations between the two summands in the parenthesis. In particular, all contributions which are quadratic in creation and annihilation operators cancel exactly. After some algebraic manipulations, we find the identity

$$\begin{aligned} i\varepsilon\, \partial_t \langle \xi_{N,t}, \mathcal{N}\, \xi_{N,t}\rangle = -4i\text{Im}\, \frac{1}{N} \int dx dy V(x-y) \\ \times \big\langle\, \xi_{N,t}, \big\{ a^*(u_{N,t,y}) a^*(\bar{v}_{N,t,y}) a^*(\bar{v}_{N,t,x}) a(\bar{v}_{N,t,x}) \\ + a^*(u_{N,t,x}) a(u_{N,t,x}) a(\bar{v}_{N,t,y}) a(u_{N,t,y}) \\ + a(u_{N,t,x}) a(\bar{v}_{N,t,x}) a(\bar{v}_{N,t,y}) a(u_{N,t,y}) \big\}\, \xi_{N,t} \big\rangle \end{aligned} \tag{20}$$

where $u_{N,t} = 1 - \omega_{N,t}$ and $v_{N,t}$ is constructed from $\omega_{N,t}$ as explained after (10) (we use here the notation $u_{N,t,x}(z) = u_{N,t}(x,z)$ and similarly for $\bar{v}_{N,t,x}$).

In order to apply Gronwall's inequality, we need to bound the r.h.s. of (20) in terms of $\langle \xi_{N,t}, \mathcal{N}\xi_{N,t}\rangle$. Let us consider, for example, the contribution of the last term in the parenthesis. Expanding the potential in a Fourier integral, we find

$$\begin{aligned}\frac{1}{N}\int dxdy\, V(x-y)\,\langle \xi_{N,t}, a(u_{N,t,x})a(\bar{v}_{N,t,x})a(\bar{v}_{N,t,y})a(u_{N,t,y})\xi_{N,t}\rangle \\ = \frac{1}{N}\int dp\, \widehat{V}(p)\Big\langle \int dr_1 ds_1 (\bar{v}_{N,t}e^{ip\cdot x}u_{N,t})(r_1,s_1)a^*_{r_1}a^*_{s_1}\,\xi_{N,t}, \\ \int dr_2 ds_2 (\bar{v}_{N,t}e^{ip\cdot x}u_{N,t})(r_2,s_2)a_{r_2}a_{s_2}\,\xi_{N,t}\Big\rangle.\end{aligned} \tag{21}$$

To estimate the r.h.s. of (21), we observe that, for any operator A on $L^2(\mathbb{R}^3)$ with integral kernel $A(x,y)$, we have the inequality

$$\left\|\int dxdy\, A(x,y)\, a^\sharp_x a^\sharp_y \psi\right\| \leq \|A\|_{\mathrm{HS}}\|(\mathcal{N}+1)^{1/2}\psi\|$$

where $a^\sharp$ is either an annihilation operator a or a creation operator a^*. Applying this bound to (21), we conclude that

$$\begin{aligned}\left|\frac{1}{N}\int dxdy V(x-y)\,\langle \xi_{N,t}, a(u_{N,t,x})a(\bar{v}_{N,t,x})a(\bar{v}_{N,t,y})a(u_{N,t,y})\xi_{N,t}\rangle\right| \\ \leq \frac{1}{N}\int dp\,|\widehat{V}(p)|\,\|\bar{v}_{N,t}\,e^{ip\cdot x}u_{N,t}\|^2_{\mathrm{HS}}\,\|(\mathcal{N}+1)^{1/2}\,\xi_{N,t}\|^2.\end{aligned} \tag{22}$$

Since the operator norm of $u_{N,t}$ is bounded by one, we find

$$\|\bar{v}_{N,t}\,e^{ip\cdot x}u_{N,t}\|^2_{\mathrm{HS}} \leq \|\bar{v}_{N,t}\|^2_{\mathrm{HS}} = \mathrm{tr}\;\omega_{N,t} = N. \tag{23}$$

Hence, assuming (11), we obtain

$$\begin{aligned}\left|\frac{1}{N}\int dxdy V(x-y)\,\langle \xi_{N,t}, a(u_{N,t,x})a(\omega_{N,t,x})a(\omega_{N,t,y})a(u_{N,t,y})\xi_{N,t}\rangle\right| \\ \leq C\langle\, \xi_{N,t}, (\mathcal{N}+1)\xi_{N,t}\rangle.\end{aligned}$$

However, because of the factor $\varepsilon = N^{-1/3}$ on the l.h.s. of (20), this bound is not sufficient, yet. Instead, we have to squeeze out an additional factor of ε. To this end, we notice that, using the orthogonality $\bar{v}_{N,t}u_{N,t} = 0$, the estimate (23) can be improved to

$$\begin{aligned}\|\bar{v}_{N,t}\,e^{ip\cdot x}u_{N,t}\|^2_{\mathrm{HS}} &= \|\bar{v}_{N,t}\,[e^{ip\cdot x}, u_{N,t}]\|^2_{\mathrm{HS}} = \|\bar{v}_{N,t}\,[e^{ip\cdot x}, \omega_{N,t}]\|^2_{\mathrm{HS}} \\ &\leq \|[e^{ip\cdot x}, \omega_{N,t}]\|^2_{\mathrm{HS}} \leq \mathrm{tr}\,|[e^{ip\cdot x}, \omega_{N,t}]| \\ &\leq C(1+|p|)\,\mathrm{tr}\,|[x,\omega_{N,t}]|.\end{aligned} \tag{24}$$

At time $t=0$, the r.h.s. of (24) is bounded, according to the first semiclassical bound in (12), by $C(1+|p|)N\varepsilon$ (and hence it is smaller than (23) by a factor ε, as desired). Using also the second semiclassical bound in (12) for the initial density ω_N, it is possible to propagate these estimates along the solution of the Hartree-Fock equation, showing in particular that, for any $t\in\mathbb{R}$,

$$\mathrm{tr}\,|[x,\omega_{N,t}]| \leq CN\varepsilon\exp(c|t|)\,.$$

Inserting this inequality in (24) and then plugging the result in (22), we conclude that

$$\left|\frac{1}{N}\int dxdyV(x-y)\,\langle\xi_{N,t},a(u_{N,t,x})a(\bar{v}_{N,t,x})a(\bar{v}_{N,t,y})a(u_{N,t,y})\xi_{N,t}\rangle\right|$$
$$\leq C\varepsilon\exp(c|t|)\langle\xi_{N,t},\mathcal{N}\xi_{N,t}\rangle.$$

This estimate, together with similar bounds for the other terms on the r.h.s. of (20), implies that

$$\left|\frac{d}{dt}\langle\xi_{N,t}\mathcal{N}\xi_{N,t}\rangle\right|\leq Ce^{c|t|}\langle\xi_{N,t}(\mathcal{N}+1)\xi_{N,t}\rangle.$$

From Gronwall's inequality, we find

$$\langle\xi_{N,t}(\mathcal{N}+1)\xi_{N,t}\rangle\leq C\exp(c\exp(c|t|)).$$

With (19), this implies the claim (13).

Bibliography

[1] V. Bach. Error bound for the Hartree-Fock energy of atoms and molecules. *Comm. Math. Phys.* **147** (1992), no. 3, 527–548.

[2] C. Bardos, F. Golse, A. D. Gottlieb, and N. J. Mauser. Mean field dynamics of fermions and the time-dependent Hartree-Fock equation. *J. Math. Pures Appl. (9)* **82** (2003), no. 6, 665–683.

[3] N. Benedikter, M. Porta, B. Schlein. Mean field evolution of fermionic systems. *Comm. Math. Phys.* **331** (2014) 1087–1131.

[4] N. Benedikter, M. Porta, B. Schlein. Mean-field dynamics of fermions with relativistic dispersion. *J. Math. Phys.* **55**, 021901 (2014).

[5] A. Elgart, L. Erdős, B. Schlein, and H.-T. Yau. Nonlinear Hartree equation as the mean field limit of weakly coupled fermions. *J. Math. Pures Appl. (9)* **83** (2004), no. 10, 1241–1273.

[6] J. Fröhlich and A. Knowles. A microscopic derivation of the time-dependent Hartree-Fock equation with Coulomb two-body interaction, *J. Stat. Phys.* **145** (2011), no. 1, 23–50.

[7] G.M. Graf and J.P. Solovej. A correlation estimate with applications to quantum systems with Coulomb interactions. *Rev. Math. Phys.* **6** (1994), 977–997.

[8] H. Narnhofer and G. L. Sewell. Vlasov hydrodynamics of a quantum mechanical model. *Comm. Math. Phys.* **79** (1981), no. 1, 9–24.

[9] H. Spohn. On the Vlasov hierarchy. *Math. Methods Appl. Sci.* **3** (1981), no. 4, 445–455.

PART C

Contributed talks

On the ground state energy of the multipolaron in the strong coupling limit

I. Anapolitanos

Universität Stuttgart, Fachbereich Mathematik, 70550 Stuttgart, Germany

B. Landon

Department of Mathematics, Harvard University, 1 Oxford Street, Cambridge MA 02138 USA

We discuss two models for the multipolaron: the Fröhlich Hamiltonian and the Pekar-Tomasevich functional. The ground state energies of the multipolaron in the two models coincide to leading order in α in the so-called strong coupling limit $\alpha \to \infty$. We outline the history of this result and we explain some of the main ideas of our proof given in [2].

Keywords: Polaron, strong coupling limit.

1. A polaron

A polar crystal is a lattice of positive and negative ions that alternate as in the picture. An example of such a material is salt. If we insert an electron in the polar crystal, then it attracts the positive ions and it repels the negative ones. The motion of the ions creates a potential that acts back on the electron. The entire system of the electron interacting with the crystal is called a polaron. An interesting fact that attracted the attention of physicists is that even though electrons always repel each other, polarons may attract each other.

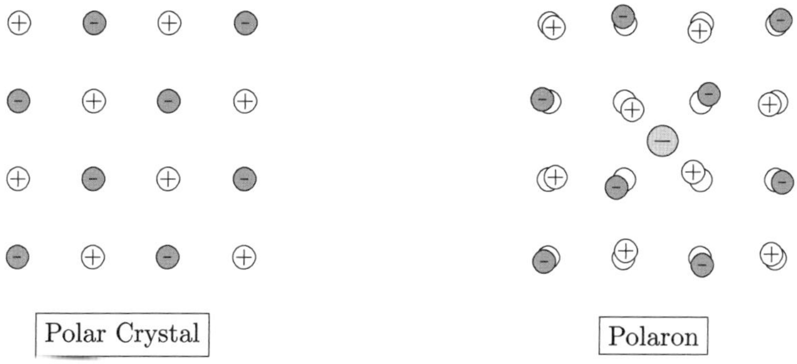

Figure 1.

The Fröhlich Hamiltonian

To model the polaron, Fröhlich assumed that the crystal is infinite and considered the limit in which the distance between neighboring ions goes to zero (the continuum limit). In this limit the crystal is described as a continuous system of harmonic oscillators. We consider the space $L^2(\mathbb{R}^3)$ accounting for the electron, and the bosonic Fock space $\mathcal{F}$ over $L^2(\mathbb{R}^3)$ accounting for the crystal. The Fröhlich Hamiltonian, which we here treat as a quadratic form, acts on a subspace of $L^2(\mathbb{R}^3) \otimes \mathcal{F}$ and is defined as

$$H_\alpha = -\Delta_x + \int a^*(k)a(k)dk - \frac{\sqrt{\alpha}}{\sqrt{2}\pi} \int_{\mathbb{R}^3} \frac{dk}{|k|} \left(e^{ikx}a(k) + e^{-ikx}a^*(k)\right). \qquad (1)$$

The first term, namely $-\Delta_x$, is the kinetic energy operator for the electron. The second term is the number operator in momentum space and gives the energy of oscillations in the crystal, which are called phonons. The last term is an interaction term. It involves the creation and annihilation operators a and a^* in momentum space. The coupling constant α quantifies the strength of the interaction between the crystal and the electrons. The expression for the interaction term can be derived from a classical model by quantizing Poisson's equation $\Delta V = -\rho$, where V and ρ are, respectively, the potential and change of ion charge density caused by the oscillations of the ions. For more details we refer the reader to Chapter 6 of [4].

The ground state energy of H_α is defined as

$$E(\alpha) = \inf_{\|\Phi\|_{L^2(\mathbb{R}^3)\otimes\mathcal{F}}=1, \Phi\in\mathcal{Q}} \langle\Phi, H_\alpha\Phi\rangle,$$

where $\mathcal{Q}$, is the intersection of the form domains of $-\Delta_x$ and $N = \int dka^*(k)a(k)$. We note that it is not trivial, but it is well known that $E(\alpha) > -\infty$, this has been proven by Lieb and Thomas [10] . From their proof method, it follows that the interaction term of the Hamiltonian H_α, is form bounded with respect to $-\Delta_x + N$, with relative bound zero. It follows then from the KLMN theorem, we refer for example to [12] Theorem X.17, that this quadratic form is associated to a self adjoint operator. We are interested in studying $E(\alpha)$. One reason is that it is associated to the problem of attraction of polarons mentioned above. This will be discussed below in some detail. Of course the Laplacian and the number operator are both very well understood and the difficulty in studying $E(\alpha)$ arises from the interaction term. In some materials α is small and in this case $E(\alpha)$ is well studied with perturbation theory, see [8]. However, in some materials α is large and this note is devoted to this case.

The Pekar functional in relation to the Fröhlich Hamiltonian

To study the ground state energy $E(\alpha)$, Pekar started with the ansatz that there is an approximate energy minimizer of the tensor product form $\Phi = \psi \otimes \eta_\psi$, we refer to the work of Pekar [11]. Here $\eta_\psi \in \mathcal{F}$ depends on $\psi \in L^2(\mathbb{R}^3)$. To choose η_ψ appropriately, he observed that there is a square term of the operators a, a^* and a cross term. Completing the square, we therefore obtain a positive operator. An optimal η_ψ would make its expectation vanish. In turns out that such an η_ψ exists, and with this choice we obtain the upper bound

$$E(\alpha) \leq \langle \Phi, H_\alpha \Phi \rangle = \left\langle \psi, \left(-\Delta - \alpha \int dy \frac{|\psi(y)|^2}{|\cdot - y|} \right) \psi \right\rangle, \tag{2}$$

where the inequality holds because Φ is a normalized test function. The first term on the right hand side contains the kinetic energy of the electron. The second term models the potential created by the crystal as a self-interaction term of the electron.

The Pekar functional is defined as

$$\mathcal{P}_\alpha[\psi] = \int_{\mathbb{R}^3} |\nabla\psi(x)|^2 dx - \alpha \int\int_{\mathbb{R}^3 \times \mathbb{R}^3} \frac{|\psi(x)|^2 |\psi(y)|^2}{|x-y|} dx dy.$$

The ground state energy of the Pekar functional is defined as $\mathcal{E}(\alpha) := \inf_{\|\psi\|=1, \psi \in H^1(\mathbb{R}^3)} \mathcal{P}_\alpha[\psi]$. Using that (2) holds for all normalized ψ it follows that

$$E(\alpha) \leq \mathcal{E}(\alpha), \ \forall \alpha > 0. \tag{3}$$

If we use the rescaling $\psi_\alpha(x) := \alpha^{\frac{3}{2}} \psi(\alpha x)$ we obtain that $\mathcal{P}_\alpha[\psi_\alpha] = \alpha^2 \mathcal{P}_1[\psi]$. This implies that $\mathcal{E}(\alpha) = \alpha^2 \mathcal{E}(1)$. Using the last equation and (3) we obtain that

$$E(\alpha) \leq \alpha^2 \mathcal{E}(1).$$

As a consequence, the Pekar ansatz gives an upper bound for the ground state energy of H_a in terms of the considerably simpler Pekar functional. Of course, Pekar wanted to obtain not only an upper bound but a lower bound as well. This begs the question of whether $E(\alpha)$ can be bounded from below in terms of $\mathcal{E}(1)$.

The positive answer was given about thirty years later, in 1983, by Donsker and Varadhan. They proved a lower bound of the form $E(\alpha) \geq \alpha^2 \mathcal{E}(1) - o(\alpha^2)$, where the relative error is negligible as $\alpha \to \infty$, but no information was given on how big the error is. In 1997 Lieb and Thomas gave a simpler proof, which also improved the upper bound to $E(\alpha) \geq \alpha^2 \mathcal{E}(1) - O(\alpha^{\frac{9}{5}})$.

A multipolaron

Now we generalize this discussion to the multipolaron or N-polaron. It is created when we insert N electrons into a polar crystal. The Fröhlich Hamiltonian can be generalized, in a straightforward way, to the case of an N-polaron. It is defined as

$$H_{U,\alpha}^{(N)} := -\sum_{j=1}^{N} \Delta_{x_j} + \sum_{i<j} \frac{U}{|x_i - x_j|} + \int a^*(k) a(k) dk$$

$$-\sum_{j=1}^{N} \frac{\sqrt{\alpha}}{\sqrt{2}\pi} \int_{\mathbb{R}^3} \frac{dk}{|k|} \left(e^{ikx_j} a(k) + e^{-ikx_j} a^*(k) \right) . \tag{4}$$

The second sum, namely the repulsion between the electrons, is essentially the only new term of this generalized Fröhlich Hamiltonian. The constant U measures the strength of the repulsion between the electrons. From a physical derivation, see [5], it follows that $U > 2\alpha$. Intuitively such an inequality must hold, because α quantifies the response of the crystal when we insert an electron into the system, and the response of the crystal can not be stronger than the input of the electron. It is convenient to define

$$\nu = \frac{U}{\alpha}.$$

The inequality for the physical regime can be rewritten as

$$\nu > 2.$$

The Pekar ansatz of an approximate minimizer of a tensor product form leads us to the Pekar-Tomasevich functional, defined as

$$\begin{aligned} \mathcal{P}_{U,\alpha}^{(N)}[\psi] = \sum_{i=1}^{N} \int_{\mathbb{R}^{3N}} |\nabla_{x_i} \psi|^2 dX + U \sum_{i<j} \int_{\mathbb{R}^{3N}} \frac{|\psi(x_1, ..., x_N)|^2}{|x_i - x_j|} dx_1 ... dx_N \\ - \alpha \int\int_{\mathbb{R}^3 \times \mathbb{R}^3} \frac{\rho_\psi(x) \rho_\psi(y)}{|x-y|} dx dy, \end{aligned}$$

where

$$\rho_\psi(x) := \sum_{j=1}^{N} \int |\psi(x_1, ..., x_{j-1}, x, x_{j+1}, ..., x_N)|^2 dx_1 ... \widehat{dx_j} ... dx_N,$$

is the electron density. The hat indicates that $\widehat{dx_j}$ is omitted in the integration. The ground state energies of the Frohlich Hamiltonian and Pekar-Tomasevich functional are denoted by $E_{U,\alpha}^{(N)}$ and $\mathcal{E}_{U,\alpha}^{(N)}$ and are defined as before. Similarly as in the case $N = 1$, we have that $\mathcal{E}_{U,\alpha}^{(N)} = \alpha^2 \mathcal{E}_\nu^{(N)}$, where $\mathcal{E}_\nu^{(N)} := \mathcal{E}_{\nu,1}^{(N)}$. Using the Pekar ansatz it follows, similarly as for $N = 1$, that $E_{U,\alpha}^{(N)} \leq \alpha^2 \mathcal{E}_\nu^{(N)}$. The natural question of whether there is a lower bound of $E_{U,\alpha}^{(N)}$ arises again.

2. Main result and related results

Now we are ready to state the main result of this note. For simplicity we consider here only the physical regime $\nu > 2$, and we also state our result, in [2], in a specialized form.

Theorem 2.1 (Anapolitanos, Landon). *For any given $\nu > 2$, there exists a constant $C_\nu > 0$ such that for all $\alpha > N^4$,*

$$E_{U,\alpha}^{(N)} \geq \alpha^2 \mathcal{E}_\nu^{(N)} - C_\nu \alpha^{9/5} N^{9/5}.$$

This result generalizes the result of Lieb and Thomas mentioned above. Results associated to ours have been proven recently. One of them is by Griesemer and Wellig, see [7], giving a similar bound for $N = 1$ but with very general external electric and magnetic fields. Wellig, see [14], has generalized our result by including some types of external electric and magnetic fields, including periodic ones. His error term estimate is slightly weaker than ours. Unfortunately, none of the results have taken into account the Fermionic statistics of the electrons. In 2010 it was proven by Frank, Lieb, Seiringer and Thomas, see [6], that the ground state energy $\mathcal{E}_\nu^{(N)}$ of the Pekar-Tomasevich functional is, for fixed ν, asymptotically proportional to N. In fact we used this to prove Theorem 2.1. It is unsatisfactory that the remainder has a higher order in N than the leading term and the unsatisfactory assumption $\alpha > N^4$ is associated to this issue.

As mentioned earlier, even though electrons always repel, polarons may attract. In this regard, our result has the following application. Lewin proved in 2010 that if the ratio $\nu = \frac{U}{\alpha}$ is in the physical regime but close to the threshold value 2, then there is binding of polarons in the Pekar-Tomasevich model. More precisely, he proved, in [9] , that

Theorem 2.2 (Lewin 2010). *For all $N > 0$ there exists $\epsilon_N > 0$ so that for $2 < \nu < 2 + \epsilon_N$, we have the strict inequality*

$$\mathcal{E}_\nu^{(N)} < \mathcal{E}_\nu^{(k)} + \mathcal{E}_\nu^{(N-k)}, \quad k = 1, \ldots, N-1.$$

Since, due to our theorem, the Pekar-Tomasevich and Fröhlich ground state energies agree to leading order in α, using the result of Lewin we obtain a corollary. It says that under the additional assumption that α is large, we have binding in the Fröhlich model. More precisely,

Corollary 2.1. *For all $N > 0$ there exists $\epsilon_N > 0$ such that for $2 < \nu < 2 + \epsilon_N$, and α large enough we have the strict inequality*

$$E_{U,\alpha}^{(N)} < E_{U,\alpha}^{(k)} + E_{U,\alpha}^{(N-k)}, \quad k = 1, \ldots, N-1.$$

In 2013 the result of Lewin was generalized by Griesemer and one of the authors, see [1] , to include a constant magnetic field. This result together with the result of Wellig [14] mentioned above, gives a similar corollary in the presence of a constant magnetic field. As well, the work of Wellig and Griesemer [7] implies a similar corollary for two polarons in a large class of electric and magnetic fields.

3. Main ideas of the proof

In the rest of the note we explain some main ideas of the proof of Theorem 2.1. In most of the proof we follow closely the proof of Lieb and Thomas in [10] , so we will focus on explaining what has changed with respect to their proof. With the help of the IMS localization it follows that

$$E_{U,\alpha}^{(N)} \geq \inf_{\psi \in \mathcal{B}_R} \langle \psi | H_U^{(N)} | \psi \rangle - \frac{3N\pi^2}{R^2},$$

where

$$\mathcal{B}_R := \{\psi \in L^2(\mathbb{R}^{3N}; \mathcal{F}) |\ \mathrm{supp}\psi \subset C_1 \times C_2 \times ... \times C_N, \text{ with } C_j \text{ a cube in } \mathbb{R}^3 \text{ of size } R\}.$$

This estimate was shown by Lieb and Thomas in [10] for $N = 1$ but can be shown for any N by a trivial adaption of their proof. Its meaning is that the support of the electrons can be localized into some cubes of size R after paying an energy cost of $\frac{3N\pi^2}{R^2}$.

The additional difficulty arising when $N > 1$ is that a priori we do not know whether or not these cubes are close to each other. To handle this difficulty we split the electron cubes $C_1, ..., C_N$ into l-clusters, where $l > 0$, as follows:

We say that two cubes are in the same l-cluster, if they are connected with a path where consecutive cubes have distance less than l. We refer also to the picture below. For cubes that are in the same l-cluster, we know that they are not too far from each other, and it turns out that the methods of Lieb and Thomas are directly applicable. The cubes that are in different l-clusters are pairwise more than a distance l from each other, which allows us to control the interaction between them, with an appropriate choice of l. The picture on the right gives an example of a system with two different l-clusters. The lines connecting the cubes have length less than l.

To control the interaction we generalize a lemma of Frank, Lieb, Seiringer and Thomas in [6] , which they proved for $N = 2$ in order to control the interaction of two electrons that are sufficiently far from each other. More precisely we proceed as follows:

We define $E_{C_1,\dots,C_N}$ to be the ground state energy of the Fröhlich Hamiltonian with the electrons localized in the cubes C_j, namely

$$E_{C_1,\dots,C_N} = \inf_{\mathrm{supp}\psi \subset C_1 \times C_2 \times \cdots \times C_N} \langle \psi | H^{(N)}_{U,\alpha} | \psi \rangle.$$

It is known, see for example [13] , that

$$E_{C_1,\dots,C_N} = -\lim_{T\to\infty} \frac{1}{T} \log Z_{C_1,\dots,C_N}(T), \tag{5}$$

where the partition function $Z_{C_1,\dots,C_N}(T)$ is given by a Feynmann-Kac formula, namely

$$\begin{aligned} &Z_{C_1,\dots,C_N}(T) \\ &= \int_{C_1} dx_1 \dots \int_{C_N} dx_N \int dW^T_{x_1}(\omega_1)\dots dW^T_{x_N}(\omega_N) \chi_{C_1}(\omega_1)\dots\chi_{C_N}(\omega_N) \\ &\times \exp\left[\int_{\mathbb{R}} ds \frac{e^{-|s|}}{2} \sum_{i,j=1}^{N} \int_0^T \frac{\alpha dt}{|\omega_i(t)-\omega_j(t+s)|} - \sum_{i\neq j} \int_0^T \frac{(U/2)dt}{|\omega_i(t)-\omega_j(t)|}\right]. \end{aligned} \tag{6}$$

Here $dW^T_{x_j}$ denotes the conditional Wiener measure for the Brownian paths ω_j starting at point x_j at time 0 and ending again at point x_j at time T. Moreover, $\chi_{C_j}(\omega_j)$ is the characteristic function of the paths ω_j that stay inside the cube C_j for all times. In the exponential we recognize in the second sum the repulsion terms between the electrons and the first term has attraction terms between electrons. Since the attraction is mediated by the crystal there are self-interaction terms as well, namely the terms $i = j$ included in the summation. If we split the interaction terms in the exponential to interaction terms within l-clusters and interaction terms between different l-clusters, the former will give the sum of energies of the l-clusters. This is of course because the formulae (5) and (6) apply when as a system we consider only one cluster.

To control the interaction terms between different l-clusters, namely the terms

$$I_{ij} := \int_{\mathbb{R}} ds \frac{e^{-|s|}}{2} \int_0^T \frac{\alpha dt}{|\omega_i(t)-\omega_j(t+s)|} - \int_0^T \frac{(U/2)dt}{|\omega_i(t)-\omega_j(t)|},$$

where i, j are in different clusters, we proceed as follows: we observe that we are in the physical regime so that $U/2 > \alpha$. So if choose the distance l of the clusters to be large enough, then the ratio of the denominators $|\omega_i(t)-\omega_j(t)|, |\omega_i(t)-\omega_j(t+s)|$ is close enough to 1 for all times due to the constraint that the paths must lie in the C_j's. This implies that $\frac{\alpha}{|\omega_i(t)-\omega_j(t+s)|} < \frac{(U/2)}{|\omega_i(t)-\omega_j(t)|}$ and since moreover $\int_{\mathbb{R}} ds \frac{e^{-|s|}}{2} = 1$, we obtain that $I_{ij} < 0$. This implies that the repulsive interaction terms dominate the attractive interaction terms, so that the total interaction between the clusters is positive. Therefore, the energy of the system is at least as big as the sum of the energies of the clusters. Using this and applying the methods of Lieb and Thomas to each cluster we obtain Theorem 2.1.

Acknowledgements The authors are grateful to Marcel Griesemer and Robert Seiringer for suggesting to them this problem. They acknowledge numerous inspiring discussions with Marcel Griesemer, Robert Seiringer and David Wellig. They would also like to thank Marcel Griesemer and David Wellig for access to an early version of their work [7]. I.A. thanks Ulrich Linden, for useful comments on the current text and Fumio Hiroshima for useful remarks. The research of I.A. was supported by the German Science Foundation (DFG), under Grant No. GR 3213/1-1. The research of B.L was partly supported by NSERC.

Bibliography

[1] I. Anapolitanos, M. Griesemer: *Multipolarons in a constant magnetic field.* To appear in Ann. Henri Poincare.

[2] I. Anapolitanos, B. Landon: *The ground state energy of the multipolaron in the strong coupling limit.* Lett. Math. Phys. **103** no. 12, 1347-1366 (2013).

[3] M.D. Donsker, S.R.S. Varadhan: *Asymptotics for the polaron.* Comm. Pure Appl. Math., **36** no. 4, 505-528 (1983).

[4] R. Feynmann: *Statistical mechanics. A set of lectures.* Westview Press (1972).

[5] H. Fröhlich: *Theory of electrical breakdown in ionic crystals.* Proc. R. Soc. Lond. A, **160**, 230-241 (1937).

[6] R. Frank, E.H. Lieb, R. Seiringer, L.E. Thomas: *Stability and absence of binding for multi-polaron systems.* Publ. Math. IHES **113** no. 1, 39 - 67 (2011).

[7] M. Griesemer, D. Wellig: *Strong coupling polarons in electromagnetic fields.* J. Phys. A: Math. Theor. **46** no. 42, 5202-5212 (2013).

[8] T.D. Lee, F. Low, D. Pines: *The motion of slow electrons in a polar crystal.* Phys. Rev. **90**, 297-302 (1953).

[9] M. Lewin: *Geometric methods for nonlinear many-body quantum systems.* J. Funct. Anal, **260** 3535-3595 (2011).

[10] E.H. Lieb, L.E. Thomas: *Exact ground state energy of the strong-coupling polaron.* Commun. Math. Phys., **183**, 511-519 (1997). Erratum: ibid., **188**, 499-500 (1997).

[11] S.I. Pekar: *Untersuchungen über die Elektronentheorie der Kristalle.* Akademie Verlag, Berlin (1954).

[12] M. Reed, B. Simon: *Methods of Modern Mathematical Physics II: Fourier Analysis and selfadjointness* Academic Press, London (1975).

[13] G. Roepstorff: *Path Integral Approach to Quantum Physics.* Springer, Berlin-Heidelberg-New York (1994).

[14] D. Wellig: *On the Strong Coupling Limit of Many-Polaron Systems in Electromagnetic Fields.* arXiv:1308.5118.

A variation on Smilansky's model

D. Barseghyan

Fachbereich Mathematik, Universität Erlangen-Nürnberg,
Cauerstrasse 11, 91058 Erlangen, Germany
E-mail: barseghyan@math.fau.de

P. Exner

Doppler Institute for Mathematical Physics and Applied Mathematics,
Czech Technical University, Břehová 7, 11519 Prague, Czechia and
Department of Theoretical Physics, NPI, Academy of Sciences,
25068 Řež near Prague, Czechia
E-mail: exner@ujf.cas.cz

We discuss a model which modifies the system studied first by Smilansky. In our case a singular potential 'channel' is replaced by a regular, below unbounded potential which shrinks as it becomes deeper. We demonstrate that, similarly to the original model, such a system exhibits a spectral transition with respect to the coupling constant, and determine the critical value above which a new spectral branch opens. The result is generalized to situations with multiple potential 'channels'.

Keywords: Smilansky model, Schrödinger operators, spectral transition.

1. Introduction

In an inspirative paper [1] Uzy Smilansky discussed a simple example of quantum dynamics which could exhibit a behavior of two substantial different types controlled by the value of the coupling constant. The model in which this can be demonstrated allows for interpretation in different ways, as a one-dimensional system coupled to a heat bath, as a simple quantum graph to the vertices of which one or more one-dimensional harmonic oscillators are attached with a position-dependent coupling strength, or as a two-dimensional quantum system described by the Hamiltonian

$$H_{\mathrm{Sm}} = -\frac{\partial^2}{\partial x^2} + \frac{1}{2}\left(-\frac{\partial^2}{\partial y^2} + y^2\right) + \lambda y \delta(x)\,. \tag{1}$$

The harmonic oscillator frequency can be, of course, changed by scaling to any positive ω^2, what is important is the last term. It was argued [1] that the behavior of the system depends crucially on the coupling parameter: if $|\lambda| > \sqrt{2}$ the particle can escape to infinity along the singular 'channel' in the y direction.

The claim can be made mathematically rigorous in terms of the spectral properties of such an operator: one can prove that that for $|\lambda|$ exceeding the critical value the operator has an additional branch of absolutely continuous spectrum which is not bounded from below [2]. The model was subsequently generalized [3, 4] to the situation when one has more than a single singular 'channel' and its further properties were studied, in particular, the discrete spectrum in the subcritical case.

Moreover, it appeared that there can be other motivations to investigate such systems. Recently Guarneri has used the Smilansky model — or rather its modification in which the motion in the x direction is restricted to a finite interval with periodic boundary conditions — to describe quantum measurements [5]; he studied the time evolution in such a situation identifying the escape along a particular 'channel' with the appropriate reduction of the wave packet.

The mentioned paper [5] concludes with expressing the hope that 'similar behavior may be reproducible with smoother interaction potentials and also in purely classical models'. The aim of this contribution is to demonstrate that this is indeed the case; we summarize here our original paper [7] to which we refer for details of the proofs which are sketched here. We are going to investigate a model in which the δ interaction with y-dependent strength is replaced by a smooth potential channel of increasing depth, and to show that it exhibits the analogous spectral transition as the coupling parameter exceeds a critical value.

To be specific, we are going to study the two-dimensional system the Hamiltonian of which is the partial differential operator on $L^2(\mathbb{R}^2)$ acting as

$$H = -\frac{\partial^2}{\partial x^2} - \frac{\partial^2}{\partial y^2} + \omega^2 y^2 - \lambda y^2 V(xy)\chi_{\{|x|\le a\}}(x), \tag{2}$$

where ω, a are positive constants, $\chi_{\{|y|\le a\}}$ is the indicator function of the interval $(-a, a)$, and the potential V with $\operatorname{supp} V \subset [-a, a]$ is a nonnegative function with bounded first derivative. By Faris-Lavine theorem [6, Thms. X.28 and X.38] the above operator is essentially self-adjoint on $C_0^\infty(\mathbb{R}^2)$; the same is true for its generalization,

$$H = -\frac{\partial^2}{\partial x^2} - \frac{\partial^2}{\partial y^2} + \omega^2 y^2 - \sum_{j=1}^{N} \lambda_j y^2 V_j(xy)\chi_{\{|x-b_j|\le a_j\}}(x) \tag{3}$$

with a finite number of potential channels, where the functions V_j are positive with bounded first derivative, with the supports contained in the intervals $(b_j - a_j, b_j + a_j)$ and such that $\operatorname{supp} V_j \cap \operatorname{supp} V_k = \emptyset$ holds for $j \neq k$.

Our aim is to demonstrate existence of a critical coupling separating two different situations: on one side the spectrum is bounded from below (and discrete in its negative part) while on the other it covers the whole real line. We note that the results discussed here depend substantially on the asymptotic behavior of the potential channels and would not change if the potential is modified in the vicinity of the x-axis, for instance, by replacing the cut-off functions in (2) and (3) with $\chi_{|y|\geq a}(y)$ and $\chi_{|y|\geq a_j}(y)$, respectively. It is also not important that in contrast to the original model with the Hamiltonian (1) we assume that the potential channels depth increases in both directions parallel to the y-axis. Finally, let us notice that if we replace the potential V by the family $V_\varepsilon : V_\varepsilon(u) = \frac{1}{\varepsilon}V(\frac{\cdot}{\varepsilon})$ the quadratic forms of the corresponding operators H_ε converge to the form of operator $H_{\rm Sm}$ as $\varepsilon \to 0$, and because of this one can understand the model described by Hamiltonian H as the regular modification of $H_{\rm Sm}$.

2. Subcritical case

To state the result we will employ a one-dimensional comparison operator

$$L = -\frac{\mathrm{d}^2}{\mathrm{d}x^2} + \omega^2 - \lambda V(x) \tag{4}$$

on $L^2(\mathbb{R})$ with the domain $H^2(\mathbb{R})$. The important property will be the sign of its spectral threshold; since V is supposed to be nonnegative, the latter is a monotonous function of λ and there is a $\lambda_{\rm crit} > 0$ at which the sign changes.

We shall first focus on the subcritical-coupling case.

Theorem 2.1. *Under the stated assumptions, the spectrum of operator H given by (2) is bounded from below provided the operator L is non-negative.*

Sketch of proof. It is obviously sufficient to prove the claim for $\lambda = 1$. We employ Neumann bracketing. Let h_n and $\widetilde{h}_n$ be respectively the restrictions of operator H to the strips $G_n = \mathbb{R} \times \{y : \ln n < y \leq \ln(n+1)\}$ and $\widetilde{G}_n = \mathbb{R} \times \{y : -\ln(n+1) < y \leq -\ln n\}$, $n = 1, 2, \ldots$, with Neumann boundary conditions. Then we have the inequality

$$H \geq \bigoplus_{n=1}^{\infty} (h_n \oplus \widetilde{h}_n), \tag{5}$$

hence straightforward estimates of the spectra $\sigma(h_n)$ and $\sigma(\widetilde{h}_n)$ and their behavior as $n \to \infty$ in combination with the minimax principle yield the result. □

By a straightforward modification of the argument we get the following claim.

Corollary 2.1. *Let H be the operator on $(-c, c) \times \mathbb{R}$ for some $c \geq a$ given the differential expression (2) with Dirichlet (Neumann, periodic) boundary conditions* *in the variable x. The spectrum of H is bounded from below if $L \geq 0$ holds, where L is the operator (4) on $L^2(-c, c)$ with Dirichlet (respectively, Neumann or periodic) boundary conditions.*

3. Supercritical case

Let us turn to the case where the 'escape to infinity' is possible.

Theorem 3.1. *Under our hypotheses, $\sigma(H) = \mathbb{R}$ holds if $\inf \sigma(L) < 0$.*

Sketch of proof. To prove that any real number μ belongs to essential spectrum of operator H we use Weyl's criterion [6, Thm. VII.12]: we find a sequence $\{\psi_k\}_{k=1}^{\infty} \subset D(H)$ such that $\|\psi_k\| = 1$ which contains no convergent subsequence and the norm $\|H\psi_k - \mu\psi_k\| \to 0$ as $k \to \infty$.

Since the claim is invariant under scaling transformations we can suppose without loss of generality that $\inf \sigma(L) = -1$. The spectral threshold is easily seen to be a simple isolated eigenvalue; we denote the corresponding normalized eigenfunction of L by h. Our aim is to show first that $0 \in \sigma_{\mathrm{ess}}(H)$.

We fix a positive ε and choose a natural number $k = k(\varepsilon)$ with which we associate a function $\chi_k \subset C_0^2(1, k)$ satisfying the following conditions

$$\int_1^k \frac{1}{z}\chi_k^2(z)\,\mathrm{d}z = 1 \quad \text{and} \quad \int_1^k z(\chi_k'(z))^2\,\mathrm{d}z < \varepsilon. \tag{6}$$

To give an example, consider the function

$$\tilde{\chi}_k(z) = \frac{8\ln^3 z}{\ln^3 k}\chi_{\{1\leq z\leq\sqrt{k}\}}(z) + \frac{2\ln k - 2\ln z}{\ln k}\chi_{\{\sqrt{k}+1\leq z\leq k-1\}}(z)$$

$$+g_k(z)\chi_{\{\sqrt{k}<z<\sqrt{k}+1\}}(z) + q_k(z)\chi_{\{k-1<z\leq k\}}(z),$$

where g_k and q_k are interpolating functions chosen in such a way that $\tilde{\chi}_k \in C_0^2(1, k)$. The first integral in (6) is positive for $\tilde{\chi}_k$, in fact we have $\int_1^{\sqrt{k}} \frac{1}{z}\tilde{\chi}_k^2(z)\,\mathrm{d}z \geq \frac{1}{4}$, hence we can define $\chi_k(z) = \left(\int_1^k \frac{1}{z}\tilde{\chi}_k^2(z)\,\mathrm{d}z\right)^{-1/2} \tilde{\chi}_k(z)$. This function satisfies by definition the first condition of (6) and one can check that it also satisfies the second one provided k is sufficiently large; this follows from the fact that $\int_1^k z(\chi_k'(z))^2\,\mathrm{d}z = \mathcal{O}\left(\frac{1}{\ln k}\right)$ holds as $k \to \infty$.

Such functions allow us to construct the Weyl sequence we seek. Given a function χ_k with the described properties, we define

$$\psi_k(x, y) := h(xy)\,e^{iy^2/2}\chi_k\left(\frac{y}{n_k}\right) + \frac{f(xy)}{y^2}\,e^{iy^2/2}\chi_k\left(\frac{y}{n_k}\right), \tag{7}$$

where $f(t) := -\frac{i}{2}\, t^2 h(t)$, $t \in \mathbb{R}$, and $n_k \in \mathbb{N}$ is a positive integer. Then straightforward estimates [7] show that choosing n_k large enough for a given k one can achieve that $\|\psi_k\|_{L^2(\mathbb{R}^2)} \geq \frac{1}{2}$ and $\|H\psi_k\|^2_{L^2(\mathbb{R}^2)} < c\varepsilon$ with a fixed c holds for $k = k(\varepsilon)$.

To complete the proof we fix a sequence $\{\varepsilon_j\}_{j=1}^{\infty}$ such that $\varepsilon_j \searrow 0$ holds as $j \to \infty$ and to any j we construct a function $\psi_{k(\varepsilon_j)}$ with the corresponding numbers chosen in such a way that $n_{k(\varepsilon_j)} > k(\varepsilon_{j-1}) n_{k(\varepsilon_{j-1})}$. The norms of $H\psi_{k(\varepsilon_j)}$ satisfy the inequality $\|H\psi_k\|^2_{L^2(\mathbb{R}^2)} < c\varepsilon$, and since the supports of $\psi_{k(\varepsilon_j)}$, $j = 1, 2, \ldots$, do not intersect each other by construction, their sequence converges weakly to zero. This yields the sought Weyl sequence for zero energy; for any nonzero real number μ we use the same procedure replacing the above ψ_k with

$$\psi_k(x,y) = h(xy)\, e^{i\epsilon_\mu(y)} \chi_k\left(\frac{y}{n_k}\right) + \frac{f(xy)}{y^2}\, e^{i\epsilon_\mu(y)} \chi_k\left(\frac{y}{n_k}\right),$$

where $\epsilon_\mu(y) := \int_{\sqrt{|\mu|}}^{y} \sqrt{t^2+\mu}\, \mathrm{d}t$, and furthermore, the functions f, χ_k in this expression are defined in the same way as above. □

In a similar way one can treat the 'longitudinally restricted' case [7]:

Theorem 3.2. *Let H be the operator on $L^2(-c,c)\otimes L^2(\mathbb{R})$ for some $c > 0$ acting as (2) with Dirichlet condition at $x = \pm c$ and denote by L the corresponding Dirichlet operator (4) on $L^2(-c,c)$. If the spectral threshold of L is negative, the spectrum of H covers the whole real axis.*

Observing the domains of the quadratic forms associated with such operators we can extend the last result in the following way.

Corollary 3.1. *The claim of Theorem 3.2 remains valid if the Dirichlet boundary conditions at $x = \pm c$ are replaced by any other self-adjoint boundary conditions.*

Finally, these results allow us also through bracketing arguments to answer the question about spectral transition for the model with multiple singular channels.

Theorem 3.3. *Let H be the operator (2) with the potentials satisfying the stated assumptions, namely the functions V_j are positive with bounded first derivative and $\operatorname{supp} V_j \cap \operatorname{supp} V_k = \emptyset$ holds for $j \neq k$. Denote by L_j the operator (4) on $L^2(R)$ with the potential V_j and $t_V := \min_j \inf \sigma(L_j)$. Then H is bounded from below if and only if $t_V \geq 0$ and in the opposite case its spectrum covers the whole real axis.*

Acknowledgments

The research was supported by the Czech Science Foundation within the project 14-06818S. D.B. is grateful to Mittag-Leffler Institute for the hospitality and to H. Schulz-Baldes for a useful discussion.

Bibliography

[1] U. Smilansky: Irreversible quantum graphs, *Waves Random Media* **14** (2004), 143–153.
[2] M. Solomyak: On a differential operator appearing in the theory of irreversible quantum graphs, *Waves Random Media* **14** (2004), 173–185.
[3] S. Naboko, M. Solomyak: On the absolutely continuous spectrum in a model of an irreversible quantum graph, *Proc. Lond. Math. Soc.*, **92** (2006), 251–272.
[4] W.D. Evans, M. Solomyak: Smilansky's model of irreversible quantum graphs: I. The absolutely continuous spectrum, II .The point spectrum, *J. Phys. A: Math. Gen.*, **38** (2005), 4611–4627, 7661–7675.
[5] I. Guarneri: Irreversible behaviour and collapse of wave packets in a quantum system with point interactions, *J. Phys. A: Math. Theor.*, **44** (2011), 485304
[6] M. Reed, B. Simon: Methods of Modern Mathematical Physics, I. Functional Analysis. II. Fourier Analysis. Self-Adjointness, *Academic Press*, New York 1981, 1975.
[7] D. Barseghyan, P. Exner: A regular version of Smilansky model, *J. Math. Phys.*, **55** (2014), 042194.

Deriving the Gross-Pitaevskii equation

N. Benedikter

Institute of Mathematics, University of Zurich,
Winterthurerstrasse 190, 8057 Zurich, Switzerland
E-mail: niels.benedikter@tofoq.de

In experiments, Bose-Einstein condensates are prepared by cooling a dilute Bose gas in a trap. After the phase transition has been reached, the trap is switched off and the evolution of the condensate observed. The evolution is macroscopically described by the Gross-Pitaevskii equation. On the microscopic level, the dynamics of Bose gases are described by the N-body Schrödinger equation. We review our article [1] in which we construct a class of initial data in Fock space which are energetically close to the ground state and prove that their evolution approximately follows the Gross-Pitaevskii equation. The key idea is to model two-particle correlations with a Bogoliubov transformation.

Keywords: Bose-Einstein condensate, dilute Bose gas, Gross-Pitaevskii equation, correlations, many-body systems, Bogoliubov transformation, squeezed coherent states.

1. Bose-Einstein condensates in the Gross-Pitaevskii regime

In dilute gases of bosonic particles, at very low temperatures a phase transition occurs and a macroscopic number of particles occupies the same one-particle state. Bose and Einstein theoretically predicted this state of matter in 1924, considering non-interacting bosons. The experimental confirmation took 70 years and was rewarded with a Nobel prize in 2001, and the theoretical study of interacting Bose-Einstein condensates still poses challenging problems.

A dilute gas of N bosons can be described with the Hamilton operator

$$H_N^U = \sum_{j=1}^{N} (-\Delta_j + U(x_j)) + \sum_{i<j}^{N} N^2 V(N(x_i - x_j))$$

acting on the Hilbert space $L^2_s(\mathbb{R}^{3N})$, the subspace of $L^2(\mathbb{R}^{3N})$ symmetric with respect to permutation of the N particles. The external potential U models the trap that confines the particles. The interaction potential V is repulsive ($V \geq 0$) and spherically symmetric; the Gross-Pitaevskii scaling $N^2V(N.)$ models rare but strong collisions. We are interested in large N (in experiments, $N \gg 10^3$).

A central role is played by the solution f to the zero-energy scattering equation

$$\Big(-\Delta+\frac{1}{2}V\Big)f=0,\quad \text{with boundary condition } f(x)\to 1\quad (|x|\to\infty).$$

Its solution has the form $f(x)\simeq 1-a_0/|x|$ for large x, where $a_0:=(8\pi)^{-1}\int fV\mathrm{d}x$ is the scattering length of V. By scaling, $f(N.)$ solves the zero-energy scattering equation with scaled potential $N^2V(N.)$.

It was proven [2] that the ground state energy E_N of the Hamiltonian H_N^U satisfies

$$\lim_{N\to\infty}\frac{E_N}{N}=\min_{\varphi\in L^2(\mathbb{R}^3),\,\|\varphi\|_2=1}\mathcal{E}_{\mathrm{GP}}(\varphi),\tag{1}$$

with the Gross-Pitaevskii (GP) energy functional

$$\mathcal{E}_{\mathrm{GP}}(\varphi):=\int\mathrm{d}x\,\big(|\nabla\varphi|^2+U|\varphi|^2+4\pi a_0|\varphi|^4\big)\,.\tag{2}$$

The ground state ψ_N^{gs} exhibits [3] complete Bose-Einstein condensation, in the sense

$$\gamma^{(1)}_{\psi_N^{\mathrm{gs}}}\xrightarrow{\text{in trace norm}}|\varphi_{\mathrm{GP}}\rangle\langle\varphi_{\mathrm{GP}}|\quad(N\to\infty),$$

where $|\varphi_{\mathrm{GP}}\rangle\langle\varphi_{\mathrm{GP}}|$ is the projection on the minimizer φ_{GP} of the GP functional (2), and $\gamma^{(1)}_{\psi_N^{\mathrm{gs}}}$ is the one-particle reduced density matrix associated with ψ_N^{gs}, i.e. the trace-class operator on $L^2(\mathbb{R}^3)$ defined through the integral kernel

$$\gamma^{(1)}_{\psi_N^{\mathrm{gs}}}(x;y):=\int\mathrm{d}x_2\ldots\mathrm{d}x_N\,\psi_N^{\mathrm{gs}}(x,x_2,\ldots,x_N)\,\overline{\psi_N^{\mathrm{gs}}}(y,x_2,\ldots,x_N).$$

Since we generally assume $\|\psi\|_2=1$, we have $\operatorname{tr}\gamma^{(1)}_\psi=1$.

When the traps are switched off ($U=0$), the system starts to evolve, following the Schrödinger equation

$$i\partial_t\psi_{N,t}=H_N^0\psi_{N,t},\quad \psi_{N,0}=\psi_N^{\mathrm{gs}}.$$

It was proven [4, 5] that $\gamma^{(1)}_{\psi_{N,t}}\to|\varphi_t\rangle\langle\varphi_t|$ as $N\to\infty$ (for any fixed $t>0$), where φ_t is the solution to the non-linear GP equation (here with initial data $\varphi=\varphi_{\mathrm{GP}}$)

$$i\partial_t\varphi_t=-\Delta\varphi_t+8\pi a_0|\varphi_t|^2\varphi_t,\quad \varphi_0=\varphi.\tag{3}$$

In our analysis we generalize the system to Fock space, with the advantage that we can use initial data that are superpositions of states with different numbers of particles. We introduce the bosonic Fock space $\mathcal{F}:=\bigoplus_{j=0}^\infty L^2_{\mathrm{s}}(\mathbb{R}^{3j})$ and creation/annihilation operators (more precisely operator-valued distributions) a_x^*, a_x, which create/annihilate a particle at $x\in\mathbb{R}^3$. They satisfy the canonical commutation relations $[a_x,a_y^*]=\delta(x-y)$, $[a_x^*,a_y^*]=0=[a_x,a_y]$. We introduce the number of particles operator $\mathcal{N}=\int\mathrm{d}x\,a_x^*a_x$ and the vacuum vector $\Omega=(1,0,\ldots)\in\mathcal{F}$. On Fock space $\mathcal{F}$ we define the Hamiltonian

$$\mathcal{H}_N:=\int\mathrm{d}x\,\nabla_x a_x^*\nabla_x a_x+\frac{1}{2N}\int\mathrm{d}x\mathrm{d}y\,N^3V(N(x-y))a_x^*a_y^*a_ya_x.\tag{4}$$

The restriction of $\mathcal{H}_N$ to $L^2_s(\mathbb{R}^{3N})$ coincides with the Hamiltonian H^0_N.

For $g \in L^2(\mathbb{R}^3)$ we define the Weyl operator

$$W(g) := \exp\left(\int \mathrm{d}x\, a^*_x g(x) - \text{h.c.}\right),$$

and for integral kernels $k \in L^2(\mathbb{R}^3 \times \mathbb{R}^3)$ we introduce the Bogoliubov transformation

$$T(k) := \exp\left(\frac{1}{2}\int \mathrm{d}x\mathrm{d}y\, k(x;y) a^*_x a^*_y - \text{h.c.}\right). \tag{5}$$

The Weyl operators have the important property of shifting the operators,

$$W^*(g)a^*_x W(g) = a^*_x + \bar{g}(x), \quad W^*(g) a_x W(g) = a_x + g(x), \tag{6}$$

whereas the Bogoliubov transformation $T(k)$ acts by

$$T^*(k) a^*_x T(k) = \int \mathrm{d}y\, \left(a^*_y \cosh(k)(y;x) + a_y \sinh(k)(y;x)\right). \tag{7}$$

Here $\cosh(k)(y;x)$ and $\sinh(k)(y;x)$ are the integral kernels defined by the power series (in k) of the hyperbolic cosine/sine, with product in the sense of operators.

We will use a Weyl operator to generate a condensate. (The coherent state $W(g)\Omega$ describes a condensate with approximately $\|g\|_2^2$ particles in the one-particle state $g/\|g\|_2$.) A Bogoliubov transformation, on the other hand, is used to implement correlations among the particles. Our main result [1] is the following theorem.

Theorem 1.1. *Let $V \geq 0$ and $V \in L^1 \cap L^3(\mathbb{R}^3, (1+|x|^6)\mathrm{d}x)$. Let $\varphi \in H^4(\mathbb{R}^3)$ with $\|\varphi\|_2 = 1$. Let $k_0(x;y) := -N\left(1 - f\left(N\left(x-y\right)\right)\right)\varphi(x)\varphi(y)$. Let $\chi \in \mathcal{F}$, possibly depending on N but s. t. $\langle \chi, \left(\mathcal{N} + 1 + \mathcal{N}^2/N + \mathcal{H}_N\right)\chi\rangle$ is bounded uniform in N. We consider $\psi_{N,t} := e^{-i\mathcal{H}_N t} W(\sqrt{N}\varphi) T(k_0)\chi$, the solution to the Schrödinger equation in Fock space, i. e. $i\partial_t \psi_{N,t} = \mathcal{H}_N \psi_{N,t}$. Then there exist constants $C, c > 0$ s. t.*

$$\operatorname{tr}\left|\gamma^{(1)}_{\psi_{N,t}} - |\varphi_t\rangle\langle\varphi_t|\right| \leq \frac{C}{\sqrt{N}} \exp(c\exp(c|t|)),$$

where φ_t solves the GP equation (3) with initial data $\varphi_0 = \varphi$.

The vector χ in the initial data describes small deviations from the squeezed coherent state $W(\sqrt{N}\varphi)T(k_0)\Omega$. The correlation structure is inserted already in the initial data; our proof keeps this structure static, showing its approximate stability. Our initial data arises naturally as an approximation to the ground state since

$$\langle W(\sqrt{N}\varphi)T(k_0)\chi, (\mathcal{H}_N + \int \mathrm{d}x\, U(x) a^*_x a_x) W(\sqrt{N}\varphi)T(k_0)\chi\rangle = N\mathcal{E}_{\text{GP}}(\varphi) + \mathcal{O}(N^{1/2}).$$

In our article [1] we also discuss initial data with exact number of particles.

2. Strategy: Modeling correlations by a Bogoliubov transformation

The approach is inspired by the method of coherent states [6], developed for studying the mean-field regime. However, coherent states cannot provide a good approximation in the GP regime because they describe completely uncorrelated states. To take into account the correlations we use Bogoliubov transformations.

For technical reasons we will compare the solution of the many-body Schrödinger equation first to the solution of the modified GP equation

$$i\partial_t \varphi_t^{(N)} = -\Delta\varphi_t^{(N)} + \left(N^3 f(N.)V(N.) * |\varphi_t^{(N)}|^2\right)\varphi_t^{(N)}, \quad \varphi_0^{(N)} = \varphi. \tag{8}$$

Since $N^3 f(N.)V(N.) \to 8\pi a_0 \delta$, it is easy to compare $\varphi_t^{(N)}$ with the solution φ_t of the GP equation (3). With f the solution to the zero-energy scattering equation, let

$$k_t(x;y) := -N\left(1 - f\left(N\left(x-y\right)\right)\right)\varphi_t^{(N)}(x)\,\varphi_t^{(N)}(y).$$

We try to approximate the full evolution $\psi_{N,t} = e^{-i\mathcal{H}_N t} W(\sqrt{N}\varphi)T(k_0)\chi$ with the (up to the small deviation given by χ) squeezed coherent state $W(\sqrt{N}\varphi_t^{(N)})T(k_t)\chi$. Thus, inspired by [6], we introduce the fluctuation dynamics

$$\mathcal{U}_N(t) := T^*(k_t)W^*(\sqrt{N}\varphi_t^{(N)})e^{-i\mathcal{H}_N t}W(\sqrt{N}\varphi)T(k_0).$$

We find the estimate

$$\operatorname{tr}\left|\gamma_{\psi_{N,t}}^{(1)} - |\varphi_t^{(N)}\rangle\langle\varphi_t^{(N)}|\right| \leq \frac{C}{\sqrt{N}}\langle \mathcal{U}_N(t)\chi, \mathcal{N}\mathcal{U}_N(t)\chi\rangle. \tag{9}$$

Hence, to show the convergence of the many-body dynamics to the GP equation, the central task is to bound the number of fluctuations $\langle \mathcal{U}_N(t)\chi, \mathcal{N}\mathcal{U}_N(t)\chi\rangle$ uniformly in N. In the next section we explain how to obtain such a bound.

3. Controlling the number of fluctuations

In this section we use the shorthands $W_t := W(\sqrt{N}\varphi_t^{(N)})$, $T_t := T(k_t)$ and $\langle . \rangle_t := \langle \mathcal{U}_N(t)\chi, . \,\mathcal{U}_N(t)\chi\rangle$.

We intend to use Grönwall's lemma. Hence we compute the derivative

$$\partial_t\langle\mathcal{N}\rangle_t = \langle [i\mathcal{L}_N(t), \mathcal{N}]\rangle_t, \tag{10}$$

with $\mathcal{L}_N(t)$ the time-dependent generator of $\mathcal{U}_N(t)$. Explicitly

$$\mathcal{L}_N(t) = (i\partial_t T_t^*)T_t + T_t^*\left((i\partial_t W_t^*)W_t + W_t^*\mathcal{H}_N W_t\right)T_t =: (i\partial_t T_t^*)T_t + T_t^*\mathcal{L}_N^{(0)}(t)T_t.$$

The term $(i\partial_t T_t^*)T_t$ is harmless. Let us focus on the second term. In $\mathcal{L}_N^{(0)}(t)$ we have

$$(i\partial_t W_t^*)W_t = -\sqrt{N}\int \mathrm{d}x\, a_x^*\, i\partial_t\varphi_t^{(N)}(x) + \text{h.c.}\ (+\text{ irrelevant scalar}).$$

For $W_t^* \mathcal{H}_N W_t$ we use (6) and expand. We get summands which are linear in creation/annihilation operators and formally of order $N^{1/2}$; moreover quadratic summands of order one, cubics of order $N^{-1/2}$ and quartics of order N^{-1}.

Unlike in the mean-field regime [6], where the Hartree equation implies complete cancellation of the linear terms in $W_t^* \mathcal{H}_N W_t$ with $(i\partial_t W_t^*)W_t$, the modified GP equation (8) leaves us with the linear, large remainder

$$N^{1/2} \int \mathrm{d}x \left(N^3 V(N.)\,(1 - f(N.)) * |\varphi_t^{(N)}|^2\right)(x)\, \varphi_t^{(N)}(x)\, a_x^* + \mathrm{h.\,c.} \qquad (11)$$

The key observation is that by conjugating $\mathcal{L}_N^{(0)}(t)$ with T_t, using (7) and expanding, we get (among many other terms) cubic terms which are not normal-ordered. Normal-ordering them by the canonical commutation relations gives rise to a linear term which cancels (11).

Similarly we get cancellations between quadratic and quartic terms of $\mathcal{L}_N^{(0)}(t)$: we conjugate them with T_t and expand the product; then normal-ordering of quartic terms gives rise to extra quadratic terms. Using the zero-energy scattering equation we now find a cancellation of otherwise problematic quadratic terms. (For identifying this cancellation think of $\sinh(k_t)(x;y)$ as $k_t(x;y)$ and of $\cosh(k_t)(x;y)$ as $\delta(x-y)$.)

These cancellations are crucial; they allow us to prove the operator inequality

$$[i\mathcal{L}_N(t), \mathcal{N}] \leq \mathcal{H}_N + C_t\left(\mathcal{N}^2/N + \mathcal{N} + 1\right).$$

(We denote by C_t varying constants which may grow exponentially in t (since we use bounds of the form $\|\varphi_t^{(N)}\|_{H^n} \leq Ce^{K|t|}$), but are independent of N.)

Next observe that $\mathcal{L}_N(t) = \mathcal{H}_N +$ other terms, where 'other terms' can be bounded above and below by $\varepsilon\mathcal{H}_N$ (any $\varepsilon > 0$) and the number operator. Thus

$$\mathcal{H}_N \leq C_t\left(\mathcal{L}_N(t) + \mathcal{N}^2/N + \mathcal{N} + 1\right). \qquad (12)$$

Thus we obtain $[i\mathcal{L}_N(t), \mathcal{N}] \leq C_t\left(\mathcal{L}_N(t) + \mathcal{N}^2/N + \mathcal{N} + 1\right)$. It is possible to control $\langle \mathcal{N}^2/N\rangle_t$ by $\langle(\mathcal{N}+1)^2/N\rangle_{t=0}$ combined with $\langle\mathcal{N}\rangle_t$. This implies

$$\partial_t\langle\mathcal{N}\rangle_t \leq C_t\langle\mathcal{N} + 1 + \mathcal{L}_N(t)\rangle_t + C_t\langle(\mathcal{N}+1)^2/N\rangle_0.$$

To close the scheme of Grönwall's lemma, we need to control the growth of $\langle\mathcal{L}_N(t)\rangle_t$. Similar to the above estimates we find

$$\partial_t\langle\mathcal{L}_N(t)\rangle_t = \langle\dot{\mathcal{L}}_N(t)\rangle_t \leq C_t\langle\mathcal{L}_N(t) + \mathcal{N} + 1\rangle_t + C_t\langle(\mathcal{N}+1)^2/N\rangle_0.$$

Combining the last two bounds, we obtain (for some fixed D_t to be chosen later)

$$\partial_t\langle D_t(\mathcal{N}+1) + \mathcal{L}_N(t)\rangle_t \leq C_t\langle D_t(\mathcal{N}+1) + \mathcal{L}_N(t)\rangle_t + C_t\langle(\mathcal{N}+1)^2/N\rangle_0.$$

Thus, Grönwall's lemma implies that for some $C, c > 0$

$$\langle\mathcal{L}_N(t) + D_t(\mathcal{N}+1)\rangle_t \leq C\exp(c\exp(ct))\langle\mathcal{L}_N(0) + \mathcal{N} + 1 + \mathcal{N}^2/N\rangle_0.$$

By (12) there exists a $C_t > 0$ such that $\mathcal{L}_N(t) + C_t(\mathcal{N}^2/N + \mathcal{N}) \geq 0$. Choosing $D_t := C_t + 1$ we obtain

$$\langle \mathcal{N} \rangle_t \leq \langle \mathcal{L}_N(t) + D_t(\mathcal{N}^2/N + \mathcal{N}) \rangle_t \leq C \exp(c \exp(ct)),$$

which by (9) completes the proof of the main result.

Acknowledgments

The author would like to thank Benjamin Schlein and Gustavo de Oliveira for useful comments on a previous version of this review.

Bibliography

[1] N. Benedikter, G. de Oliveira and B. Schlein, *Preprint arXiv:1208.0373* (2012).
[2] E. H. Lieb, R. Seiringer and J. Yngvason, *Phys. Rev. A* **61**, p. 043602 (2000).
[3] E. H. Lieb and R. Seiringer, *Phys. Rev. Lett.* **88**, p. 170409 (2002).
[4] L. Erdős, B. Schlein and H.-T. Yau, *Ann. of Math. (2)* **172**, 291 (2010).
[5] P. Pickl, *Preprint arXiv:1001.4894* (2010).
[6] I. Rodnianski and B. Schlein, *Comm. Math. Phys.* **291**, 31 (2009).

Boundary triplets approach for Dirac operator

A. A. Boitsev

Petersburg National Research University of Information Technologies, Mechanics and Optics, 49 Kronverkskiy, St. Petersburg, 197101, Russia
E-mail: boitsevanton@gmail.com

Boundary triplets approach is applied to the construction of self-adjoint extensions of the Dirac operator. Previous results on the extension of tensor product of operators are discussed.

1. Introduction

The operators having representation of sum of tensor products appear in quantum physics. In order to describe the correct interaction between two systems the self-adjoint extension theory is used. We consider an operator S having form $S = A \otimes I_T + I_A \otimes T$, where A is densely defined closed symmetric operator in Hilbert space $\mathcal{H}_A$ and T is bounded and self-adjoint operator.

The results for these types of operators are obtained in paper [9]. We are going to show the application of the obtained results to a particular operator - Dirac operator of the form

$$S := -ic\frac{d}{dx} \otimes I_T + I \otimes \begin{pmatrix} \frac{c^2}{2} & 0 \\ 0 & -\frac{c^2}{2} \end{pmatrix}. \qquad (1)$$

2. Preliminaries

2.1. *Linear relations*

A linear relation Θ in $\mathcal{H}$ is a closed linear subspace of $\mathcal{H} \oplus \mathcal{H}$. The set of all linear relations in $\mathcal{H}$ is denoted by $\widetilde{\mathcal{C}}(\mathcal{H})$. Denote also by $\mathcal{C}(\mathcal{H})$ the set of all closed linear (not necessarily densely defined) operators in $\mathcal{H}$. Identifying each operator $T \in \mathcal{C}(\mathcal{H})$ with its graph $\mathrm{gr}\,(T)$ we regard $\mathcal{C}(\mathcal{H})$ as a subset of $\widetilde{\mathcal{C}}(\mathcal{H})$.

The role of the set $\widetilde{\mathcal{C}}(\mathcal{H})$ in extension theory becomes clear from Proposition 2.1. However, its role in the operator theory is substantially motivated by the following circumstances: in contrast to $\mathcal{C}(\mathcal{H})$, the set $\widetilde{\mathcal{C}}(\mathcal{H})$ is closed with respect to taking inverse and adjoint relations Θ^{-1} and Θ^*. The latter are given by: $\Theta^{-1} = \{\{g, f\} : \{f, g\} \in \Theta\}$ and

$$\Theta^* = \left\{ \begin{pmatrix} k \\ k' \end{pmatrix} : (h', k) = (h, k') \text{ for all } \begin{pmatrix} h \\ h' \end{pmatrix} \in \Theta \right\}.$$

A linear relation Θ is called symmetric if $\Theta \subset \Theta^*$ and self-adjoint if $\Theta = \Theta^*$.

2.2. *Boundary triplets and proper extensions*

Let us briefly recall some basic fact on boundary triplets. Let S be a densely defined closed symmetric operator with equal deficiency indices $n_\pm(S) := \dim(\mathfrak{N}_{\pm i})$, $\mathfrak{N}_z := \ker(S^* - z)$, $z \in \mathbb{C}_\pm$, acting on some separable Hilbert space $\mathfrak{H}$.

Definition 2.1.

1) A closed extension $\widetilde{S}$ of S is called proper if $\operatorname{dom}(S) \subset \operatorname{dom}(\widetilde{S}) \subset \operatorname{dom}(S^*)$.

2) Two proper extensions $\widetilde{S}'$, $\widetilde{S}$ are called disjoint if $\operatorname{dom}(\widetilde{S}') \cap \operatorname{dom}(\widetilde{S}) = \operatorname{dom}(S)$ and transversal if in addition $\operatorname{dom}(\widetilde{S}') + \operatorname{dom}(\widetilde{S}) = \operatorname{dom}(S^*)$.

Denote by Ext_S the set of all proper extensions of S completed by the non-proper extensions S and S^*. Any self-adjoint or maximal dissipative (accumulative) extension is proper.

Definition 2.2 ([5]). *A triplet $\Pi = \{\mathcal{H}, \Gamma_0, \Gamma_1\}$, where $\mathcal{H}$ is an auxiliary Hilbert space and $\Gamma_0, \Gamma_1 : \operatorname{dom}(S^*) \to \mathcal{H}$ are linear mappings, is called a boundary triplet for S^* if the "abstract Green's identity"*

$$(S^* f, g) - (f, S^* g) = (\Gamma_1 f, \Gamma_0 g) - (\Gamma_0 f, \Gamma_1 g), \quad f, g \in \operatorname{dom}(S^*), \tag{2}$$

fied and the mapping $\Gamma := (\Gamma_0, \Gamma_1)^\top : \operatorname{dom}(S^*) \to \mathcal{H} \oplus \mathcal{H}$ is surjective, i.e. $\operatorname{ran}(\Gamma) = \mathcal{H} \oplus \mathcal{H}$. ◇

A boundary triplet $\Pi = \{\mathcal{H}, \Gamma_0, \Gamma_1\}$ for S^* always exists whenever $n_+(S) = n_-(S)$. Note also that $n_\pm(S) = \dim(\mathcal{H})$ and $\ker(\Gamma_0) \cap \ker(\Gamma_1) = \operatorname{dom}(S)$.

With any boundary triplet Π one associates two canonical self-adjoint extensions $S_j := S^* \upharpoonright \ker(\Gamma_j)$, $j \in \{0, 1\}$. Conversely, for any extension $S_0 = S_0^* \in \mathrm{Ext}_S$ there exists a (non-unique) boundary triplet $\Pi = \{\mathcal{H}, \Gamma_0, \Gamma_1\}$ for S^* such that $S_0 := S^* \upharpoonright \ker(\Gamma_0)$.

Using the concept of boundary triplets one can parameterize all proper extensions of A in the following way.

Proposition 2.1 ([3, 6]). *Let $\Pi = \{\mathcal{H}, \Gamma_0, \Gamma_1\}$ be a boundary triplet for S^*. Then the mapping*

$$\mathrm{Ext}_S \ni \widetilde{S} \to \Gamma \mathrm{dom}\,(\widetilde{S}) = (\Gamma_0 f, \Gamma_1 f)^\top : f \in \mathrm{dom}\,(\widetilde{S})\} =: \Theta \in \widetilde{\mathcal{C}}(\mathcal{H}) \tag{3}$$

establishes a bijective correspondence between the sets Ext_S *and* $\widetilde{\mathcal{C}}(\mathcal{H})$. *We write* $\widetilde{S} = S_\Theta$ *if* $\widetilde{S}$ *corresponds to* Θ *by* (3). *Moreover, the following holds:*

(i) $S_\Theta^* = S_{\Theta^*}$, *in particular,* $S_\Theta^* = S_\Theta$ *if and only if* $\Theta^* = \Theta$.

(ii) S_Θ *is symmetric (self-adjoint) if and only if* Θ *is symmetric (self-adjoint).*

(iii) The extensions S_Θ *and* S_0 *are disjoint (transversal) if and only if there is a closed (bounded) operator* B *such that* $\Theta = \mathrm{gr}\,(B)$. *In this case* (3) *takes the form*

$$S_\Theta := S_{\mathrm{gr}\,(B)} = S^* \upharpoonright \ker\,(\Gamma_1 - B\Gamma_0). \tag{4}$$

In particular, $S_j := S^* \upharpoonright \ker\,(\Gamma_j) = S_{\Theta_j}$, $j \in \{0,1\}$, where $\Theta_0 := \{0\} \times \mathcal{H}$ and $\Theta_1 := \mathcal{H} \times \{0\} = \mathrm{gr}\,(\mathbb{O})$ where $\mathbb{O}$ denotes the zero operator in $\mathcal{H}$. Note also that $\widetilde{\mathcal{C}}(\mathcal{H})$ contains the trivial linear relations $\{0\} \times \{0\}$ and $\mathcal{H} \times \mathcal{H}$ parameterizing the extensions S and S^*, respectively, for any boundary triplet Π.

2.3. *γ-field and Weyl function*

It is well known that Weyl functions are an important tool in the direct and inverse spectral theory of Sturm-Liouville operators. In [2, 3] the concept of Weyl function was generalized to the case of an arbitrary symmetric operator S with $n_+(S) = n_-(S) \leq \infty$. Following [3] we briefly recall basic facts on Weyl functions and γ-fields associated with a boundary triplet Π.

Definition 2.3 ([2, 3]). Let $\Pi = \{\mathcal{H}, \Gamma_0, \Gamma_1\}$ be a boundary triplet for S^* and $S_0 = S^* \upharpoonright \ker\,(\Gamma_0)$. The operator valued functions $\gamma(\cdot) : \rho(S_0) \to [\mathcal{H}, \mathfrak{H}]$ and $M(\cdot) : \rho(S_0) \to [\mathcal{H}]$ defined by

$$\gamma(z) := \left(\Gamma_0 \upharpoonright \mathfrak{N}_z\right)^{-1} \quad \text{and} \quad M(z) := \Gamma_1 \gamma(z), \quad z \in \rho(S_0), \tag{5}$$

are called the γ-field and the Weyl function, respectively, corresponding to the boundary triplet Π.

Clearly, the Weyl function can equivalently be defined by

$$M(z)\Gamma_0 f_z = \Gamma_1 f_z, \qquad f_z \in \mathfrak{N}_z, \quad z \in \rho(S_0).$$

The γ-field $\gamma(\cdot)$ and the Weyl function $M(\cdot)$ in (5) are well defined. Moreover, both $\gamma(\cdot)$ and $M(\cdot)$ are holomorphic on $\rho(S_0)$.

2.4. *Krein-type formula for resolvents*

Let $\Pi = \{\mathcal{H}, \Gamma_0, \Gamma_1\}$ be a boundary triplet for S^*, $M(\cdot)$ and $\gamma(\cdot)$ the corresponding Weyl function and γ-field, respectively. For any proper (not necessarily self-adjoint) extension $\widetilde{S}_\Theta \in \mathrm{Ext}_S$ with non-empty resolvent set $\rho(\widetilde{S}_\Theta)$ the following Krein-type formula holds (cf. [2–4])

$$(S_\Theta - z)^{-1} - (S_0 - z)^{-1} = \gamma(z)(\Theta - M(z))^{-1}\gamma^*(\overline{z}), \quad z \in \rho(S_0) \cap \rho(S_\Theta). \tag{6}$$

Formula (6) extends the known Krein formula for canonical resolvents to the case of any $S_\Theta \in \mathrm{Ext}_S$ with $\rho(S_\Theta) \neq \emptyset$. Moreover, due to relations (3), (4) and (5) formula (6) is connected with the boundary triplet Π. Emphasize, that this connection makes it possible to apply the Krein-type formula (6) to boundary value problems.

2.5. *Results on bounded self-adjoint operator*

Let A be a densely defined symmetric operator with equal deficiency indices acting in the separable Hilbert space $\mathfrak{H}_A$ and let T be bounded self-adjoint operator acting on the separable Hilbert space $\mathcal{H}_T$. The operator $S := A \otimes I_T + I_A \otimes T$ is considered. The operator S is defined as a closure of the operator $A \odot I_T + I_A \odot T$, so that

$$\mathrm{dom}\,(A \odot I_T + I_A \odot T) := \{f = \sum_{k=1}^{n} g_k \otimes h_k : g_k \in \mathrm{dom}\,(A), h_k \in \mathcal{H}_T\} \tag{7}$$

and

$$(A \odot I_T + I_A \odot T)f = \sum_{k=1}^{n} (Ag_k \otimes h_k + g_k \otimes Th_k), f \in \mathrm{dom}\,(A \odot I_T + I_A \odot T). \tag{8}$$

The operator S is a well defined closed symmetric operator (see [9]).

Let us recall the main results of [9].

Proposition 2.2. *If $\Pi_A = \{\mathcal{H}_A, \Gamma_0^A, \Gamma_1^A\}$ is a boundary triplet for A^*, then $\Pi = \{\mathcal{H}, \Gamma_0, \Gamma_1\}$ is a boundary triplet for S^*, where*

$$\mathcal{H} := \mathcal{H}_A \otimes \mathcal{H}_T, \Gamma_0^S := \Gamma_0^A \otimes I, \Gamma_1^S := \Gamma_1^A \otimes I. \tag{9}$$

For the γ_S-field and Weyl function corresponding to Π_S the following statements hold:

Proposition 2.3. *Let Π_A be a boundary triplet for A^* with γ-field $\gamma_A(z)$. If Π_S is the boundary triplet of Proposition 2.2 of S^*, then the γ-field $\gamma_S(\cdot)$ of Π_S admits the representation*

$$\gamma_S(z) = \int_a^b d\widehat{E}_T(\lambda)\, \gamma_A(z - \lambda) \otimes I_{\mathcal{H}_T} = \int_a^b \gamma_A(z - \lambda) \otimes I_{\mathcal{H}_T}\, d\widehat{E}_T(\lambda) \tag{10}$$

$z \in \mathbb{C}_\pm$ where $\sigma(T) \subset [a, b)$.

Proposition 2.4. *Let Π_A be a boundary triplet for A^* with Weyl function $M_A(\cdot)$. If Π_S is the boundary triplet of Proposition 2.2 of S^*, then the Weyl function $M_S(\cdot)$ of Π_S admits the representation*

$$\begin{aligned} M_S(z) &= \int_a^b d\widehat{E}_T(\lambda)\, M_A(z-\lambda) \otimes I_{\mathcal{H}_T} \\ &= \int_a^b M_A(z-\lambda) \otimes I_{\mathcal{H}_T}\, d\widehat{E}_T(\lambda), \end{aligned} \tag{11}$$

$z \in \mathbb{C}_\pm$ where $\sigma(T) \subset [a,b)$. In particular, if $n_\pm(A) = 1$, then $M_A(\cdot)$ is scalar, $\mathcal{H}_S = \mathcal{H}_T$ and

$$M_S(z) = M_A(z-T), \qquad z \in \mathbb{C}_\pm. \tag{12}$$

3. Application

Let us consider Dirac operator of the form

$$S := -ic\frac{d}{dx} \otimes I_T + I \otimes \begin{pmatrix} \frac{c^2}{2} & 0 \\ 0 & -\frac{c^2}{2} \end{pmatrix} \tag{13}$$

defined on $\mathcal{H}_S = L_2(\mathbb{R}) \otimes \mathbb{C}^2$. It's symmetric restriction is given by restricting operator A_0 so that $f(0) = 0$. We can construct the γ-field and the Weyl function corresponding to the operator S^* in terms of the γ-field and the Weyl function corresponding to the operator A^*. Let us obtain the boundary triplet Π_A, corresponding to the operator A^*.

$$(A^*f,g) - (f,A^*g) = ic\left(f(0+)\overline{g}(0+) - f(0-)\overline{g}(0-)\right). \tag{14}$$

Let us choose the boundary operators Γ_0^A, Γ_1^A in the form:

$$\Gamma_0^A f = -i\sqrt{2c}\left(\frac{f(0-)-f(0+)}{2}\right), \Gamma_1^A f = -\sqrt{2c}\left(\frac{f(0+)+f(0-)}{2}\right). \tag{15}$$

The condition of surjectivity is obviously obeyed.

In order to obtain the γ-field we need to construct a deficiency element. Obviously, deficiency indices of A are $(1;1)$ and the deficiency elements we can choose as:

a) If $\Im(z) > 0$

$$f_+(x) = \begin{cases} e^{\frac{izx}{c}}, & x \geqslant 0 \\ 0, & x < 0 \end{cases} \tag{16}$$

b) If $\Im(z) < 0$

$$f_-(x) = \begin{cases} 0, & x \geqslant 0 \\ e^{\frac{izx}{c}}, & x < 0 \end{cases} \tag{17}$$

In accordance with definition, we get:

$$\gamma_A(z) = \begin{cases} -i\sqrt{\frac{2}{c}}, & \Im(z) > 0 \\ i\sqrt{\frac{2}{c}}, & \Im(z) < 0 \end{cases} \tag{18}$$

The Weyl function $M_A(z)$ is easily obtained:

$$M_A(z) = \begin{cases} i, & \Im(z) > 0 \\ -i, & \Im(z) < 0 \end{cases} \tag{19}$$

Now, in accordance with the results mentioned above,

$$M_S(z) = M_A(z - T) \tag{20}$$

Acknowledgments

This work was partially financially supported by the Government of the Russian Federation (grant 074-U01), by Ministry of Science and Education of the Russian Federation (GOSZADANIE 2014/190, Project 14.Z50.31.0031), by grants of the President of Russia (state contracts 14.124.13.2045-MK and 14.124.13.1493-MK).

Bibliography

[1] H. Baumgärtel and M. Wollenberg. *Mathematical scattering theory*, volume 59 of *Mathematical Textbooks and Monographs, Part II: Mathematical Monographs.* Akademie-Verlag, Berlin, 1983.

[2] V. A. Derkach and M. M. Malamud. On the Weyl function and Hermite operators with lacunae. *Dokl. Akad. Nauk SSSR*, 293(5):1041–1046, 1987.

[3] V. A. Derkach and M. M. Malamud. Generalized resolvents and the boundary value problems for Hermitian operators with gaps. *J. Funct. Anal.*, 95(1):1–95, 1991.

[4] V. A. Derkach and M. M. Malamud. The extension theory of Hermitian operators and the moment problem. *J. Math. Sci.*, 73(2):141–242, 1995. Analysis. 3.

[5] V. I. Gorbachuk and M. L. Gorbachuk. *Boundary value problems for operator differential equations*, volume 48 of *Mathematics and its Applications (Soviet Series).* Kluwer Academic Publishers Group, Dordrecht, 1991.

[6] M. M. Malamud. Some classes of extensions of a Hermitian operator with lacunae. *Ukraïn. Mat. Zh.*, 44(2):215–233, 1992.

[7] Mark Malamud and Hagen Neidhardt. Sturm-liouville boundary value problems with operator potentials and unitary equivalence. *J. Differ. Equations*, 252(11):5875–5922, 2012.

[8] Konrad Schmüdgen. *Unbounded self-adjoint operators on Hilbert space*, volume 265 of *Graduate Texts in Mathematics*. Springer, Dordrecht, 2012.

[9] A.A.Boitsev, H.Neidhardt, I.Yu.Popov Weyl function for sum of operators tensor product Nanosystems: physics, chemistry, mathematics, 4(6):747–759, 2013

Bose-Einstein condensation on quantum graphs

J. Bolte

Department of Mathematics, Royal Holloway, University of London, Egham, TW20 0EX, United Kingdom
E-mail: jens.bolte@rhul.ac.uk

J. Kerner

Institut für Analysis, Dynamik und Modellierung, Universität Stuttgart, 70565 Stuttgart, Germany
E-mail: Joachim.Kerner@mathematik.uni-stuttgart.de

We present results on Bose-Einstein condensation (BEC) on general compact quantum graphs, i.e., one-dimensional systems with a (potentially) complex topology. We first investigate non-interacting many-particle systems and provide a complete classification of systems that exhibit condensation. We then consider models with interactions that consist of a singular part as well as a hardcore part. In this way we obtain generalisations of the Tonks-Girardeau gas to graphs. For this we find an absence of phase transitions which then indicates an absence of BEC.

Keywords: Quantum graphs, many-particle quantum graphs, Bose-Einstein condensation, Tonks-Girardeau gas.

1. Introduction

We present results on Bose-Einstein condensation (BEC) in many-particle systems on compact quantum graphs [1–3] Quantum graphs are models of particles moving along the edges of a metric graph. They hence combine the simplicity of a one-dimensional model with the complexity of a graph. One of the major findings about quantum graphs was that the one-particle spectra display correlations that are well described by random-matrix theory [4, 5]. For that reason quantum graphs have become popular models in quantum chaos.

Bose-Einstein condensation, on the other hand, is a quantum mechanical phenomenon of many-particle systems that can be well described within the framework of quantum statistical mechanics. Originally, this condensation was predicted by Einstein for a gas of non-interacting bosons in three dimensions [6]. The conclusion was that below some critical temperature T_{crit}, or above some critical particle density ρ_{crit}, all additional particles brought into the system must condense into the one-particle ground state. In other words, the one-particle ground state becomes macroscopically occupied. As recognised later, [7] the macroscopic occupation of a *single-particle* state is indeed the underlying mechanism that gives rise to the condensation, also in interacting many-particle system.

We here introduce many-particle systems on graphs and study BEC in non-interacting systems as well as in systems with hardcore interactions, thus generalising the Tonks-Girardeau gas [8] to graphs.

2. Preliminaries

In this section we briefly summarise relevant concepts of BEC as well as of many-particle quantum graphs. For more details on BEC see [7, 9], on quantum graphs see [5, 10–12], and on many-particle quantum graphs see [1, 2].

A definition of BEC that also applies to a system of interacting bosons was given in [7] and employs the reduced one-particle density matrix obtained from the canonical density matrix (at inverse temperature $\beta = \frac{1}{T}$),

$$\rho_N(x,y) = \frac{1}{Z_N(\beta)} \sum_n e^{-\beta E_n^N} \bar{\Psi}_n(x)\Psi_n(y) \ , \tag{1}$$

by tracing out all degrees of freedom except one. Here $\Psi_n(x)$ is the n-th eigenfunction of the N-particle system with eigenvalues E_n^N, and $Z_N(\beta) = \sum_n e^{-\beta E_n^N}$ is the canonical partition function. Condensation is defined to occur when the largest eigenvalue of the reduced density matrix is asymptotically of order one as $N \to \infty$. Note that this limit (the thermodynamic limit) is accompanied by the limit $V \to \infty$, where V is the volume of the one-particle configuration space, such that the particle density remains fixed.

Unfortunately, it is in general very difficult to prove (or disprove) BEC in an interacting system in the sense of Penrose-Onsager. Instead we shall employ the connection between BEC and phase transitions in the case of interacting bosons on a graph.

The classical configuration space for a particle on a graph is a compact metric graph, i.e., a finite, connected graph $\Gamma = (\mathcal{V}, \mathcal{E})$ with vertex set $\mathcal{V} = \{v_1, \ldots, v_V\}$ and edge set $\mathcal{E} = \{e_1, \ldots, e_E\}$. The edges are identified with intervals $[0, l_e]$, $e = 1, \ldots, E$, thus assigning lengths to intervals. This then introduces a metric on the graph. Note that a graph is called *compact* when all lengths are finite.

For the one-particle quantum system the Hilbert space is

$$\mathcal{H}_1 = L^2(\Gamma) = \bigoplus_{e=1}^{E} L^2(0, l_e). \tag{2}$$

The quantum dynamics shall be generated by a self-adjoint realisation of the Laplacian $-\Delta_1$. As a differential expression this acts on each edge-component f_e of a function $F = (f_1, \dots, f_E) \in C_0^\infty(\Gamma)$ as the negative second derivative, $-\Delta_1 F = (-f_1'', \dots, -f_E'')$. As shown in [11], one way to arrive at all possible self-adjoint realisations of the Laplacian uses quadratic forms,

$$Q_1[F] = \sum_{e=1}^{E} \int_0^{l_e} |f'(x)|^2 \mathrm{d}x - \langle F_{bv}, L_1 F_{bv} \rangle_{\mathbb{C}^{2E}} \ , \tag{3}$$

with domain

$$\mathcal{D}_{Q_1} = \{F \in H^1(\Gamma); P_1 F_{bv} = 0\} \ . \tag{4}$$

Here $F_{bv} = (f_1(0), \dots, f_E(0), f_1(l_1), \dots, f_E(l_E))^T \in \mathbb{C}^{2E}$ is the vector of boundary values, P_1 is an orthogonal projection on $\mathbb{C}^{2E}$ and L_1 a self-adjoint endomorphism of $\ker P_1$.

The N-particle Hilbert space is the N-fold tensor product $\mathcal{H}_N = \mathcal{H}_1 \otimes \cdots \otimes \mathcal{H}_1$,

$$\mathcal{H}_N = L^2(\Gamma_N) = \bigoplus_{e_1 e_2 \dots e_N} L^2(D_{e_1 e_2 \dots e_N}), \tag{5}$$

where $D_{e_1 e_2 \dots e_N} = (0, l_{e_1}) \times \cdots \times (0, l_{e_N})$ are N-dimensional hypercubes. In order to introduce contact interactions that are formally given by a Hamiltonian

$$H_N = -\Delta_N + \alpha \sum_{i \neq j} \delta(x_{e_i} - x_{e_j}) \tag{6}$$

we also need to dissect the N-particle configuration space further by cutting the hypercubes along all hypersurplanes defined by $x_{e_i} = x_{e_j}$ when $e_i = e_j$. We denote (5) as $L^2(\Gamma_N^*)$ when taking this further dissection into account. The bosonic subspace is denoted as $L_B^2(\Gamma_N^*)$.

As in the one-particle case, we require the N-particle Hamiltonian to be a self-adjoint realisation of the Laplacian $-\Delta_N$. As a differential expression it acts as

$$(-\Delta_N \Psi)_{e_1 \dots e_N} = -\left(\frac{\partial^2}{\partial x_{e_1}^2} + \cdots + \frac{\partial^2}{\partial x_{e_N}^2} \right) \psi_{e_1 \dots e_N} \ , \tag{7}$$

on functions $\Psi = (\psi_{e_1 \dots e_N}) \in C_0^\infty(\Gamma_N)$. Self-adjoint (bosonic) realisations of $-\Delta_N$ are obtained through suitable quadratic forms. For this, we first define vectors of boundary values

$$\Psi_{bv}(\boldsymbol{y}) = \begin{pmatrix} \sqrt{l_{e_2} \dots l_{e_N}} \, \psi_{e_1 \dots e_N}(0, l_{e_2} y_1, \dots, l_{e_N} y_{N-1}) \\ \sqrt{l_{e_2} \dots l_{e_N}} \, \psi_{e_1 \dots e_N}(l_{e_1}, l_{e_2} y_1, \dots, l_{e_N} y_{N-1}) \end{pmatrix}, \tag{8}$$

with $\boldsymbol{y} \in [0,1]^{N-1}$. The desired quadratic form now reads

$$\begin{aligned} Q_B^{(N)}[\Psi] = N \sum_{e_1 \dots e_N} \int_0^{l_{e_1}} \dots \int_0^{l_{e_N}} |\psi_{e_1 \dots e_N, x_{e_1}}(x_{e_1}, \dots, x_{e_N})|^2 \, \mathrm{d}x_{e_N} \dots \mathrm{d}x_{e_1} \\ - N \int_{[0,1]^{N-1}} \langle \Psi_{bv}, L_N(\boldsymbol{y}) \Psi_{bv} \rangle_{\mathbb{C}^{2E^N}} \, \mathrm{d}\boldsymbol{y}, \end{aligned} \tag{9}$$

and is defined on

$$\mathcal{D}_{Q_B^{(N)}} = \{\Psi \in H^1_{0,B}(\Gamma_N^*);\ P_N(\boldsymbol{y})\Psi_{bv}(\boldsymbol{y}) = 0 \text{ for a.e. } \boldsymbol{y} \in [0,1]^{N-1}\}. \tag{10}$$

Here $\Psi \in H^1_{0,B}(\Gamma_N^*) \subset H^1_B(\Gamma_N^*)$ if each component $\psi_{e_1 e_2 \dots e_N}$ is in H^1 and vanishes on the hyperplanes $x_{e_i} = x_{e_j}$. Moreover, $P_N, L_N : [0,1]^{N-1} \to M(2E^N, \mathbb{C})$ are bounded and measurable maps such that $P_N(\boldsymbol{y})$ is an orthogonal projection and $L_N(\boldsymbol{y})$ is a self-adjoint endomorphism of $\ker P_N(\boldsymbol{y})$.

The self-adjoint realisation of $-\Delta_N$ associated with $Q_B^{(N)}[\cdot]$ is denoted as $(-\Delta_N, \mathcal{D}_N^{\infty}(P_N, L_N))$, see [2]. Due to the Dirichlet conditions along the hyperplanes $x_{e_i} = x_{e_j}$ this operator is a rigorous version of the Hamiltonian (6) in the limit $\alpha \to \infty$. Therefore it represents *hardcore interactions*. It is important to note that the coordinate dependence of the maps P_N and L_N leads to (additional) singular many-particle interactions that are localised in the vertices of the graph [1].

3. BEC in non-interacting systems many-particle

In this section we provide a complete classification of non-interacting many-particle systems on general compact quantum graph in terms of the presence or absence of BEC. Since non-interacting Hamiltonians are entirely determined by corresponding one-particle Hamiltonians $(-\Delta_1, \mathcal{D}_1(P_1, L_1))$, it is sufficient to refer to the latter only.

Definition 3.1. Let Γ be a compact, metric graph with edge lengths $l_1, ..., l_E$. The thermodynamic limit (TL) consists of the scaling $l_e \mapsto \eta l_e$ and taking the limit $\eta \to \infty$.

As a first result we obtain the following [3].

Lemma 3.1. *If the one-particle Laplacian $(-\Delta_1, \mathcal{D}_1(P_1, L_1))$ is such that L_1 is negative semi-definite, no BEC occurs in the corresponding free Bose gas at finite temperature ($T > 0$) in the thermodynamic limit.*

The proof uses the formalism of the grand-canonical ensemble and, via a bracketing construction, compares the given systems with free Bose gases with Dirichlet- and

Neumann boundary conditions in the vertices. The number of particles is not fixed and the particle density is adjusted via the chemical potential μ.

Our main result in this section is now the following [3].

Theorem 3.1. *Let a free Bose gas be given on a quantum graph with a one-particle Laplacian $(-\Delta_1, \mathcal{D}_1(P_1, L_1))$ such that L_1 has at least one positive eigenvalue. Then, in the thermodynamical limit, there is a critical temperature $T_c > 0$ such that BEC occurs below T_c.*

In order to prove this statement one shows that the one-particle ground state energy (which is negative since L_1 has a positive eigenvalue, compare (3)) remains negative in the thermodynamic limit. Indeed, one uses the lower bound for the one-particle ground state energy of [10] as well as a suitable Rayleigh quotient to prove that the ground state energy converges to $-L_{max}^2$, where $L_{max} > 0$ is the largest positive eigenvalue of L_1. One also uses that the number of negative eigenstates is bounded from above [10] (even in the thermodynamic limit) and applies the trace formula for quantum graphs [12]. With standard arguments the Theorem then follows.

4. Interacting many-particle systems

In this final section we consider systems of bosons interacting via singular interactions localised at the vertices of the graph as well as hardcore contact interactions, i.e., we consider (bosonic) self-adjoint realisations $(-\Delta_N, \mathcal{D}_N^\infty(P_N, L_N))$ of the N-particle Laplacian. Our goal is to prove that no phase transitions (in the free energy density) are present, indicating an absence of condensation. For this recall that the Landau free energy density of a sequence of N-particle Hamiltonians H_N with discrete spectra $\{E_n^N\}$ is defined as

$$f(\beta,\mu) = -\lim_{V\to\infty} \frac{1}{\beta V} \log \sum_{N=1}^{\infty} e^{N\beta\mu} \mathrm{Tr}_{\mathcal{H}_N} e^{-\beta H_N} \ . \tag{11}$$

The vanishing of functions in the domain $\mathcal{D}_N^\infty(P_N, L_N)$ along the hyperplanes $x_{e_i} = x_{e_j}$ allows to define a Fermi-Bose mapping, relating the bosonic free-energy density to a fermionic one (which is known explicitly). This leads to the following result [3].

Theorem 4.1. *Let $(-\Delta_N, \mathcal{D}_N^\infty(P_N, L_N))_{N\in\mathbb{N}}$ be a family of bosonic Laplacians with repulsive hardcore interactions, indexed by the particle number N. Assume that for this family the operator of multiplication with $L_N(\cdot)$ on $L^2([0,1]^{N-1})$ is uniformly bounded with respect to N. Then the bosonic Landau free-energy density $f(\beta,\mu)$ is given by*

$$f(\beta,\mu) = \frac{1}{\pi\beta} \int_0^\infty \log\left(1 + e^{-\beta(k^2-\mu)}\right) \mathrm{d}k \ . \tag{12}$$

This function is smooth and, hence, there exists no phase transition.

It is important to note that Theorem 4.1 holds *independently* of the singular interactions in the vertices, i.e., independently of the maps P_N and L_N. Hence, even when BEC occurs without the hardcore interactions, switching on the latter destroys the condensation.

Bibliography

[1] J. Bolte and J. Kerner, *J. Phys. A: Math. Theor.* **46**, p. 045206 (2013).
[2] J. Bolte and J. Kerner, *J. Phys. A: Math. Theor.* **46**, p. 045207 (2013).
[3] J. Bolte and J. Kerner, *preprint arXiv:1309.6091* (2013), to appear in *J. Math. Phys.*
[4] T. Kottos and U. Smilansky, *Physical Review Letters* **79**, 4794 (1997).
[5] S. Gnutzmann and U. Smilansky, *Taylor and Francis. Advances in Physics* **55**, 527 (2006).
[6] A. Einstein, *Sitzber. Kgl. Preuss. Akadm. Wiss.* , p. 3 (1925).
[7] O. Penrose and L. Onsager, *Physical Review* **104**, 576 (1956).
[8] M. D. Girardeau, *Journal of Mathematical Physics* **1**, 516 (1960).
[9] M. A. Cazalilla, R. Citro, T. Giamarchi, E. Orignac and M. Rigol, *Rev. Mod. Phys.* **83**, 1405 (2011).
[10] V. Kostrykin and R. Schrader, *Contemporary Mathematics* **415**, 201 (2006).
[11] P. Kuchment, *Waves Random Media* **14**, S107 (2004).
[12] J. Bolte and S. Endres, *Ann. H. Poincaré* **10**, 189 (2009).

Description of quantum and classical dynamics via Feynman formulae

Ya. A. Butko

Bauman Moscow State Technical University,
105005, 2nd Baumanskaya str., 5, Moscow, Russia;
Saarland University, P.O. Box 15 11 50, 66041 Saarbrücken, Germany
E-mail: yanabutko@yandex.ru, kinderknecht@math.uni-sb.de

A method to describe classical and quantum evolution systems is presented. The method is based on the so-called Feynman formulae, i.e. on representations of solutions of evolution equations as limits of n-fold iterated integrals of some (usually elementary) functions when n tends to infinity. Such formulae are suitable for direct calculations and numeric modeling. On the other hand, these formulae are related to some functional (or path) integrals with respect to probability measures and Feynman type pseudomeasures. We outline some results related to classical and quantum systems on different geometrical structures.

Keywords: Feynman formulae, approximations of evolution semigroups.

1. Introduction

Feynman formula is a representation of a solution of an evolution equation by a limit of iterated n-fold integrals when n tends to infinity. One succeeds to obtain such representations for solutions of different classes of evolution equations on various geometrical structures (see, e.g. [1]–[10] and references therein). Sometimes one succeeds to get Feynman formulae containing only integrals of elementary functions. Such Feynman formulae allow to calculate solutions of evolution equations directly and, consequently, to model the corresponding dynamics numerically.

The limits of iterated integrals in Feynman formulae often coincide with functional integrals (i.e. integrals over infinite dimensional path spaces) with respect to probability measures (such integrals are called Feynman–Kac formulae, cf. [4, 5, 10]) or with respect to Feynman type pseudomeasures (such integrals are called Feynman path integrals, cf. [2, 3]). First, the representations of solutions of evolution equations by Feynman–Kac formulas allow to investigate the considered dynamics by means of stochastic analysis, e.g., by the method of Monte–Carlo. And the approach of Feynman formulae provides a new way to establish Feynman–Kac formulae. Second, the limits of iterated finite dimensional integrals in some Feynman formulae can be used to define rigorously several Feynman path integrals over paths both in configuration and in phase space. And, on the other hand, the iterated integrals themselves reveal, in turn, the tool to calculate Feynman path integrals. Besides, different types of Feynman formulae for the same evolution equation allow to connect some Feynman path integrals with functional integrals with respect to probability measures. This gives another way to calculate Feynman path integrals and, moreover, develops deeper connections between quantum mechanics and stochastic analysis. Furthermore, the approach of Feynman formulae allows to distinguish different procedures of quantization in the language of Feynman path integrals (cf. [2, 3]).

2. The method of Feynman formulae

Let $(X, \|\cdot\|_X)$ be a Banach space, $\mathcal{L}(X)$ be the Banach space of all bounded linear operators on X with the operator norm $\|\cdot\|$, Id be the identity operator in X, $\mathrm{Dom}(L)$ denote the domain of a linear operator $L : \mathrm{Dom}(L) \subset X \to X$. A one-parameter family $(T_t)_{t\geq 0}$ of bounded linear operators $T_t : X \to X$ is called a strongly continuous semigroup, if $T_0 = \mathrm{Id}$, $T_{s+t} = T_s \circ T_t$ for all $s, t \geq 0$ and $\lim_{t\to 0} \|T_t\varphi - \varphi\|_X = 0$ for all $\varphi \in X$. If $(T_t)_{t\geq 0}$ is a strongly continuous semigroup on a Banach space $(X, \|\cdot\|_X)$, then the generator L of $(T_t)_{t\geq 0}$ is defined by $L\varphi := \lim\limits_{t\to 0} t^{-1}(T_t\varphi - \varphi)$ with domain $\mathrm{Dom}(L) := \left\{\varphi \in X \;\middle|\; \lim\limits_{t\to 0} t^{-1}(T_t\varphi - \varphi) \quad \text{exists in } X\right\}$. Consider for some operator L an evolution equation $\frac{\partial f}{\partial t} = Lf$. If L is the generator of a strongly continuous semigroup $(T_t)_{t\geq 0}$ on a Banach space $(X, \|\cdot\|_X)$, then the (mild) solution of the Cauchy problem for this equation with the initial value $f(0) = f_0 \in X$ is given by $f(t) = T_t f_0$ for all $f_0 \in X$. Therefore, solving the evolution equation $\frac{\partial f}{\partial t} = Lf$ means to construct a semigroup $(T_t)_{t\geq 0}$ with the given generator L. If the desired semigroup is not known explicitly it can be approximated, e.g. with the help of the Chernoff Theorem (cf. [11]).

Theorem 2.1 (Chernoff). *Let X be a Banach space, $F : [0, \infty) \to \mathcal{L}(X)$ be a (strongly) continuous mapping such that $F(0) = \mathrm{Id}$ and $\|F(t)\| \leq e^{at}$ for some $a \in [0, \infty)$ and all $t \geq 0$. Let D be a linear subspace of $\mathrm{Dom}(F'(0))$ such that the restriction of the operator $F'(0)$ to this subspace is closable. Let $(L, \mathrm{Dom}(L))$ be this closure. If $(L, \mathrm{Dom}(L))$ is the generator of a strongly continuous semigroup $(T_t)_{t\geq 0}$, then for any $t_0 > 0$ the sequence $(F(t/n))^n)_{n\in\mathbb{N}}$ converges to $(T_t)_{t\geq 0}$ as $n \to \infty$ in the strong operator topology, uniformly with respect to $t \in [0, t_0]$, i.e.*

$$T_t = \lim_{n\to\infty} \left[F(t/n)\right]^n . \tag{1}$$

A family of operators $(F(t))_{t\geq 0}$ suitable for the formula (1) is called *Chernoff equivalent* to the semigroup $(T_t)_{t\geq 0}$, i.e. this family satisfies all the assertions of the Chernoff theorem with respect to this semigroup. The identity (1) we call *Feynman*

Formula. In many cases the operators $F(t)$ are integral operators and hence we have a limit of iterated integrals on the right hand side of the equality (1).

3. Some examples of Feynman formulae

Theorem 3.1 (Theo. 5.1 in [6]). *Let X be a Banach space with a norm $\|\cdot\|_X$. Let $(T_k(t))_{t\geq 0}$, $k=1,\dots,m$, be strongly continuous semigroups on X with generators $(A_k, \mathrm{Dom}(A_k))$ respectively. Assume that $A = A_1 + \dots + A_m$ with $\mathrm{Dom}(A) = \cap_{k=1}^m \mathrm{Dom}(A_k)$ is closable and that the closure is the generator of a strongly continuous semigroup $(T(t))_{t\geq 0}$ on X. Let $(F_k(t))_{t\geq 0}$, $k=1,\dots,m$, be families of operators in X which are Chernoff equivalent to the semigroups $(T_k(t))_{t\geq 0}$ respectively, i.e. for each $k \in \{1,\dots,m\}$ we have $F_k(0) = \mathrm{Id}$, $\|F_k(t)\| \leq e^{a_k t}$ for some $a_k > 0$ and there is a set $D_k \subset \mathrm{Dom}(A_k)$, which is a core for A_k, such that $\lim_{t\to 0} \left\|\frac{F_k(t)\varphi-\varphi}{t} - A_k\varphi\right\|_X = 0$ for each $\varphi \in D_k$. Assume that there exists a set $D \subset \cap_{k=1}^m D_k$ which is a core for A. Then the family $(F(t))_{t\geq 0}$, where $F(t) = F_1(t)\circ\dots\circ F_m(t)$ is Chernoff equivalent to the semigroup $(T(t))_{t\geq 0}$ and, hence, the Feynman formula $T_t = \lim_{n\to\infty} \left[F(t/n)\right]^n$ is valid in the strong operator topology locally uniformly with respect to $t \geq 0$.*

Example 3.1. Let $X = L^2(\mathbb{R}^d)$. Consider $L_0 := \frac{i}{2}\Delta$, $\mathrm{Dom}(L_0) = C_c^\infty(\mathbb{R}^d)$. The closure of $(L_0, \mathrm{Dom}(L_0))$ generates a (semi)group $(T_t^0 \equiv e^{\frac{it}{2}\Delta})_{t\geq 0}$ resolving the Cauchy problem for the Schrödinger equation $i\frac{\partial f}{\partial t} = -\frac{1}{2}\Delta f$:

$$T_t^0\varphi(q) = (2\pi i t)^{(-d/2)} \int_{\mathbb{R}^d} e^{i\frac{|q-y|^2}{2t}} \varphi(y)dy,$$

where the right hand side is understood in the regularized sense (cf. [12], IX.31). Let $F_1(t) := T_t^0$. Let $B \in C_b^1(\mathbb{R}^d)$. Consider a family of operators $(F_2(t))_{t\geq 0}$ defined by the formula

$$F_2(t)\varphi(q) := \varphi(q - tb(q)), \quad q \in Q, \ \varphi \in L^2(\mathbb{R}^d).$$

There exists $t_0 > 0$ such that for all $t \in [0, t_0]$ the operators $F_2(t)$ are continuous linear operators on $L^2(\mathbb{R}^d)$ and $\|F_2(t)\| \leq e^{ct}$ for some constant $c > 0$. For all $t > t_0$ set $F_2(t) := F_2(t_0)$. Then the family $(F_2(t))_{t\geq 0}$ is Chernoff equivalent to the semigroup $e^{-tB\nabla}$, generated by $(-B\nabla, C_c^\infty(\mathbb{R}^d))$ with $B\nabla\varphi(q) := B(q)\nabla\varphi(q)$. let $C \in C_b(\mathbb{R}^d)$. Let the semigroup $(F_3(t))_{t\geq 0}$ be defined by $F_3(t)\varphi(q) = e^{-itC(q)}\varphi(q)$. The closure of $(L, C_c^\infty(\mathbb{R}^d))$ with $L := \frac{i}{2}\Delta - B\nabla - iC$ generates (by Theo. X.22 in [1?]) a strongly continuous (semi)group $(T_t)_{t\geq 0}$. And by Theorem 3.1 the family $(F_3(t)\circ F_2(t)\circ F_1(t))_{t\geq 0}$ is Chernoff equivalent to the semigroup $(T_t)_{t\geq 0}$, i.e. the following Feynman formula holds

$$T_t\varphi(q_0) = \lim_{n\to\infty}\left[e^{-i\frac{t}{n}C}\circ F_2(t/n)\circ T^0_{t/n}\right]^n\varphi(q_0)$$

$$= \lim_{n\to\infty}(2\pi i t)^{(-dn/2)}\int\limits_{\mathbb{R}^d}\dots\int\limits_{\mathbb{R}^d} e^{-i\frac{t}{n}\sum\limits_{k=1}^{n}C(q_{k-1})}\, e^{i\sum\limits_{k=1}^{n}\frac{\left|q_{k-1}-q_k-\frac{t}{n}B(q_{k-1})\right|^2}{2t/n}}\varphi(q_n)dq_1\dots dq_n.$$

Note, that the expression in the last line can be interpreted as a Feynman path integral over trajectories in configuration space of the system, whose quantization is described by the Hamiltonian $H = -\frac{1}{2}\Delta - iB\nabla + C$:

$$T_t\varphi(q_0) \equiv e^{-itH}\varphi(q_0) = \int e^{-i\int\limits_0^t[C(\xi(s))-\frac{1}{2}|B(\xi(s))|^2]ds}\, e^{i\int\limits_0^t B(\xi(s))\cdot d\xi(s)}\varphi(\xi(t))\Phi^{q_0}(d\xi).$$

Here Φ^{q_0} is a Feynman pseudomeasure concentrated on starting at q_0 paths in configuration space and heuristic "stochastic integral" $\int\limits_0^t B(\xi(s))\cdot d\xi(s)$ is similar to the construction of [13].

Let now Q be a metric space, $X = C_b(Q)$ be the Banach space of bounded continuous functions on Q with the supremum norm $\|f\|_\infty = \sup\limits_{q\in Q}|f(q)|$. Let $(T_t)_{t\geq 0}$ be a strongly continuous semigroup on X with generator $(L, \mathrm{Dom}(L))$. Consider a function $A : Q \to [\underline{\alpha}, \overline{\alpha}] \subset (0,\infty)$ which is continuous on Q. Define an operator $\widetilde{L}$ as $\widetilde{L}\varphi(q) := A(q)(L\varphi)(q)$, $q \in Q$, with $\mathrm{Dom}(\widetilde{L}) = \mathrm{Dom}(L)$. Then $\widetilde{L}$ is also the generator of a strongly continuous semigroup which we denote by $(\widetilde{T}_t)_{t\geq 0}$. We call the operator $\widetilde{L}$ a multiplicative perturbation of the generator L, and the semigroup $(\widetilde{T}_t)_{t\geq 0}$ the semigroup with the multiplicatively perturbed (by the function A) generator.

Theorem 3.2 (Theo. 4 in [1], cf. Theo. 4.1 in [6]). *Let the family $(F(t))_{t\geq 0}$ of bounded linear operators on X be Chernoff equivalent to the semigroup $(T_t)_{t\geq 0}$. Consider a family of operators $(\widetilde{F}(t))_{t\geq 0}$ defined by the formula*

$$\widetilde{F}(t)\varphi(q) = (F(A(q)t)\varphi)(q). \tag{2}$$

Then the family $(\widetilde{F}(t))_{t\geq 0}$ acts in X and is Chernoff equivalent to the semigroup $(\widetilde{T}_t)_{t\geq 0}$ with the generator $\widetilde{L}$. Therefore, the following Feynman formula is valid locally uniformly with respect to $t \geq 0$ in $\mathcal{L}(X)$: $\widetilde{T}_t = \lim_{n\to\infty}\left[\widetilde{F}(t/n)\right]^n$.

Remark 3.1. Note that the metric space Q in Theorem 3.2 is not supposed to be locally compact or linear. Theorem 3.2 remains true also if $X = C_\infty(Q) := \left\{\varphi \in C_b(Q) : \lim\limits_{\rho(q,q_0)\to\infty}\varphi(q) = 0\right\}$, where q_0 is an arbitrary fixed point of Q and the metric space Q is supposed to be unbounded.

Example 3.2. Consider a star graph Γ with vertex v and $n \in \mathbb{N}$ external edges $l_1, \dots, l_n$. Let ρ be the metric on Γ induced by the isomorphism $l_k \cong [0, +\infty)$; $\Gamma^o = \Gamma \setminus \{v\} = \sqcup_{k=1}^n l_k^o$, $l_k^o \cong (0, +\infty)$; $\xi \in l_k^o \Rightarrow \xi = (k, x)$ where $x = \rho(\xi, v) > 0$; $\varphi_k(x) := \varphi(\xi)\big|_{\xi \in l_k^o}$. Let $\int\limits_\Gamma \varphi(\xi)d\xi := \sum\limits_{k=1}^n \int\limits_0^\infty \varphi_k(x)dx$. Let $C^2_\infty(\Gamma) = \{\varphi \in C_\infty(\Gamma) : \varphi \in C^2_\infty(\Gamma^o), \varphi'' \text{ extends to } \Gamma \text{ as a function in } C_\infty(\Gamma)\}$. Let δ_v be the Dirac delta-measure concentrated at the vertex v. Let $\rho_v(\xi, \eta) := \rho(\xi, v) + \rho(v, \eta)$ for all $\xi, \eta \in \Gamma$. Let $1_k(\xi) = 1$ if $\xi \in l_k^o$, $1_k(\xi) = 0$ if $\xi \notin l_k^o$. Let $g(t, z) = (2\pi t)^{-1/2} \exp\left\{\frac{-z^2}{2t}\right\}$. Define $p(t, \xi, \eta) = \sum\limits_{k=1}^n 1_k(\xi)1_k(\eta)g(t, \rho(\xi, \eta))$, $p_v(t, \xi, \eta) = \sum\limits_{k=1}^n 1_k(\xi)1_k(\eta)g(t, \rho_v(\xi, \eta))$, $p^D(t, \xi, \eta) = p(t, \xi, \eta) - p_v(t, \xi, \eta)$. Let a, c, $b_k \in [0, 1]$, $k = 1, \dots, n$, $a \neq 1$ and $a + c + \sum\limits_{k=1}^n b_k = 1$. Consider an operator L_1 on $C_\infty(\Gamma)$ with $\mathrm{Dom}(L_1) = \{\varphi \in C^2_\infty(\Gamma) : a\varphi(v) + \frac{c}{2}\varphi''(v) = \sum\limits_{k=1}^n b_k \varphi_k'(v)\}$ and $L_1\varphi = \frac{1}{2}\varphi''$ for all $\varphi \in \mathrm{Dom}(L_1)$. Then $(L_1, \mathrm{Dom}(L_1))$ is the generator of a strongly continuous semigroup $(T_t^1)_{t\geq 0}$ on the space $C_\infty(\Gamma)$ and $T_t^1\varphi(\xi) = \int\limits_\Gamma \varphi(\eta)P(t, \xi, d\eta)$ for each $\varphi \in C_\infty(\Gamma)$. The transition kernel $P(t, \xi, d\eta)$ is given explicitly by the following formulae [14] : for the case $a + c \in (0, 1)$ with $w_k = \frac{b_k}{1-a-c}$, $\beta = \frac{a}{1-a-c}$, $\gamma = \frac{c}{1-a-c}$ and

$$g_{\beta,\gamma}(t, z) = \frac{1}{\gamma^2}(2\pi t)^{-1/2} \int\limits_0^t \frac{s + \gamma z}{(t-s)^{3/2}} \exp\left\{ - \frac{(s+\gamma z)^2}{2\gamma^2(t-s)} \right\} e^{-\beta s/\gamma} ds,$$

$$\begin{aligned} &P(t, \xi, d\eta) \qquad (3)\\ &= p^D(t, \xi, \eta)d\eta + \sum_{k,m=1}^n 1_k(\xi)1_m(\eta)2w_m g_{\beta,\gamma}(t, \rho_v(\xi, \eta))d\eta + \gamma g_{\beta,\gamma}(t, \rho(\xi, v))\delta_v(d\eta); \end{aligned}$$

for the case $a + c = 0$ with $w_k = b_k$

$$P(t, \xi, \eta) = p^d(t, \xi, \eta)d\eta + \sum_{k,m=1}^n 1_k(\xi)1_m(\eta)2w_m g(t, \rho_v(\xi, \eta))d\eta; \qquad (4)$$

for the case $a + c = 1$ with $a = \frac{\beta}{1+\beta}$, $\quad c = \frac{1}{1+\beta}$

$$P(t, \xi, \eta) = p^d(t, \xi, \eta)d\eta - \left(\int\limits_0^t e^{-\beta(t-s)} \frac{\rho(\xi, v)}{\sqrt{2\pi s^3}} \exp\left\{ -\frac{\rho(\xi, v)^2}{2s} \right\} ds, \right) \delta_v(d\eta). \qquad (5)$$

The heat kernel $P(t, \xi, d\eta)$ in (3) is the transition kernel of the process of Brownian motion on Γ constructed by killing (after an exponential holding time with the rate β at the vertex) the Walsh process (the analogue of the reflected Brownian motion) with sticky vertex with stickness parameter γ (see [14] for the detailed exposition).

Let $A(\cdot) \in C(\Gamma)$, let $\exists \underline{\alpha}, \overline{\alpha} \in (0, +\infty)$ such that $\underline{\alpha} \leq A(\xi) \leq \overline{\alpha}$ for all $\xi \in \Gamma$. Let $B(\cdot) \in C_b(\Gamma)$, $B(v) = 0$. Let $C(\cdot) \in C_b(\Gamma)$. As before let a, c, $b_k \in [0, 1]$,

$k = 1, \dots, n$ with $a \neq 1$ and $a + c + \sum_{k=1}^{n} b_k = 1$. Consider an operator L such that for all $\varphi \in \mathrm{Dom}(L) := \left\{ \varphi \in C^2_\infty(\Gamma) : a\varphi(v) + \frac{c}{2}\varphi''(v) = \sum_{k=1}^{n} b_k \varphi_k'(v) \right\}$ one has

$$L\varphi(\xi) := A(\xi)\varphi''(\xi) + B(\xi)\varphi'(\xi) + C(\xi)\varphi(\xi).$$

Then the operator $(L, \mathrm{Dom}(L))$ is the generator of a strongly continuous semigroup $(T_t)_{t\geq 0}$ on the space $C_\infty(\Gamma)$.

Proposition 3.1. *Consider a family $(F_1(t))_{t\geq 0}$ on $C_\infty(\Gamma)$ defined by*

$$F_1(t)\varphi(\xi) := \widetilde{T^1_t}\varphi(\xi) \equiv \int\limits_\Gamma \varphi(\eta) P(A(\xi)t, \xi, d\eta).$$

Consider a family $(F_2(t))_{t\geq 0}$ on $C_\infty(\Gamma)$ defined by

$$F_2(t)\varphi(\xi) := \varphi(\xi + tB(\xi)) := \begin{cases} \varphi_k(x + tB_k(x)), & \xi = (k, x),\ x + tB_k(x) > 0, \\ \varphi(v), & \xi = (k, x),\ x + tB_k(x) \leq 0. \end{cases}$$

Consider a family $(F_3(t))_{t\geq 0}$ with $F_3(t)\varphi(\xi) = e^{tC(\xi)}\varphi(\xi)$. Then by Theorem 3.1 and Theorem 3.2 the family $(F(t))_{t\geq 0}$ with $F(t) := F_3(t) \circ F_2(t) \circ F_1(t)$, i.e.

$$F(t)\varphi(\xi) = e^{tC(\xi)} \int\limits_\Gamma \varphi(\eta) P(A(\xi + tB(\xi))t, \xi + tB(\xi), d\eta),$$

is Chernoff equivalent to the semigroup $(T_t)_{t\geq 0}$ on the space $C_\infty(\Gamma)$ generated by $(L, \mathrm{Dom}(L))$. And hence the Feynman formula (1) is valid locally uniformly with respect to $t \geq 0$ in $\mathcal{L}(C_\infty(\Gamma))$.

Acknowledgments

Financial support by the Ministry of education and science of Russian Federation through the project 14.B37.21.0370, by the President of Russian Federation through the projects MK-4255.2012.1 and MK-943.2010.1, by DFG, DAAD, Erasmus Mundus and RFBR is gratefully acknowledged.

Bibliography

[1] Butko Ya.A.: Feynman formulae for evolution semigroups (in Russian), *Electronic scientific and technical periodical "Science and education"*, N3 (2014), DOI: 10.7463/0314.0701581.

[2] Butko, Ya. A., Grothaus, M., Smolyanov, O.G.: Feynman formulae and phase space Feynman path integrals for tau-quantization of some Lévy-Khintchine type Hamilton functions, preprint arXiv:1311.2883 [math-ph], 2013.

[3] Böttcher, B., Butko, Ya.A., Schilling, R.L., Smolyanov, O.G.: Feynman formulae and path integrals for some evolutionary semigroups related to τ-quantization, *Rus. J. Math. Phys.*, **18** N 4 (2011), 387–399.

[4] Butko, Ya. A.: Feynman formulas and functional integrals for diffusion with drift in a domain on a manifold, *Math. Notes.* **83** (2008), 301–316.

[5] Butko, Ya., Grothaus, M., Smolyanov, O.G.: Lagrangian Feynman Formulae for Second Order Parabolic Equations in Bounded and Unbounded Domains, *Inf. Dim. Anal. Quant. Probab. Rel. Top.* **13** (2010), 377–392.

[6] Butko, Ya.A., Schilling, R.L., Smolyanov, O.G.: Lagrangian and Hamiltonian Feynman formulae for some Feller semigroups and their perturbations, *Inf. Dim. Anal. Quant. Probab. Rel. Top.* **15** N 3 (2012), 26 p.

[7] Plyashechnik, A.S.: Feynman Formula for Schrödinger-type equations with time- and space-dependent coefficients, *Rus. J. Math. Phys.* **19** N 3 (2012), 340–359.

[8] Sakbaev, V.G., Smolyanov, O.G.: Dynamics of a Quantum Particle with Discontinuous Position-Dependent Mass, *Doklady Math.* **82** (2010), 630–634.

[9] Smolyanov, O.G., Shamarov, N.N.: Feynman and Feynman-Kac formulae for evolution equations with Vladimirov operator, *Doklady Math.* **77** (2008), 345–349.

[10] Smolyanov, O.G., Weizsäcker, H.v., Wittich, O.: Chernoff's Theorem and Discrete Time Approximations of Brownian Motion on Manifolds, *Potential Analysis* **26** (2007), 1–29.

[11] Chernoff, P.: Product formulas, nonlinear semigroups and addition of unbounded operators, *Mem. Am. Math. Soc.* **140** (1974).

[12] Reed, M., Simon, B.: Methods of Modern Mathematical Physics. V. II. Academic Press, 1980.

[13] Albeverio, S., Brzezniak, Z.: Oscillatory integrals on Hilbert spaces and Schrödinger equation with magnetic fields, *J. Math. Phys.* **36** N5 (1995), 2135–2156.

[14] Kostrykin, V., Potthof, Ju., Schrader, R.: Construction of the paths of Brownian motions on star graphs II, *Commun.Stoch.Anal.* **6** N 2 (2012), 247-261.

Asymptotic observables, propagation estimates and the problem of asymptotic completeness in algebraic QFT

W. Dybalski

Institut für Theoretische Physik, ETH Zürich,
8093 Zürich, Switzerland
and
Zentrum Mathematik, Technische Universität München,
D-85747 Garching, Germany
E-mail: dybalski@ma.tum.de

We review recent results on the existence of asymptotic observables in algebraic QFT. The problem of asymptotic completeness is discussed from this perspective.

Keywords: Local quantum field theory, Haag–Ruelle scattering theory, Araki–Haag detectors, asymptotic completeness.

1. Introduction

The problem of existence of asymptotic observables has a long tradition in algebraic quantum field theory (AQFT), starting with a seminal work of Araki and Haag [1]. These authors introduced certain time-dependent families of observables, and showed that their limits as time tends to infinity (*i.e.* certain asymptotic observables), behave as idealized particle detectors[a]. However, in [1] the convergence was only shown on certain domains of scattering states. The existence of such asymptotic observables on arbitrary vectors of bounded energy has remained an open problem for over four decades. It can be expected from quantum mechanics that a solution of this problem is a key to asymptotic completeness in AQFT.

In an ongoing project with C. Gérard we gave a solution of this problem for a certain class of detectors [4, 5]. Moreover, we demonstrated that the linear span of the ranges of these detectors coincides with the subspace of scattering states. This weak variant of asymptotic completeness has a simple physical interpretation: given any initial state (think of a container with hydrogen gas used as a proton source at the LHC), after a typical particle physics experiment one obtains a configuration of

[a] We consider only the case of outgoing asymptotic observables, since the incoming case is analogous.

independent particles in terms of which the measurement results are interpreted. We show that in the context of massive theories this empirical fact can be derived from the basic assumptions of locality and positivity of energy in a model-independent manner.

2. Framework

We work in the standard framework of AQFT, which is given by:

(1) A net of local von Neumann algebras $\mathcal{O} \mapsto \mathfrak{A}(\mathcal{O}) \subset B(\mathcal{H})$, labelled by open bounded regions of Minkowski spacetime $\mathcal{O} \subset \mathbb{R}^{d+1}$. $\mathfrak{A}(\mathcal{O})$ is interpreted as the algebra of all observables measurable in $\mathcal{O}$.

(2) The global algebra of this net $\mathfrak{A} := \overline{\bigcup_{\mathcal{O}\subset\mathbb{R}^{d+1}} \mathfrak{A}(\mathcal{O})}$.

(3) A strongly continuous unitary representation of translations $\mathbb{R}^{d+1} \ni (t,x) \mapsto U(t,x)$ acting on $\mathcal{H}$. Its generators (H,P) have the interpretation of the total energy and momentum operators. Their joint spectrum is denoted $\mathcal{S}p\, U$ and their spectral projection on a Borel set Δ by $\mathbb{1}_\Delta(U)$.

On these objects we impose the Haag–Kastler postulates:

(1) (isotony) $\mathcal{O}_1 \subset \mathcal{O}_2 \Rightarrow \mathfrak{A}(\mathcal{O}_1) \subset \mathfrak{A}(\mathcal{O}_2)$.

(2) (locality) $\mathcal{O}_1 \times \mathcal{O}_2 \Rightarrow [\mathfrak{A}(\mathcal{O}_1), \mathfrak{A}(\mathcal{O}_2)] = 0$, where $\times$ means spacelike separation.

(3) (covariance) $U(t,x)\mathfrak{A}(\mathcal{O})U(t,x)^* = \mathfrak{A}(\mathcal{O} + (t,x))$.

(4) (positivity of energy) $\mathcal{S}p\, U = \{0\} \cup H_m \cup G_{2m}$.

(5) (uniqueness and cyclicity of the vacuum) $\mathbb{1}_{\{0\}}(U) = |\Omega\rangle\langle\Omega|$ and $\overline{\mathfrak{A}\Omega} = \mathcal{H}$.

We adopt a restrictive variant of the positivity of energy assumption (4), suitable for massive theories, which requires that $\mathcal{S}p\, U$ consists of $\{0\}$, corresponding to the vacuum vector Ω, a mass hyperboloid $H_m := \{ (E,p) \in \mathbb{R}^{d+1} \,|\, E = \sqrt{p^2+m^2} \}$ carrying single-particle states and the multiparticle spectrum $G_{2m} := \{ (E,p) \in \mathbb{R}^{d+1} \,|\, E \geq \sqrt{p^2+(2m)^2} \}$. We recall that there exist interacting quantum field theories which satisfy the above assumptions, for example $\lambda\phi_2^4$ at small λ [6].

We remark that our results still hold (after minor modifications) if (4) is relaxed to $\mathcal{S}p\, U = \{0\} \cup H_m \cup G_M$, where $G_M \subset \{ (E,p) \in \mathbb{R}^{d+1} \,|\, E \geq \sqrt{p^2+M^2} \}$, $m < M \leq 2m$. This form of the spectrum allows for bound states below the $2m$ threshold.

3. Generalized creation/annihilation operators

In concrete models, the conventional building blocks of physical observables are creation and annihilation operators. To construct their counterparts in our abstract setting, we need to control the energy-momentum transfer of elements of the algebra of observables $\mathfrak{A}$. The energy-momentum transfer (or the Arveson spectrum) of an observable $B \in \mathfrak{A}$ is denoted $\mathrm{supp}\,\widehat{B}$ and defined as the support of the operator-valued distribution:

$$\tilde{B}(E,p) := (2\pi)^{-\frac{d+1}{2}} \int dtdx\, \mathrm{e}^{-\mathrm{i}Et+\mathrm{i}px} B(t,x), \tag{1}$$

where $B(t,x) := U(t,x)BU(t,x)^*$. It has the expected properties, in particular

$$\mathrm{supp}\,\widehat{B^*} = -\mathrm{supp}\,\widehat{B}, \tag{2}$$

$$B1\!\!1_\Delta(U)\mathcal{H} \subset 1\!\!1_{\overline{\Delta+\mathrm{supp}\,\widehat{B}}}(U)\mathcal{H}. \tag{3}$$

To use B^* as a creation operator of a single-particle state, we would like $\mathrm{supp}\,\widehat{B^*}$ to be a small neighbourhood of a point on the mass hyperboloid H_m. An operator with such an energy-momentum transfer cannot be strictly local, *i.e.* contained in $\mathfrak{A}(\mathcal{O})$ for some open bounded region $\mathcal{O}$. However, it can be *almost local i.e.* it can be approximated in norm by operators localized in double-cones, centred at zero, with increasing radius, up to an error vanishing faster than any inverse power of this radius. A typical example of an almost local operator is

$$B = \int dtdx\, A(t,x)f(t,x), \quad A \in \mathfrak{A}(\mathcal{O}), f \in \mathcal{S}(\mathbb{R}^{d+1}) \tag{4}$$

for some open bounded $\mathcal{O}$. Clearly, $\mathrm{supp}\,\widehat{B} \subset \mathrm{supp}\hat{f}$, so the energy-momentum transfer of B^* can be a compact set. Also, for such operators the function $(t,x) \mapsto B(t,x)$ is smooth in norm and all its derivatives are again almost local. We denote by $\mathcal{L}_0 \subset \mathfrak{A}$ the subspace of operators of the form (4) whose energy-momentum transfers are compact sets supported outside of the future lightcone[b]. The elements of $\mathcal{L}_0$ are 'annihilation operators' in the sense that they annihilate the vacuum, but not in the sense of canonical commutation relations.

The above construction of generalized creation and annihilation operators is well known since early days of AQFT. We conclude this section with a more recent concept of 'improper' annihilation operators, introduced in our recent work [4]. They are defined as maps

$$a_B : \mathcal{H}_{\mathrm{c}} \mapsto \mathcal{H} \otimes L^2(\mathbb{R}^d), \tag{5}$$

$$(a_B\Psi)(x) = B(x)\Psi, \tag{6}$$

where $B \in \mathcal{L}_0$ and $\mathcal{H}_{\mathrm{c}}$ is the domain of vectors $\Psi \in \mathcal{H}$ s.t. $\Psi = 1\!\!1_\Delta(U)\Psi$ for some compact set Δ. The fact that $x \mapsto B(x)\Psi$ is square-integrable is non-trivial, but it

[b]This definition is slightly more restrictive than the one from [4, 5].

follows from Lemma 2.2 of [2]. Similarly, for a family $B_1, \dots, B_n$ of operators from $\mathcal{L}_0$, we define

$$a_{B_1,\dots,B_n} : \mathcal{H}_c \to \mathcal{H} \otimes L^2(\mathbb{R}^{nd}), \tag{7}$$
$$(a_{B_1,\dots,B_n}\Psi)(x_1, \dots, x_n) = B_1(x_1)\dots B_n(x_n)\Psi. \tag{8}$$

4. Propagation observables and asymptotic observables

We recall that in non-relativistic scattering theory, time-dependent estimates on the propagation properties of solutions of evolution equations are usually called *propagation estimates*, and time-dependent observables used to derive them are called *propagation observables*. Asymptotic observables are limits (as time $t \to \infty$) of propagation observables evolved in the Heisenberg picture.

Let us first consider the classical phase space $T^*\mathbb{R}^d = \mathbb{R}^d \times (\mathbb{R}^d)'$, $h \in \mathcal{S}(T^*\mathbb{R}^d)$ and a classical propagation observable

$$t \mapsto h_t \in \mathcal{S}(T^*\mathbb{R}^d), \tag{9}$$

where $h_t(x, \xi) := h(x/t, \xi)$. We can elevate it to a quantum-mechanical propagation observable with the help of the Weyl quantization:

$$t \mapsto h_t^{\mathrm{w}} \in B(L^2(\mathbb{R}^d)), \tag{10}$$

where

$$(h_t^{\mathrm{w}}u)(x) = (2\pi)^{-d} \int \mathrm{e}^{\mathrm{i}(x-y)\cdot\xi} h_t\Big(\frac{x+y}{2}, \xi\Big)u(y)dyd\xi, \; u \in L^2(\mathbb{R}^d). \tag{11}$$

Next, using the map a_B, $B \in \mathcal{L}_0$, we define a quantum-field-theoretical propagation observable:

$$t \mapsto a_B^*(1_{\mathcal{H}} \otimes h_t^{\mathrm{w}})a_B, \tag{12}$$

which is a family of (unbounded) operators on $\mathcal{H}_c$.

Given the propagation observable (12), the corresponding asymptotic observable is approximated (as $t \to \infty$) by:

$$C_t := \mathrm{e}^{\mathrm{i}tH} a_B^*(1_{\mathcal{H}} \otimes h_t^{\mathrm{w}})a_B \mathrm{e}^{-\mathrm{i}tH}. \tag{13}$$

We mention as an aside that in the case of h independent of momentum ξ the above formula gives

$$C_t = \int dx\, h(x/t)(B^*B)(t, x), \tag{14}$$

which is the standard Araki–Haag detector [1, 2].

Our main result, stated in Theorem 4.1 below, concerns the strong convergence as $t \to \infty$ of approximating sequences of the form

$$t \mapsto C_{1,t} \dots C_{n,t} 1_\Delta(U), \tag{15}$$

where $\Delta \subset G_{2m}$ is an open set, whose extension is small compared to m, and $C_{i,t}$, $i = 1, \dots, n$ are defined as in (13). So far we can handle detectors which satisfy the following admissibility conditions:

(a) The energy-momentum transfers of B_i^* are small neighbourhoods of distinct points p_i on the mass hyperboloid H_m s.t. $p_1 + \cdots + p_n \in \Delta$.

(b) The functions $h_{i,t} \in \mathcal{S}(T^*\mathbb{R}^d)$ have the form

$$h_{i,t}(x,\xi) = h_{0,i}(x/t)\chi(x/t - \nabla\omega(\xi)),$$

where $h_{0,i} \in C_0^\infty(\mathbb{R}^d)$ have mutually disjoint supports, $\chi \in C_0^\infty(\mathbb{R}^d)$ is supported in a small neighbourhood of zero and $\omega(\xi) := \sqrt{\xi^2 + m^2}$.

The first assumption above says that B_i^* are 'creation operators' of single-particle states with energy-momentum vectors centered around p_i. The disjointness of supports of $h_{0,i}$ in the second assumption ensures that the corresponding detectors are localized in spacelike separated regions for large t, which helps to exploit locality. The function χ keeps the average velocity x/t close to the instantaneous velocity $\nabla\omega(\xi)$ in accordance with the expected ballistic motion of a particle at asymptotic times.

Theorem 4.1. *Let $\Delta \subset G_{2m}$ be a small open subset, B_i, h_i, $i = 1, \dots, n$, be admissible as specified in* (a), (b) *above and let $C_{i,t}$ be given by (*13*). Then there exists the limit*

$$Q_n^+(\Delta) := \mathrm{s-}\lim_{t\to\infty} C_{1,t} \dots C_{n,t} \mathbb{1}_\Delta(U). \tag{16}$$

5. Outline of the proof of Theorem 4.1

The proof of Theorem 4.1 given in [5] relies on the method of propagation estimates combined with the Haag–Ruelle scattering theory. Here we outline a different argument (also obtained jointly with C. Gérard) which does not use the Haag–Ruelle theory.

We note that for $n = 1$ condition (a) is not compatible with $\Delta \subset G_{2m}$, hence the theorem does not provide any information on the convergence of one detector. Let us consider the first interesting case which is $n = 2$: Making use of locality and disjointness of supports of $h_{0,1}$, $h_{0,2}$ (condition (b)) we can write

$$C_{1,t}C_{2,t}\mathbb{1}_\Delta(U) = \mathrm{e}^{\mathrm{i}tH} a^*_{B_1,B_2}(\mathbb{1}_{\mathcal{H}} \otimes h^{\mathrm{w}}_{1,t} h^{\mathrm{w}}_{2,t}) a_{B_1,B_2} \mathrm{e}^{-\mathrm{i}tH} \mathbb{1}_\Delta(U) + O(t^{-\infty}), \tag{17}$$

where $O(t^{-\infty})$ denotes a term which vanishes in norm faster than any inverse power of t. Thus it is enough to show strong convergence of

$$f(t) := \mathrm{e}^{\mathrm{i}tH} a^*_{B_1,B_2}(\mathbb{1}_{\mathcal{H}} \otimes h^{\mathrm{w}}_{1,t} h^{\mathrm{w}}_{2,t}) a_{B_1,B_2} \mathrm{e}^{-\mathrm{i}tH} \mathbb{1}_\Delta(U) \tag{18}$$

as $t \to \infty$. The key observation, which allows us to transport methods from quantum-mechanical scattering theory to the present context, is that the time-derivative of f can be expressed by the Heisenberg derivative of the quantum-mechanical propagation observable $t \mapsto h^{\mathrm{w}}_{1,t} h^{\mathrm{w}}_{2,t}$. More precisely

$$\partial_t f(t) = \mathrm{e}^{\mathrm{i}tH} a^*_{B_1,B_2} (1\!\!1_{\mathcal{H}} \otimes \mathcal{D}(h^{\mathrm{w}}_{1,t} h^{\mathrm{w}}_{2,t})) a_{B_1,B_2} \mathrm{e}^{-\mathrm{i}tH} 1\!\!1_{\Delta}(U) + O(t^{-\infty}), \qquad (19)$$

where $\mathcal{D} = \partial_t + \mathrm{i}[\tilde{\omega}, \cdot\,]$ and $\tilde{\omega} = \omega(-\mathrm{i}\nabla_{x_1}) + \omega(-\mathrm{i}\nabla_{x_2})$. Thus the asymptotic time-evolution of a relativistic QFT is governed by a quantum-mechanical Hamiltonian $\tilde{\omega}$. In the case of $n = 2$, using relation (19) and the standard phase space propagation estimates (see *e.g.* [3]) one can show the convergence of $t \mapsto f(t)$ [4]. In this case functions χ appearing in condition (b) above are not needed and one can prove the convergence of a product of two conventional Araki–Haag detectors (14). For $n > 2$ things become more complicated due to our limited understanding of quantum mechanical *dispersive systems*, that is systems of particles with non-quadratic dispersion relations. In this case the phase space propagation estimate is not available and we had to derive a new propagation estimate to control the convergence of $t \mapsto f(t)$. As explained in more detail in [5], it requires the presence of the functions χ in the propagation observables.

Let us conclude this section with a few remarks about the key property (19): The term involving $\partial_t(h^{\mathrm{w}}_{1,t} h^{\mathrm{w}}_{2,t})$ is self-explanatory. Let us indicate how $\partial_t \mathrm{e}^{\mathrm{i}Ht}$ and $\partial_t \mathrm{e}^{-\mathrm{i}Ht}$ give rise to the commutator with $\tilde{\omega}$: We note the identities

$$B_1(x_1)B_2(x_2)1\!\!1_{\Delta}(U) = 1\!\!1_{\{0\}}(U)B_1(x_1)1\!\!1_{H_m}(U)B_2(x_2)1\!\!1_{\Delta}(U), \qquad (20)$$

$$\begin{aligned} B_1(x_1)B_2(x_2)H1\!\!1_{\Delta}(U) = {} & HB_1(x_1)B_2(x_2)1\!\!1_{\Delta}(U) + [B_1(x_1),[B_2(x_2),H]]1\!\!1_{\Delta}(U) \qquad (21) \\ & + [B_1(x_1),H]B_2(x_2)1\!\!1_{\Delta}(U) + [B_2(x_2),H]B_1(x_1)1\!\!1_{\Delta}(U), \end{aligned}$$

where (20) follows from condition (a) and (3), and (21) is a simple computation. The first term on the r.h.s. of (21) vanishes due to (20) and translation invariance of the vacuum. The term involving the double commutator contributes to $O(t^{-\infty})$ on the r.h.s. of (19) due to locality and the disjointness of supports of $h_{1,0}$ and $h_{2,0}$. The last two terms on the r.h.s. (21) give rise to the quantum-mechanical Hamiltonian $\tilde{\omega}$. In fact, keeping (20) in mind, we note that

$$\begin{aligned} 1\!\!1_{\{0\}}(U)[B_i(x_i),H]1\!\!1_{H_m}(U) &= 1\!\!1_{\{0\}}(U)B_i(x_i)\omega(P)1\!\!1_{H_m}(U) \\ &= \omega(-\mathrm{i}\nabla_{x_i})1\!\!1_{\{0\}}(U)B_i(x_i)1\!\!1_{H_m}(U), \end{aligned} \qquad (22)$$

where the last step follows by taking the Fourier transform of ω and exploiting the invariance of the vacuum.

6. The problem of asymptotic completeness

We recall that under the assumptions from Section 2 the Haag–Ruelle scattering theory [8, 10] gives a canonical wave operator whose range will be denoted by $\mathcal{H}^+$.

$\mathcal{H}^+$ can equivalently be seen as a subspace of $\mathcal{H}$ spanned by (outgoing) scattering states, including the vacuum (0-particle state) and $\mathcal{H}_1 := 1\!\!1_{H_m}(U)\mathcal{H}$ (single-particle states). The property of asymptotic completeness requires that $\mathcal{H}^+ = \mathcal{H}$ *i.e.* that all states in the physical Hilbert space can be interpreted as configurations of particles. It is well known that this property does not follow from the Haag–Kastler postulates: Firstly, it could happen that not all the superselection sectors of the theory are accommodated in $\mathcal{H}$. Single-particle states of charged particles from the omitted sectors would then be missing in $\mathcal{H}$ and consequently their scattering states would not belong to $\mathcal{H}^+$. But scattering configurations of such particles with total charge zero would clearly belong to $\mathcal{H}$ thus violating asymptotic completeness. Secondly, even if all the superselection sectors are accommodated in $\mathcal{H}$, the physical Hilbert space may still contain pathological states with too many local degrees of freedom which do not admit any particle interpretation. Such states appear *e.g.* in certain generalized free fields [7].

One approach to the problem of asymptotic completeness in AQFT is to amend the Haag–Kastler postulates with some physically motivated *a priori* conditions which should characterize theories with a reasonable particle interpretation. This approach, pioneered in [9], resulted in a family of *phase space conditions* which brought many important insights in AQFT, but did not have much impact on its scattering theory. Our strategy is different: we try to get as close as possible to proving complete particle interpretation without adopting additional assumptions. *A posteriori* our results can be reformulated as a condition for asymptotic completeness.

To illustrate this strategy, let us discuss the following result from [5] which complements Theorem 4.1:

Theorem 6.1. *Under the assumptions of Theorem 4.1, the range of each asymptotic observable $Q_n^+(\Delta)$ belongs to the subspace of scattering states $1\!\!1_\Delta(U)\mathcal{H}^+$. Moreover, $1\!\!1_\Delta(U)\mathcal{H}^+$ is spanned by the ranges of $Q_n^+(\Delta)$, $n \in \mathbb{N}$.*

We note that the second part of this result ensures that 'sufficiently many' asymptotic observables, constructed in Theorem 4.1 are non-zero. This part is relatively easy to prove proceeding similarly as in [1]. The essential part of Theorem 6.1 is the first statement, which says that a certain subspace of $\mathcal{H}$ (namely the span of the ranges of the asymptotic observables $Q_n^+(\Delta)$), which *a priori* has nothing to do with scattering states, is in fact contained in the subspace of scattering states $\mathcal{H}^+$. Thus the requirement that the ranges of $Q_n^+(\Delta)$ (together with the vacuum vector and single-particle states) span the entire Hilbert space $\mathcal{H}$, is a condition for asymptotic completeness.

If we are ready to interpret $Q_n^+(\Delta)$ as particle detectors, Theorem 6.1 has a simple physical interpretation. It says that for any initial state $\Psi \in \mathcal{H}$, the state resulting from the measurement $Q_n^+(\Delta)$ will be a vector in $\mathcal{H}^+$, that is a configuration of particles. Theorem 6.1 also says that if Ψ belongs to the orthogonal complement of $\mathcal{H}^+$ (as a neutral configuration of particles from superselection sectors omitted in $\mathcal{H}$, a state with too many local degrees of freedom or any other pathology Haag–Kastler postulates may admit) then Ψ will be annihilated by $Q_n^+(\Delta)$. This insight cannot be obtained by methods from [1], where only the existence of asymptotic observables on $\mathcal{H}^+$ is considered.

In the language of many-body scattering theory our results concern the problem of asymptotic completeness in the *free region* (*i.e.* outside of the collision planes $\{x_1 = x_2\}$, $\{x_1 = x_3\}$, etc.) In fact, condition (b) above ensures that the classical propagation observables, corresponding to (15), vanish near the collision planes. A natural future direction is to drop condition (b) and understand the asymptotic dynamics of AQFT close to the collision planes. As this condition enters already in the first step of our analysis (see (17)), this goal surely requires new ideas. Taking quantum mechanics as a guide, some relativistic counterpart of the Mourre theory may be needed here. One can speculate that the covariance under Lorentz boosts, which we did not exploit in our analysis so far, will provide this missing ingredient.

Bibliography

[1] H. Araki and R. Haag: *Collision cross sections in terms of local observables.* Commun. Math. Phys. **4**, (1967) 77–91.

[2] D. Buchholz: *Harmonic analysis of local operators.* Commun. Math. Phys. **129**, (1990) 631–641.

[3] J. Dereziński and C. Gérard: *Scattering theory of classical and quantum N-particle systems.* Springer, 1997.

[4] W. Dybalski and C. Gérard: *Towards asymptotic completeness of two-particle scattering in local relativistic QFT.* Commun. Math. Phys. **326**, (2014) 81–109.

[5] W. Dybalski and C. Gérard: *A criterion for asymptotic completeness in local relativistic QFT.* Accepted for publication in Commun. Math. Phys. Preprint arXiv:1308.5187 [math-ph].

[6] J. Glimm, A. Jaffe and T. Spencer: *The particle structure of the weakly coupled $P(\phi)_2$ model and other applications of high temperature expansions: Part I. Physics of quantum field models. Part II. The cluster expansion.* In: Constructive quantum field theory. (Erice, 1973), G. Velo, A. S. Wightman (eds.). Berlin, Heidelberg, New York: Springer 1973.

[7] O. W. Greenberg: *Generalized free fields and models of local field theory.* Ann. of Phys. **16**, (1961) 158–176.

[8] R. Haag: *Quantum field theories with composite particles and asymptotic conditions.* Phys. Rev. **112**, (1958) 669–673.

[9] R. Haag and J.A. Swieca: *When does a quantum field theory describe particles?* Commun. Math. Phys. **1**, (1965) 308–320.

[10] D. Ruelle: *On the asymptotic condition in quantum field theory.* Helv. Phys. Acta **35**, (1962) 147–163.

Recent probabilistic results on covariant Schrödinger operators on infinite weighted graphs

B. Güneysu

Institut für Mathematatik, Humboldt-Universität zu Berlin, 12489 Berlin, Germany
E-mail: gueneysu@math.hu-berlin.de

O. Milatovic

Department of Mathematics and Statistics, University of North Florida, Jacksonville, FL 32224, USA
E-mail: omilatov@unf.edu

We review recent probabilistic results on covariant Schrödinger operators on vector bundles over (possibly locally infinite) weighted graphs, and explain applications like semi-classical limits. We also clarify the relationship between these results and their formal analogues on smooth (possibly noncompact) Riemannian manifolds.

Keywords: Infinite graph, vector bundle, Feynman-Kac formula, partition function.

1. A review of covariant Schrödinger operators on smooth Riemannian manifolds

Let us start by taking a look at covariant Schrödinger-type operators on Riemannian manifolds: Assume that $E \to M$ is a smooth finite dimensional Hermitian vector bundle over a possibly noncompact smooth Riemannian manifold M without boundary, with a Hermitian covariant derivative ∇ on E, which means that ∇ is a complex linear map

$$\nabla : \Gamma_{\mathsf{C}^\infty}(M,E) \longrightarrow \Omega^1_{\mathsf{C}^\infty}(M,E), \text{ such that for all } \Psi_1, \Psi_2 \in \Gamma_{\mathsf{C}^\infty}(M,E) \text{ one has:}$$
$$\mathrm{d}(\Psi_1,\Psi_2) = (\nabla\Psi_1,\Psi_2) + (\Psi_1,\nabla\Psi_2). \tag{1}$$

Then the symmetric nonnegative sesquilinear form

$$\tilde{Q}_{\nabla,0}(\Psi_1,\Psi_2) = \int_M \big(\nabla\Psi_1(x),\nabla\Psi_2(x)\big)_x \mathrm{vol}(\mathrm{d}x),\ \Psi_1,\Psi_2 \in \Gamma_{\mathsf{C}^\infty_c}(M,E), \tag{2}$$

is closable in the Hilbert space $\Gamma_{\mathsf{L}^2}(M,E)$, as this form corresponds to a non-negative symmetric operator. If $Q_{\nabla,0} := \overline{\tilde{Q}_{\nabla,0}}$, and if $V : M \to \mathrm{End}(E)$ is a potential

(= a measurable self-adjoint section in $\mathrm{End}(E) \to M$) with $V = V^{+} - V^{-}$ for some potentials $V^{\pm} \geq 0$ such that $|V^{+}| \in \mathsf{L}^{1}_{\mathrm{loc}}(M)$ and $|V^{-}| \in \mathscr{K}(M)$ (the Kato class on M), then the form $Q_{\nabla,V} := Q_{\nabla,0} + Q_{V}$ (where Q_V is the maximally defined form given by V), is symmetric, densely defined, closed, and semibounded (from below), and we can consider the corresponding self-adjoint operator $H_{\nabla,V}$. In this full generality, these operators appear as the nonrelativistic Hamilton operators corresponding to atoms in a magnetic field, possibly with the electron's spin taken into account [4]. A fundamental result in this context is the *Feynman-Kac formula*: Namely, the integral kernel of the underlying covariant Schrödinger semigroup $(\mathrm{e}^{-tH_{\nabla,V}})_{t\geq 0}$ is given [3] by the well-defined path integral formula

$$\mathrm{e}^{-tH_{\nabla,V}}(x,y) = p(t,x,y)\mathbb{E}^{x}\left[\mathscr{A}_t^{\nabla,V} //_t^{\nabla,-1} \middle| \mathbb{X}_t = y\right] \in \mathrm{Hom}(E_y, E_x), \tag{3}$$

where the process $\mathbb{X}$ is a (scaled) Brownian motion on M under $\mathbb{P}^{x}$ with transition density $p(t,x,y)$, where

$$//_t^{\nabla} \in \mathrm{Hom}(E_{\mathbb{X}_0}, E_{\mathbb{X}_t})$$

is the (unitary) stochastic ∇-parallel transport along the Brownian paths, and

$\mathscr{A}_t^{\nabla,V} \in \mathrm{End}(E_{\mathbb{X}_0})$ is the path ordered exponential

$$\mathscr{A}_t^{\nabla,V} = \mathbf{1}_{\mathbb{X}_0} + \sum_{n=1}^{\infty}(-1)^n \int_{t\sigma_n} //_{s_1}^{\nabla,-1} V(\mathbb{X}_{s_1}) //_{s_1}^{\nabla} \cdots //_{s_n}^{\nabla,-1} V(\mathbb{X}_{s_n}) //_{s_n}^{\nabla} \,\mathrm{d}s_1 \cdots \mathrm{d}s_n,$$

with $t\sigma_n \subset \mathbb{R}^n$ the t-scaled standard n-simplex. Along with many other applications, the Feynman-Kac formula can be used to prove the following quantum mechanical result: One has *a semiclassical behaviour of the quantum partition function*, in the sense that for all $\beta > 0$ one has [2, 10]

$$\frac{\mathrm{tr}\left(\mathrm{e}^{-\beta\hbar H_{\nabla,V/\hbar}}\right)}{(2\pi)^{-\dim(M)}\hbar^{-\dim(M)/2}\int_M \int_{\mathrm{T}^*_x M} \mathrm{tr}_{E_x}\left(\mathrm{e}^{-\beta(|v|^2 + V(x))}\right) \mathrm{d}v \ \mathrm{vol}(\mathrm{d}x)} \xrightarrow{\hbar\to 0+} 1, \tag{4}$$

which is in fact equivalent to

$$(2\pi\hbar)^{\dim(M)/2}\mathrm{tr}\left(\mathrm{e}^{-\beta\hbar H_{\nabla,V/\hbar}}\right) \xrightarrow{\hbar\to 0+} \int_M \mathrm{tr}_{E_x}\left(\mathrm{e}^{-\beta V(x)}\right)\mathrm{vol}(\mathrm{d}x). \tag{5}$$

In (4), $\mathrm{d}v$ stands for the Lebesgue measure on $\mathrm{T}^*_x M$, and it should be noted that in general this result requires some additional global Golden-Thompson control [10] on V, like

$$\int_M p(\beta\hbar, x, x)\mathrm{e}^{-\beta \min\{\text{eigenvalues of } V(x)\}}\mathrm{vol}(\mathrm{d}x) < \infty.$$

The aim of this note is to translate all of the above data into the setting of infinite graphs, and to present recently obtained weighted graph-analogues of these results.

2. Covariant Schrödinger operators on infinite graphs: Recent results

Let (X, b, m) be an arbitrary weighted graph, that is, X is a countable set, b is a symmetric function

$$b : X \times X \longrightarrow [0, \infty) \text{ which satisfies } b(x,x) = 0, \sum_{y \in X} b(x,y) < \infty \text{ for all } x \in X,$$

and $m : X \to (0, \infty)$ is an arbitrary function. We shall interpret b as an edge weight function and write $x \sim_b y$, if $b(x, y) > 0$, and m is interpreted as a vertex weight function. In this setting, a *complex vector bundle* $F \to X$ (over the countable set X) with $\mathrm{rank}(F) = \nu \in \mathbb{N}$ is given by a family $F = \bigsqcup_{x \in X} F_x$ of ν-dimensional complex linear spaces, with the corresponding space of sections

$$\Gamma(X, F) = \{f \big| f : X \to F, f(x) \in F_x\},$$

which is a $\mathsf{C}(X)$-module, with $\mathsf{C}(X)$ the complex algebra of functions on X. If additionally each fiber F_x comes equipped with a complex scalar product $(\bullet, \bullet)_x = (\bullet, \bullet)_x^F$, then $F \to X$ is referred to as a *Hermitian vector bundle*, and the norm and operator norm corresponding to $(\bullet, \bullet)_x$ will be denoted with $| \bullet |_x$.

Definition 2.1. Let $F \to X$ be a complex vector bundle with rank $\nu \in \mathbb{N}$.
(i) An assignment Φ which assigns to any $x \sim_b y$ an isomorphism of complex vector spaces $\Phi_{x,y} : F_x \to F_y$ is called a *b-connection* on $F \to X$, if one has $\Phi_{y,x} = \Phi_{x,y}^{-1}$ for all $x \sim_b y$.
(ii) If $F \to X$ is Hermitian, then a b-connection Φ on $F \to X$ is called *unitary*, if $\Phi_{x,y}^* = \Phi_{x,y}^{-1}$ for all $x \sim_b y$.

We fix a Hermitian vector bundle $F \to X$ of rank $\nu \in \mathbb{N}$, with a unitary b-connection Φ defined on it. These data determine the sesquilinear form

$$\tilde{Q}_{\Phi,0}(\Psi_1, \Psi_2) = \frac{1}{2} \sum_{x \sim_b y} b(x,y) \big(\Psi_1(x) - \Phi_{y,x}\Psi_1(y), \Psi_2(x) - \Phi_{y,x}\Psi_2(y)\big)_x$$

in the Hilbert space $\Gamma_{\ell^2_m}(X, F)$ given by the sections $\Psi \in \Gamma(X, F)$ such that

$$\|\Psi\|_m^2 := \sum_{x \in X} |\Psi(x)|_x^2 m(x) < \infty,$$

with domain of $\tilde{Q}_{\Phi,0}$ consisting of finitely supported sections $\Gamma_c(X, F)$. Clearly, $\tilde{Q}_{\Phi,0}$ is densely defined, symmetric, and nonnegative, and in fact it is closable (although, in contrast to the Riemannian setting, $\tilde{Q}_{\Phi,0}$ need not come from a symmetric operator). Furthermore, we point out that in general $\tilde{Q}_{\Phi,0}$ is not bounded. However, with

$$C(b, m) := \sup_{x \in X} \frac{1}{m(x)} \sum_{y \in X} b(x, y)$$

one always has

$$\tilde{Q}_{\Phi,0}(f,f) \leq 2C(b,m)\|f\|_m^2,$$

and one often has $C(b,m) < \infty$ in applications (cf. [5] and the references therein for more details on these facts). Let us explain the analogy of $\tilde{Q}_{\Phi,0}$ to the Riemannian case (2): Firstly, we have the edge vector bundle

$$F^b := \bigsqcup_{(x,y)\in X^b} F^b_{(x,y)} := \bigsqcup_{(x,y)\in X^b} F_x \longrightarrow X^b := \{b > 0\} \subset X \times X,$$

which with $\Omega^1(X,F;b) := \Gamma(X^b, F^b)$ carries the canonical Hermitian structure

$$\big(\alpha_1(x,y), \alpha_2(x,y)\big)_{(x,y)} := \big(\alpha_1(x,y), \alpha_2(x,y)\big)^F_x, \quad \alpha_1, \alpha_2 \in \Omega^1(X,F;b).$$

Then Φ induces the complex linear map

$$\nabla_\Phi : \Gamma(X,F) \longrightarrow \Omega^1(X,F;b), \quad \nabla_\Phi\Psi(x,y) := \Phi_{y,x}\Psi(y) - \Psi(x),$$

which satisfies the Leibniz rule

$$\nabla_\Phi(f\Psi)(x,y) = \mathrm{d}f(x,y)\Psi(x) + f(y)\nabla_\Phi\Psi(x,y), \quad f \in \mathsf{C}(X),$$
$$\text{where } \mathrm{d} : \mathsf{C}(X) \longrightarrow \Omega^1(X;b) := \mathsf{C}(X^b), \ \mathrm{d}f(x,y) := f(y) - f(x)$$

is the covariant derivative corresponding to the identity connection on the trivial complex line bundle $X \times \mathbb{C} \to X$ over X. Furthermore, the unitarity of Φ implies that ∇_Φ is Hermitian in the sense that for all $\Psi_1, \Psi_2 \in \Gamma(X,F)$ one has

$$\mathrm{d}\big(\Psi_1, \Psi_2\big)(x,y) = \big(\nabla_\Phi\Psi_1(x,y), \Psi_2(x)\big)_x - \big(\Psi_1(y), \nabla_{\Phi^{-1}}\Psi_2(x,y)\big)_y,$$

which clearly is a discrete analogue to (1). Finally, for $\Psi_1, \Psi_2 \in \Gamma_c(X,F)$ one has

$$\tilde{Q}_{\Phi,0}(\Psi_1, \Psi_2) = \frac{1}{2}\sum_{x\sim_b y} b(x,y)\big(\nabla_\Phi\Psi_1(x,y), \nabla_\Phi\Psi_2(x,y)\big)_{(x,y)},$$

which is of type (2) in the situation of "unweighted edges" $b(x,y) \in \{0,1\}$.

Let $Q_{\Phi,0} := \overline{\tilde{Q}_{\Phi,0}}$, and note that the canonical scalar regular Dirichlet form Q on $\ell^2_m(X)$ and its associated operator H arise as special cases of the above construction, upon taking the identity connection on $X\times\mathbb{C} \to X$. Given a *potential* V on $F \to X$, that is, $V \in \Gamma(X, \mathrm{End}(F))$ is pointwise self-adjoint, we can define a symmetric sesqui-linear form Q_V in $\Gamma_{\ell^2_m}(X,F)$ by

$$Q_V(\Psi_1,\Psi_2) = \sum_{x\in X}\big(V(x)\Psi_1(x), \Psi_2(x)\big)_x m(x), \quad \mathsf{D}(Q_V) = \Gamma_{\ell^2_m \cap \ell^2_{|V|\cdot m}}(X,F).$$

Let $\mathscr{K}(Q) \supset \ell^\infty(X)$ be the Kato class corresponding to Q, that is, $w : X \to \mathbb{C}$ is in $\mathscr{K}(Q)$, if and only if

$$\lim_{t\to 0+}\sup_{x\in X}\int_0^t\int_X \mathrm{e}^{-sH}(x,y)|w(y)|m(y)\mathrm{d}s = 0.$$

Definition 2.2 (and Proposition). *V is called* Kato decomposable, *if it admits a decomposition $V = V^+ - V^-$ into potentials $V^\pm \geq 0$ such that $|V^-| \in \mathscr{K}(Q)$. In this situation, for any $\hbar > 0$ the form $Q_{\Phi,V/\hbar} := Q_{\Phi,0} + Q_{V/\hbar}$ is densely defined, symmetric, closed and semi-bounded, and the self-adjoint operator corresponding to $Q_{\Phi,V/\hbar}$ will be denoted with $H_{\Phi,V/\hbar}$.*

We fix a Kato decomposable potential V on $F \to X$, where we refer the reader to [9] for essential self-adjointness properties of $H_{\Phi,V}$. Let us now prepare the ingredients for the Feynman-Kac formula: As Q is a regular Dirichlet form on a nice space, we can associate a reversible strong right-Markov process to it. A convenient version

$$\mathbb{X} : [0,\tau) \times \Omega \longrightarrow X, \text{ with lifetime } \tau : \Omega \longrightarrow [0,\infty],$$

of this process has been constructed in [6] and the references therein, on a filtered probability space $(\Omega, \mathscr{F}, \mathscr{F}_*, \mathbb{P})$. Let $\tau_n : \Omega \to [0,\infty]$, $n \in \mathbb{N}_0$, denote the n-th jump time of $\mathbb{X}$, and let $N(t) : \Omega \to \mathbb{N}_0 \cup \{\infty\}$ be its number of jumps until $t \geq 0$. Many path properties of this process have been derived in [6, 7]. We shall need in the sequel that $\tau = \sup_n \tau_n$ and

$$\mathbb{P}\big(b(\mathbb{X}_{\tau_n}, \mathbb{X}_{\tau_{n+1}}) > 0 \text{ for all } n \in \mathbb{N}_0\big) = 1, \ \{N(t) < \infty\} = \{t < \tau\} \ \text{ for all } t \geq 0. \quad (6)$$

In particular, the Φ-parallel transport along the paths of $\mathbb{X}$ is well-defined by

$$//^{\Phi} : [0,\tau) \times \Omega \longrightarrow F \boxtimes F^* = \bigsqcup_{(x,y) \in X \times X} \mathrm{Hom}(F_y, F_x)$$

$$//_t^{\Phi} := \begin{cases} \mathbf{1}_{\mathbb{X}_0}, & \text{if } N(t) = 0 \\ \Phi_{\mathbb{X}_{\tau_{N(t)-1}}, \mathbb{X}_{\tau_{N(t)}}} \cdots \Phi_{\mathbb{X}_{\tau_0}, \mathbb{X}_{\tau_1}} & \text{else} \end{cases} \in \mathrm{Hom}(F_{\mathbb{X}_0}, F_{\mathbb{X}_t}),$$

which gives a pathwise unitary process, and we can also define the process

$$\mathscr{A}^{\Phi,V} : [0,\tau) \times \Omega \longrightarrow \mathrm{End}(F)$$

as the path ordered exponential

$$\mathscr{A}_t^{\Phi,V} - \mathbf{1}_{\mathbb{X}_0}$$
$$= \sum_{n=1}^{\infty} (-1)^n \int_{t\sigma_n} //_{s_1}^{\Phi,-1} V(\mathbb{X}_{s_1}) //_{s_1}^{\Phi} \cdots //_{s_n}^{\Phi,-1} V(\mathbb{X}_{s_n}) //_{s_n}^{\Phi} \, \mathrm{d}s_1 \cdots \mathrm{d}s_n \in \mathrm{End}(F_{\mathbb{X}_0}).$$

Generalizing the scalar magnetic situation from [6], the following Feynman-Kac formula, which is our discrete analogue of (3), has been proven in [7]:

Theorem 2.1. *With $\mathbb{P}^x := \mathbb{P}(\bullet | \mathbb{X}_0 = x)$, the integral kernel*

$$[0,\infty) \times X \times X \ni (t,x,y) \longmapsto \mathrm{e}^{-tH_{\Phi,V}}(x,y) \in \mathrm{Hom}(F_y, F_x) \subset F \boxtimes F^*,$$

is given by

$$\mathrm{e}^{-tH_{\Phi,V}}(x,y) = m(y)^{-1} \mathbb{P}^x(\mathbb{X}_t = y) \mathbb{E}^x\big[\mathscr{A}_t^{\Phi,V} //_t^{\Phi,-1} \big| \mathbb{X}_t = y\big], \quad (7)$$

in other words, one has the representation

$$e^{-tH_{\Phi,V}}\Psi(x)=\sum_{y\in X}e^{-tH_{\Phi,V}}(x,y)\Psi(y)m(y),\quad t\geq 0,\ \Psi\in\Gamma_{\ell^2_m}(X,F),\ x\in X.$$

And the following semiclassical limit is our discrete variant of (5):

Theorem 2.2. *Assume that there is a scalar Kato decomposable function $w : X \to \mathbb{R}$ with $V \geq w$. Then for all $\beta > 0$ with $\sum_{x\in X} e^{-\beta w(x)} < \infty$, one has*

$$\mathrm{tr}(e^{-\beta\hbar H_{\Phi,V/\hbar}})\leq\sum_{x\in X}\mathrm{tr}_x(e^{-\beta V(x)})<\infty, \tag{8}$$

$$\mathrm{tr}(e^{-\hbar\beta H_{\Phi,V/\hbar}})\xrightarrow{\hbar\to 0+}\sum_{x\in X}\mathrm{tr}_x(e^{-\beta V(x)}). \tag{9}$$

Let us remark here that (X, b, m) is completely arbitrary in these results (in particular, (X, b) may be locally infinite, and we allow $\inf m = 0$). Furthermore, if (X, b) does not support a symmetry which is respected by m and $\mathbb{P}^\bullet$ appropriately, then the Golden-Thompson bound (8) does not follow directly from (7), but rather from a combination of (7) for $V = 0$ with the abstract operator variant of the Golden-Thompson bound [5, 6] (and a combination of geometric and functional analytic approximation arguments). The proof of (9) uses semigroup domination and the corresponding result in the scalar "nonmagnetic" situation, which itself makes full use of the path properties of $\mathbb{X}$. Finally, we would like to point out that by combining the above results with the fact that Brownian motion on (noncompact) Riemannian manifolds can be approximated in law by time continuous geodesic random walks (cf. [8], where even the situation of time-dependent Riemannian metrics is considered), it should be possible to approximate spectral data of the covariant Schrödinger operators from Section 1 by the discrete ones of this Section. This should be possible under very general assumptions on the underlying data. We refer the reader to [1, 11] for special cases in the flat Euclidean space.

Bibliography

[1] S. Albeverio, R. Høegh-Krohn, H. Holden and T. Kolsrud, A covariant Feynman–Kac formula for unitary bundles over Euclidean space, in *Stochastic Partial Differential Equations and Applications II (Trento, 1988)*, Lecture Notes in Math., Vol. 1390 (Springer, Berlin, 1989), pp. 1–12.

[2] C. Bär and F. Pfäffle, Asymptotic heat kernel expansion in the semi-classical limit, *Comm. Math. Phys.* **294**, 731–744 (2010).

[3] B. Güneysu, On generalized Schrödinger semigroups, *J. Funct. Anal.* **262**, 4639–4674 (2012).

[4] B. Güneysu, Nonrelativistic hydrogen type stability problems on nonparabolic 3-manifolds, *Ann. Henri Poincaré* **13**, 1557–1573 (2012).

[5] B. Güneysu, Semiclassical limits of quantum partition functions on infinite graphs, arXiv:1402.2452.
[6] B. Güneysu, M. Keller and M. Schmidt, A Feynman–Kac–Itô formula for magnetic Schrödinger operators on graphs, arXiv:1301.1304v2.
[7] B. Güneysu, O. Milatovic and F. Truc, Generalized Schödinger semigroups on infinite graphs, *Potential Anal.* **41**, 517–541 (2014).
[8] K. Kuwada, Convergence of time-inhomogeneous geodesic random walks and its application to coupling methods, *Ann. Probab.* **40**, 1945–1979 (2012).
[9] O. Milatovic and F. Truc, Essential self-adjointness of Schrödinger operators on vector bundles over infinite graphs, arXiv:1307.1213.
[10] B. Simon, *Functional Integration and Quantum Physics*, 2nd edn. (AMS Chelsea Publishing, Providence, 2005).
[11] V. S. Varadarajan and D. Weisbart, Convergence of quantum systems on grids, *J. Math. Anal. Appl.* **336**, 608–624 (2007).

Resolvent expansion for the discrete one-dimensional Schrödinger operator

K. Ito

Graduate School of Pure and Applied Sciences, University of Tsukuba,
1-1-1 Tennodai, Tsukuba Ibaraki, 305-8571 Japan
E-mail: ito-ken@math.tsukuba.ac.jp

A. Jensen

Department of Mathematical Sciences, Aalborg University,
Fredrik Bajers Vej 7G, DK-9220 Aalborg Ø, Denmark
E-mail: matarne@math.aau.dk

We describe a classification of threshold properties for one-dimensional discrete Schrödinger operators and associated asymptotic expansions of the resolvent for a general class of interactions.

Keywords: Resolvent expansion, discrete Schrödinger operator, threshold properties.

1. Introduction

We consider the discrete one-dimensional Schrödinger operator $H = H_0 + V$, where $(H_0x)[n] = -(x[n+1] + x[n-1] - 2x[n])$ and V is a self-adjoint operator on $\ell^2(\mathbb{Z})$ with a decay property given by V extending to a compact operator from $\ell^{\infty,-\beta}(\mathbb{Z})$ to $\ell^{1,\beta}(\mathbb{Z})$ for some $\beta \geq 1$. Let $H = H_0 + V$. We give a description of the solutions to $Hx = 0$, and $Hx = 4x$, $x \in \ell^{\infty,-\beta}(\mathbb{Z})$. Using this description we give asymptotic expansions of the resolvent of H at the two thresholds 0 and 4.

The precise results are quite complicated and long to state, so here we will give an outline of the results. The full results can be found in [2]. The proofs use the technique introduced in [4].

2. Outline of results

The operator H_0 on $\mathcal{H} = \ell^2(\mathbb{Z})$ has a purely absolutely continuous spectrum equal to $[0, 4]$. The points 0 and 4 are called the thresholds. We state results for the point 0. The results for the point 4 follow from these results due to the following observation.

It is well known that if we define for any sequence x the operator $(Jx)[n] = (-1)^n x[n]$, then J satisfies $J^2 = 1$ and is a bounded self-adjoint and unitary operator on $\mathcal{H}$. Let $V_J = JVJ^{-1}$. Furthermore, we have $J(H_0+V)J^{-1} = -(H_0 - V_J - 4)$ and

$$J(H_0 + V - z)^{-1}J^{-1} = -(H_0 - V_J - (4 - z))^{-1}.$$

We introduce for $s \in \mathbb{R}$ the spaces

$$\mathcal{L}^s = \ell^{1,s}(\mathbb{Z}) = \Big\{x\colon \mathbb{Z} \to \mathbb{C};\ \|x\|_{1,s} = \sum_{n\in\mathbb{Z}} (1+n^2)^{s/2}|x[n]| < \infty\Big\},$$

$$(\mathcal{L}^s)^* = \ell^{\infty,-s}(\mathbb{Z}) = \Big\{x\colon \mathbb{Z} \to \mathbb{C};\ \|x\|_{\infty,-s} = \sup_{n\in\mathbb{Z}}(1+n^2)^{-s/2}|x[n]| < \infty\Big\}.$$

We use the factored perturbation technique. Bounded operators and compact operators between Banach spaces are denoted $\mathcal{B}(\mathcal{X},\mathcal{Y})$ and $\mathcal{C}(\mathcal{X},\mathcal{Y})$, respectively.

Assumption 2.1. *Let $V \in \mathcal{B}(\mathcal{H},\mathcal{H})$ be self-adjoint, and assume that there exist a real number $\beta \geq 1$, a Hilbert space $\mathcal{K}$, an injective operator $v \in \mathcal{B}(\mathcal{K},\mathcal{L}^\beta) \cap \mathcal{C}(\mathcal{K},\mathcal{L}^1)$ and a self-adjoint unitary operator $U \in \mathcal{B}(\mathcal{K})$ such that*

$$V = vUv^* \in \mathcal{B}((\mathcal{L}^\beta)^*, \mathcal{L}^\beta) \cap \mathcal{C}((\mathcal{L}^1)^*, \mathcal{L}^1).$$

This class of interactions has the advantage that it is additive, i.e. a sum of two interactions satisfying Assumption 2.1 will also satisfy this assumption. Sums of multiplicative interactions and finite rank interactions are included, provided they satisfy the decay conditions in the assumption.

Under Assumption 2.1 the perturbed Schrödinger operator $H = H_0 + V$ is bounded and self-adjoint on $\mathcal{H}$, and we can expand the perturbed resolvent $R(z) = (H-z)^{-1}$ as follows. For $z \in \mathbb{C} \setminus [0,\infty)$ we take the determination of the square root with $\operatorname{Im}\sqrt{z} > 0$. We use the notation $\mathcal{B}^n = \mathcal{B}(\mathcal{L}^n, (\mathcal{L}^n)^*)$.

Theorem 2.1. *Suppose $\beta \geq 4$ in Assumption 2.1, and let $N \in [-2, \beta - 6]$ be any integer. Then as $z \to 0$ in $\mathbb{C}\setminus[0,\infty)$ the resolvent $R(z)$ has the asymptotic expansion in the uniform topology of $\mathcal{B}^{N+4}$:*

$$R(z) = \sum_{j=-2}^{N} z^{j/2} G_j + \mathcal{O}(z^{(N+1)/2}), \quad G_j \in \mathcal{B}^{j+3}, \tag{1}$$

and the coefficients G_j can be computed explicitly.

To describe the leading coefficients in the expansion given in (1) we define the generalized eigenspace associated with the threshold $\lambda = 0$ as

$$\widetilde{\mathcal{E}} = \{\Psi \in (\mathcal{L}^\beta)^*;\ H\Psi = 0\}, \quad \widetilde{d} = \dim \widetilde{\mathcal{E}}.$$

This eigenspace is finite-dimensional and the (generalized) eigenfunctions have a particular asymptotics at infinity. Define the sequences $\mathbf{1}, \boldsymbol{\sigma} \in (\mathcal{L}^0)^*$ and $\mathbf{n}, |\mathbf{n}| \in (\mathcal{L}^1)^*$ by

$$\mathbf{1}[n] = 1, \quad \boldsymbol{\sigma}[n] = \begin{cases} \pm 1 & \text{if } \pm n > 0, \\ 0 & \text{if } \quad n = 0, \end{cases} \quad \mathbf{n}[n] = n, \quad |\mathbf{n}|[n] = |n|,$$

respectively.

Theorem 2.2. *Suppose $\beta \geq 1$ in Assumption 2.1. Then*

$$\widetilde{\mathcal{E}} \subset \mathbb{C}\mathbf{n} \oplus \mathbb{C}|\mathbf{n}| \oplus \mathbb{C}\mathbf{1} \oplus \mathbb{C}\boldsymbol{\sigma} \oplus \mathcal{L}^{\beta-2}, \quad \widetilde{d} < \infty.$$

We use the following subspaces to classify the solutions to $H\Psi = 0$.

$$\mathcal{E} = \widetilde{\mathcal{E}} \cap \left(\mathbb{C}\mathbf{1} \oplus \mathbb{C}\boldsymbol{\sigma} \oplus \mathcal{L}^{\beta-2}\right), \quad d = \dim \mathcal{E}; \tag{2}$$

$$E = \widetilde{\mathcal{E}} \cap \mathcal{L}^{\beta-2}, \quad d_0 = \dim E. \tag{3}$$

We use the following terminology in the classification. It is the same terminology as was introduced in [3].

Definition 2.1. The threshold $\lambda = 0$ is said to be

(1) a *regular point*, if $\mathcal{E} = E = \{0\}$;

(2) an *exceptional point of the first kind*, if $\mathcal{E} \supsetneq E = \{0\}$;

(3) an *exceptional point of the second kind*, if $\mathcal{E} = E \supsetneq \{0\}$;

(4) an *exceptional point of the third kind*, if $\mathcal{E} \supsetneq E \supsetneq \{0\}$.

These classifications *characterize and are characterized by* the expansion coefficients G_{-2} and G_{-1} of the singular part in (1):

Theorem 2.3. *Suppose $\beta \geq 4$ in Assumption 2.1. Then $\widetilde{d} = d_0 + 2$ and there exist bases $\Psi_j \in E$, $j = 1, \ldots, d_0$, and $\Psi_j \in \mathcal{E}/E$, $j = d_0 + 1, \ldots, d$, such that*

$$G_{-2} = -\sum_{j=1}^{d_0} \langle \Psi_j, \cdot \rangle \Psi_j, \quad G_{-1} \equiv i \sum_{j=d_0+1}^{d} \langle \Psi_j, \cdot \rangle \Psi_j \mod \langle E, \cdot \rangle E, \tag{4}$$

where $\langle E, \cdot \rangle E \subset \mathcal{B}((\mathcal{L}^{\beta-2})^*, \mathcal{L}^{\beta-2})$ is the subspace spanned by the operators of the form $\langle \Psi, \cdot \rangle \Psi'$ with $\Psi, \Psi' \in E$. Furthermore, one can choose $\Psi_j \in E$, $j = 1, \ldots, d$, to be orthonormal, and hence $-G_{-2}$ is the orthogonal projection onto E.

It is possible to give a very precise and complete characterization of the coefficients in Theorem 2.3, see [2].

3. Examples of interactions

The class of interactions that satisfy Assumption 2.1 is very large. We will here give some examples. Obviously *non-local* operators of finite rank are included, provided the decay conditions are satisfied. More generally, we can state the following result.

Proposition 3.1. *Let $\beta \geq 1$ be any real number, and $v_j \in \mathcal{L}^\beta$, $j = 1, 2, \ldots$, be at most a countable number of linearly independent vectors with*

$$\sum_j \|v_j\|^2_{\mathcal{L}^\beta} < \infty. \tag{5}$$

Then for any $\sigma_j \in \{\pm 1\}$ the operator series

$$V = \sum_j \sigma_j \langle v_j, \cdot \rangle v_j$$

converge in the uniform topology of $\mathcal{B}((\mathcal{L}^\beta)^*, \mathcal{L}^\beta)$ and satisfy Assumption 2.1 with the same β.

To prove this result we just take an abstract Hilbert space $\mathcal{K}$ of dimension $\#\{v_j\}$. An operator V is *local*, if it does not expand the support of sequences. Note that H_0 is not local, in contrast to the continuous case.

Since the base space is discrete, a local operator V is always identifiable with a multiplication operator by some function, which we also denote by V. A local V satisfies Assumption 2.1, if

$$\sum_{n \in \mathbb{Z}} (1 + n^2)^\beta |V[n]| < \infty, \quad \beta \geq 1.$$

For local interactions zero cannot be an eigenvalue and we can have at most one linearly independent threshold resonance function, see [2].

We give an example of a finite rank interaction, where zero is an exceptional point of the third kind.

Example 3.1. We define the following sequences:

$$\phi_j[n] = \begin{cases} -1, & n = 4j, \\ 1, & n = 4j + 1, \\ 0, & \text{otherwise}, \end{cases} \qquad j = 0, 1, 2,$$

$$u_j[n] = \begin{cases} 1 & n \leq 4j, \\ -1 & n > 4j, \end{cases} \qquad j = 0, 1, 2.$$

Then we define

$$Vx = -\sum_{j=0}^{2} \langle \phi_j, x \rangle \phi_j.$$

With these definitions it is straightforward to verify that

$$(H_0 + V)u_j = 0, \quad j = 0, 1, 2.$$

If we define

$$w[n] = \begin{cases} 1 & n = 1, 2, 3, 4, \\ 0 & \text{otherwise}, \end{cases}$$

then we have $u_2 = \frac{1}{2}u_0 + \frac{1}{2}u_1 + w$.

In this example we see that it is possible to have two linearly independent threshold resonance functions and an ℓ^2 eigenfunction. The example can be modified to get a zero eigenvalue of any finite multiplicity.

4. Comments on the literature

Resolvent expansions for one-dimensional discrete Schrödinger operators with multiplicative potentials was treated in the generic case in [5] and in the general case in [1]. Besides [2] we are not aware of any complete treatments of the resolvent expansion problem and the associated complete classification of the threshold (generalized) eigenspaces for a general class of interactions.

Bibliography

[1] S. Cuccagna, *L^p continuity of wave operators in $\mathbb{Z}$*, J. Math. Anal. Appl. **354** (2009), 594–605.

[2] K. Ito and A. Jensen, *A complete classification of threshold properties for one-dimensional discrete Schrödinger operators*, submitted, arXiv:1312.1396.

[3] A. Jensen and T. Kato, *Spectral properties of Schrödinger operators and time-decay of the wave functions*, Duke Math. J. **46** (1979), 583–611.

[4] A. Jensen and G. Nenciu, *A unified approach to resolvent expansions at thresholds*, Rev. Math. Phys. **13** (2001), no. 6, 717–754.

[5] D. E. Pelinovsky and A. Stefanov, *On the spectral theory and dispersive estimates for a discrete Schrödinger equation in one dimension*, J. Math. Phys. **49** (2008), 113501.

Spectral asymptotics for a δ' interaction supported by an infinite curve

M. Jex

Doppler Institute for Mathematical Physics and Applied Mathematics, Czech Technical University in Prague, Břehová 7, 11519 Prague, and Department of Physics, Faculty of Nuclear Sciences and Physical Engineering, Czech Technical University in Prague, Břehová 7, 11519 Prague, Czech Republic
E-mail: jexmicha@fjfi.cvut.cz

We consider a generalized Schrödinger operator in $L^2(\mathbb{R}^2)$ describing an attractive δ' interaction in a strong coupling limit. δ' interaction is characterized by a coupling parameter β and it is supported by a C^4-smooth infinite asymptotically straight curve Γ without self-intersections. It is shown that in the strong coupling limit, $\beta \to 0_+$, the eigenvalues for a non-straight curve behave as $-\frac{4}{\beta^2} + \mu_j + \mathcal{O}(\beta|\ln\beta|)$, where μ_j is the j-th eigenvalue of the Schrödinger operator on $L^2(\mathbb{R})$ with the potential $-\frac{1}{4}\gamma^2$ where γ is the signed curvature of Γ.

Keywords: δ' interaction, quantum graphs, spectral theory.

1. Introduction

The quantum mechanics describing the particle confined to various manifolds is studied quite extensively. It is very useful for describing various nanostructures in physics but it also offers a large variety of interesting problems from the purely mathematical point of view. Systems where the confinement is realized by a singular attractive potential, so called 'leaky' quantum graphs [1], have the advantage that they take quantum tunneling effects into account in contrast to quantum graphs [2]. The confining potential is often taken to be of the δ type. One can think also about more singular types of potentials namely δ' type based on the concept of δ' interaction in one dimension [3].

We are interested in the spectrum of the operator which can be formally written as

$$H = -\Delta - \beta^{-1}\delta'(\cdot - \Gamma)$$

where δ' interaction is supported by an infinite curve Γ in $\mathbb{R}^2$. We are interested in the strong coupling regime which corresponds to small values of the parameter β. We derive spectral asymptotics of discrete and essential spectra. As a byproduct we obtain that for a non-straight curve the bound state arises for sufficiently small β in an alternative way to one presented in [4].

2. Formulation of the Problem and Results

We consider a curve Γ parameterized by its arc length

$$\Gamma : \mathbb{R} \to \mathbb{R}^2, \quad s \mapsto (\Gamma_1(s), \Gamma_2(s)),$$

where $\Gamma_1(s), \Gamma_2(s) \in C^4(\mathbb{R})$ are component functions. We denote signed curvature as $\gamma(s) := (\Gamma_1''\Gamma_2' - \Gamma_1'\Gamma_2'')(s)$. We introduce several conditions for the curve Γ as:
(Γ1) Γ is C^4 smooth curve,
(Γ2) Γ has no "near self-intersections", i.e. there exists its strip neighborhood of a finite thickness which does not intersect with itself,
(Γ3) Γ is asymptotically straight in the sense that $\lim_{|s|\to\infty} \gamma(s) = 0$ and
(Γ4) Γ is not a straight line.

The operator, we are interested in, acts as a free Laplacian outside of the interaction support

$$(H_\beta\psi)(x) = -(\Delta\psi)(x)$$

for $x \in \mathbb{R}^2 \setminus \Gamma$ with the domain which can be written as $\mathcal{D}(H_\beta) = \{\psi \in H^2(\mathbb{R}^2 \setminus \Gamma) \mid \partial_{n_\Gamma}\psi(x) = \partial_{-n_\Gamma}\psi(x) = \psi'(x)|_\Gamma, -\beta\psi'(x)|_\Gamma = \psi(x)|_{\partial_+\Gamma} - \psi(x)|_{\partial_-\Gamma}\}$. The vector n_Γ denotes the normal to Γ and $\psi(x)|_{\partial_\pm\Gamma}$ are the appropriate traces of the function ψ. We also need to introduce the operator defined on the line as

$$S = -\frac{\partial^2}{\partial s^2} - \frac{1}{4}\gamma(s)^2, \tag{1}$$

with the domain $\mathcal{D}(S) = H^2(\mathbb{R})$. The eigenvalues of the operator S are denoted by μ_j with the multiplicity taken into account. Now we are ready to write down the main results of our paper.

Theorem 2.1. *Let an infinite curve Γ satisfy conditions (Γ1)–(Γ3), then $\sigma_{\mathrm{ess}}(H_\beta) \subseteq [\epsilon(\beta), \infty)$, where $\epsilon(\beta) \to -\frac{4}{\beta^2}$ holds as $\beta \to 0_+$.*

Theorem 2.2. *Let an infinite curve Γ satisfy assumptions (Γ1)–(Γ4), then H_β has at least one isolated eigenvalue below the threshold of the essential spectrum for all sufficiently small $\beta > 0$, and the j-th eigenvalue behaves in the strong coupling limit $\beta \to 0_+$ as*

$$\lambda_j = -\frac{4}{\beta^2} + \mu_j + \mathcal{O}(-\beta\ln(\beta)).$$

3. Bracketing Estimates

For the purpose of the proofs we introduce curvelinear coordinates (s, u) along the curve in the same way as done in [6], i.e.

$$(x, y) = \left(\Gamma_1(s) + u\Gamma_2'(s), \Gamma_2(s) - u\Gamma_1'(s)\right) . \tag{2}$$

The map $\Gamma(s) : \mathbb{R} \to \mathbb{R}^2$ is injective, because we have curve Γ parameterized by its arc length. The conditions (Γ1) and (Γ3) imply that the curvature $\gamma(s)$ is bounded. As a result the coordinates (2) are injective in the neighborhood of the point $\Gamma(s)$ for $u < 1/\gamma(s)$. Due to the condition (Γ2) a strip neighborhood of thickness $2l$ which does not intersect with itself exists and as a result the map (2) is injective for all u small enough. We denote d as a maximum of u for which the map (2) is injective. d can be estimated from above by $\min\{min_{s\in\mathbb{R}}(\gamma(s)^{-1}), l\}$. A strip neighborhood around Γ of thickness $2a < 2d$ is denoted by $\Omega_a := \{x \in \mathbb{R}^2 : \text{dist}\,(x, \Gamma) < a\}$.

The quadratic form associated with the operator H_β was derived in [5] and it can be written as

$$h_\beta[\psi] = \|\nabla\psi\|^2 - \beta^{-1} \int_{\mathbb{R}} |\psi(s, 0_+) - \psi(s, 0_-)|^2 \, \mathrm{d}s \, .$$

where we used the curvelinear coordinates in the strip neighborhood of the curve Γ for the functions $\psi \in H^1(\mathbb{R}^2 \setminus \Gamma)$ as $\psi(s, u)$.

For the proofs of both theorems we will need estimates of our operator H_β via Dirichlet and Neumann bracketing as done in [6, 7]. We introduce the operators with added either Dirichlet or Neummann boundary conditions at the boundary of the strip neighborhood Ω_a of Γ. We introduce quadratic forms $h_{D,\beta}$ and $h_{N,\beta}$ on the strip neighborhood of Γ which can be written as

$$h_{D/N,\beta}[\psi] = \|\nabla\psi\|^2 - \beta^{-1} \int_{\mathbb{R}} |\psi(s, 0_+) - \psi(s, 0_-)|^2 \, \mathrm{d}s$$

with the domains $\mathcal{D}(h_{N,\beta}) = H^1(\Omega_a \setminus \Gamma)$ and $\mathcal{D}(h_{D,\beta}) = \tilde{H}_0^1(\Omega_a \setminus \Gamma)$, where $\tilde{H}_0^1(\Omega_a \setminus \Gamma)$ is set of all functions in $H^1(\Omega_a \setminus \Gamma)$ vanishing at the boundary of Ω_a. The operators associated with the quadratic forms $h_{D,\beta}$ and $h_{N,\beta}$ are denoted by $H_{D,\beta}$ and $H_{N,\beta}$, respectively. With the help of Dirichlet-Neumann bracketing we are able to write the following inequality

$$-\Delta^N_{\mathbb{R}^2\setminus\Omega_a} \oplus H_{N,\beta} \le H_\beta \le -\Delta^D_{\mathbb{R}^2\setminus\Omega_a} \oplus H_{D,\beta} \, , \tag{3}$$

where $-\Delta^{N/D}_{\mathbb{R}^2\setminus\Omega_a}$ denotes either Neumann or Dirichlet Laplacian on $\mathbb{R}^2 \setminus \Omega_a$ respectively. The Neumann Laplacian and the Dirichlet Laplacian are positive and as a result all the information about the negative spectrum, which we are interested in, is encoded in the operators $H_{D/N,\beta}$.

Now we rewrite the quadratic forms $h_{D/N,\beta}$ in the curvelinear coordinates (2). We obtain expressions analogous to those obtained in [6].

Lemma 3.1. *Quadratic forms $h_{D,\beta}$, $h_{N,\beta}$ are unitarily equivalent to quadratic forms $q_{D,\beta}$ and $q_{N,\beta}$ which can be written as*

$$q_{D,\beta}[f] = \|\frac{\partial_s f}{g}\|^2 + \|\partial_u f\|^2 + (f, Vf) - \beta^{-1}\int_{\mathbb{R}} |f(s,0_+) - f(s,0_-)|^2 \,\mathrm{d}s + \frac{1}{2}\int_{\mathbb{R}} \gamma(s)\big(|f(s,0_+)|^2 - |f(s,0_-)|^2\big)\,\mathrm{d}s$$

$$q_{N,\beta}[g] = q_{D,\beta}[g] - \int_{\mathbb{R}} \frac{\gamma(s)}{2(1+a\gamma(s))}|f(s,a)|^2\,\mathrm{d}s + \int_{\mathbb{R}} \frac{\gamma(s)}{2(1-a\gamma(s))}|f(s,-a)|^2\,\mathrm{d}s$$

defined on $\tilde{H}_0^1(\mathbb{R}\times((-a,0)\cup(0,a)))$ and $H^1(\mathbb{R}\times((-a,0)\cup(0,a)))$, respectively. The geometrically induced potential in these formulæ is given by

$$V(s,u) = \frac{u\gamma''}{2g^3} - \frac{5(u\gamma')^2}{4g^4} - \frac{\gamma^2}{4g^2}$$

with $g(s,u) := 1 + u\gamma(s)$.

The proof of this lemma can be done step by step as done in [6, Lemma 3.1] so we omit the details.

We will also need cruder estimates by quadratic forms $q^{\pm}_{a,\beta}[f]$ which satisfy $q^{-}_{a,\beta}[f] \leq q_{N,\beta}[f] \leq h_\beta[f] \leq q_{D,\beta}[f] \leq q^{+}_{a,\beta}[f]$. The quadratic form $q^{+}_{a,\beta}[f]$ can be written as

$$q^{+}_{a,\beta}[f] = \|\partial_u f\|^2 + (1-a\gamma_+)^{-2}\|\partial_s f\|^2 + (f, V^{(+)}f) - \beta^{-1}\int_{\mathbb{R}} |f(s,0_+) - f(s,0_-)|^2\mathrm{d}s + \frac{1}{2}\int_{\mathbb{R}} \gamma(s)\big(|f(s,0_+)|^2 - |f(s,0_-)|^2\big)\,\mathrm{d}s$$

where $V^{(+)} := \frac{a(\gamma'')_+}{2(1-a\gamma_+)^3} - \frac{\gamma^2}{4(1+a\gamma_+)^2}$ and $f_+ := \max_{s\in\mathbb{R}}|f|$ denotes maximum of $|f|$. The quadratic form $q^{-}_{a,\beta}[f]$ can be written as

$$q^{-}_{a,\beta}[f] = \|\partial_u f\|^2 + (1+a\gamma_+)^{-2}\|\partial_s f\|^2 + (f, V^{(-)}f) - \beta^{-1}\int_{\mathbb{R}} |f(s,0_+) - f(s,0_-)|^2\,\mathrm{d}s - \frac{1}{2}\int_{\mathbb{R}} \gamma(s)\big(|f(s,0_+)|^2 - |f(s,0_-)|^2\big)\,\mathrm{d}s - \gamma_+\int_{\mathbb{R}} |f(s,a)|^2\,\mathrm{d}s - \gamma_+\int_{\mathbb{R}} |f(s,-a)|^2\,\mathrm{d}s$$

where $V^{(-)} = -\frac{a(\gamma'')_+}{2(1-a\gamma_+)^3} - \frac{5(a(\gamma')_+)^2}{4(1-a\gamma_+)^4} - \frac{\gamma^2}{4(1-a\gamma_+)^2}$. The operators $Q^{\pm}_{a,\beta}$ associated with $q^{\pm}_{a,\beta}[f]$ can be written as $Q^{\pm}_{a,\beta} = U^{\pm}_a \otimes I + \int_{\mathbb{R}}^{\oplus} T^{\pm}_{a,\beta}(s)\,\mathrm{d}s$ where $U^{\pm}_a$ corresponds to the longitudinal variable s and $T^{\pm}_{a,\beta}(s)$ corresponds to the transversal variable u.

The operators $T^{\pm}_{a,\beta}(s)$ act as $T^{\pm}_{a,\beta}(s)f = -f''$ with the domains

$$\mathcal{D}(T^{+}_{a,\beta}(s)) = \{f \in H^2((-a,a)\setminus\{0\}) \mid f(a) = f(-a) = 0\,,$$
$$f'(0_-) = f'(0_+) = -\beta^{-1}(f(0_+) - f(0_-)) + \frac{1}{2}\gamma(s)(f(0_+) + f(0_-))\}$$
$$\mathcal{D}(T^{-}_{a,\beta}(s)) = \{f \in H^2((-a,a)\setminus\{0\}) \mid \mp\gamma_+ f(\pm a) = f'(\pm a)\,,$$
$$f'(0_-) = f'(0_+) = -\beta^{-1}(f(0_+) - f(0_-)) + \frac{1}{2}\gamma(s)(f(0_+) + f(0_-))\}\,.$$

The operators $U^{\pm}_a$ act as $U^{\pm}_a f = -(1\mp a\gamma_+)^{-2}f'' + V^{(\pm)}f$ with the domain $\mathcal{D}(U^{\pm}_a) = H^2(\mathbb{R})$. The operators $T^{\pm}_{a,\beta}(s)$ depend on the variable s, however, their negative spectrum is independent of s. Now we state two lemmata estimating the eigenvalues of operators $T^{\pm}_{a,\beta}(s)$ and $U^{\pm}_a$. Their proofs can be found in [6, Lemmata 3.2-3.4].

Lemma 3.2. *Each of the operators $T^{\pm}_{a,\beta}(s)$ has exactly one negative eigenvalue $t_{\pm}(a,\beta)$, respectively, which is independent of s provided that $\frac{a}{\beta} > 2$ and $\frac{2}{\beta} > \gamma_+$. For all $\beta > 0$ sufficiently small these eigenvalues satisfy the inequalities*

$$-\frac{4}{\beta^2} - \frac{16}{\beta^2}\exp\left(-\frac{4a}{\beta}\right) \le t_-(a,\beta) \le -\frac{4}{\beta^2} \le t_+(a,\beta) \le -\frac{4}{\beta^2} + \frac{16}{\beta^2}\exp\left(-\frac{4a}{\beta}\right)\,.$$

Lemma 3.3. *There is a positive C independent of a and j such that*

$$|\mu^{\pm}_j(a) - \mu_j| \le Caj^2$$

holds for $j \in \mathbb{N}$ and $0 < a < \frac{1}{2\gamma_+}$, where $\mu^{\pm}_j(a)$ are the eigenvalues of $U^{\pm}_a$, respectively, with the multiplicity taken into account.

Now we are ready to prove our main theorems.

4. Proof of Theorem 2.1

First we prove the trivial case for the straight line. By separation of variables the spectrum is $\sigma(H_\beta) = \sigma_{\mathrm{ess}}(H_\beta) = \left[-\frac{4}{\beta^2}, \infty\right)$.

The case for non-straight curve is done similarly as for the singular interaction supported by nonplanar surfaces in [8, 9]. The inclusion $\sigma_{\mathrm{ess}}(H_\beta) \subseteq [\epsilon(\beta), \infty)$ can be rewritten as

$$\inf \sigma_{\mathrm{ess}}(H_\beta) \ge \epsilon(\beta)\,.$$

The inequality $H_\beta \ge H_{N,\beta} \oplus -\Delta^N_{\mathbb{R}^2\setminus\Omega_d}$ implies that it is sufficient to check $\inf\sigma_{\mathrm{ess}}(H_{N,\beta}) \ge \epsilon(\beta)$ in $L^2(\Omega_a)$ for $a < d$ because the operator $-\Delta^N_{\mathbb{R}^3\setminus\Omega_d}$ is positive. Next we divide the curve Γ into two parts. First part is defined as $\Gamma^{\mathrm{int}}_\tau := \{\Gamma(s) | s < \tau\}$ and the second one $\Gamma^{\mathrm{ext}}_\tau := \Gamma \setminus \overline{\Gamma^{\mathrm{int}}_\tau}$. The corresponding strip neighborhoods are defined as $\Omega_a^{\rm int} := \{x(s,u) \in \Omega_a | s < \tau\}$ and $\Omega_a^{\rm ext} := \{x(s,u) \in \Omega_a | s > \tau\}$. We introduce Neumann decoupled operators on $\Omega_a^{\rm int,ext}$ as

$$H_{\beta,\tau}^{N,\rm int} \oplus H_{\beta,\tau}^{N,\rm ext}.$$

The operators $H_{\beta,\tau}^{N,\omega}$, $\omega = \rm int, ext$, are associated with quadratic forms $h_{\beta,\tau}^{N,\omega}$ which can be written as

$$\begin{aligned} h_{\beta,\tau}^{N,\omega} = \|\frac{\partial_s f}{g}\|^2 + \|\partial_u f\|^2 + (f, Vf) - \beta^{-1} \int_{\Gamma_\tau^\omega} |f(s,0_+) - f(s,0_-)|^2 \, \mathrm{d}s \\ + \frac{1}{2} \int_{\Gamma_\tau^\omega} \gamma(s) \big(|f(s,0_+)|^2 - |f(s,0_-)|^2 \big) \, \mathrm{d}s \\ - \int_{\Gamma_\tau^\omega} \frac{\gamma(s)}{2(1+a\gamma(s))} |f(s,a)|^2 \, \mathrm{d}s + \int_{\Gamma_\tau^\omega} \frac{\gamma(s)}{2(1-a\gamma(s))} |f(s,-a)|^2 \, \mathrm{d}s \end{aligned}$$

with the domains $H^1(\Omega_a^\omega)$. Neumann bracketing implies that $H_{N,\beta} \geq H_{\beta,\tau}^{N,\rm int} \oplus H_{\beta,\tau}^{N,\rm ext}$. The spectrum of the operator $H_{\beta,\tau}^{N,\rm int}$ is purely discrete [10] and as a result min-max principle implies that

$$\inf \sigma_{\rm ess}(H_{N,\beta}) \geq \inf \sigma_{\rm ess}(H_{\beta,\tau}^{N,\rm ext}).$$

We denote the following expression $V_\tau := \inf_{|s|>\tau, u\in(-a,a)} V(s,u)$. The assumption ($\Gamma 3$) gives us that

$$\lim_{\tau\to\infty} V_\tau = 0$$

With the help of Lemma 3.2 we can write the following estimates

$$\begin{aligned} h_{\beta,\tau}^{N,\rm ext}[f] \geq \|\partial_u f\|^2 + V_\tau \|f\|^2 - \beta^{-1} \int_{\Gamma_\tau^{\rm ext}} |f(s,0_+) - f(s,0_-)|^2 \, \mathrm{d}s \\ + \frac{1}{2} \int_{\Gamma_\tau^{\rm ext}} \gamma(s) \big(|f(s,0_+)|^2 - |f(s,0_-)|^2 \big) \, \mathrm{d}s \\ - \int_{\Gamma_\tau^{\rm ext}} \frac{\gamma(s)}{2(1+a\gamma(s))} |f(s,a)|^2 \, \mathrm{d}s + \int_{\Gamma_\tau^{\rm ext}} \frac{\gamma(s)}{2(1-a\gamma(s))} |f(s,-a)|^2 \, \mathrm{d}s \\ \geq \left(V_\tau - \frac{4}{\beta^2} - \frac{16}{\beta^2} \exp\left(-\frac{4a}{\beta} \right) \right) \|f\|^2 \end{aligned}$$

Because we can choose τ arbitrarily large we obtain the desired result.

5. Proof of Theorem 2.2

For the proof of the second theorem we use the inequalities (3) and Lemmata 3.2 and 3.3. First we put $a(\beta) = -\frac{3}{4}\beta \ln \beta$. Now with the explicit form of $Q_{a,\beta}^{\pm}$ in mind and the fact that $T_{a,\beta}^{\pm}(s)$ have exactly one negative eigenvalue we have that the spectra of $Q^{\pm}_{a,\beta}$ can be written as $t_{\pm}(d(\beta),\beta) + \mu_j^{\pm}(d(\beta))$. Using Lemmata 3.2 and 3.3 we obtain

$$t_{\pm}(a(\beta),\beta) + \mu_j^{\pm}(a(\beta)) = -\frac{4}{\beta^2} + \mu_j + \mathcal{O}(\beta|\ln\beta|)\,.$$

The min-max principle along with the inequality (3) completes the proof.

Acknowledgments

The research was supported by the Czech Science Foundation within the project 14-06818S and by by Grant Agency of the Czech Technical University in Prague, grant No. SGS13/217/OHK4/3T/14.

Bibliography

[1] P. Exner: *Leaky quantum graphs: a review*, Proceedings of the Isaac Newton Institute programme "Analysis on Graphs and Applications", AMS "Proceedings of Symposia in Pure Mathematics" Series, vol. 77, Providence, R.I., 2008; pp. 523–564.

[2] G. Berkolaiko, P. Kuchment: *Introduction to Quantum Graphs*, Amer. Math. Soc., Providence, R.I., 2013.

[3] S. Albeverio, F. Gesztesy, R. Høegh-Krohn, H. Holden: *Solvable Models in Quantum Mechanics*, 2nd edition with an appendix by P. Exner, AMS Chelsea Publishing, Providence, R.I., 2005.

[4] J. Behrndt, P. Exner, V. Lotoreichik: *Schrödinger operators with δ- and δ'-interactions on Lipschitz surfaces and chromatic numbers of associated partitions*, `arXiv:1307.0074 [math-ph]`

[5] J. Behrndt, M. Langer, V. Lotoreichik: *Schrödinger operators with δ and δ'-potentials supported on hypersurfaces*, *Ann. Henri Poincaré* **14** (2013), 385–423.

[6] P. Exner, M. Jex: *Spectral asymptotics of a strong δ' interaction on a planar loop*, *J. Phys. A: Math. Theor.* **46** (2013), 345201.

[7] P. Exner, K. Yoshitomi: *Asymptotics of eigenvalues of the Schrödinger operator with a strong δ-interaction on a loop*, *J. Geom. Phys.* **41** (2002), 344–358.

[8] P. Exner, S. Kondej: *Bound state due to a strong δ interaction supported by a curved surface*, *J. Phys. A: Math. Gen.* **36** (2003), 443–457.

[9] P. Exner, M. Jex: *Spectral asymptotics of a strong δ' interaction supported by a surface*, `arXiv:1402.6117 [math-ph]`.

[10] E.B. Davies: *Spectral Theory and Differential Operators*, Cambridge: Cambridge University Press, (1998).

Asymptotically predefined spectral gaps for the Neumann Laplacian in periodic domains

A. Khrabustovskyi

Department of Mathematics, Karlsruhe Institute of Technology,
Karlsruhe, 76131, Germany
E-mail: andrii.khrabustovskyi@kit.edu
www.math.kit.edu/grk1294/~Khrabustovskyi

Let Ω be a periodic unbounded domain in $\mathbb{R}^n$. It is well-known that the spectrum of the Neumann Laplacian in $L_2(\Omega)$ is a locally finite union of compact intervals called bands. The neighbouring bands may overlap, otherwise we have a gap in the spectrum. Our goal is to construct such periodic domain Ω that the spectrum of the corresponding Neumann Laplacian has at least m gaps ($m \in \mathbb{N}$ is a given number) and these gaps are close to predefined intervals.

Keywords: Periodic domains, Neumann Laplacian, spectrum, gaps, asymptotic analysis.

Introduction

The problems we are going to solve belong to the spectral theory of periodic self-adjoint differential operators in unbounded periodic domains. It is known (see, e.g, [17]) that usually the spectrum of such operators is a locally finite union of compact intervals called bands. In general the bands may overlap. The open interval $(\alpha, \beta) \subset \mathbb{R}$ is called a *gap* if it has an empty intersection with the spectrum, but its ends belong to it.

In general the presence of gaps in the spectrum is not guaranteed (for example, the spectrum of the Laplacian in $L_2(\mathbb{R}^n)$ has no gaps: $\sigma(-\Delta_{\mathbb{R}^n}) = [0, \infty)$). Therefore one of the central questions arising here is to understand whether a given operator really has a gap in its spectrum. The importance of such questions is motivated by various applications, since the presence of gaps is important for the description of wave processes which are governed by differential operators under consideration. Namely, if the wave frequency belongs to a gap, then the corresponding wave cannot propagate in the medium. This feature is a dominant requirement for so-called photonic crystals, which are materials with periodic dielectric structure attracting much attention in recent years. The electromagnetic waves of certain frequencies cannot propagate in them, which is caused by gaps in the spectrum of the Maxwell operator or related scalar operators. We refer to book [4] concerning mathematical problems arising in this field.

The problem of existence of spectral gaps for various periodic operators attracts a lot attention in the last twenty years. Various examples were presented in [6, 11, 12, 21] for periodic divergence type elliptic operators in $\mathbb{R}^n$, in [10] for periodic magnetic Schrödinger operators, in [3, 5, 9, 19] for Laplace-Beltrami operators on periodic Riemannian manifolds, in [7, 8] for Maxwell operators with periodic coefficients in $\mathbb{R}^n$, in [18] for the Laplacians in periodically perforated domains. There are also many papers devoted to the same problem for periodic operators in unbounded domains with a waveguide geometry (see, e.g., [1, 2, 20]).

The second natural question arising here is how to control the edges of the gaps via a suitable choice of the coefficients of the operators or/and via a suitable choice of the geometry of the medium. Such problem was studied by the author in [13] for the Laplace-Beltrami operators on periodic Riemannian manifolds and in [14] for periodic elliptic operators in divergent form in $\mathbb{R}^n$.

In the current work we present the results for the Neumann Laplacians in unbounded periodic domains. We construct a family $\{\Omega^\varepsilon\}_\varepsilon$ of periodic domains in $\mathbb{R}^n$ such that the spectrum of the Neumann Laplacian in $L_2(\Omega^\varepsilon)$ has at least m gaps when ε is small enough ($m \in \mathbb{N}$ is a given number), and moreover the first m gaps converge to predefined intervals as $\varepsilon \to 0$. The complete proofs can be found in preprint [16]. Some preliminary results (the case $m = 1$) were obtained in [15].

1. The construction of the family of periodic domains

Let $\varepsilon > 0$ be a small parameter and let $n \in \mathbb{N} \setminus \{1\}$. Let $B_j \subset \mathbb{R}^n$, $j = 1, \dots m$ be arbitrary open domains with Lipschitz boundaries satisfying the following conditions:

(b_1) $\overline{B_j} \cap \overline{B_k} = \varnothing$ for $j \neq k$,

(b_2) $\bigcup\limits_{j=1}^{m} \overline{B_j} \subset Y$, where $Y = \{x = (x_1, \dots, x_n) \in \mathbb{R}^n : \ |x_i| < 1/2, \forall i\}$,

(b_3) for any $j = 1, \dots, m$ the boundary of ∂B_j has a flat subset, namely there exists $\tilde{x}^j \in \partial B_j^\varepsilon$ and $r_j > 0$ such that $B_{r_j}(\tilde{x}^j) \cap \partial B_j$ belong to a $(n-1)$-dimensional hyperplain. Here by $B_{r_j}(\tilde{x}^j)$ we denote the ball with the center at the point $\tilde{x}^j$ and the radius r_j.

For $j = 1, \dots, m$ we denote

$$S_j^\varepsilon = \partial B_j \setminus \left(\bigcup_{j=1}^{m} D_j^\varepsilon \right),$$

where $D_j^\varepsilon = \{x \in \partial B: |x - \tilde{x}^j| < d_j^\varepsilon\}$ with

$$d_j^\varepsilon = \begin{cases} a_j \varepsilon^{\frac{2}{n-2}}, & n > 2, \\ \varepsilon^{-1}\exp(-\frac{1}{a_j\varepsilon^2}), & n = 2. \end{cases}, \qquad a_j > 0 \text{ are constants.}$$

Finally we set

$$\Omega^\varepsilon = \mathbb{R}^n \setminus \left(\bigcup_{i\in\mathbb{Z}^n} \bigcup_{j=1}^{m} S_{ij}^\varepsilon \right), \text{ where } S_{ij}^\varepsilon = \varepsilon(S_j^\varepsilon + i).$$

We denote by $\mathcal{A}^\varepsilon$ the Neumann Laplacian in Ω^ε. More precisely $\mathcal{A}^\varepsilon$ is the operator which acts in $L_2(\Omega^\varepsilon)$ and is generated by the sesquilinear form

$$\eta^\varepsilon[u,v] = \int_{\Omega^\varepsilon} (\nabla u, \nabla \bar{v})\, dx, \ \operatorname{dom}(\eta^\varepsilon) = H^1(\Omega^\varepsilon). \tag{1}$$

By $\sigma(\mathcal{A}^\varepsilon)$ we denote the spectrum of $\mathcal{A}^\varepsilon$. Our goal is to describe the behaviour of $\sigma(\mathcal{A}^\varepsilon)$ as $\varepsilon \to 0$.

2. The convergence theorem

We denote:

$$\sigma_j = \begin{cases} \dfrac{\varkappa a_j^{n-2}}{4b_j}, & n > 2, \\ \dfrac{\pi a_j}{2b_j}, & n = 2. \end{cases} \tag{2}$$

Here b_j is the volume of B_j, $\varkappa$ is the capacity of the disc $\{x = (x_1, \ldots, x_n) \in \mathbb{R}^n: |x| < 1, \ x_n = 0\}$. It is supposed that the numbers a_j and b_j are such that $\sigma_j < \sigma_{j+1}$, $j = 1, \ldots m-1$.

Let us consider the following equation (with unknown $\lambda \in \mathbb{C}$):

$$1 + \sum_{j=1}^{m} \frac{\sigma_j b_j}{(1 - \sum\limits_{i=1}^{m} b_i)(\sigma_j - \lambda)} = 0. \tag{3}$$

It is easy to show that if $\sigma_j < \sigma_{j+1}$ then equation (3) has exactly m roots $\{\mu_j\}_{j=1}^m$, they are real and moreover they interlace with σ_j, i.e.

$$\sigma_j < \mu_j < \sigma_{j+1}, \ j = \overline{1, m-1}, \quad \sigma_m < \mu_m < \infty.$$

Now we can formulate the main result on the behaviour of $\sigma(\mathcal{A}^\varepsilon)$ as $\varepsilon \to 0$.

Theorem 2.1. *Let $L > 0$ be an arbitrarily large number. Then the spectrum $\sigma(\mathcal{A}^\varepsilon)$ of the operator $\mathcal{A}^\varepsilon$ has the following structure in $[0, L]$ when ε is small enough:*

$$\sigma(\mathcal{A}^\varepsilon) \cap [0, L] = [0, L] \setminus \bigcup_{j=1}^{m} (\sigma_j^\varepsilon, \mu_j^\varepsilon), \tag{4}$$

where the intervals $(\sigma_j^\varepsilon, \mu_j^\varepsilon)$ satisfy

$$\forall j = 1, \ldots, m: \quad \lim_{\varepsilon \to 0} \sigma_j^\varepsilon = \sigma_j, \quad \lim_{\varepsilon \to 0} \mu_j^\varepsilon = \mu_j. \tag{5}$$

3. Control of the edges of the gaps

Our final goal is to find such $a_j > 0$ and domains B_j satisfying (b_1)-(b_3) that the intervals (σ_j, μ_j) will coincide with predefined intervals.

First of all we note that for an arbitrary numbers $b_j > 0$ $(j = 1, \ldots, m)$ satisfying $\sum\limits_{j=1}^{m} b_j < 1$ one can easily construct such domains B_j satisfying (b_1)-(b_3) that the volumes of B_j are equal to b_j. Also we note that the numbers σ_j, μ_j depend on the volumes of B_j, but not on their geometry. Therefore our goal can be reformulated as follows: for arbitrary intervals (α_j, β_j) to find such $a_j > 0$ and $b_j > 0$ with $\sum\limits_{j=1}^{m} b_j < 1$ that (σ_j, μ_j) coincide with (α_j, β_j). One has the following result.

Theorem 3.1. *Let (α_j, β_j) $(j = 1, \ldots, m)$ be arbitrary intervals satisfying*

$$0 < \alpha_1, \quad \alpha_j < \beta_j < \alpha_{j+1}, \; j = \overline{1, m-1}, \quad \alpha_m < \beta_m < \infty. \tag{6}$$

We denote

$$\rho_j = \left(\frac{\beta_j - \alpha_j}{\alpha_j}\right) \prod_{i = \overline{1,m} | i \neq j} \left(\frac{\beta_i - \alpha_j}{\alpha_i - \alpha_j}\right)$$

and choose the parameters a_j, b_j by the formulae

$$a_j = \begin{cases} \left(\dfrac{4\sigma_j \rho_j}{\kappa(1 + \sum_{i=1}^{m} \rho_i)}\right)^{\frac{1}{n-2}}, \\ \dfrac{2\sigma_j \rho_j}{\pi(1 + \sum_{i=1}^{m} \rho_i)}, \end{cases} \qquad b_j = \frac{\rho_j}{1 + \sum\limits_{i=1}^{m} \rho_i}.$$

Then

$$\sigma_j = \alpha_j, \; \mu_j = \beta_j.$$

We note that since the intervals (α_j, β_j) satisfy (6) then

$$\forall j: \; \beta_j > \alpha_j, \quad \forall i \neq j: \; \mathrm{sign}(\beta_i - \alpha_j) = \mathrm{sign}(\alpha_i - \alpha_j) \neq 0.$$

Therefore $\rho_j > 0$ and thus the choice of a_j and b_j is correct.

4. Sketch of the proof

In this section we give a brief outline of the proof of Theorem 2.1. The complete analysis can be found in preprint [16] (for the case $m = 1$ see [15]). Theorem 3.1 is proved in the same way as Theorem 4.1 from [13] or Theorem 2.2 from [14].

We denote by A^ε the Neumann Laplacian in the domain $\varepsilon^{-1}\Omega^\varepsilon$. It is clear that $\sigma(\mathcal{A}^\varepsilon) = \varepsilon^{-2}\sigma(A^\varepsilon)$. Let $[a_k^-(\varepsilon), a_k^+(\varepsilon)]$ be the k-th band of $\sigma(A^\varepsilon)$, that is

$$\sigma(\mathcal{A}^\varepsilon) = \bigcup_{k=1}^{\infty}[\varepsilon^{-2}a_k^-(\varepsilon), \varepsilon^{-2}a_k^+(\varepsilon)],\ a_k^-(\varepsilon), a_k^+(\varepsilon) \underset{k\to\infty}{\nearrow} \infty. \tag{7}$$

We denote $Y^\varepsilon = Y \setminus \bigcup_{j=1}^{m} S_j^\varepsilon$. By $A^{N,\varepsilon}$ (respectively, $A^{D,\varepsilon}$, $A^{p,\varepsilon}$, $A^{a,\varepsilon}$) we denote the operators in $L_2(Y^\varepsilon)$ defined by the operation $-\Delta$ in Y^ε, the Neumann boundary conditions on $\bigcup_{j=1}^{m} S_j^\varepsilon$ and the Neumann (respectively, Dirichlet, periodic, antiperiodic boundary conditions on ∂Y). By $\{\lambda_k^{N,\varepsilon}\}_{k\in\mathbb{N}}$, $\{\lambda_k^{D,\varepsilon}\}_{k\in\mathbb{N}}$, $\{\lambda_k^{p,\varepsilon}\}_{k\in\mathbb{N}}$, $\{\lambda_k^{a,\varepsilon}\}_{k\in\mathbb{N}}$ we denote the corresponding sequences of eigenvalues written in the increasing order and repeated according to their multiplicity. One has the following enclosures:

$$\lambda_k^{p,\varepsilon}, \lambda_k^{a,\varepsilon} \in [a_k^-(\varepsilon), a_k^+(\varepsilon)] \quad \text{and} \quad [a_k^-(\varepsilon), a_k^+(\varepsilon)] \subset [\lambda_k^{N,\varepsilon}, \lambda_k^{D,\varepsilon}]. \tag{8}$$

At first, it is easy to show that for each $k \in \mathbb{N}$

$$\lim_{\varepsilon\to 0} \lambda_k^{a,\varepsilon} = \lambda_k^a, \tag{9}$$

where λ_k^a is the k-th eigenvalues of the operator $A_a = \left(-\Delta_{B_{m+1}}^a\right) \oplus_{j=1}^m \left(-\Delta_{B_j}\right)$. Here $B_{m+1} = Y \setminus \bigcup_{j=1}^{m} \overline{B_j}$, $-\Delta_{B_{m+1}}^a$ is the Laplacian in B_{m+1} with the Neumann boundary conditions on $\partial B_{m+1} \setminus \partial Y$ and the antiperiodic boundary conditions on ∂Y, $-\Delta_{B_j}$ is the Neumann Laplacian in B_j ($j = \overline{1, m}$). It is clear that $\lambda_1^a = \cdots = \lambda_m^a = 0$, while $\lambda_{m+1}^a > 0$. Thus, in view of (8), (9) ,

$$\liminf_{\varepsilon\to 0} a_{m+1}^+(\varepsilon) > 0 \tag{10}$$

(hence, due to (7), $\mathcal{A}^\varepsilon$ has at most m gaps in $[0, L]$, when ε is less than some ε_L).

The remaining part of our analysis is devoted to the proof of the following:

$$\begin{aligned} \lim_{\varepsilon\to 0} \varepsilon^{-2}\lambda_k^{D,\varepsilon} &= \lim_{\varepsilon\to 0} \varepsilon^{-2}\lambda_k^{a,\varepsilon} = \sigma_k,\ k = 1, \ldots m, \\ \lim_{\varepsilon\to 0} \varepsilon^{-2}\lambda_k^{N,\varepsilon} &= \lim_{\varepsilon\to 0} \varepsilon^{-2}\lambda_k^{p,\varepsilon} = \mu_{k-1},\ k = 2, \ldots m+1 \end{aligned} \tag{11}$$

(it is clear that $\lambda_1^{N,\varepsilon} = \lambda_1^{p,\varepsilon} = 0$). The assertions of Theorem 2.1 follow directly from (7), (8), (10), (11).

In order to prove (11) we construct suitable approximations for the corresponding eigenfunctions. We explain briefly the ideas focusing on the Neumann eigenvalues only. The proof consists of several steps.

I. Let $\{u_k^{N,\varepsilon}\}_{k\in\mathbb{N}}$ be a system of eigenfunctions corresponding to $\{\lambda_k^{N,\varepsilon}\}_{k\in\mathbb{N}}$ and

$$u_k^{N,\varepsilon} \text{ are real}, \quad (u_k^{N,\varepsilon}, u_l^{N,\varepsilon})_{L_2(Y^\varepsilon)} = \delta_{kl}, \quad \int\limits_{Y^\varepsilon} u_k^{N,\varepsilon}(x)dx \geq 0. \tag{12}$$

Using the Cauchy inequality and (12) we obtain

$$\left(\int\limits_{B_j} u_k^{N,\varepsilon}(x)dx\right)^2 \leq |B_j| \left\|u_k^{N,\varepsilon}\right\|^2_{L_2(B_j)} \leq |B_j|$$

and therefore one can extract a subsequence (still denoted by ε) such that for $k = 1,\ldots,m+1$, $j = 1,\ldots,m+1$

$$|B_j|^{-1} \int\limits_{B_j} u_k^{N,\varepsilon}(x)dx \to \mathbf{c}_k^j \text{ as } \varepsilon \to 0, \quad \mathbf{c}_k^j \text{ are constants.}$$

One can proof using the minimax principle that

$$\lambda_k^{N,\varepsilon} \leq C\varepsilon^2, \ k = 1,\ldots,m+1 \tag{13}$$

(the infimum limit of the next eigenvalue $\lambda_{m+2}^{N,\varepsilon}$ is positive, it can be shown in the same way as (10)). Then, due to (13) and the Poincaré inequality, we get

$$\text{for } k = 1,\ldots m+1, \ j = 1,\ldots,m+1: \ u_k^{N,\varepsilon} \to \mathbf{c}_k^j \text{ in } L_2(B_j) \text{ as } \varepsilon \to 0. \tag{14}$$

II. We construct an approximation $v_k^{N,\varepsilon}$ for the eigenfunction $u_k^{N,\varepsilon}$ in the following way: $v_k^{N,\varepsilon}$ is equal to $\mathbf{c}_k^j$ in $B_j \setminus \widehat{D}_j^\varepsilon$ ($j = 1,\ldots m+1$), where $\widehat{D}_j^\varepsilon$ is some small neighborhood of D_j^ε, while in $\widehat{D}_j^\varepsilon$ we construct it using some harmonic and cut-off functions. It has the following properties as $\varepsilon \to 0$:

$$\|\nabla v_k^{N,\varepsilon}\|^2_{L_2(Y^\varepsilon)} \sim \varepsilon^2 F(\mathbf{c}_k^1,\ldots,\mathbf{c}_k^{m+1}), \quad v_k^{N,\varepsilon} \to \mathbf{c}_k^j \text{ in } L_2(B_j) \text{ as } \varepsilon \to 0, \tag{15}$$

where $F(c^1,\ldots,c^{m+1}) = \sum\limits_{j=1}^{m} \sigma_j(c^j - c^{m+1})^2$. The proof of the asymptotic equality

$$\lambda_k^{N,\varepsilon} = \|\nabla u_k^{N,\varepsilon}\|^2_{L_2(Y^\varepsilon)} \sim \|\nabla v_k^{N,\varepsilon}\|^2_{L_2(Y^\varepsilon)} \text{ as } \varepsilon \to 0 \tag{16}$$

is based on the minimax principle and (14), (15).

III. It is clear that the first eigenfunction $u_1^{N,\varepsilon}$ is equal to 1, thus $\mathbf{c}_1^j = 1$, $\forall j$. Using the minimax principle and (15)-(16) one can prove that the next constants

$\mathbf{c}_k^j$, $k = 2, \ldots, m+1$, can be found by solving the following recurrent problem:

$$F(\mathbf{c}_k^1, \ldots, \mathbf{c}_k^{m+1}) = \min_{H_k} F(c^1, \ldots, c^{m+1}), \ k = 2, \ldots, m+1, \text{ where}$$

$$H_k = \left\{ (c^1, \ldots, c^{m+1}) \in \mathbb{R}^{m+1} : \sum_{j=1}^{m+1} (c^j)^2 |B_j| = 1, \ \sum_{j=1}^{m+1} c^j \mathbf{c}_l^j |B_j| = 0 \text{ for } l = \overline{1, k-1} \right\}.$$

It is difficult to solve this problem for large m, but it is possible to prove that the numbers $F(\mathbf{c}_k^1, \ldots, \mathbf{c}_k^{m+1})$, $k = 2, \ldots m+1$ (i.e., the minimums of this problem) are the roots of equation (3). Using this, (15), (16) and the definition of the numbers μ_j we conclude that $\lim_{\varepsilon \to 0} \varepsilon^{-2} \lambda_k^{N,\varepsilon} = \mu_{k-1}$, $k = 2, \ldots m+1$.

Acknowledgements

This work is supported by DFG through Research Training Group 1294 "Analysis, Simulation and Design of Nanotechnological Processes".

Bibliography

[1] D. Borisov, K. Pankrashkin, Quantum waveguides with small periodic perturbations: gaps and edges of Brillouin zones, *J. Phys. A* **46** (2013).

[2] G. Cardone, S. Nazarov, C. Perugia, A gap in the essential spectrum of a cylindrical waveguide with a periodic perturbation of the surface, *Math. Nachr.* **283**, 1222 (2010).

[3] E. B. Davies, E. M. II Harrell, Conformally flat Riemannian metrics, Schroedinger operators, and semiclassical approximation, *J. Differ. Equ.* **66**, 165 (1987).

[4] W. Dorfler, A. Lechleiter, M. Plum, G. Schneider, C. Wieners, *Photonic Crystals. Mathematical Analysis and Numerical Approximation* (Springer, Berlin, 2011).

[5] P. Exner, O. Post, Convergence of spectra of graph-like thin manifolds, *J. Geom. Phys.* **54**, 77 (2005).

[6] A. Figotin, P. Kuchment, Band-gap structure of the spectrum of periodic dielectric and acoustic media. I. Scalar model, *SIAM J. Appl. Math.* **56**, 68 (1996).

[7] A. Figotin, P. Kuchment, Band-gap structure of the spectrum of periodic dielectric and acoustic media. II. Two-dimensional photonic crystals, *SIAM J. Appl. Math.* **56**, 1561 (1996).

[8] N. Filonov, Gaps in the spectrum of the Maxwell operator with periodic coefficients, *Commun. Math. Phys.* **240**, 161 (2003).

[9] E.L. Green, Spectral theory of Laplace-Beltrami operators with periodic metrics, *J. Differ. Equ.* **133**, 15 (1997).

[10] R. Hempel, I. Herbst, Strong magnetic fields, Dirichlet boundaries, and spectral gaps, *Commun. Math. Phys.* **169**, 237 (1995).

[11] R. Hempel, K. Lienau, Spectral properties of periodic media in the large coupling limit, *Commun. Partial Differ. Eq.* **25**, 1445 (2000).

[12] V. Hoang, M. Plum, C. Wieners, A computer-assisted proof for photonic band gaps, *Z. Angew. Math. Phys.* **60**, 1035 (2009).

[13] A. Khrabustovsky, Periodic Riemannian manifold with preassigned gaps in spectrum of Laplace-Beltrami operator, *J. Differ. Equ.* **252**, 2339 (2012).

[14] A. Khrabustovskyi, Periodic elliptic operators with asymptotically preassigned spectrum, *Asymp. Anal.* **82**, 1 (2013).

[15] A. Khrabustovskyi, E. Khruslov, Gaps in the spectrum of the Neumann Laplacian generated by a system of periodically distributed traps, *Math. Meth. Appl. Sci.* (2014), DOI: 10.1002/mma.3046.

[16] A. Khrabustovskyi, Opening up and control of spectral gaps of the Laplacian in periodic domains, (2014), arXiv:1308.4091.

[17] P. Kuchment, *Floquet Theory For Partial Differential Equations* (Birkhauser, Basel, 1993).

[18] S.A. Nazarov, K. Ruotsalainen, J. Taskinen, Spectral gaps in the Dirichlet and Neumann problems on the plane perforated by a doubleperiodic family of circular holes, *J. Math. Sci.* **181**, 164 (2012).

[19] O. Post, Periodic manifolds with spectral gaps, *J. Differ. Equ.* **187**, 23 (2003).

[20] K. Yoshitomi, Band gap of the spectrum in periodically curved quantum waveguides. *J. Differ. Equ.* **142**, 123 (1998).

[21] V. Zhikov, On spectrum gaps of some divergent elliptic operators with periodic coefficients, *St. Petersb. Math. J.* **16**, 773 (2005).

Graph model for the Stokes flow

M. O. Kovaleva

Department of Higher Mathematics,
ITMO University,
Kronverkskiy pr. 49, St. Petersburg, 197101, Russian Federation
E-mail: vega14@mail.ru

I. Yu. Popov

Department of Higher Mathematics,
ITMO University,
Kronverkskiy pr. 49, St. Petersburg, 197101, Russian Federation
E-mail: popov1955@gmail.com

Quasi-one-dimensional model for the Stokes flow concentrated near a network is suggested. It is analogous to the quantum graph model in quantum mechanics. A version of the Harnack's inequality for the flow is obtained.

Keywords: Metric graph, Stokes flow, spectrum.

1. Introduction

Quantum graph is a graph considered as a one-dimensional singular variety and equipped with a self-adjoint differential operator. Reasons for studying quantum graphs arise as simplified (due to the reduced dimensions) models in physics, chemistry and nanotechnology, when one considers propagation of waves of different nature through a quasi-one-dimensional system that looks like a thin neighborhood of a graph. The intensive investigation of quantum graphs started in 1980th ([1, 2]). As for the state of the art in the field, see, e.g., [3]. The model has attractive features. From the one side, the method allows one to obtain explicit solutions of the model problem, from the other side, it gives one good approximation for "realistic" physical solutions. It is a reason for seeking of new applications for the model. In this paper we suggest a version of the metric graph model for the Stokes flow. The Stokes approximation is often used in the cases when one can omit the inertial terms in the Navier-Stokes equation, i.e. when the Reynolds number is small. More precisely, our model is related with the Stokes flow through narrow channels. A channel is considered as a "narrow" one if the ratio of its width and length is essentially less than one. We can note that real sizes of channels for different problems differ considerably. Particularly, the Stokes approximation is used for the description of flows through nanotubes (see, e.g., [4–6], blood flow [7–9] and geophysical flows [10, 11]. The problem becomes more complicated due to the viscosity variation and to the fact that we have to deal with a network of channels. In the present paper we try to construct a simplified quasi-one-dimensional model for the flow.

There are several ways to construct quasi-one-dimensional model for the Stokes flow through a network. One way is based on the investigation of the Stokes equations (2D or 3D) in a system of narrow tubes and the consideration of the limiting procedure for the tubes widths tending to zero. This approach is similar to so-called model of thick quantum graphs (see, e.g., [12, 13]). We use another way initially suggested in the diffraction theory to describe waves concentrated near lines (see, e.g., [14, 15]). As for the procedure, it is as follows. We look for a solution in the form of formal asymptotic expansion in small parameter (channel width). By substitution of this series in the equation and collecting terms of identical orders, we get a chain of 1D problems for the coefficients of this expansion, particularly, a metric graph problem for the main term of the asymptotic expansion. Keeping in mind the background problem, one can see that the term "Stokes graph" is appropriate for this metric graph model.

2. Model construction

Consider the plane flow with varying viscosity and density, i.e. the system of coupled 2D Stokes and continuity equations. In the Cartesian coordinates $(\widetilde{x},\widetilde{y})$ it has the form

$$2\widetilde{\eta}\frac{\partial^2 v_x}{\partial \widetilde{x}^2}+2\frac{\partial\widetilde{\eta}}{\partial\widetilde{x}}\frac{\partial v_x}{\partial\widetilde{x}}+\widetilde{\eta}\frac{\partial^2 v_x}{\partial\widetilde{y}^2}+\widetilde{\eta}\frac{\partial^2 v_y}{\partial\widetilde{y}\partial\widetilde{x}}+\frac{\partial\widetilde{\eta}}{\partial\widetilde{y}}\frac{\partial v_x}{\partial\widetilde{y}}+\frac{\partial\widetilde{\eta}}{\partial\widetilde{y}}\frac{\partial v_y}{\partial\widetilde{x}}-\frac{\partial P}{\partial\widetilde{x}}=-\widetilde{\rho}G_x, \quad (1)$$

$$\widetilde{\eta}\frac{\partial^2 v_y}{\partial \widetilde{x}^2}+\widetilde{\eta}\frac{\partial^2 v_x}{\partial\widetilde{y}\partial\widetilde{x}}+\frac{\partial\widetilde{\eta}}{\partial\widetilde{x}}\frac{\partial v_x}{\partial\widetilde{y}}+\frac{\partial\widetilde{\eta}}{\partial\widetilde{x}}\frac{\partial v_y}{\partial\widetilde{x}}+2\frac{\partial\widetilde{\eta}}{\partial\widetilde{y}}\frac{\partial v_y}{\partial\widetilde{y}}+2\widetilde{\eta}\frac{\partial^2 v_y}{\partial\widetilde{y}^2}-\frac{\partial P}{\partial\widetilde{y}}=-\widetilde{\rho}G_y, \quad (2)$$

$$\frac{\partial(\widetilde{\rho}v_x)}{\partial\widetilde{x}}+\frac{\partial(\widetilde{\rho}v_y)}{\partial\widetilde{y}}=0. \quad (3)$$

Here (v_x, v_y) is the flow velocity, $\widetilde{\eta}=\widetilde{\eta}(\widetilde{x},\widetilde{y})$ is the viscosity, P is the pressure, $\widetilde{\rho}=\widetilde{\rho}(\widetilde{x},\widetilde{y})$ is the density, (G_x, G_y) is the gravitational force (we use here geophysical terminology, it can be another driving (external) force). Note that (3) is the continuity equation.

At first, we consider the flow concentrated near the axis. More precisely, we deal with the case of smooth functions $\widetilde{\eta}$ and $\widetilde{\rho}$ having essentially greater values outside the strip $\{(\widetilde{x},\widetilde{y}) : \widetilde{x}\in(-\infty,\infty), \widetilde{y}\in(-\varepsilon,\varepsilon)\}$ than inside it:

$$\widetilde{\eta}(\widetilde{x},\widetilde{y})=\eta(\widetilde{x})\widetilde{\eta}_2(\widetilde{y}), \widetilde{\eta}_2(\widetilde{y})=\begin{cases}\eta_c, & |\widetilde{y}|\le\varepsilon,\\ \eta_w(\widetilde{y}), & |\widetilde{y}|>2\varepsilon,\end{cases}, \eta_w\gg\eta_c,$$

$$\widetilde{\rho}(\widetilde{x},\widetilde{y}) = \rho(\widetilde{x})\widetilde{\rho}_2(\widetilde{y}), \widetilde{\rho}_2(\widetilde{y}) = \begin{cases} \rho_c, & |\widetilde{y}| \leq \varepsilon, \\ \rho_w(\widetilde{y}), & |\widetilde{y}| > 2\varepsilon, \end{cases}, \rho_w \gg \rho_c.$$

Here ε is a small parameter. We will assume (for simplicity) that $\eta_w(\widetilde{y})$ and $\rho_w(\widetilde{y})$ grows exponentially when $|\widetilde{y}| \to \infty$. It means that, really, the flow is only inside the strip. We introduce new coordinates $x = \widetilde{x}, \quad y = \widetilde{y}/\varepsilon$. and seek the solutions in the form of formal asymptotic series in ε :

$$v_x = v_x^0 + v_x^1\varepsilon + ..., \quad v_y = v_y^0 + v_y^1\varepsilon + ..., \quad P = P^0 + P^1\varepsilon + ...$$

Let us insert the series into equations (1), (2), (3) and collect terms of the same powers of ε. We get the chain of equations for the series coefficients. Terms of order ε^{-2} are in the Stokes equations (in new variables) only:

$$\widetilde{\eta}\frac{\partial^2 v_x^0}{\partial y^2} + \frac{\partial\widetilde{\eta}}{\partial y}\frac{\partial v_x^0}{\partial y} = 0, \quad \widetilde{\eta}\frac{\partial^2 v_y^0}{\partial y^2} + \frac{\partial\widetilde{\eta}}{\partial y}\frac{\partial v_y^0}{\partial y} = 0.$$

These simple equations give us

$$v_x^0 = \frac{f_x(x)}{\eta(x)}\int\frac{dy}{\widetilde{\eta}_2(y)} = u(x)\int\frac{dy}{\widetilde{\eta}_2(y)}, \quad v_y^0 = \frac{f_y(x)}{\eta(x)}\int\frac{dy}{\widetilde{\eta}_2(y)}. \tag{4}$$

Here $u(x), f_y(x)$ are some functions of one variable. They should be determined later. Terms of order ε^{-1} are in three equations. The continuity equation gives us: $v_y^0 = g_y(x)/\widetilde{\rho}$. To have no contradiction with (4) we conclude that $v_y^0 = 0$, i.e. the series for v_y starts from another power than for v_x. It is a conventional situation (see, e.g., [16]). The Stokes equations give us the following relations for the terms of this order:

$$\widetilde{\eta}\frac{\partial^2 v_x^1}{\partial y^2} + \frac{\partial\widetilde{\eta}}{\partial y}\frac{\partial v_x^1}{\partial y} + \widetilde{\eta}\frac{\partial^2 v_y^0}{\partial y\partial x} + \frac{\partial\widetilde{\eta}}{\partial y}\frac{\partial v_y^0}{\partial x} = 0,$$

$$2\widetilde{\eta}\frac{\partial^2 v_y^1}{\partial y^2} + 2\frac{\partial\widetilde{\eta}}{\partial y}\frac{\partial v_y^1}{\partial y} + \widetilde{\eta}\frac{\partial^2 v_x^0}{\partial y\partial x} + \frac{\partial\widetilde{\eta}}{\partial x}\frac{\partial v_x^0}{\partial y} - \frac{\partial P^0}{\partial y} = 0.$$

These equations are solvable in respect to v_x^1, v_y^1. The pressure term P^0 should be determined later.

Consider the terms of zero order. We have the following three equations:

$$\frac{\partial\widetilde{\rho}}{\partial x}v_x^0 + \widetilde{\rho}\frac{\partial v_x^0}{\partial x} + \frac{\partial\widetilde{\rho}}{\partial y}v_y^1 + \widetilde{\rho}\frac{\partial v_y^1}{\partial y} = 0,$$

$$2\widetilde{\eta}\frac{\partial^2 v_x^0}{\partial x^2} + 2\frac{\partial\widetilde{\eta}}{\partial x}\frac{\partial v_x^0}{\partial x} + \widetilde{\eta}\frac{\partial^2 v_x^2}{\partial y^2} + \widetilde{\eta}\frac{\partial^2 v_y^1}{\partial y\partial x} + \frac{\partial\widetilde{\eta}}{\partial y}\frac{\partial v_x^2}{\partial y} + \frac{\partial\widetilde{\eta}}{\partial y}\frac{\partial v_y^1}{\partial x} - \frac{\partial P^0}{\partial x} = -\widetilde{\rho}G_x,$$

$$\widetilde{\eta}\frac{\partial^2 v_x^1}{\partial y\partial x} + \frac{\partial\widetilde{\eta}}{\partial x}\frac{\partial v_x^1}{\partial y} + 2\frac{\partial\widetilde{\eta}}{\partial y}\frac{\partial v_y^2}{\partial y} + 2\widetilde{\eta}\frac{\partial^2 v_y^2}{\partial y^2} - \frac{\partial P^1}{\partial y} = -\widetilde{\rho}G_y.$$

At present, we confine our attention by the first term of the asymptotic series for the velocity. The pressure is determined at the next stage (by taking into account the solvability condition ($\frac{\partial^2 P^0}{\partial y \partial x} = \frac{\partial P^0}{\partial x \partial y}$)). Here we insert $v_y^0 = 0$. The last equation does not contain an information about v^0. But the first two equations give us an interesting correlation. Taking into account the expressions for $\widetilde{\eta}, \widetilde{\rho}, v_x^0$, one can see that it leads to the following equation for the function $u(x)$:

$$u'' - \frac{\eta'}{\eta}\frac{\rho'}{\rho}u = -\frac{\rho}{\eta}G_x. \tag{5}$$

Particularly, in the homogeneous case ($G_x = 0$)

$$u'' - \frac{\eta'}{\eta}\frac{\rho'}{\rho}u = 0. \tag{6}$$

we have the 1D Schrödinger equation with a specific potential for the function $u(x)$ corresponding to zero energy (the analogous problem for the quantum graph is known as the threshold resonance, see, e.g., [19]).

The obtained correlation opens an intriguing perspective related with graphs. Let us consider a metric graph Γ with the Schrödinger operator H on each edge:

$$H = -\frac{d^2}{dx^2} + \frac{\eta'}{\eta}\frac{\rho'}{\rho}. \tag{7}$$

The domain of the operator consists of functions from the Sobolev space H^2 at each edge satisfying boundary conditions at vertices (see below). Keeping in mind the above treatment of the Stokes flow, we will refer to the graph as to the Stokes graph. It is necessary to determine boundary conditions at vertices. Consider a vertex (say zero point) with n output edges. From physical conditions (see, e.g., [10]) one has $\rho_1(0) = \rho_2(0) = ...\rho_n(0) = \rho(0)$ and $u_1'(+0) = u_2'(+0) = ...u_n'(+0) = u'(0)$. The last condition is related with the most natural physical condition of the pressure continuity [17, 18]. Here $u_j'(+0)$ is the derivative in the outgoing direction at the vertex 0. The continuity equation gives us (for this vertex under the assumption that $\sum\limits_{j=1}^{n} \rho_j'(+0) \neq 0$):

$$\sum_{j=1}^{n} u_j(+0) = -\left(\rho(0)\left(\sum_{j=1}^{n} \rho_j'(+0)\right)^{-1}\right) u'(0). \tag{8}$$

It is well-known δ'-coupling condition for a quantum graph (see, e.g., [20–22]).

Discussion

To show usefulness of the model we present two physically sound theorems. First, introduce two conventional definitions [17]. We consider a graph Γ and assume the Dirichlet or Neumann conditions at its boundary ($u|_{\partial\Gamma} = 0$. or $u'|_{\partial\Gamma} = 0$.) As for the relation with the flow properties, the Dirichlet condition corresponds to the absence of the flow through the boundary vertices and the Neumann condition, in contrast, corresponds to free flow through these vertices.

Definition. We will say that the operator H preserve the sign of a solution on Γ if for any subgraph $\Gamma_0 \subseteq \Gamma$ $(\Gamma_0 \neq \varnothing)$ there is no solution u of the equation $Hu = 0$ with condition (8) at the vertices such that $u(x)$ changes sign inside Γ_0.

Note that this property is analogous to non-oscillation property from [17]. The difference is that the authors of [17] deal with continuous functions on Γ. Correspondingly, changing of sign means that there is a point where $u(x) = 0$. In our case the continuity is broken at vertices.

Definition. Function u is a solution of the differential inequality $Hu \geq 0$ if u is a solution of the equation $Hu = f$ for some non-negative $f \in C(\Gamma)$.

Inequality analogous to the Harnack's inequality for elliptic problem on manifold was obtained in [17] for continuous on the graph solutions of the differential inequality. We proved similar inequality for the Stokes graph, i.e. for the case of another condition at vertices corresponding to the absence of the continuity at these points (8). To prove the inequality in this case we assume additionally that the following condition takes place at any vertex:

$$\sum_{j=1}^{n} \rho_j'(+0) > 0. \tag{9}$$

Theorem. Let the operator H preserve the sign of a solution on Γ. If a solution u of the differential inequality $Hu \geq 0$ is non-negative on Γ then for any locally compact (in respect to Γ) subgraph $\Gamma_0, \Gamma_0 \subset \Gamma$, the following inequality takes place:

$$\max_{x\in\Gamma_0} u(x) \leq \gamma \min_{x\in\Gamma_0} u(x),$$

where the constant γ depends on Γ_0 only.

The structure of the proof is analogous to that in [17]. It will be published in another paper. We can mention that constant γ is estimated with using of the Green function of the operator H.

This theorem gives us an interesting property of the Stokes flow velocity in this case. Namely, Let us consider the inhomogeneous equation (5). The solution (velocity), naturally, depends on the right hand side of the equation (i.e. on the driving forces). The theorem shows an interesting fact that the ratio of the maximal and the minimal values of the velocity (over any subgraph Γ_0, $\Gamma_0 \subseteq \Gamma$, possibly, over the whole graph Γ if it is finite) is bounded by a constant which depends only on Γ_0 and does not depend on the inhomogeneous term.

The following theorem describes a critical case when 0 is an eigenvalue of the operator H.

Theorem. Let the operator H preserve the sign of a solution on any subgraph $\Gamma_0, \Gamma_0 \subset \Gamma$ which does not coincide with Γ but do not preserve the sign of a solution on the whole Γ. Then each solution u of the differential inequality $Hu \geq 0, \quad u|_{\partial\Gamma} \geq 0$, is a solution of the problem $Hu = 0, \quad u|_{\partial\Gamma} = 0$.

Spectral problem is one more source of interesting results for fluid mechanics. We will describe it in another publication.

Acknowledgments

This work was partially financially supported by the Government of the Russian Federation (grant 074-U01), by Ministry of Science and Education of the Russian Federation (GOSZADANIE 2014/190, Project 14.Z50.31.0031), by grant of Russian Foundation for Basic Researches and grants of the President of Russia (state contracts 14.124.13.2045-MK and 14.124.13.1493-MK).

Bibliography

[1] N. Gerasimenko, B. Pavlov, *Theor. Math. Phys.* **74** 230 (1988).

[2] P. Exner, P. Seba, *Rep. Math. Phys.* **28** 7 (1989).

[3] P. Exner, J.P. Keating, P. Kuchment, T. Sunada, A. Teplyaev (eds.), *Analysis on Graphs and Its Applications. Proc. Symp. Pure Math. 77* (Amer. Math. Soc., Providence, RI, 2008).

[4] M. B. Belonenko, S. A. Chivilikhin, V. V. Gusarov, I. Yu. Popov, O. A. Rodygina, *Europhys. Lett.* **101** (6) 66001 (2013).

[5] I. Yu. Popov, *Phys. Scripta* **83** 045601/1-3 (2011).

[6] T.-D. Li, J. Gao, R. Szoszkeiwicz, U. Landman, E. Riedo, *Phys. Rev. B.* **75** 115415/1-6 (2007).

[7] M. Sharan, A. S. Popel, *Biorheology* **38** 415 (2001).

[8] T. W. Secomb, A. V. El-Kareh, *Biorheology* **32** 169 (1995).

[9] V. P. Srivastava, *Int. J. Applications and Appl. Math.* **2** (1) 51 (2007).

[10] T. Gerya, *Introduction to Numerical Geodynamic Modelling* (Cambridge University Press, Cambridge, 2010).

[11] A. Ismail-Zadeh, P. Tackley, *Computational methods in Geodynamics* (Cambridge University Press, Cambridge, 2010).

[12] O. Post, *J. Phys. A: Math. Gen.* **38** 4917 (2005).

[13] S. Molchanov, B. Vainberg, *Comm. Math.Phys.* **273** 533 (2007).

[14] V. M. Babich, V. S. Buldyrev, *Asymptotic Methods in Short-wavelength Diffraction Theory (Alpha Science Series on Wave Phenomena)* (Alpha Science Intl Ltd., Oxford, 2009).

[15] A. I. Popov, V. M. Babich, *J. Math. Sci.* **173** (3) 243 (2011).

[16] M. Van Dyke, *Perturbation Methods in Fluid Mechanics* (Parabolic Press, Stanford, 1975).

[17] Yu. V. Pokornyi, O. M. Penkin, V. L. Pryadev, A. V. Borovskikh, K. P. Lazarev, S. A. Shabrov, *Differential Equations on Geometrical Graphs* (Moscow, Fizmatlit, 2004).
[18] A. V. Gudzovskii, *Doklady RAS* **358** (6) 765 (1998).
[19] C. Cacciapuoti, P. Exner, *J. Phys. A.* **40** F511 (2007).
[20] S. Albeverio, F. Gesztesy, R. Hoegh-Krohn, H. Holden, with an appendix by P. Exner, *Solvable Models in Quantum Mechanics:Second Edition* (AMS Chelsea Publishing. Providence, R.I. 2005).
[21] P. Kuchment, *Waves Random Media* **14** S107 (2004).
[22] A. Mikhailova, B. Pavlov, I. Popov, T. Rudakova, A. Yafyasov, *Math. Nachr.* **235** 101 (2002).

Point contacts and boundary triples

V. Lotoreichik

Institute of Computational Mathematics,
Graz University of Technology,
Steyrergasse 30, Graz, 8010, Austria
E-mail: lotoreichik@math.tugraz.at

H. Neidhardt

Weierstrass Institute for Applied Analysis and Stochastics,
Mohrenstrasse 39, Berlin, 10117, Germany
E-mail: hagen.neidhardt@wias-berlin.de

I. Yu. Popov

Department of Higher Mathematics,
St. Petersbsurg National Research University of
IT, Mechanics and Optics,
Kronverkskiy pr. 49, St. Petersburg, 197101, Russian Federation
E-mail: popov1955@gmail.com

We suggest an abstract approach for point contact problems in the framework of boundary triples. Using this approach we obtain the perturbation series for a simple eigenvalue in the discrete spectrum of the model self-adjoint extension with weak point coupling. An example of a two-level quantum model is provided.

Keywords: Boundary triples, Weyl function, point contacts, weak coupling, perturbation series.

1. Introduction

Let H_0 be a self-adjoint operator with an isolated simple eigenvalue λ_0. Further let V be a bounded or unbounded self-adjoint operator such that the family of operators $H(\varkappa) := H_0 + \varkappa V$ is well-defined and self-adjoint for sufficiently small coupling constants $\varkappa \in \mathbb{R}$. If V is relatively compact with respect to H_0, then there is a smooth function $\lambda(\varkappa)$ such that $\lambda(\varkappa)$ is a simple eigenvalue of $H(\varkappa)$ for each $\varkappa$ and $\lim_{\varkappa\to 0} \lambda(\varkappa) = \lambda_0$ holds, cf. [1]. Since the function $\lambda(\varkappa)$ is smooth it admits a Taylor-type expansion of the form

$$\lambda(\varkappa) = \lambda_0 + a\varkappa + b\varkappa^2 + O(\varkappa^3). \tag{1}$$

The problem is to compute the coefficients a and b of this perturbation series in terms of the operators H_0 and V.

In a slightly modified form, similar problem appears for point contacts in quantum mechanics. Typically one considers two quantum systems which do not interact, where one of them has a simple isolated eigenvalue λ_0. If both systems are coupled by a point contact, then the eigenvalue λ_0 can move either along the real axis or become a pole of the analytic continuation of the resolvent of the coupled system in the lower complex half-plane. In the last case one speaks of resonances. The eigenvalue case realizes if the second system has no spectrum around λ_0 while the resonance case appears if the second system has continuous spectrum around λ_0, that is, if λ_0 is an embedded eigenvalue for the decoupled systems. If the point interaction depends on a parameter $\varkappa$ such that for $\varkappa \to 0$ the coupled system converges to the decoupled one, then again a perturbation series is expected either for the eigenvalues or for the resonances. In the following we focus on the eigenvalue case.

Perturbation series for point interactions were perhaps first studied by B. S. Pavlov in [2, 3] for a model of point interactions with an inner structure, where the first order coefficient a was computed. A direct sum of two three-dimensional Schrödinger operators coupled by a point contact was considered by P. Exner in [4]. In this paper he was able to compute the first and second order coefficients a and b. See also related work [5] on spin-dependent point interactions and [6] for perturbation of eigenvalues at threshold in point contact models. A survey on the resonance case can be found in [7], see also references therein. Point contact models are often used in other areas of mathematical physics. In [8] a model of a small window in the screen is studied. In [9] Maxwell and Schrödinger operators are coupled via a point contact and in [10] a model of a three-dimensional Helmholtz resonator is constructed via point coupling.

In the following we consider an abstract point contact model and are interested in the perturbation series for its eigenvalues. In particular, let $\widetilde{A}$ and $\widehat{A}$ be two densely defined closed symmetric operators in the Hilbert spaces $\widetilde{\mathcal{H}}$ and $\widehat{\mathcal{H}}$, respectively, both having equal finite deficiency indices (d, d). Let us consider the direct sum $A := \widetilde{A} \oplus \widehat{A}$ which is also a densely defined closed symmetric operator in $\widetilde{\mathcal{H}} \oplus \widehat{\mathcal{H}}$ with deficiency indices $(2d, 2d)$. Further, let $\widetilde{A}_{[\alpha]}$ and $\widehat{A}_{[\beta]}$ be self-adjoint extensions of $\widetilde{A}$ and $\widehat{A}$, respectively. The Hamiltonian of the decoupled system is given by $H_0 := \widetilde{A}_{[\alpha]} \oplus \widehat{A}_{[\beta]}$, which is a self-adjoint extension of A. As usual the Hamiltonian of the point coupled system is given by another self-adjoint extension H of A, which can not be decomposed into the orthogonal sum with respect to the decomposition $\widetilde{\mathcal{H}} \oplus \widehat{\mathcal{H}}$. The family $H(\varkappa)$ from above is now replaced by a one-parametric family of point contacts, that means, by a family of self-adjoint extensions $H(\varkappa)$ of A.

To make the problem precise we use the framework of boundary triples. In this framework a subfamily A_Λ of self-adjoint extensions of A are labeled by Hermitian matrices Λ in $\mathbb{C}^{2d}$ via an abstract boundary condition involving Λ. In particular, there is a Hermitian matrix Λ_0 such that $H_0 = A_{\Lambda_0}$. Let us assume that λ_0 is an isolated eigenvalue of $\widehat{A}_{[\beta]}$, but a resolvent point of $\widetilde{A}_{[\alpha]}$. Moreover, let $\Lambda(\varkappa)$ be a one-parametric sufficiently regular family of Hermitian matrices in $\mathbb{C}^{2d}$, which converges to Λ_0 as $\varkappa \to 0$. Setting $H(\varkappa) := A_{\Lambda(\varkappa)}$ one gets a family of self-adjoint extensions $H(\varkappa)$ of A which converges in an appropriate sense to H_0 as $\varkappa \to 0$ and in the discrete spectra of the operators $H(\varkappa)$ exists a branch of the type (1). The goal is to compute the coefficients a and b for this branch in terms of the Taylor coefficients of $\Lambda(\varkappa)$ and the abstract Weyl function $M(\cdot)$ which is an important ingredient of the boundary triple approach. In general this problem can not be reduced to the investigation of holomorphic operator families of the types (A) or (B) in the sense of Kato which are thoroughly discussed in [1]. Only in some special cases such a reduction can be done.

We solve this problem for arbitrary $d \in \mathbb{N}$ for the first order coefficient a. In the special case of $d = 1$ we obtain first and second order coefficients a and b which is beyond Pavlov [2, 3] and covers [4]. In general, it would be also possible to compute b for arbitrary finite deficiency indices, however, we have not included that for the purpose to avoid tedious computations.

Our abstract results are illustrated with a vector two-level quantum model, which is a generalization of the analogous scalar model considered in [4, 7].

2. Boundary triples and Weyl functions

The reader may consult with [11–16] for the theory of boundary triples and its applications. In this note we use this concept only in the case of finite deficiency indices. Throughout this section the following hypothesis is employed.

Hypothesis I. *Let A be a closed, symmetric, densely defined operator in a Hilbert space $\mathcal{H}$ with equal finite deficiency indices (d, d).*

Definition 2.1. Assume that Hypothesis I holds. The triple $\{\mathbb{C}^d, \Gamma_0, \Gamma_1\}$ with $\Gamma_0, \Gamma_1 \colon \operatorname{dom} A^* \to \mathbb{C}^d$ is a boundary triple for A^* if the following conditions hold: the mapping $\Gamma := (\Gamma_0, \Gamma_1)^\top$ is surjective onto $\mathbb{C}^{2d}$ and the abstract Green's identity $(A^*f, g)_\mathcal{H} - (f, A^*g)_\mathcal{H} = (\Gamma_1 f, \Gamma_0 g)_{\mathbb{C}^d} - (\Gamma_0 f, \Gamma_1 g)_{\mathbb{C}^d}$ holds for all $f, g \in \operatorname{dom} A^*$.

Boundary triples is an efficient tool to parametrize self-adjoint extensions of a symmetric operator.

Proposition 2.1 ([14, Proposition 1.4], [16, Proposition 14.7]). *Assume that Hypothesis I holds. Let $\{\mathbb{C}^d, \Gamma_0, \Gamma_1\}$ be a boundary triple for A^*. Then for each self-adjoint extension $\widetilde{A}$ of A there is a unique self-adjoint relation Θ in $\mathbb{C}^d$ such that $\widetilde{A} = A_\Theta := A^* \upharpoonright \{f \in \operatorname{dom} A^* : \Gamma f \in \Theta\}$.*

Remark 2.1. The self-adjoint extension $A_0 := A^* \upharpoonright \ker \Gamma_0$ is distinguished. It corresponds to the self-adjoint relation $\Theta_\infty := \left\{\begin{pmatrix} 0 \\ h \end{pmatrix} : h \in \mathbb{C}^d\right\}$. If Θ is the graph of a Hermitian matrix Λ in $\mathbb{C}^d$, i.e $\Theta = \operatorname{graph}(\Lambda)$, then one easily checks that

$$A_{\operatorname{graph}(\Lambda)} = A_\Lambda := A^* \upharpoonright \{f \in \operatorname{dom} A^* : \Gamma_1 f = \Lambda \Gamma_0 f\}. \tag{2}$$

The operator Λ is called the boundary operator with respect to the boundary triple $\{\mathbb{C}^d, \Gamma_0, \Gamma_1\}$.

One can associate γ-fields and Weyl functions with boundary triples.

Definition 2.2. Assume that Hypothesis I holds. Let $\{\mathbb{C}^d, \Gamma_0, \Gamma_1\}$ be a boundary triple for A^*. The function $\gamma\colon \rho(A_0) \to \mathcal{B}(\mathbb{C}^d, \mathcal{H})$ defined as

$$\gamma(\lambda) := \big(\Gamma_0 \upharpoonright \ker(A^* - \lambda)\big)^{-1}, \qquad \lambda \in \rho(A_0),$$

is called the *γ-field*. The function $M\colon \rho(A_0) \to \mathbb{C}^{d\times d}$ defined as $M(\lambda) := \Gamma_1 \gamma(\lambda)$ is called the *Weyl function*.

Proposition 2.2. *Assume that Hypothesis I holds. Let M be the Weyl function associated with a boundary triple $\{\mathbb{C}^d, \Gamma_0, \Gamma_1\}$ for A^*. Let the self-adjoint operator A_Λ in $\mathcal{H}$ be as in* (2)*. Then the following statements hold.*

(i) *The function $M(\cdot)$ is holomorphic on $\rho(A_0)$.*
(ii) *For $\lambda \in \rho(A_0)$ the relation* $\dim\ker(A_\Lambda - \lambda) = \dim\ker(\Lambda - M(\lambda))$ *holds.*
(iii) *If $\lambda_0 \in \rho(A_0)$ is a simple eigenvalue of A_Λ, then the function*

$$D_\Lambda(\lambda) := \det(\Lambda - M(\lambda)), \qquad \lambda \in \rho(A_0),$$

has a simple zero at $\lambda = \lambda_0$. In particular, $D'_\Lambda(\lambda_0) \neq 0$ holds.

Proof. All the statements of this proposition are known. Item (i) can be found in [12, Proposition 1.21], see also [16, Proposition 14.15 (iv)] and item (ii) is given in [12, Theorem 1.36 (1)], see also [16, Proposition 14.17 (ii)]. For item (iii) see [15, Corollary 4.4, Proposition 5.1 (iii)]. □

3. Abstract point contact and its weak coupling regime

In this section we present an abstract treatment of point contacts in the framework of boundary triples and obtain the perturbation series of the simple eigenvalue in the weak coupling regime. We make use of the following hypothesis.

Hypothesis II. *Let $\widetilde{\mathcal{H}}$ and $\widehat{\mathcal{H}}$ be separable Hilbert spaces. Let $\widetilde{A}$ and $\widehat{A}$ be closed, densely defined, symmetric operators in $\widetilde{\mathcal{H}}$ and $\widehat{\mathcal{H}}$, respectively, both with deficiency indices (d, d). Let $\{\mathbb{C}^d, \widetilde{\Gamma}_0, \widetilde{\Gamma}_1\}$ and $\{\mathbb{C}^d, \widehat{\Gamma}_0, \widehat{\Gamma}_1\}$ be boundary triples for $\widetilde{A}^*$ and $\widehat{A}^*$, respectively.*

The next lemma appears to be useful in what follows.

Lemma 3.1 ([12, Section 1.4.4]). *Assume that Hypothesis II holds. Then the operator $\widetilde{A} \oplus \widehat{A}$ is closed, densely defined and symmetric in the Hilbert space $\widetilde{\mathcal{H}} \oplus \widehat{\mathcal{H}}$ with deficiency indices $(2d, 2d)$ and $\{\mathbb{C}^{2d}, \widetilde{\Gamma}_0 \oplus \widehat{\Gamma}_0, \widetilde{\Gamma}_1 \oplus \widehat{\Gamma}_1\}$ is a boundary triple for $(\widetilde{A} \oplus \widehat{A})^*$.*

Our model operator A_Λ in the Hilbert space $\widetilde{\mathcal{H}} \oplus \widehat{\mathcal{H}}$ is defined as

$$\begin{aligned} A_\Lambda(\widetilde{f} \oplus \widehat{f}) &:= \widetilde{A}^*\widetilde{f} \oplus \widehat{A}^*\widehat{f}, \\ \operatorname{dom} A_\Lambda &:= \left\{ \widetilde{f} \oplus \widehat{f} \in \operatorname{dom} \widetilde{A}^* \oplus \operatorname{dom} \widehat{A}^* : \Lambda \begin{pmatrix} \widetilde{\Gamma}_0 \widetilde{f} \\ \widehat{\Gamma}_0 \widehat{f} \end{pmatrix} = \begin{pmatrix} \widetilde{\Gamma}_1 \widetilde{f} \\ \widehat{\Gamma}_1 \widehat{f} \end{pmatrix} \right\}, \end{aligned}$$

with a Hermitian $2d \times 2d$ matrix of the form

$$\Lambda := \begin{pmatrix} \alpha I_d & \omega I_d \\ \overline{\omega} I_d & \beta I_d \end{pmatrix}, \qquad \alpha, \beta \in \mathbb{R},\ \omega \in \mathbb{C}. \tag{3}$$

Proposition 3.1. *The operator A_Λ, defined as above, is self-adjoint in the Hilbert space $\widetilde{\mathcal{H}} \oplus \widehat{\mathcal{H}}$.*

Proof. The statement of this proposition is a straightforward consequence of the structure of the matrix Λ, Proposition 2.1, Remark 2.1 and Lemma 3.1. □

The next theorem contains the main results of this note: the two terms expansion of a bound state of A_Λ for small coupling parameter $|\omega|$ in the case of arbitrary $d \in \mathbb{N}$ and the three terms analogous expansion in the special case $d = 1$. In its formulation we use self-adjoint operators

$$\begin{aligned} \widetilde{A}_{[\alpha]} &:= \widetilde{A}^* \upharpoonright \ker(\widetilde{\Gamma}_1 - \alpha\widetilde{\Gamma}_0), \qquad & \widetilde{A}_0 &:= \widetilde{A}^* \upharpoonright \ker \widetilde{\Gamma}_0, \\ \widehat{A}_{[\beta]} &:= \widehat{A}^* \upharpoonright \ker(\widehat{\Gamma}_1 - \beta\widehat{\Gamma}_0), \qquad & \widehat{A}_0 &:= \widehat{A}^* \upharpoonright \ker \widehat{\Gamma}_0. \end{aligned} \tag{4}$$

Let L be a $d \times d$-matrix. In the following we use the notion of the adjugate matrix $\operatorname{adj}(L)$, cf. [17, 18]. Notice that the adjugate of a matrix is quite different from the adjoint one L^*.

Theorem 3.1. *Assume that Hypothesis II holds with some $d \in \mathbb{N}$. Let $\widetilde{M}$ and $\widehat{M}$ be the Weyl functions associated with boundary triples from that hypothesis. Let the self-adjoint operators $\widetilde{A}_{[\alpha]}$ and $\widehat{A}_{[\beta]}$ be as above. Assume that the real value λ_0 satisfies $\lambda_0 \in \rho(\widetilde{A}_0) \cap \rho(\widehat{A}_0) \cap \rho(\widetilde{A}_{[\alpha]})$ and λ_0 is a simple isolated eigenvalue of $\widehat{A}_{[\beta]}$.*

(i) *Then for sufficiently small $|\omega|$ in the discrete spectrum of A_Λ there is a branch*

$$\lambda(|\omega|^2) = \lambda_0 + a|\omega|^2 + O(|\omega|^4), \qquad |\omega| \to 0+, \tag{5}$$

with

$$a := \frac{\operatorname{tr}\Big(\operatorname{adj}\big(\beta I_d - \widehat{M}(\lambda_0)\big)\big(\widetilde{M}(\lambda_0) - \alpha I_d\big)^{-1}\Big)}{\operatorname{tr}\Big(\operatorname{adj}\big(\beta I_d - \widehat{M}(\lambda_0)\big)\widehat{M}'(\lambda_0)\Big)}, \tag{6}$$

where $\operatorname{adj}(\beta I_d - \widehat{M}(\lambda_0))$ *is the adjugate matrix.*

(ii) *Suppose that $d = 1$. Then the expansion (5) can be extended as*

$$\lambda(|\omega|^2) = \lambda_0 + a|\omega|^2 + b|\omega|^4 + O(|\omega|^6), \qquad |\omega| \to 0+, \tag{7}$$

with

$$a := \frac{1}{(\widetilde{M}(\lambda_0) - \alpha)\widehat{M}'(\lambda_0)} \tag{8}$$

and

$$b := \frac{1}{((\widetilde{M}(\lambda_0) - \alpha)\widehat{M}'(\lambda_0))^2}\left(\frac{\widetilde{M}'(\lambda_0)}{\alpha - \widetilde{M}(\lambda_0)} - \frac{1}{2}\frac{\widehat{M}''(\lambda_0)}{\widehat{M}'(\lambda_0)}\right). \tag{9}$$

Proof. (i) The proof of this item is carried out in three steps.
Step I. For sufficiently small $\varepsilon > 0$ the interval $I := (\lambda_0 - \varepsilon, \lambda_0 + \varepsilon)$ is contained in the set $\rho(\widetilde{A}_0) \cap \rho(\widehat{A}_0)$. By Proposition 2.2 (i) the following matrix-valued function

$$T(\lambda) := (\alpha I_d - \widetilde{M}(\lambda))(\beta I_d - \widehat{M}(\lambda))$$

is well-defined and C^∞-smooth on I. Next we introduce the scalar-valued function

$$F\colon I \times \mathbb{R} \to \mathbb{R}, \quad F(\lambda, x) := \det\big(T(\lambda) - xI_d\big), \tag{10}$$

which is C^∞-smooth on $I \times \mathbb{R}$.
Step II. The following two functions

$$\widetilde{D}_\alpha(\lambda) := \det\big(\alpha I_d - \widetilde{M}(\lambda)\big) \quad \text{and} \quad \widehat{D}_\beta(\lambda) := \det\big(\beta I_d - \widehat{M}(\lambda)\big)$$

are well-defined and C^∞-smooth on I. Jacobi's formula [18, 19] and the identity $\operatorname{adj}(L_1 L_2) = \operatorname{adj}(L_2)\operatorname{adj}(L_1)$ imply

$$F_x(\lambda_0, 0) = -\operatorname{tr}\Big(\operatorname{adj}\big(\beta I_d - \widehat{M}(\lambda_0)\big)\operatorname{adj}\big(\alpha I_d - \widetilde{M}(\lambda_0)\big)\Big).$$

In view of $\lambda_0 \in \rho(\widetilde{A}_{[\alpha]})$ and of Proposition 2.2 (ii) the matrix $\alpha I_d - \widetilde{M}(\lambda_0)$ is invertible. For any invertible matrix L the identity $\mathrm{adj}\,(L) = \det\,(L)\,L^{-1}$ holds. Hence, we arrive at

$$F_x(\lambda_0, 0) = \widetilde{D}_\alpha(\lambda_0)\mathrm{tr}\Big(\mathrm{adj}\,\big(\beta I_d - \widehat{M}(\lambda_0)\big)\big(\widetilde{M}(\lambda_0) - \alpha I_d\big)^{-1}\Big). \tag{11}$$

Note that $F(\lambda, 0) = \widetilde{D}_\alpha(\lambda)\widehat{D}_\beta(\lambda)$, where the identity $\det\,(L_1L_2) = \det\,(L_1)\det\,(L_2)$ is used. In view of $\lambda_0 \in \sigma_{\mathrm{d}}(\widehat{A}_{[\beta]})$ and of Proposition 2.2 (ii) we get $\widehat{D}_\beta(\lambda_0) = 0$, which implies $F(\lambda_0, 0) = 0$. Next we compute F_λ at the point $(\lambda_0, 0)$

$$\begin{aligned} F_\lambda(\lambda_0, 0) &= \Big(\frac{d}{d\lambda}F(\lambda, 0)\Big)\Big|_{\lambda=\lambda_0} \\ &= \widetilde{D}'_\alpha(\lambda_0)\widehat{D}_\beta(\lambda_0) + \widetilde{D}_\alpha(\lambda_0)\widehat{D}'_\beta(\lambda_0) = \widetilde{D}_\alpha(\lambda_0)\widehat{D}'_\beta(\lambda_0). \end{aligned} \tag{12}$$

Since the eigenvalue λ_0 is simple in the spectrum of $\widehat{A}_{[\beta]}$, by Proposition 2.2 (iii) $\widehat{D}'_\beta(\lambda_0) \neq 0$ holds. Similarly $\widetilde{D}_\alpha(\lambda_0) \neq 0$ because of $\lambda_0 \in \rho(\widetilde{A}_{[\alpha]})$. Hence we obtain that $F_\lambda(\lambda_0, 0) \neq 0$. Recall that $F(\lambda_0, 0) = 0$ and that F is C^∞-smooth. Therefore, by the classical implicit function theorem [20, Theorem 3.3.1] there exists the C^∞-smooth function $\lambda(\cdot)$ defined on a sufficiently small neighborhood of the origin such that $\lambda(0) = \lambda_0$ and that $F(\lambda(x), x) = 0$ holds pointwise. The derivative of $\lambda(\cdot)$ is given as usual by

$$\lambda'(x) = -\frac{F_x(\lambda(x), x)}{F_\lambda(\lambda(x), x)}. \tag{13}$$

Again using Jacobi's formula we get

$$\widehat{D}'_\beta(\lambda_0) = -\mathrm{tr}\Big(\mathrm{adj}\,\big(\beta I_d - \widehat{M}(\lambda_0)\big)\widehat{M}'(\lambda_0)\Big). \tag{14}$$

Substituting (11), (12) and (14) into (13) we arrive at $\lambda'(0) = a$ with a given by (6). Hence we obtain that

$$\lambda(x) = \lambda_0 + ax + O(x^2), \qquad x \to 0. \tag{15}$$

Step III. By Proposition 2.2 (ii) a point $\lambda \in \rho(\widetilde{A}_0) \cap \rho(\widehat{A}_0)$ satisfying

$$\det\begin{pmatrix} \alpha I_d - \widetilde{M}(\lambda) & \omega I_d \\ \overline{\omega} I_d & \beta I_d - \widehat{M}(\lambda) \end{pmatrix} = 0$$

is in the discrete spectrum of A_Λ. By [21, Theorem 3] one gets that

$$\det\begin{pmatrix} \alpha I_d - \widetilde{M}(\lambda) & \omega I_d \\ \overline{\omega} I_d & \beta I_d - \widehat{M}(\lambda) \end{pmatrix} = \det((\alpha I_d - \widetilde{M}(\lambda))(\beta I_d - \widehat{M}(\lambda)) - |\omega|^2 I_d).$$

That is $\lambda \in \rho(\widetilde{A}_0) \cap \rho(\widehat{A}_0)$ satisfying $F(\lambda, |\omega|^2) = 0$ with F as in (10) belongs to the discrete spectrum of A_Λ. Hence for sufficiently small $|\omega|^2$ we have $\lambda(|\omega|^2) \in \sigma_{\mathrm{d}}(A_\Lambda)$ with $\lambda(\cdot)$ defined by Step II. Finally, the expansion (15) implies (5) in the formulation of the theorem.

(ii) The proof of this item goes along the lines of the proof of (i) and we indicate only the differences. Let F be defined as in (10). In this special case ($d = 1$) we have

$$F(\lambda, x) = \big(\alpha - \widetilde{M}(\lambda)\big)\big(\beta - \widehat{M}(\lambda)\big) - x.$$

The 1st and 2nd order partial derivatives of F are computed below

$$\begin{aligned} F_x(\lambda, x) &= -1, \quad F_{\lambda x}(\lambda, x) = 0, \quad F_{xx}(\lambda, x) = 0, \\ F_\lambda(\lambda, x) &= -\widetilde{M}'(\lambda)(\beta - \widehat{M}(\lambda)) - (\alpha - \widetilde{M}(\lambda))\widehat{M}'(\lambda), \\ F_{\lambda\lambda}(\lambda, x) &= -\widetilde{M}''(\lambda)(\beta - \widehat{M}(\lambda)) + 2\widetilde{M}'(\lambda)\widehat{M}'(\lambda) - (\alpha - \widetilde{M}(\lambda))\widehat{M}''(\lambda). \end{aligned} \tag{16}$$

In particular, we have at the point $(\lambda_0, 0)$

$$\begin{aligned} F_\lambda(\lambda_0, 0) &= (\widetilde{M}(\lambda_0) - \alpha)\widehat{M}'(\lambda_0), \\ F_{\lambda\lambda}(\lambda_0, 0) &= 2\widetilde{M}'(\lambda_0)\widehat{M}'(\lambda_0) + (\widetilde{M}(\lambda_0) - \alpha)\widehat{M}''(\lambda_0), \end{aligned} \tag{17}$$

where we used that $\beta - \widehat{M}(\lambda_0) = 0$, which is true in view of $\lambda_0 \in \sigma_{\rm d}(\widehat{A}_{[\beta]})$. Similarly as on Step II in the proof of (i) we get that $F(\lambda_0, 0) = 0$ and $F_\lambda(\lambda_0, 0) \neq 0$. Hence, there exists the C^∞-smooth function $\lambda(\cdot)$ defined on a sufficiently small neighborhood of the origin such that $\lambda(0) = \lambda_0$, that $F(\lambda(x), x) = 0$ holds pointwise and that $\lambda'(x)$ is as in (13). Substituting the identity $F_x(\lambda(x), x) = -1$ into (13) we obtain that

$$\lambda'(x) = \frac{1}{F_\lambda(\lambda(x), x)}, \tag{18}$$

and further substituting (17) into the above formula we get $\lambda'(0) = a$ with a as in (8). Taking the derivative in (18) we get

$$\lambda''(x) = -\frac{F_{\lambda\lambda}(\lambda(x), x)\lambda'(x) + F_{\lambda x}(\lambda(x), x)}{(F_\lambda(\lambda(x), x))^2}.$$

Plugging (16) and (17) into the above formulae we obtain $\lambda''(0) = 2b$ with b as in (9). Hence we arrive at the expansion

$$\lambda(x) = \lambda_0 + ax + bx^2 + O(x^3), \qquad x \to 0,$$

which implies (7) similarly as on Step III in the proof of (i) the expansion (15) implied the formula (5). □

Remark 3.1. The roles of the operators $\widetilde{A}_{[\alpha]}$ and $\widehat{A}_{[\beta]}$ in the above theorem can be interchanged.

Remark 3.2. Note that $\operatorname{adj}(0) = 1$ and in the special case $d = 1$ the formula (6) reduces to (8).

4. An example

Let the operator A in the Hilbert space $L^2(\mathbb{R}^3) \otimes \mathbb{C}^d$ with $d \in \mathbb{N}$ be defined as

$$Af := (-\Delta + Q)f, \qquad \operatorname{dom} A := \{f \in H^2(\mathbb{R}^3) \otimes \mathbb{C}^d \colon f(0,0,0) = \mathbf{0}\}, \tag{19}$$

where $Q = Q^*$ is a $d \times d$ matrix. The operator A is closed, symmetric, densely defined with deficiency indices (d, d), cf. [22–24] and [25, Section 3]. The adjoint of the above symmetric operator can be characterized [23, 24] as

$$\begin{aligned}\operatorname{dom} A^* &= \left\{f = f_0 + \vec{a}\tfrac{e^{-|x|}}{|x|} + \vec{b}e^{-|x|} \colon f_0 \in \operatorname{dom} A, \vec{a}, \vec{b} \in \mathbb{C}^d\right\},\\ A^* f &= -\Delta f_0 - \vec{a}\tfrac{e^{-|x|}}{|x|} - \vec{b}\Big(e^{-|x|} - \tfrac{2e^{-|x|}}{|x|}\Big) + Qf.\end{aligned}$$

The triple $\{\mathbb{C}^d, \Upsilon_0, \Upsilon_1\}$ with $\Upsilon_0, \Upsilon_1 \colon \operatorname{dom} A^* \to \mathbb{C}^d$, where

$$\Upsilon_0 f := \sqrt{4\pi}\vec{a} \quad \text{and} \quad \Upsilon_1 f := \sqrt{4\pi} \lim_{|x|\to 0} \big(f(x) - \tfrac{\vec{a}}{|x|}\big)$$

is a boundary triple for A^*. By [24, Proposition 4.1 (iii)] and [25, Proposition 3.3] the Weyl function associated with the boundary triple $\{\mathbb{C}^d, \Upsilon_0, \Upsilon_1\}$ is given by $M(\lambda) = i\sqrt{\lambda - Q}$, $\lambda \in \rho(A_0)$. Let us assume that Q has only simple eigenvalues $q_1 < \ldots < q_k < \ldots < q_d$. It turns out that $\sigma(A_0) = [q_1, \infty)$. Consider the self-adjoint extension $A_{[\alpha]} = A^* \upharpoonright \ker(\Upsilon_1 - \alpha\Upsilon_0)$ of A with $\alpha < 0$. The extension $A_{[\alpha]}$ has below the threshold q_1 at most d eigenvalues. Moreover, $\lambda \in (-\infty, q_1)$ is an eigenvalue if and only if $\lambda + \alpha^2 \in \sigma(Q)$. Hence, $\lambda_k := q_k - \alpha^2$ is an eigenvalue of $A_{[\alpha]}$ below the threshold q_1 if and only if $q_k - \alpha^2 < q_1$. In particular, we have $A_{[\alpha]} \geq q_1 - \alpha^2$.

Suppose that symmetric operators $\widetilde{A}$ and $\widehat{A}$ are as in (19) with $Q = \widetilde{Q}$ and $Q = \widehat{Q}$, respectively. Let us assume that $\{\widetilde{q}_1, \widetilde{q}_2, \ldots, \widetilde{q}_d\}$ and $\{\widehat{q}_1, \widehat{q}_2, \ldots, \widehat{q}_d\}$ are the simple eigenvalues of $\widetilde{Q}$ and $\widehat{Q}$, respectively, ordered increasingly. We denote the instances of the triple $\{\mathbb{C}^d, \Upsilon_0, \Upsilon_1\}$ for $\widetilde{A}^*$ by $\{\mathbb{C}^d, \widetilde{\Gamma}_0, \widetilde{\Gamma}_1\}$ and for $\widehat{A}^*$ by $\{\mathbb{C}^d, \widehat{\Gamma}_0, \widehat{\Gamma}_1\}$. The corresponding Weyl functions are clearly given by

$$\widetilde{M}(\lambda) = i\sqrt{\lambda - \widetilde{Q}} \quad \text{and} \quad \widehat{M}(\lambda) = i\sqrt{\lambda - \widehat{Q}}.$$

Note that the triple $\{\mathbb{C}^{2d}, \Gamma_0, \Gamma_1\}$ with $\Gamma_i = \widetilde{\Gamma}_i \oplus \widehat{\Gamma}_i$, $i = 0, 1$, is a boundary triple for $(\widetilde{A} \oplus \widehat{A})^*$. The self-adjoint extensions $\widetilde{A}_{[\alpha]}$, $\widehat{A}_{[\beta]}$ $(\alpha, \beta < 0)$ and $\widetilde{A}_0$, $\widehat{A}_0$ are defined as in (4). The self-adjoint extension $\widetilde{A}_{[\alpha]}$ of $\widetilde{A}$ satisfies $\widetilde{A}_{[\alpha]} \geq \widetilde{q}_1 - \alpha^2$. The self-adjoint extension $\widehat{A}_{[\beta]}$ of $\widehat{A}$ has the spectrum $\sigma(\widehat{A}_{[\beta]}) = \big(\cup_{k=1}^{l} \{\widehat{q}_k - \beta^2\}\big) \cup [\widehat{q}_1, +\infty)$, where the eigenvalues are simple and l is the greatest integer $l \in \{1, \ldots, d\}$ satisfying $\widehat{q}_l - \beta^2 < \widehat{q}_1$. If the condition

$$\widehat{\lambda}_k := \widehat{q}_k - \beta^2 < \widetilde{q}_1 - \alpha^2$$

holds for some $k \leq l$, then $\widehat{\lambda}_k$ is a simple isolated eigenvalue of $\widehat{A}_{[\beta]}$ and simultaneously a resolvent point of $\widetilde{A}_{[\alpha]}$. Consider the self-adjoint extension $A_\Lambda := (\widetilde{A} \oplus \widehat{A})^* \upharpoonright$

$\ker(\Gamma_1 - \Lambda\Gamma_0)$ of $\widetilde{A} \oplus \widehat{A}$ with Λ as in (3). Simple computations give us

$$\widehat{M}'(\widehat{\lambda}_k) = \frac{i}{2}\big(\widehat{\lambda}_k - \widehat{Q}\big)^{-1/2} = \frac{1}{2}\big(\widehat{Q} - \widehat{\lambda}_k\big)^{-1/2}.$$

With the above formula in hands we get using Theorem 3.1 (i) that in the discrete spectrum of A_Λ exists a branch with the expansion

$$\widehat{\lambda}_k(|\omega|^2) = \widehat{\lambda}_k - 2\frac{\operatorname{tr}\Big(\operatorname{adj}\big(\sqrt{\widehat{Q}-\widehat{\lambda}_k}+\beta\big)\big(\sqrt{\widetilde{Q}-\widehat{\lambda}_k}+\alpha\big)^{-1}\Big)}{\operatorname{tr}\Big(\operatorname{adj}\big(\sqrt{\widehat{Q}-\widehat{\lambda}_k}+\beta\big)\big(\widehat{Q}-\widehat{\lambda}_k\big)^{-1/2}\Big)}|\omega|^2 + O(|\omega|^4), \quad \omega \to 0.$$

Notice that $\widehat{Q} - \widehat{\lambda}_1 \geq 0$ and $\widetilde{Q} - \widehat{\lambda}_1 \geq 0$. Since $\sqrt{\widetilde{Q} - \widehat{\lambda}_1} + \alpha > 0$ and furthermore $\operatorname{adj}\big(\sqrt{\widehat{Q} - \widehat{\lambda}_1} + \beta\big) \geq 0$ we get that $\widehat{\lambda}_1(|\omega|^2) < \widehat{\lambda}_1$ for sufficiently small ω.

Next we consider the special case $d = 1$. Setting $\widetilde{q} := \widetilde{q}_1$ and $\widehat{q} := \widehat{q}_1$ we get $\sigma(\widetilde{A}_{[\alpha]}) = \{\widetilde{q} - \alpha^2\} \cup [\widetilde{q}, +\infty)$ and $\sigma(\widehat{A}_{[\beta]}) = \{\widehat{q} - \beta^2\} \cup [\widehat{q}, +\infty)$. Let $\lambda_0 := \widehat{q} - \beta^2$. If $\widehat{q} - \beta^2 < \widetilde{q}$ and $\widetilde{q} - \alpha^2 \neq \widehat{q} - \beta^2$, then $\lambda_0 \in \rho(\widetilde{A}_{[\alpha]})$.

Let $E := \widehat{q} - \widetilde{q}$. Notice that $\beta^2 - E > 0$. Simple computations give us

$$\widetilde{M}(\lambda_0) = -\sqrt{\beta^2 - E}, \qquad \widetilde{M}'(\lambda_0) = \frac{1}{2\sqrt{\beta^2 - E}},$$
$$\widehat{M}'(\lambda_0) = \frac{1}{2|\beta|}, \qquad \widehat{M}''(\lambda_0) = \frac{1}{4|\beta|^3}.$$

Hence, according to Theorem 3.1 (ii) in the discrete spectrum of A_Λ exists a branch with the expansion

$$\lambda(|\omega|^2) = \lambda_0 + \frac{2\beta}{\sqrt{\beta^2 - E} + \alpha}|\omega|^2 + \frac{1}{\big(\sqrt{\beta^2 - E} + \alpha\big)^3}\frac{\beta^2 + E - \alpha\sqrt{\beta^2 - E}}{\sqrt{\beta^2 - E}}|\omega|^4 + O(|\omega|^6)$$

as $\omega \to 0$. The latter result is consistent with [4, Theorem 3.1], see also [7].

Acknowledgment

The work of VL was supported by the Austrian Science Fund (FWF), project P 25162-N26. VL thanks Weierstrass Institute for Applied Analysis and Stochastics for hospitality. The work of IYP was supported by the Government of Russian Federation (grant 074-U01) and by State contract of the Russian Ministry of Education and Science. IYP thanks TU Graz for hospitality. Jussi Behrndt and Jonathan Rohleder are acknowledged for interesting and fruitful discussions. The authors are very grateful to the anonymous referee for useful suggestions.

Bibliography

[1] T. Kato, *Perturbation theory for linear operators* (Springer-Verlag, Berlin, 1995).

[2] B. S. Pavlov, *Theor. Math. Phys.* **59**, 544 (1984).

[3] B. S. Pavlov, *Russian Math. Surveys* **42**, 127 (1987).

[4] P. Exner, *Helv. Phys. Acta* **64**, 592 (1991).

[5] C. Cacciapuoti, R. Carlone and R. Figari, *J. Phys. A* **42**, p. 035202 (2009).

[6] C. Cacciapuoti, R. Carlone and R. Figari, *J. Phys. A* **43**, p. 474009 (2010).

[7] P. Exner, in *Mathematical physics, spectral theory and stochastic analysis*, (Basel: Birkhäuser/Springer, 2013) pp. 165–227.

[8] I. Y. Popov, *Russ. Acad. Sci. Sb. Math.* **75**, 285 (1992).

[9] I. Y. Popov, *J. Math. Phys.* **53**, p. 063505 (2012).

[10] I. Y. Popov, *Z. Anal. Anwendungen* **32**, 155 (2013).

[11] J. F. Brasche, M. M. Malamud and H. Neidhardt, *Integr. Equ. Oper. Theory* **43**, 264 (2002).

[12] J. Brüning, V. Geyler and K. Pankrashkin, *Rev. Math. Phys.* **20**, 1 (2008).

[13] V. Derkach and M. Malamud, *J. Funct. Anal.* **95**, 1 (1991).

[14] V. Derkach and M. Malamud, *J. Math. Sci.* **73**, 141 (1995).

[15] M. M. Malamud and H. Neidhardt, *Russ. J. Math. Phys.* **21**, 55 (2014).

[16] K. Schmüdgen, *Unbounded self-adjoint operators on Hilbert space.* (Dordrecht: Springer, 2012).

[17] S. Bosch, *Lineare Algebra., 4th revised ed.* (Berlin: Springer, 2008).

[18] J. R. Magnus and H. Neudecker, *Matrix differential calculus with applications in statistics and econometrics. Revised edition.* (Chichester: Wiley, 1999).

[19] M. A. Goldberg, *Amer. Math. Monthly* **43**, 1124 (1972).

[20] S. Krantz and H. Parks, *The implicit function theorem. History, theory, and applications. Reprint of the 2003 hardback edition.* (New York, NY: Birkhauser/Springer, 2013).

[21] J. R. Silvester, *The Mathematical Gazette* **84**, 460 (2000).

[22] S. Albeverio, F. Gesztesy, R. Hoegh-Krohn and H. Holden, *Solvable models in quantum mechanics* (AMS Chelsea Publishing, 2005).

[23] J. Behrndt, M. Malamud and H. Neidhardt, *Proc. Lond. Math. Soc. (3)* **97**, 568 (2008).

[24] N. Goloshchapova, M. Malamud and V. Zastavnyi, *Math. Nachr.* **285**, 1839 (2012).

[25] A. A. Boitsev, H. Neidhardt and I. Y. Popov, *Nanosystems: Phys. Chem. Math.* **4**, p. 747 (2013).

Trace formulas for singular and additive non-selfadjoint perturbations

M. M. Malamud

Institute of Applied Mathematics and Mechanics,
National Academy of Science of Ukraine,
R. Luxemburg str. 74
Donetsk 83114, Ukraine
E-mail: mmm@telenet.dn.ua

H. Neidhardt

Weierstrass Institute for Applied Analysis and Stochastics,
Mohrenstr. 39,
D-10117 Berlin, Germany
E-mail: hagen.neidhardt@wias-berlin.de

The paper gives an overview on trace formulas for pairs of self-adjoint, maximal dissipative and other types of resolvent comparable operators. In particular, recent results on trace formulas for singular pairs of resolvent comparable operators are presented. The results improve and generalize certain classical results of M.G. Krein for pairs of self-adjoint and dissipative operators, the results of A. Rybkin for such pairs, as well as the results of V. Adamyan, B. Pavlov, and M.G. Krein for pairs $\{A, A^*\}$ with a maximal dissipative operator A.

Keywords: Symmetric operators, perturbation determinants, trace formulas, spectral shift functions.

1. Introduction

Trace formulas play an important role in spectral theory. I.M. Lifshitz was the first who considered such formulas in [1] considering the following situation. Let H_0 be a self-adjoint operator in $\mathfrak{H}$ and let $V = V^*$ be a finite dimensional operator in $\mathfrak{H}$. He proved that for a wide class of functions $\varphi(\cdot) : \mathbb{R} \longrightarrow \mathbb{R}$ the following formula holds

$$\operatorname{tr}(\varphi(H) - \varphi(H_0)) = \int_{\mathbb{R}} \varphi'(t)\xi(t)dt \tag{1}$$

where $H = H_0 + V$. The operator H_0 might have arbitrary spectrum. I.M. Lifshitz applied formula (1) to compute the free energy of a crystal in the case of presence of an impurity at a lattice point. Since that time formulas of type (1) are called trace formulas. The function $\xi(\cdot)$ is called a *spectral shift function* (in short SSF).

Outgoing from [1], Krein extended formula (1) in [2] to trace class perturbations $V = \overline{H - H_0} \in \mathfrak{S}_1$. Moreover, for the rigorous justification of the existence of a SSF of a pair $\{H, H_0\}$ he introduced a concept of perturbation determinant $\Delta_{H/H_0}(\cdot)$ and proved the inversion formula

$$\xi(t) = \frac{1}{\pi} \lim_{y\downarrow 0} \mathrm{Im}\left(\log(\Delta_{H'/H}(t+iy))\right) \quad \text{for a.e.} \quad t \in \mathbb{R},$$

expressing $\xi(\cdot)$ by means of $\Delta_{H/H_0}(\cdot)$. Here the branch of the logarithm is fixed by the condition $\lim_{y\to+\infty} \log(\Delta_{H'/H}(t+iy)) = 0$. Such treatment has allowed him to show that there exists a unique SSF satisfying $\xi(\cdot) \in L^1(\mathbb{R}; dt)$. Krein introduced the class $\mathcal{K}(\mathbb{R})$ of functions in [3] admitting the integral representation

$$\Phi(t) = i \int_{\mathbb{R}} \frac{e^{-ist} - 1}{s}\, dp(s), \quad t \in \mathbb{R}, \tag{2}$$

with $p(\cdot) : \mathbb{R} \longrightarrow \mathbb{C}$ being of bounded variation, and proved (1) for $\Phi \in \mathcal{K}(\mathbb{R})$. Clearly, the class $\mathcal{K}(\mathbb{R})$ consists of absolutely continuous functions $\Phi(\cdot)$ with the derivatives $\Phi'(t) = \int_{\mathbb{R}} e^{-ist}\, dp(s)$ being the Fourier-Stieltjes transform. The widest known class of functions for which (1) holds was found by V. Peller in [4] where he proved (1) for functions $\Phi \in B^1_{\infty 1}(\mathbb{R})$ of the so-called Besov class.

In subsequent publications [3, 5] Krein extended (1) to a pair $\{H, H_0\}$ of self-adjoint *resolvent comparable operators*, i.e., operators satisfying

$$(H - z)^{-1} - (H_0 - z)^{-1} \in \mathfrak{S}_1, \quad z \in \rho(H) \cap \rho(H_0). \tag{3}$$

This extension has been motivated by applications to Schrödinger operators $H = H_0 + q$ and other differential operators. Clearly, H is not a trace class perturbation of $H_0 = -\Delta$ while the pair $\{H, H_0\}$ satisfies (3) for certain classes of decaying potentials q. For pairs satisfying (3) a spectral shift function $\xi(\cdot) = \overline{\xi(\cdot)}$ exists and belongs to $L^1(\mathbb{R}; \frac{dt}{1+t^2})$ which is unique only up to an additive real constant.

A first generalization of trace formulas to pairs of non-selfadjoint and non-unitary operators goes back to the paper by Langer in [6].

Pairs $\{H, H^*\}$ with a *m-dissipative* operator H were treated by different methods in [7, 8]. Recall that a densely defined operator T is called accumulative if $\mathrm{Im}\,(Tf, f) \leq 0$, $f \in \mathrm{dom}\,(T)$. It is called a maximal accumulative (m-accumulative) if T does not admit any accumulative extension. An operator T' is called dissipative (maximal dissipative or m-dissipative) if $T := -T'$ is accumulative (m-accumulative). Using the functional model and assuming that the pair $\{H, H^*\}$ is resolvent comparable,

Adamyan and Pavlov proved in [7] the following trace formula

$$\begin{aligned}&\operatorname{tr}(\Phi(H)-\Phi(H^*))\\&=\sum_{z_k\in\sigma_p(H)} m_k(\Phi(z_k)-\Phi(\overline{z}_k))+\frac{i}{\pi}\int_{\mathbb{R}}\Phi'(t)d\mu(t)-ia\operatorname{res}_\infty(\Phi)\end{aligned}\tag{4}$$

where Φ is holomorphic in a neighborhood of $\sigma(H)\cup\sigma(H^*)$. Here μ is a Borel measure on $\mathbb{R}$ satisfying $\int_{\mathbb{R}}\frac{d\mu}{1+t^2}<\infty$, $\{z_k\}_k\subset\mathbb{C}_+$ the set of eigenvalues of H, m_k the algebraic multiplicity of z_k, $a\geq 0$ and $\operatorname{res}_\infty(\Phi)$ is the residuum of Φ at infinity. Earlier L. Sakhnovich in [8] using the triangular model of M.S. Livsic proved (4) for bounded H with $H-H^*\in\mathfrak{S}_1$ and rational Φ.

A. Rybkin considered a pair $\{H,H_0\}$ consisting of a m-dissipative operator H and $H_0=H_0^*$ in [9, 10]. Assuming (3) he proved the following trace formula

$$\operatorname{tr}(\Phi(H)-\Phi(H_0))=\sum_{z_k\in\sigma(H)} m_k\Phi(z_k)+\int_{\mathbb{R}}\Phi'(t)\omega(t)dt.\tag{5}$$

Here $\Phi(\cdot)$ is holomorphic in a neighborhood of $\sigma(H)\cup\sigma(H_0)$ and decays sufficiently fast at infinity. The meaning of the integral in (5) was clarified by A. Rybkin in the subsequent publication [11].

Krein extended formula (4) to pairs $\{H^*,H\}$ with *not necessarily dissipative* $H:=H_0+iG$ in [12] assuming only that $H_0=H_0^*$ and $G=G^*\in\mathfrak{S}_1$. He proved (cf. [12], Theorem 8.4) that the trace formula

$$\operatorname{tr}(\Phi(H)-\Phi(H^*))=\sum_{z_k\in\sigma_p(H)} m_k\left(\Phi(z_k)-\Phi(\overline{z}_k)\right)+i\int_{\mathbb{R}}\Phi'(t)d\omega(t)\tag{6}$$

holds with $\omega(\cdot)=\overline{\omega(\cdot)}$ being a function of bounded variation on $\mathbb{R}$. Since now $\operatorname{res}_\infty(\Phi)=0$, (6) coincides with (4) if $\omega(\cdot)$ is non-decreasing. However as distinguished from (4) in the case of additive perturbations $\omega(\cdot)$ is bounded.

Investigating the pairs $\{H,H_0\}$ with $H_0=H_0^*$, $H:=H_0-iG$, and $G\geq 0$, Krein introduced the class $\mathcal{K}_+(\mathbb{R})$ of functions holomorphic in $\mathbb{C}_-$ and admitting integral representations (2) with functions $p(\cdot):\mathbb{R}\longrightarrow\mathbb{C}$ being of bounded variation and supported on $\mathbb{R}_+$. He proved in [12], Theorem 9.2, that if $\Phi\in\mathcal{K}_+(\mathbb{R})$, then $\Phi(H)-\Phi(H_0)\in\mathfrak{S}_1$, and instead of (1) the following trace formula holds

$$\operatorname{tr}(\Phi(H)-\Phi(H_0))=-i\int_{\mathbb{R}}\Phi'(t)\,d\omega_K(t).\tag{7}$$

Here $\omega_K(\cdot)=\overline{\omega_K(\cdot)}$ is a bounded non-decreasing function.

Finally, pairs $\{H,H_0\}$ with $H_0=H_0^*$ and $H:=H_0-iG$ where $G\geq 0$, and $G\log(G)\in\mathfrak{S}_1$, were studied by V. Adamyan and one of the authors in [13]. It is proved in [13] that under the assumption $G\log(G) \in \mathfrak{S}_1$ there exists a *real-valued* function $\xi(\cdot) \in L^1(\mathbb{R}, \frac{dt}{1+t^2})$ such that in place of (7) the following formula holds

$$\mathrm{tr}\,(\Phi(H) - \Phi(H_0)) = \int_{\mathbb{R}} \Phi'(t)\;\xi(t)\;dt$$

for Φ from a certain class of holomorphic functions which is narrower than $\mathcal{K}_+(\mathbb{R})$.

2. Trace formulas for pairs of resolvent comparable operators

In this section we present recent generalizations of the above results obtained in [14], see also [15]. The results were obtained by applying the technique of boundary triplets to the computation of perturbation determinants, cf. [16].

We start with a pair of m-accumulative operators. Denote by $\mathcal{F}_+(\widetilde{A}, \widetilde{A}')$ a class of holomorphic function such that the Riesz-Dunford calculus is applicable to the pair $\{\widetilde{A}', \widetilde{A}\}$ in the upper half plane.

Theorem 2.1. *Let $\{\widetilde{A}', \widetilde{A}\}$ be a pair of m-accumulative resolvent comparable operators. Assume that $\rho(\widetilde{A}) \cap \mathbb{C}_- \neq \emptyset$ and $\Phi \in \mathcal{F}_+(\widetilde{A}, \widetilde{A}')$. Then $\Phi(\widetilde{A}') - \Phi(\widetilde{A}) \in \mathfrak{S}_1$ and the following statements hold:*

(i) *There exists a complex-valued function $\omega(\cdot) \in L^1(\mathbb{R}; \frac{dt}{1+t^2})$ (a spectral shift function of the pair $\{\widetilde{A}', \widetilde{A}\}$) such that the following trace formula holds*

$$\mathrm{tr}\,(\Phi(\widetilde{A}') - \Phi(\widetilde{A})) = \frac{1}{\pi}\int_{\mathbb{R}} \Phi'(t)\omega(t)dt. \tag{8}$$

A complex-valued function $\widetilde{\omega} \in L^1(\mathbb{R}; \frac{dt}{1+t^2})$ is also a SSF of the pair $\{\widetilde{A}', \widetilde{A}\}$, i.e. (8) holds with $\widetilde{\omega}$ in place of ω, if and only if $\widetilde{\omega}(\cdot) - \omega(\cdot) \in H^1_-(\mathbb{R}, \frac{dt}{1+t^2})$.

(ii) *If in addition, the imaginary part of ω satisfies the Zygmund condition, i.e.*

$$\int_{\mathbb{R}} |\omega_I(t)| \log(1 + |\omega_I(t)|)\frac{dt}{1+t^2} < \infty, \qquad \omega_I(\cdot) := \mathrm{Im}\,(\omega(\cdot)),$$

then there exists a real-valued SSF $\xi(\cdot) \in L^1\left(\mathbb{R}; \frac{dt}{1+t^2}\right)$. *The latter happens if, in particular, $\omega(\cdot) \in L^2\left(\mathbb{R}; \frac{dt}{(1+t^2)^{\alpha/2}}\right)$ for some $\alpha \in [0, 2]$. Moreover, if $\alpha \in (0, 1)$, then there is real SSF $\xi(\cdot)$ satisfying $\xi(\cdot) \in L^2\left(\mathbb{R}; \frac{dt}{(1+t^2)^{\alpha/2}}\right)$.*

(iii) *If $\widetilde{A} = \widetilde{A}^*$ (resp. $\widetilde{A}' = \widetilde{A}'^*$), then there exists a spectral shift function $\omega(\cdot)$ of $\{\widetilde{A}', \widetilde{A}\}$ satisfying $\mathrm{Im}\,(\omega(t)) \leq 0$ (resp. $\mathrm{Im}\,(\omega(t)) \geq 0$) for a.e. $t \in \mathbb{R}$.*

This theorem substantially generalizes and complements Krein's result (cf. formula (7)) in several directions:

- The measure ω_K in (7) is absolutely continuous.
- The unperturbed operator $\widetilde{A}$ is admitted to be m-accumulative instead of self-adjoint.
- The operator $\widetilde{A}'$ is not necessarily an additive perturbation of $\widetilde{A}$.

Theorem 2.2. *Let the assumption of Theorem 2.1 be satisfied.*

(i) *If for some* $\lambda_0 \in \rho(\widetilde{A}') \cap \rho(\widetilde{A}) \cap \rho(A_0) \cap \mathbb{R}$ *the inequality*

$$\mathrm{Re}\,((\widetilde{A}' - \lambda_0)^{-1}) \leq \mathrm{Re}\,((\widetilde{A} - \lambda_0)^{-1}) \tag{9}$$

holds, then there exists a complex-valued SSF $\omega(\cdot)$ *for the pair* $\{\widetilde{A}, \widetilde{A}'\}$ *satisfying* $\mathrm{Re}\,(\omega(t)) \geq 0$ *for a.e.* $t \in \mathbb{R}$.

(ii) *If for some* $\lambda_0 \in \rho(\widetilde{A}') \cap \rho(\widetilde{A}) \cap \rho(A_0) \cap \mathbb{R}$ *the inequality*

$$\mathrm{Im}\,((\widetilde{A}' - \lambda_0)^{-1}) \leq \mathrm{Im}\,((\widetilde{A} - \lambda_0)^{-1}) \tag{10}$$

holds, then there exists a complex-valued SSF $\omega(\cdot)$ *for the pair* $\{\widetilde{A}, \widetilde{A}'\}$ *such that* $\mathrm{Im}\,(\omega(t)) \geq 0$ *for a.e.* $t \in \mathbb{R}$.

(iii) *If both conditions* (9) *and* (10) *are satisfied, then there exists* $\omega(\cdot)$ *for the pair* $\{\widetilde{A}, \widetilde{A}'\}$ *such that* $\mathrm{Re}\,(\omega(t)) \geq 0$ *and* $\mathrm{Im}\,(\omega(t)) \geq 0$ *for a.e.* $t \in \mathbb{R}$.

Next we extend Theorem 2.1 to the case of a pair $\{\widetilde{A}', \widetilde{A}\}$ with m-accumulative $\widetilde{A}$ and arbitrary $\widetilde{A}'$.

Theorem 2.3. *Let* $\{\widetilde{A}', \widetilde{A}\}$ *be a pair of resolvent comparable operators. Assume also that* $\widetilde{A}$ *is* m*-accumulative,* $\rho(\widetilde{A}) \cap \mathbb{C}_- \neq \emptyset$, *and* $\Phi \in \mathcal{F}_+(\widetilde{A}, \widetilde{A}')$. *Then* $\Phi(\widetilde{A}') - \Phi(\widetilde{A}) \in \mathfrak{S}_1$ *and the following trace formula holds*

$$\mathrm{tr}\,(\Phi(\widetilde{A}') - \Phi(\widetilde{A})) = \sum_k m_k(\Phi(z_k) - \Phi(\overline{z}_k)) + \frac{1}{\pi}\int_{\mathbb{R}} \Phi'(t)d\nu(t) + i\alpha\,\mathrm{res}_\infty(\Phi).$$

Here $\{z_k\}_k := \sigma_p(\widetilde{A}') \cap \mathbb{C}_+$, m_k *is the algebraic multiplicity of* z_k, $\alpha \geq 0$, $d\nu(\cdot) := \omega(\cdot)dt + id\mu(\cdot)$, *where* $\mu(\cdot)$ *is a non-negative Borel measure satisfying* $\int \frac{1}{1+t^2}d\mu(t) < \infty$ *and* $\omega(\cdot) \in L^1(\mathbb{R}; \frac{dt}{1+t^2})$ *is complex-valued.*

Theorem 2.4. *Let* $\{\widetilde{A}, \widetilde{A}^*\}$ *be a resolvent comparable pair and* $\Phi \in \mathcal{F}_+(\widetilde{A}, \widetilde{A}')$. *Then* $\Phi(\widetilde{A}) - \Phi(\widetilde{A}^*) \in \mathfrak{S}_1$ *and the following trace formula holds*

$$\mathrm{tr}\,(\Phi(\widetilde{A}) - \Phi(\widetilde{A}^*)) = \sum_n m_n(\Phi(z_n) - \Phi(\overline{z}_n)) + \frac{i}{\pi}\int_{\mathbb{R}} \Phi'(t)d\mu(t) + i\alpha\,\mathrm{res}_\infty(\Phi).$$

Here $\{z_n\}_n := \sigma_p(\widetilde{A}) \cap (\mathbb{C} \setminus \mathbb{R})$ *and* m_n *denotes the algebraic multiplicity of* z_n, μ *is* a real-valued measure *on* $\mathbb{R}$ *satisfying* $\int_{\mathbb{R}} \frac{1}{1+t^2}|d\mu(t)| < \infty$, *and* $\alpha \in \mathbb{R}$.

If $\widetilde{A}$ is m-dissipative, this trace formula is just formula (4) by Adamyan and Pavlov in [7]. On the other hand, it extends Krein's formula (6) to the case of non-additive perturbations with a not necessarily m-dissipative operator $\widetilde{A}$.

Finally, we consider a pair $\{\widetilde{A}', \widetilde{A}\}$ with a m-dissipative $\widetilde{A}$ and arbitrary $\widetilde{A}'$.

Theorem 2.5. *Let the pairs $\{\widetilde{A}', \widetilde{A}\}$ and $\{\widetilde{A}', \widetilde{A}^*\}$ be resolvent comparable. Assume also that $\widetilde{A}$ is m-dissipative, $\rho(\widetilde{A}') \cap \rho(\widetilde{A}^*) \cap \rho(\widetilde{A}) \cap \mathbb{C}_+ \neq \emptyset$, and $\Phi \in \mathcal{F}_+(\widetilde{A}, \widetilde{A}')$. Then $\Phi(\widetilde{A}') - \Phi(\widetilde{A}) \in \mathfrak{S}_1$ and the following trace formula holds*

$$\begin{aligned}\operatorname{tr}(\Phi(\widetilde{A}') - \Phi(\widetilde{A})) = &\sum_k m_k'(\Phi(z_k') - \Phi(\overline{z}_k')) - \sum_l m_l(\Phi(z_l) - \Phi(\overline{z}_l)) \\ &+ \frac{1}{\pi}\int_{\mathbb{R}} \Phi'(t)d\mu(t) + i\alpha \operatorname{res}_\infty(\Phi).\end{aligned}$$

Here $\{z_k'\}_k = \sigma(\widetilde{A}') \cap \mathbb{C}_+ = \sigma_p(\widetilde{A}') \cap \mathbb{C}_+$ and $\{z_l\}_l = \sigma(\widetilde{A}) \cap \mathbb{C}_+ = \sigma_p(\widetilde{A}) \cap \mathbb{C}_+$, m_k' and m_l are the algebraic multiplicities of the eigenvalues z_k' and z_l, respectively, μ is a complex-valued Borel measure on $\mathbb{R}$ satisfying $\int_{\mathbb{R}} \frac{1}{1+t^2}|d\mu(t)| < \infty$ and $\alpha \in \mathbb{R}$.

3. Trace formulas for additive perturbations

The results of Section 2 can be improved for pairs $\{H', H\}$ where H' is an additive perturbation of H, i.e. $H' := H + V$ and $V \in \mathfrak{S}_1$.

Theorem 3.1. *Let H be a m-accumulative operator, $\Phi \in \mathcal{K}_+(\mathbb{R})$, $V \in \mathfrak{S}_1$, and*

$$\operatorname{Im}(Vf, f) \leq -\operatorname{Im}(Hf, f), \quad f \in \operatorname{dom}(H).$$

(i) *$H' := H + V$ is also m-accumulative and $\Phi(H') - \Phi(H) \in \mathfrak{S}_1$.*

(ii) *There exists a summable complex-valued SSF $\omega(\cdot)$ for $\{H', H\}$, i.e. $\omega(\cdot) \in L^1(\mathbb{R}; dt)$, such that the trace formula (8) holds. Moreover,*

$$\operatorname{tr}(V) = \frac{1}{\pi}\int_{\mathbb{R}} \omega(t)dt.$$

(iii) *If $V = iV_1$, $V_1 = V_1^*$, then there is a purely imaginary SSF $\omega(\cdot) = i\omega_1(\cdot)$, $\omega_1(\cdot) = \overline{\omega_1(\cdot)}$. Moreover, if $V_1 \leq 0$, then ω_1 can be chosen to satisfy $\omega_1(\cdot) \leq 0$.*

(iv) *If $V = V^*$, then there is a summable real SSF $\omega(\cdot)$. Moreover, if in addition $V \leq 0$, then $\omega(\cdot)$ can be chosen to satisfy $\omega(\cdot) \geq 0$.*

Theorem 3.1 clarifies Theorems 2.1 and 2.2 in the case of additive perturbations.

Theorem 3.2. *Let H be a maximal accumulative operator, $V \in \mathfrak{S}_1$ and let $H' := H + V$. If $\Phi \in \mathcal{F}_+(H, H')$, then $\Phi(H') - \Phi(H) \in \mathfrak{S}_1$ and the trace formula*

$$\operatorname{tr}(\Phi(H') - \Phi(H)) = \sum_k m_k(\Phi(z_k) - \Phi(\overline{z}_k)) + \frac{i}{\pi}\int_{\mathbb{R}} \Phi'(t)d\mu(t),$$

holds. Here $\{z_k\}_k = \sigma(\widetilde{A}') \cap \mathbb{C}_+ = \sigma_p(\widetilde{A}') \cap \mathbb{C}_+$, m_k is the algebraic multiplicity of z_k, $\alpha \geq 0$, and $d\nu(\cdot) := \omega(\cdot)dt + id\mu(\cdot)$ where $\mu(\cdot)$ is a non-negative finite Borel measure on $\mathbb{R}$ and $\omega(\cdot) \in L^1(\mathbb{R}; dt)$ is complex-valued.

4. Sketch of our approach

Here we briefly outline the proof of Theorem 3.1. First we show that the perturbation determinants

$$\begin{aligned} \Delta_{H'/H_R}(z) &:= \det(I + iV_I(H_R - z)^{-1}), \\ \Delta_{H_R/H}(z) &:= \det(I + V_R(H - z)^{-1}), \end{aligned} \qquad z \in \mathbb{C}_+,$$

admit exponential representations with real and purely imaginary SSF.

Lemma 4.1 ([14, Lemma 6.5]). *Let H be a m-accumulative operator in $\mathfrak{H}$.*

(i) Let $0 \leq V_+ = V_+^* \in \mathfrak{S}_1$. Assume also that

$$(V_+ f, f) \leq -\operatorname{Im}(Hf, f), \quad f \in \operatorname{dom}(H).$$

Then the following multiplicative representation holds

$$\begin{aligned} w_+(z) &:= \det(I + iV_+(H - z)^{-1}) \\ &= \exp\left\{\frac{i}{\pi}\int_{\mathbb{R}} \frac{1}{t - z}\eta_+(t)dt\right\}, \quad z \in \mathbb{C}_+, \end{aligned} \tag{11}$$

with a non-negative $\eta_+(\cdot) \in L^1(\mathbb{R}; dt)$. Moreover, $\eta_+(\cdot)$ admits a representation

$$\eta_+(t) = -\ln(|w_+(t+i0)|), \quad w_+(t+i0) := \lim_{y\downarrow 0} w_+(t+iy), \quad \text{for a.e.} \quad t \in \mathbb{R}.$$

(ii) In addition, the following representation holds

$$w^+(z) := \det(I - iV_+(H - z)^{-1}) = \exp\left\{-\frac{i}{\pi}\int_{\mathbb{R}} \frac{1}{t - z}\eta^+(t)dt\right\}, \qquad z \in \mathbb{C}_+,$$

with a non-negative $\eta^+(\cdot) \in L^1(\mathbb{R})$ that can be recovered from

$$\eta^+(t) = \ln(|w^+(t + i0)|) := \lim_{y\downarrow 0} w^+(t + iy) \quad \text{for a.e.} \quad t \in \mathbb{R}.$$

(iii) Assume that $V = V^* \in \mathfrak{S}_1$ and Then the following representation holds

$$w(z) := \det(I + iV(H - z)^{-1}) = \exp\left\{\frac{i}{\pi}\int_{\mathbb{R}} \left(\frac{1}{t - z}\right)\eta(t)dt\right\}, \qquad z \in \mathbb{C}_+,$$

where $\eta(\cdot) = \overline{\eta(\cdot)} \in L^1(\mathbb{R}; dt)$ and $\eta(t) = -\ln(|w(t + i0)|)$ for a.e. $t \in \mathbb{R}$

To prove statement (i) we generalize Krein's procedure (see [2, 3]) by dividing the proof in three steps. The first consists in considering one-dimensional perturbations $iV_k = i\lambda_k(\cdot, e_k)e_k$, $\lambda_k > 0$. We show that the determinant $w_k(z) := \det(I + iV_k(H - z)^{-1})$, $k \in \mathbb{N}$, is a contractive holomorphic function in $\mathbb{C}_+$ satisfying $\Re w_k(z) > 0$, $z \in \mathbb{C}_+$. Hence $w_k(\cdot)$ is an outer function in $\mathbb{C}_+$ and admits a representation $w_k(z) = \kappa_+ \exp\left\{\frac{i}{\pi}\int_{\mathbb{R}}\left(\frac{1}{t-z} - \frac{t}{1+t^2}\right)\eta_k(t)dt\right\}$ with $\eta_k(\cdot) \geq 0$ and $|\kappa_+| = 1$. Then we show that it admits representation (11) with $\eta_k(\cdot) \in L^1(\mathbb{R})$ in place of $\eta_+(\cdot)$. In the second step we pass to perturbations $i\sum_1^m V_k$ of arbitrary finite rank, and in the third we extend the results to general trace class perturbations using the spectral decomposition $V_+ = \sum_{k=1}^\infty V_k = \sum_{k=1}^\infty \lambda_k(\cdot, e_k)e_k$.

Lemma 4.2 ([14, Lemma 6.3]). *Let H be a m-accumulative operator.*

(i) If $0 \leq V_+ = V_+^* \in \mathfrak{S}_1(\mathfrak{H})$, then there exists a non-negative function $\xi_+(\cdot) \in L^1(\mathbb{R}; dt)$ such that the following representation holds

$$\det\left(I + V_+(H - z)^{-1}\right) = \exp\left\{\frac{1}{\pi}\int_{\mathbb{R}}\frac{\xi_+(t)}{t-z}dt\right\}, \qquad z \in \mathbb{C}_+, \tag{12}$$

and $\operatorname{tr}(V_+) = \frac{1}{\pi}\int_{\mathbb{R}}\xi_+(t)dt$.

(ii) If $V = V^* \in \mathfrak{S}_1(\mathfrak{H})$, then there exists a real-valued function $\xi(\cdot) \in L^1(\mathbb{R}; dt)$ such that representation (12) is valid with V and $\xi(\cdot)$ in place of V_+ and $\xi_+(\cdot)$, respectively. In particular, one has

$$\operatorname{tr}(V) = \int_{\mathbb{R}}\xi(t)dt \qquad \text{and} \qquad \int_{\mathbb{R}}|\xi(t)|dt \leq \|V\|_{\mathfrak{S}_1}.$$

To prove (i) we note that H being a m-accumulative operator, admits a self-adjoint dilation $K = K^*$ acting in a larger Hilbert space $\mathfrak{K} \supseteq \mathfrak{H}$ and satisfying

$$(H - z)^{-1} = P^{\mathfrak{K}}_{\mathfrak{H}}(K - z)^{-1} \upharpoonright \mathfrak{H}, \qquad z \in \mathbb{C}_+.$$

Setting $V \upharpoonright \mathfrak{K} \ominus \mathfrak{H} = 0$ we identify V with its (trivial) continuation to $\mathfrak{K}$ and put $K' := K + V$. Clearly,

$$\Delta_{H'/H}(z) = \det(I_{\mathfrak{H}} + V(H - z)^{-1}) = \det(I_{\mathfrak{K}} + V(K - z)^{-1}) = \Delta_{K'/K}(z),$$

for $z \in \mathbb{C}_+$. It remains to apply Ref 5, Theorem 1, to obtain Lemma 4.2(ii).

By Lemmas 4.1(iii) and 4.2 (ii), the perturbation determinants $\Delta_{H'/H_R}(\cdot)$ and $\Delta_{H_R/H}(\cdot)$ admit the following multiplicative representations

$$\Delta_{H'/H_R}(z) = \exp\left\{\frac{i}{\pi}\int_{\mathbb{R}}\frac{\eta(t)}{t-z}dt\right\}, \quad \Delta_{H_R/H}(z) = \exp\left\{\frac{1}{\pi}\int_{\mathbb{R}}\frac{\xi(t)}{t-z}dt\right\}, \tag{13}$$

with real-valued functions $\eta(\cdot) \in L^1(\mathbb{R}; dt)$ and $\xi(\cdot) \in L^1(\mathbb{R}; dt)$. Applying the chain rule we derive from (13) that

$$\Delta_{H'/H}(z) = \Delta_{H'/H_R}(z)\Delta_{H_R/H}(z) = \exp\left\{\frac{1}{\pi}\int_{\mathbb{R}}\frac{\omega(t)}{t-z}dt\right\}, \quad z \in \mathbb{C}_+, \tag{14}$$

where $\omega(t) := \xi(t) + i\eta(t)$, $t \in \mathbb{R}$, is a complex valued SSF. Finally, taking the logarithmic derivative (see [17, Chap. IV, Sec. 1]) one gets the trace formula (8) for $\Phi_z(t) = (t - z)^{-1}$. For the class $\mathcal{K}_+(\mathbb{R})$ it extends in a standard fashion.

The results of Section 2 are treated in the framework of boundary triplets approach. The proofs are heavily relied on our approach to perturbation determinants developed in [14, 16]. In accordance with this approach we investigate finite dimensional perturbations of bounded accumulative boundary operators.

Bibliography

[1] I. Lifshits, *Usp. Mat. Nauk* **7**, 171 (1952).
[2] M. G. Kreĭn, *Mat. Sbornik N.S.* **33(75)**, 597 (1953).
[3] M. G. Kreĭn, Some new studies in the theory of perturbations of self-adjoint operators, in *First Math. Summer School, Part I (Russian)*, (Izdat. "Naukova Dumka", Kiev, 1964) pp. 103–187.
[4] V. V. Peller, *Funktsional. Anal. i Prilozhen.* **19**, 37 (1985).
[5] M. G. Kreĭn, *Dokl. Akad. Nauk SSSR* **144**, 268 (1962).
[6] H. Langer, *Math. Nachr.* **30**, 123 (1965).
[7] V. M. Adamjan and B. S. Pavlov, *Vestnik Leningrad. Univ. Mat. Mekh. Astronom.* **2**, 5 (1979).
[8] L. A. Sahnovič, *Trudy Moskov. Mat. Obšč.* **19**, 211 (1968).
[9] A. V. Rybkin, *Mat. Sb. (N.S.)* **125(167)**, 420 (1984).
[10] A. V. Rybkin, *Vestnik Leningrad. Univ. Mat. Mekh. Astronom.* **4**, 97 (1984).
[11] A. V. Rybkin, *Mat. Sb.* **185**, 91 (1994).
[12] M. G. Kreĭn, *J. Operator Theory* **17**, 129 (1987).
[13] V. M. Adamjan and H. Neidhardt, *J. Operator Theory* **24**, 187 (1990).
[14] M. M. Malamud and H. Neidhardt, Perturbation determinants and trace formulas for singular perturbations. Preprint, arXiv:1212.6887 [math-ph], (2012).
[15] M. M. Malamud and H. Neidhardt, Trace formulas for singular perturbations. WIAS-Preprint 1937, (2014).
[16] M. M. Malamud and H. Neidhardt, *Russ. J. Math. Phys.* **21**, 55 (2014).
[17] I. C. Gohberg and M. G. Kreĭn, *Introduction to the theory of linear nonselfadjoint operators* (American Mathematical Society, Providence, R.I., 1969).

On δ'-couplings at graph vertices

S. S. Manko

*Doppler Institute for Mathematical Physics and Applied Mathematics, Czech Technical University in Prague, Břehová 7, 11519 Prague, Czech Republic and Department of Physics, Faculty of Nuclear Science and Physical Engineering, Czech Technical University in Prague, Pohraniční 1288/1, 40501 Děčín, Czech Republic**
E-mail: stepan.manko@gmail.com

In this paper we review our recent attempt to understand the physical meaning of quantum graph models through analysis of their vertex coupling approximations. We find a family of vertex couplings, generalizing the δ'-interaction on the line, and show that with a suitable choice of the parameters they can be approximated in the norm-resolvent sense by families of operators with singularly scaled rank-one interactions. These couplings are of interest mainly because their properties contrast in a sense to those of more regular couplings, for instance, a δ'-junction is opaque at high energies. Finally, our new generalization of the δ'-interaction on the line is different from the two known ones; in contrast to those it leans on the permutation asymmetry of the approximating operators.

Keywords: Quantum graph, vertex coupling, approximation.

1. Different definitions of δ'-couplings: introduction and preliminary results

Quantum graphs attracted a lot of attention both in the physical and mathematical literature in the last decade. One of the reasons for this is that quantum graphs represent natural models for complex structures prepared from semiconductor wires, carbon nanotubes, photonic crystals, and other substances. Applications of these models also arise in dynamical systems and probability theory, spectral theory of differential operators on manifolds and in singular domains, in chemistry, for instance, in modeling electron spectra of aromatic molecules. On the other hand quantum graphs provide a useful and versatile tool to study properties of quantum dynamics in situations when the system has a nontrivial geometrical or topological structure. There are numerous papers devoted to the subject and we restrict ourselves to referring to the recent monograph [2] for an extensive bibliography and basic concepts of theory.

*On leave of absence from Pidstryhach Institute for Applied Problems of Mechanics and Mathematics, National Academy of Sciences of Ukraine, 3b Naukova str, 79060 Lviv, Ukraine

One of the central items in quantum graph models concerns the way in which the wave functions are coupled at the graph vertices. If for example n edges meet at a vertex, in the absence of external fields the requirement of self-adjointness or, in physical terms, probability current conservation leads to a condition:

$$(\mathcal{U}-\mathcal{I})\begin{pmatrix}\psi_1(0)\\ \vdots \\ \psi_n(0)\end{pmatrix}+\mathrm{i}(\mathcal{U}+\mathcal{I})\begin{pmatrix}\psi_1'(0)\\ \vdots \\ \psi_n'(0)\end{pmatrix}=0$$

coupling the vectors of boundary values of the wave functions and their derivatives, in which $\mathcal{U}$ is an $n\times n$ unitary matrix, while $\mathcal{I}$ stands for the identity matrix. To make the graph Hamiltonian self-adjoint is not a difficult task but the result leaves a lot a freedom through parameters entering those conditions the values of which have to be fixed. Different set of parameters give rise to different dynamics on the graph and the choice of the parameters should be guided by the junctions to which the graph vertices should represent an idealized description.

To indicate an example of the above parametrization, we introduce the $n\times n$ matrix $\mathcal{J}$ with all entries equal to one. It is straightforward to check that $\mathcal{U}=-\frac{n+\mathrm{i}\beta}{n-\mathrm{i}\beta}\mathcal{I}+\frac{2}{n-\mathrm{i}\beta}\mathcal{J}$ describes the δ'*-coupling*

$$\sum_{\ell=1}^{n}\psi_\ell'(0)=0, \qquad \psi_j(0)-\psi_k(0)=\frac{\beta}{n}\big(\psi_j'(0)-\psi_k'(0)\big), \quad 1\le j<k\le n, \tag{1}$$

with $\beta\in\mathbb{R}$. The case $\beta=0$ corresponds to the Kirchhoff boundary conditions, while $\beta=\infty$ implies $\mathcal{U}=\mathcal{I}$, the full Neumann decoupling. In a similar spirit, the full Neumann decoupling appears when we put $\beta=\infty$ for the matrix $\mathcal{U}=\mathcal{I}-\frac{2}{n-i\beta}\mathcal{J}$, which describes the so-called δ_s'*-coupling*,

$$\psi_j'(0)=\psi_k'(0)=:\psi'(0), \quad 1\le j<k\le n, \qquad \sum_{\ell=1}^{n}\psi_\ell(0)=\beta\psi'(0), \tag{2}$$

with $\beta\in\mathbb{R}$. The basic properties of the above couplings have been studied in papers [7, 8]. It is worth to mention here that both these couplings appeared as a generalization of the following one-dimensional point interaction – δ'*-coupling*,

$$\psi'(+0)=\psi'(-0), \qquad \psi(+0)-\psi(-0)=\beta\psi'(0). \tag{3}$$

Another possible generalization of (3) is given by the boundary conditions

$$\sum_{\ell=1}^{n}\psi_\ell'(0)=0, \qquad \psi_j(0)-\psi_k(0)=\beta(\gamma_j-\gamma_k)\sum_{\ell=1}^{n}\gamma_\ell\psi_\ell'(0), \quad 1\le j<k\le n. \tag{4}$$

If all γ_j are equal, then conditions (4) lead to the Kirchhoff boundary conditions. The same conclusion can be drawn if $\beta=0$; from this standpoint conditions (4) are similar to (1) and differ from conditions (2).

As we have said, different vertex couplings give rise to different quantum dynamics on the graph, therefore the background of a choice of n^2 parameters is the question about the physical meaning of the coupling, important without any doubt. A natural approach to this problem is to analyze various approximations of those couplings, either on the graph itself or using a tubular network shrinking to the graph skeleton. The task simplifies due to the fact that the vertex couplings, the physically interesting ones at least, are of a local nature; it is thus sufficient to solve the problem for a star-type graph with n edges meeting in a single vertex.

The purpose of this short note is to review our recent achievements in this area. First of all we want to draw attention to the new coupling at graph vertices (4). Second and more important, we address the question of physical meaning of such a coupling (suggesting an answer in terms of singularly scaled rank-one operators). Finally, in the last section we review different (new and old) approaches to the problem of approximation for the general vertex couplings.

Having introduced our purposes, let us end this section with basic properties of coupling (4). With this aim in mind, we introduce a star graph Γ consisting of n halflines connected at one vertex. Next, we consider a free spinless particle on the graph, with the Hamiltonian $\mathscr{H}_\beta$ acting on the graph edges as a negative second derivative, whose domain can be characterized by the coupling conditions (4). The following theorem gives the explicit expression of the resolvent of $\mathscr{H}_\beta$ and describes the structure of its spectrum and scattering amplitudes.

Theorem 1.1. *(i) The resolvent $(\mathscr{H}_\beta - k^2)^{-1}$ is an integral operator on $L^2(\Gamma)$ with the kernel*

$$\begin{aligned}\Xi_k(x_i, y_j) = \frac{\mathrm{i}}{2k}\left[\delta_{ij}\exp(\mathrm{i}k|x_i - y_j|) + \left(\frac{2}{n} - \delta_{ij}\right)\exp(\mathrm{i}k(x_i + y_j))\right] \\ + \Lambda_{ij}(k^2)\exp(\mathrm{i}k(x_i + y_j)), \quad i, j = 1, 2, \ldots, n,\end{aligned}$$

with $k^2 \in \rho(\mathscr{H}_\beta)$, $\Im k > 0$, and with $\Lambda_{ij}(k^2) = \beta\Pi_{ij}(1 + \mathrm{i}k\beta B)^{-1}$, where $B := \frac{1}{n}\big(\sum_{\ell=1}^n \gamma_\ell\big)^2 - \sum_{\ell=1}^n \gamma_\ell^2$ and $\Pi_{ij} := \big(\frac{1}{n}\sum_{\ell=1}^n \gamma_\ell - \gamma_i\big)\big(\frac{1}{n}\sum_{\ell=1}^n \gamma_\ell - \gamma_j\big)$.

(ii) The essential spectrum of $\mathscr{H}_\beta$ is purely absolutely continuous and covers the nonnegative real axis, while the singularly continuous spectrum is empty. If $\beta < 0$, the operator $\mathscr{H}_\beta$ has precisely one negative eigenvalue $-(\beta B)^{-2}$. If $\beta > 0$, the operator has no eigenvalues.

(iii) For any momentum $k > 0$ the on-shell scattering matrix $\mathcal{S}(k)$ for the pair $(\mathscr{H}_\beta, \mathscr{H}_0)$ takes the form $\mathcal{S}_{ij}(k) = \frac{2}{n} - \delta_{ij} - \frac{2\mathrm{i}k\beta\Pi_{ij}}{1+\mathrm{i}k\beta B}$ with Π_{ij} and B given above.

2. Approximations: vertex couplings understanding

This section is devoted to realizing of the physical meaning of coupling (4). To this aim, we are going to introduce the operator family $\mathscr{H}^\varepsilon$ on Γ, which are a rank-one perturbation of a Hamiltonian of the free particle. We demonstrate that the family of Hamiltonians $\mathscr{H}^\varepsilon$ approximates $\mathscr{H}_\beta$ in the norm-resolvent topology; moreover, the spectral and scattering quantities of the limit operator $\mathscr{H}_\beta$ may be obtained by using the limiting procedure for those of $\mathscr{H}^\varepsilon$.

Given a real-valued function V belonging to the class $L^1_{\mathrm{loc}}(\Gamma)$ with a compact support and zero mean, we introduce the following family of Schrödinger operators on $L^2(\Gamma)$ labeled by the parameter $\varepsilon \in (0,1]$,

$$\mathscr{H}^\varepsilon := -\frac{\mathrm{d}^2}{\mathrm{d}x^2} + \frac{\lambda(\varepsilon)}{\varepsilon^3} V\left(\frac{x}{\varepsilon}\right)\left\langle \cdot, V\left(\frac{\cdot}{\varepsilon}\right)\right\rangle_{L^2(\Gamma)}.$$

As we pointed out, the operator $\mathscr{H}^\varepsilon$ is a perturbation of the free Hamiltonian, and so its domain consists of those functions from the Sobolev space, which satisfy the Kirchhoff boundary conditions at the graph vertex. The function $\lambda(\cdot)$ in the above expression is supposed to be real-valued for real ε and holomorphic in the vicinity of the origin. In addition, it satisfies the condition $\lambda(\varepsilon) = \lambda_0 + \varepsilon\lambda_1 + \mathcal{O}(\varepsilon^2)$ as $\varepsilon \to 0$, where λ_0 and λ_1 are nonzero real numbers.

In 1986, Šeba [21] studied the family $\mathscr{H}^\varepsilon$ in dimension one; he proved that in the strong resolvent limit one gets the δ'-coupling (3). It was the first attempt to realize the physical meaning of this coupling and its connection with the pseudo-Hamiltonian $-\frac{\mathrm{d}^2}{\mathrm{d}x^2} + \beta\langle\cdot, \delta'(x)\rangle\delta'(x)$, where the scalar product $\langle\varphi, f\rangle$ is defined as the action of a distribution f on a test function φ. Šeba's approximation led to a nontrivial dependence of β with respect to terms λ_0 and λ_1 in the asymptotics of the coupling constant $\lambda(\varepsilon)$.

To describe the outcome of the limiting process we need the following quantities:

$$\gamma_j := \int_0^\infty x_j V(x_j)\,\mathrm{d}x_j\,, \quad j = 1, 2, \ldots, n\,,$$

and

$$A := -\sum_{j=1}^n \int_0^\infty \int_0^\infty \min\{x_j, y_j\}\, V(x_j)V(y_j)\,\mathrm{d}x_j \mathrm{d}y_j\,.$$

Using them, we define the constant β as $\frac{1}{\lambda_1 A^2}$ if $\lambda_0 A = 1$, otherwise we put $\beta = 0$. With this β our main result reads

Theorem 2.1. *(i) As $\varepsilon \to 0$, the family of Hamiltonians $\mathscr{H}^\varepsilon$ converges to $\mathscr{H}_\beta$ in the norm-resolvent sense.*

(ii) The essential spectrum of $\mathscr{H}^\varepsilon$ is purely absolutely continuous and covers the nonnegative real axis, while the singularly continuous spectrum is empty.

If $\beta < 0$, the operator $\mathscr{H}^\varepsilon$ has for all ε small enough exactly one negative eigenvalue $-\varkappa_\varepsilon^2$ with the asymptotic behavior $-\varkappa_\varepsilon^2 = -(\beta B)^{-2} + \mathcal{O}(\varepsilon)$ as $\varepsilon \to 0$, which tends to the eigenvalue of the limit operator. If $\beta > 0$, the perturbed operator has no eigenvalues. If $\beta = 0$, then there are two possibilities. If $\lambda_0 < \frac{1}{A}$, the approximating operator has for all ε small enough exactly one negative eigenvalue $-\varkappa_\varepsilon^2$ with the asymptotics $-\varkappa_\varepsilon^2 = -(A - \lambda_0^{-1})^2(\varepsilon B)^{-2} + \mathcal{O}(\varepsilon^{-1})$ as $\varepsilon \to 0$, tending to $-\infty$. In the opposite case the operator $\mathscr{H}^\varepsilon$ has no eigenvalues.

(iii) For any momentum $k > 0$ the on-shell scattering matrix $\mathcal{S}^\varepsilon(k)$ for the pair $(\mathscr{H}^\varepsilon, \mathscr{H}_0)$ converges as $\varepsilon \to 0$ to that of $(\mathscr{H}_\beta, \mathscr{H}_0)$, and moreover, there is a constant C such that $\|\mathcal{S}^\varepsilon(k) - \mathcal{S}(k)\| \le C\varepsilon$, $\varepsilon \in (0, 1]$, where $\|\cdot\|$ stands for the operator norm of the matrix.

We refer the reader to a more detailed discussion and proofs to our paper [10]. Using Šeba's approach, we introduce new (physically motivated) definition of the δ'-coupling on the graph given by (4). Interestingly enough, although Šeba proved the strong resolvent convergence, we give the proof of the norm resolvent one, which is more suitable from the viewpoint of the quantum-mechanical approximations.

3. Discussion of general vertex couplings approximations

Having discussed the δ'-couplings, we end the note with mentioning recent progress in the problem of approximation for the general vertex couplings. Historically, the problem appeared in dimension 1 (one vertex and two edges) thirty years ago. Firstly, the δ'-coupling (3) appeared as a natural counterpart of the *δ-coupling* (*δ-interaction* or *δ-potential*)

$$\psi(+0) = \psi(-0) =: \psi(0), \qquad \psi'(+0) - \psi'(-0) = \alpha\psi(0)\,.$$

This interaction could be associated with the pseudo-Hamiltonians $-\frac{\mathrm{d}^2}{\mathrm{d}x^2} + \alpha\delta(x)$ or $-\frac{\mathrm{d}^2}{\mathrm{d}x^2} + \alpha\langle\cdot, \delta(x)\rangle\delta(x)$ (see [1]). Denote the Hamiltonian corresponding to δ-interaction by $\mathscr{H}_\alpha$ and let Q be locally integrable function of compact support (including the origin), then the Hamiltonian $-\frac{\mathrm{d}^2}{\mathrm{d}x^2} + \frac{1}{\varepsilon}Q\big(\frac{x}{\varepsilon}\big)$ approximates $\mathscr{H}_\alpha$ as $\varepsilon \to 0$ in the norm resolvent sense, where $\alpha = \int_{\mathbb{R}} Q$. Another possible approximative family for $\mathscr{H}_\alpha$ is $-\frac{\mathrm{d}^2}{\mathrm{d}x^2} + \frac{1}{\varepsilon^2}\big\langle\cdot, Q\big(\frac{\cdot}{\varepsilon}\big)\big\rangle Q\big(\frac{x}{\varepsilon}\big)$. The wavefunctions of the limit Hamiltonian are continuous; generalizing this interaction to the graph case, the requirement of continuity at the vertex selects a one-parameter family corresponding to the so-called *δ-coupling*,

$$\psi_j(0) = \psi_k(0) =: \psi(0), \quad 1 \le j < k \le n, \qquad \sum_{\ell=1}^{n} \psi_\ell'(0) = \alpha\psi(0),$$

where $\alpha \in \mathbb{R}$. We can add the trivial case corresponding formally to $\alpha = \infty$, when the system decomposes into n halflines with Dirichlet endpoints. In the particular case $\alpha = 0$ we get free boundary conditions. As its one-dimensional analogue, δ-coupling at the graph vertex can be obtained as a limit of the families $-\frac{\mathrm{d}^2}{\mathrm{d}x^2}+\frac{1}{\varepsilon}Q\left(\frac{x}{\varepsilon}\right)$ and $-\frac{\mathrm{d}^2}{\mathrm{d}x^2}+\frac{1}{\varepsilon^2}\left\langle\cdot, Q\left(\frac{\cdot}{\varepsilon}\right)\right\rangle Q\left(\frac{x}{\varepsilon}\right)$ in the norm resolvent topology. For proof see paper [9] where a more general result of this type is derived. However, in the limit we get just a one-parameter subset of all the admissible matching conditions, and the other ones require a different limiting procedure, for instance, using shrinking potentials with a more singular scaling of the type $\frac{1}{\varepsilon^2}Q\left(\frac{x}{\varepsilon}\right)$. Of course, solving this problem it is also natural to consider more singular rank-one perturbation $\frac{1}{\varepsilon^3}\left\langle\cdot, Q\left(\frac{\cdot}{\varepsilon}\right)\right\rangle Q\left(\frac{x}{\varepsilon}\right)$, but this case has been already mentioned in the previous sections.

Šeba was seemingly the first to investigate the one-dimensional Hamiltonian with the potential of the type $\frac{1}{\varepsilon^2}Q\left(\frac{x}{\varepsilon}\right)$ for vertices connecting two edges, which is equivalent to generalized point interaction on the line, with the conclusion that the limit is trivial describing disconnected edges [21]. Recalling that the Schrödinger operators are quantum mechanical Hamiltonians for a particle on the line, one would have to conclude that, in dimension 1, the barrier $\frac{1}{\varepsilon^2}Q\left(\frac{\cdot}{\varepsilon}\right)$ is asymptotically opaque, i.e., that the particle cannot tunnel through it in the limit $\varepsilon \to 0$. Later it was pointed out, however, that such a claim holds only generically and a nontrivial limit may exist when the potential Q has a *zero-energy resonance* (a bounded solution ϕ_b to the equation $-\phi'' + Q\phi = 0$) – cf. [6] and subsequent papers of these authors. In [6] the numerical analysis of exactly solvable models for the potential with piece-wise constant Q of compact support was performed. Namely, the authors demonstrated that for resonant Q, the limiting value of the transmission coefficient $T_\varepsilon(k)$ of the operator $-\frac{\mathrm{d}^2}{\mathrm{d}x^2}+\frac{1}{\varepsilon^2}Q\left(\frac{x}{\varepsilon}\right)$ is different from zero; in certain cases, the limit operator was defined via the matching conditions in its domain

$$\psi(+0) = \gamma\psi(-0), \qquad \gamma\psi'(+0) = \psi'(-0)\,, \tag{5}$$

where $\gamma = \phi_b(+\infty)/\phi_b(-\infty)$ (both limits exist and do not vanish). The coupling conditions of the form (5) also appear in [18] in a realization of the pseudo-Hamiltonian $-\frac{\mathrm{d}^2}{\mathrm{d}x^2}+\alpha\delta'(x)$ by means of the distribution theory over discontinuous test functions. In the paper [17], eigenvalue and eigenfunction asymptotics as $\varepsilon \to 0$ were studied for the full-line Schrödinger operators given by the differential expression $-\frac{\mathrm{d}^2}{\mathrm{d}x^2}+\frac{1}{\varepsilon^2}Q\left(\frac{x}{\varepsilon}\right)+q(x)$, where Q is regular and of compact support and q is unbounded at infinity. The eigenfunctions were shown to satisfy in the limit the Dirichlet condition $\psi(0) = 0$ in the non-resonant case and the interface condition (5) in a resonant case, thus again exhibiting the zero-energy dichotomy (see also [16] for the convergence of the corresponding Hamiltonians). The result was further generalized, in particular, to Schrdinger operators on star graphs with three edges [19, 20] or to combinations of potentials with different scaling of the form $\frac{1}{\varepsilon}Q_1\left(\frac{\cdot}{\varepsilon}\right)+\frac{1}{\varepsilon^2}Q_2\left(\frac{\cdot}{\varepsilon}\right)$, see [15]. The latter approach leads to the more general interface conditions of the form

$$\psi(+0) = \gamma\psi(-0), \qquad \gamma\psi'(+0) - \psi'(-0) = \gamma\alpha\psi(-0)\,.$$

The same conclusion can be drawn if we take the Hamiltonian with the potential $\frac{\lambda(\varepsilon)}{\varepsilon^2}Q\big(\frac{x}{\varepsilon}\big)$, where the coupling function $\lambda(\cdot)$ also depends on the small parameter.

In our recent paper [11] we generalize this idea to the case of quantum graph to approximate new vertex couplings. To be more exact we discussed the asymptotic behaviour as $\varepsilon \to 0$ of the family of Schrödinger operators given by $-\frac{\mathrm{d}^2}{\mathrm{d}x^2} + \frac{1+\lambda\varepsilon+\mathcal{O}(\varepsilon^2)}{\varepsilon^2}Q\big(\frac{x}{\varepsilon}\big)$ on the graph. Since our approximation problem is of a local character we focus on noncompact star-shaped graphs consisting of n semi-infinite edges connected at a single vertex. As in the case of line, the limiting procedure crucially depend on the existing of the resonant solutions, i.e., bounded solutions of the equation $-\phi'' + Q\phi = 0$ on the graph. We say that a potential Q admits a zero-energy resonance of order m if there are m linearly independent resonant solutions $\phi_1, \ldots, \phi_m$. Denote by ϕ_{ij} the limit value of ϕ_i at infinity across the jth edge; the system of resonant solutions could be chosen so that $\phi_{ij} = \delta_{ij}$ for $i, j = 1, \ldots, m$, where δ_{ij} is the Kronecker symbol. Set $q_{ij} := \int_\Gamma Q\phi_i\phi_j\,\mathrm{d}x$ for $i, j = 1, \ldots, m$, then the limit operator acts as a minus second derivative on functions from the Sobolev space that obey the matching conditions

$$\psi_j(0) - \sum_{i=1}^{m} \phi_{ij}\psi_i(0) = 0, \qquad j = m+1, \ldots, n,$$
$$\psi_i'(0) + \sum_{j=m+1}^{n} \phi_{ij}\psi_j'(0) - \lambda \sum_{j=1}^{m} q_{ij}\psi_j(0) = 0, \qquad i = 1, \ldots, m.$$

If $\lambda = 0$, the above conditions do not couple function values and derivatives, and as a result, the matching conditions of the limit operator are *scale-invariant* (and generalize (5)). This means, in particular, that the corresponding operator has no eigenvalues, another manifestation of the scale-invariant character is that the scattering matrix is independent of energy.

In [5], Cheon and Shigehara proposed a formal scheme how to approximate the one-dimensional δ'-interactions (discontinuous wavefunctions) by means of a triple of δ interactions (continuous wavefunctions) with the coupling constants scaled in a nonlinear way as their distances tend to zero. The rigorous proof of the fact was given in [12], moreover, based on mentioned scheme the authors also obtained the potential-type approximation for δ'-interactions. It was a surprising result, since until that time it was believed, that no potential-type approximation to the δ'-interaction existed. The same idea was demonstrated to work in the case of graph δ'-couplings [3]. Similarly, in [14] the authors found that with a larger number of regular interactions one can deal with families described by $2n$ parameters, and changing locally the approximating graph topology one can deal with all the couplings invariant with respect to the time reversal which form an $\binom{n+1}{2}$-parameter subset. Finally, in [4] the problem of general vertex coupling approximation has been solved.

Let us finish by adding that solution to the vertex-coupling approximation problem is certainly not unique. For instance, in [13] a more natural approach was used, namely, all admissible couplings were approximated by tubular Neumann networks shrinking to the graph skeleton. Even in the latter case, however, one considers approximations on the graph as an intermediate step – cf. [13] and references therein.

Acknowledgments

The author enjoyed the pleasure of collaboration with Pavel Exner and Yuriy Golovaty. The research was supported by the European Union with the project "Support for research teams on CTU" CZ.1.07/2.3.00/30.0034.

Bibliography

[1] S. Albeverio, F. Gesztesy, R. Høegh-Krohn and H. Holden, *Solvable Models in Quantum Mechanics* (New York: American Mathematical Society, 2005) (with an appendix by P. Exner)

[2] G. Berkolaiko and P. Kuchment, *Introduction to Quantum Graphs*, (Amer. Math. Soc., Providence, R.I., 2013).

[3] T. Cheon and P. Exner, An approximation to δ' couplings on graphs, *J. Phys. A: Math. Gen.* **37**, L329 (2004).

[4] T. Cheon, P. Exner and O. Turek, Approximation of a general singular vertex coupling in quantum graphs, *Ann. Physics* **325**, 548 (2010).

[5] T. Cheon and T. Shigehara, Realizing discontinuous wave functions with renormalized short-range potentials, *Phys. Lett. A.* **243**, 111 (1998).

[6] P. Christiansen, H. Arnbak, A. Zolotaryuk, V. Ermakov and Y. Gaididei, On the existence of resonances in the transmission probability for interactions arising from derivatives of Diracs delta function, *J. Phys. A: Math. Gen.* **36**, 7589 (2003).

[7] P. Exner, Lattice Kronig-Penney models, *Phys. Rev. Lett.* **74**, 3503 (1995).

[8] P. Exner, Contact interactions on graph superlattices, *J. Phys. A: Math. Gen.* **29**, 87 (1996).

[9] P. Exner, Weakly coupled states on branching graphs, *Lett. Math. Phys.* **38**, 313 (1996).

[10] P. Exner and S. Manko, Approximations of quantum-graph vertex couplings by singularly scaled rank-one operators, *Lett. Math. Phys.* 104, 1079 (2014).

[11] P. Exner and S. Manko, Approximations of quantum-graph vertex couplings by singularly scaled potentials, *J. Phys. A: Math. Theor.* **46**, 345202 (2013).

[12] P. Exner, H. Neidhardt and V. A. Zagrebnov, Potential approximations to δ': an inverse Klauder phenomenon with norm-resolvent convergence, *Commun. Math. Phys.* **224**, 593 (2001).

[13] P. Exner and O. Post, A general approximation of quantum graph vertex couplings by scaled Schrödinger operators on thin branched manifolds, *Commun. Math. Phys.* **322**, 207 (2013).
[14] P. Exner and O. Turek, Approximations of singular vertex couplings in quantum graphs, *Rev. Math. Phys.* **19**, 571 (2007).
[15] Yu. Golovaty, 1D Schrödinger operators with short range interactions: two-scale regularization of distributional potentials, *Intgr. Eqns Oper. Theory.* **75**, 341 (2013).
[16] Yu. Golovaty and R. Hryniv, Norm resolvent convergence of singularly scaled Schrödinger operators and δ'-potentials, *Proc. R. Soc. Edinb. A.* **143**, 791 (2013).
[17] Yu. Golovaty and S. Man'ko, Solvable models for the Schrödinger operators with δ'-like potentials, *Ukr. Math. Bulletin.* **6**, 169 (2009).
[18] P. Kurasov, A. Scrinzi and N. Elander, On the δ'-interaction arising in the exterior complex scaling, *Phys. Rev. A.* **49**, 5095 (1994).
[19] S. Man'ko, On δ'-like potential scattering on star graphs, *J. Phys. A: Math. Theor.*, **43**, 445304 (2010).
[20] S. Man'ko, Schrödinger operators on star graphs with singularly scaled potentials supported near vertices, *J. Math. Phys.*, **53**, 123521 (2012).
[21] P. Šeba, Some remarks on the δ'-interaction in one dimension, *Rep. Math. Phys.* **24** 111 (1986).

Stochastic calculus and non-relativistic QED

B. Güneysu

Institut für Mathematik, Humboldt Universität zu Berlin,
Rudower Chaussee 25,
D-12489 Berlin, Germany
E-mail: gueneysu@math.hu-berlin.de

O. Matte* and J. S. Møller†

Institut for Matematik, Aarhus Universitet, Ny Munkegade 118,
DK-8000 Aarhus C, Denmark
** E-mail: matte@imf.au.dk*
† E-mail: jacob@imf.au.dk

We summarize recent results on stochastic differential equations associated with the standard model of non-relativistic quantum electrodynamics for an electron in an electrostatic potential interacting with the quantized electromagnetic field. Moreover, we present a Feynman-Kac formula for the corresponding semi-group – which is new in case the electron-spin is taken into account – and a new Feynman-Kac formula for an operator-valued integral kernel of the semi-group. Finally, we announce a proof of the norm-continuity of the integral kernel.

Keywords: Stochastic differential equation, Feynman-Kac, quantized radiation field.

1. Introduction

In his famous work on quantum electrodynamics (QED) R. P. Feynman developed in particular a path integral description for the quantum mechanical dynamics of a non-relativistic (NR) electron interacting with the quantized (relativistic) electromagnetic radiation field [2]. It is equally well-known that Feynman's formalism can be justified mathematically for various models given by semi-bounded self-adjoint Hamiltonians, provided that one considers the associated semi-groups instead of the unitary groups representing the quantum mechanical time evolution. For instance, Feynman-Kac (FK) representations of semi-groups have been extensively exploited in the study of Schrödinger operators and led to many important – physically relevant – spectral theoretic results. A natural next step, therefore, is to study FK representations of the semi-group generated by the Hamiltonian of the standard model of NRQED which is a Schrödinger operator minimally coupled to the quantized radiation field.

Employing some ideas from earlier work on polynomial quantum field models (in particular [11]), FK representations in NRQED have been derived first by F. Hiroshima [4] in the scalar case, i.e., without electron spin. A formula which takes spin into account as well has been established more recently by F. Hiroshima and J. Lőrinczi [6]. In both articles the FK formula is derived by means of repeated Trotter product expansions. In principle it should, however, also be possible to verify and analyze FK formulas in NRQED by means of a suitable version of *stochastic calculus* in Hilbert spaces [9, 10]. This approach is explored in a recent paper by the present authors [3], whose main results are summarized in this review. Before we explain its organization we also mention that various mathematical questions in NRQED (such as self-adjointness properties, ergodic properties in the scalar case, uniqueness and localization of ground states, and Gibbs measures associated with ground states) have already been investigated by means of FK representations before, mainly by F. Hiroshima and his co-workers. We refer the reader to the monograph [8] for an exhaustive exposition of these issues and appropriate references.

The model considered here is explained in Sect. 2. In Sect. 3 we introduce new representations of FK integrands in NRQED in a form resembling the *annihilation-preservation-creation processes* playing a prominent role in quantum stochastic calculus [7]. As a crucial merit of the application of the stochastic calculus, the FK integrands are identified as solutions of *stochastic differential equations (SDE's)*, which may be exploited in further investigations of NRQED. To the best of our knowledge the analysis of these novel equations is non-standard due to the appearance of non-deterministic, unbounded, non-commuting operators in their coefficients. Our main existence and uniqueness theorem for these SDE's is also stated in Sect. 3. The final Sect. 4 is devoted to various FK formulas.

2. Definition of the model

The standard model of NRQED for one electron interacting with the quantized photon field is given by a Hamiltonian acting in the Hilbert space $\mathscr{H}$ defined by

$$\mathscr{H} := L^2(\mathbb{R}^3, \hat{\mathscr{H}}) = \int_{\mathbb{R}^3}^{\oplus} \hat{\mathscr{H}}\,\mathrm{d}\boldsymbol{x}, \quad \text{with } \hat{\mathscr{H}} := \mathbb{C}^2 \otimes \mathscr{F}.$$

Here $\mathscr{F} = \oplus_{n=0}^{\infty} \mathscr{F}^{(n)}$ is the bosonic Fock space modeled over $\mathfrak{h} := L^2(\mathbb{R}^3 \times \mathbb{Z}_2, \mathrm{d}k)$, i.e., $\mathscr{F}^{(n)}$ is the n-th symmetric tensor power of $\mathfrak{h}$; $\mathrm{d}k$ denotes the product of the Lebesgue measure and the counting measure on $\mathbb{Z}_2$. We write $\mathscr{E}[\mathfrak{v}] := \{\zeta(h) : h \in \mathfrak{v}\}$ for the set of exponential vectors corresponding to some subset $\mathfrak{v} \subset \mathfrak{h}$, where

$$\zeta(h) := \big(1, ih, \dots, (n!)^{-1/2} i^n h^{\otimes_n}, \dots\big) \in \mathscr{F}, \qquad h \in \mathfrak{h},$$

and denote its complex linear hull by $\mathscr{C}[\mathfrak{v}] := \mathrm{span}_{\mathbb{C}}(\mathscr{E}[\mathfrak{v}])$. The symbols $a^\dagger(f)$ and $a(f)$ denote the usual bosonic creation and annihilation operators of a photon state $f \in \mathfrak{h}$ satisfying the following canonical commutations relations on, e.g., $\mathscr{C}[\mathfrak{h}]$,

$$[a(f), a(g)] = [a^\dagger(f), a^\dagger(g)] = 0, \quad [a(f), a^\dagger(g)] = \langle f|g\rangle\, \mathbb{1}_{\mathscr{F}}, \quad f, g \in \mathfrak{h}.$$

The field operator $\varphi(f)$ is defined as the self-adjoint closure of $(a^\dagger(f) + a(f))\restriction_{\mathscr{C}[\mathfrak{h}]}$. For $\boldsymbol{f} = (f_1, f_2, f_3) \in \mathfrak{h}^3$, we write $\varphi(\boldsymbol{f}) := (\varphi(f_1), \varphi(f_2), \varphi(f_3))$ for short. Finally, $\Gamma(T)$ denotes the second quantization of a bounded linear or anti-linear operator T defined on $\mathfrak{h}$ with norm $\leqslant 1$.

Let $\omega(\boldsymbol{k}) := (\boldsymbol{k}^2 + \mu^2)^{1/2}$, $\boldsymbol{k} \in \mathbb{R}^3$, with $\mu \geqslant 0$, and let $\boldsymbol{m} = (m_1, m_2, m_3) : \mathbb{R}^3 \to \mathbb{R}^3$ be either the identity map on $\mathbb{R}^3$ or zero. Then the differential second quantization $\mathrm{d}\Gamma(\omega)$ is interpreted as the radiation field energy and $\mathrm{d}\Gamma(\boldsymbol{m}) := (\mathrm{d}\Gamma(m_1), \mathrm{d}\Gamma(m_2), \mathrm{d}\Gamma(m_3))$ is the field momentum operator, in case $\boldsymbol{m}(\boldsymbol{k}) = \boldsymbol{k}$. We introduce the following dense subspaces of $\mathfrak{h}$,

$$\mathfrak{d} := L^2\big(\mathbb{R}^3 \times \mathbb{Z}_2, [1 + (\omega + \tfrac{1}{2}\boldsymbol{m}^2)^2]\mathrm{d}k\big), \quad \mathfrak{k} := L^2\big(\mathbb{R}^3 \times \mathbb{Z}_2, [\tfrac{1}{\omega} + (\omega + \tfrac{1}{2}\boldsymbol{m}^2)^2]\mathrm{d}k\big).$$

If we interpret a Schrödinger operator acting in $L^2(\mathbb{R}^3)$ and $\mathrm{d}\Gamma(\omega)$ as operators in $\mathscr{H}$ in the canonical way, then the Hamiltonian of NRQED is a perturbation of their sum. To explain this more precisely, we introduce the interaction terms

$$\varphi(\boldsymbol{G}) := \int_{\mathbb{R}^3}^{\oplus} \mathbb{1}_{\mathbb{C}^2} \otimes \varphi(\boldsymbol{G}_{\boldsymbol{x}})\mathrm{d}\boldsymbol{x}, \quad \boldsymbol{\sigma}\cdot\varphi(\boldsymbol{F}) := \sum_{j=1}^{3} \int_{\mathbb{R}^3}^{\oplus} \sigma_j \otimes \varphi(F_{j,\boldsymbol{x}})\mathrm{d}\boldsymbol{x},$$

where σ_1, σ_2, and σ_3 are the 2×2 Pauli spin-matrices and the coupling functions $\boldsymbol{G}_{\boldsymbol{x}} = (G_{1,\boldsymbol{x}}, G_{2,\boldsymbol{x}}, G_{3,\boldsymbol{x}})$ and $\boldsymbol{F}_{\boldsymbol{x}} = (F_{1,\boldsymbol{x}}, F_{2,\boldsymbol{x}}, F_{3,\boldsymbol{x}})$ satisfy the following:

Assumption 2.1. *The map $(\boldsymbol{x}, \boldsymbol{k}, \lambda) \mapsto (\boldsymbol{G}_{\boldsymbol{x}}, \boldsymbol{F}_{\boldsymbol{x}})(\boldsymbol{k}, \lambda)$ is measurable, $\boldsymbol{x} \mapsto \boldsymbol{G}_{\boldsymbol{x}}$ belongs to $C^2(\mathbb{R}^3, \mathfrak{h}^3)$, and $\boldsymbol{x} \mapsto \boldsymbol{F}_{\boldsymbol{x}} \in \mathfrak{h}^3$ is globally Lipschitz continuous on $\mathbb{R}^3$. All components of $\boldsymbol{G}_{\boldsymbol{x}}$, $\boldsymbol{F}_{\boldsymbol{x}}$, and $\partial_{x_\ell}\boldsymbol{G}_{\boldsymbol{x}}$ belong to $\mathfrak{k}$ and the map*

$$\mathbb{R}^3 \ni \boldsymbol{x} \longmapsto (\boldsymbol{G}_{\boldsymbol{x}}, \partial_{x_1}\boldsymbol{G}_{\boldsymbol{x}}, \partial_{x_2}\boldsymbol{G}_{\boldsymbol{x}}, \partial_{x_3}\boldsymbol{G}_{\boldsymbol{x}}, \boldsymbol{F}_{\boldsymbol{x}}) \in \mathfrak{k}^{15}$$

is bounded and continuous. Moreover, there is a conjugation $C\colon \mathfrak{h} \to \mathfrak{h}$, i.e., an anti-linear isometry with $C^2 = \mathbb{1}_{\mathfrak{h}}$, such that, for all $t \geqslant 0$, $\boldsymbol{x} \in \mathbb{R}^3$, and $\ell = 1, 2, 3$,

$$[C, e^{-t\omega + i\boldsymbol{m}\cdot\boldsymbol{x}}] = 0, \quad G_{\ell,\boldsymbol{x}}, F_{\ell,\boldsymbol{x}} \in \mathfrak{h}_C := \{f \in \mathfrak{h} : Cf = f\}. \tag{1}$$

Example 2.1. In the standard model of NRQED for one electron interacting with the quantized electromagnetic radiation field in Coulomb gauge and with a sharp ultra-violet cut-off one chooses $\omega(\boldsymbol{k}) = |\boldsymbol{k}|$, $\boldsymbol{m} = \boldsymbol{0}$, and $\boldsymbol{G}$ is given by

$$\boldsymbol{G}^{\Lambda}_{\boldsymbol{x}}(\boldsymbol{k}, \lambda) := (\alpha/2)^{1/2}(2\pi)^{-3/2}\omega(\boldsymbol{k})^{-1/2}\, 1_{\{\omega(\boldsymbol{k})\leqslant\Lambda\}}\, e^{-i\boldsymbol{k}\cdot\boldsymbol{x}}\boldsymbol{\varepsilon}(\boldsymbol{k}, \lambda),$$

where $\alpha, \Lambda > 0$. Applying a suitable unitary transformation, we may achieve that $\boldsymbol{\varepsilon}(\boldsymbol{k}, 0) = |\boldsymbol{e} \times \boldsymbol{k}|^{-1}\boldsymbol{e} \times \boldsymbol{k}$ and $\boldsymbol{\varepsilon}(\boldsymbol{k}, 1) = |\boldsymbol{k}|^{-1}\boldsymbol{k} \times \boldsymbol{\varepsilon}(\boldsymbol{k}, 0)$, for a.e. $\boldsymbol{k}$ and some unit vector $\boldsymbol{e}$ in $\mathbb{R}^3$. If the electron spin is neglected, then we choose $\boldsymbol{F}=\boldsymbol{0}$; otherwise we set $\boldsymbol{F}$ equal to $\boldsymbol{F}_{\boldsymbol{x}}^{\Lambda}(\boldsymbol{k},\lambda):=-\frac{i}{2}\boldsymbol{k}\times\boldsymbol{G}_{\boldsymbol{x}}^{\Lambda}(\boldsymbol{k},\lambda)$. The choice $\boldsymbol{m}(\boldsymbol{k}):=\boldsymbol{k}$ appears in the definition of certain fiber Hamiltonians associated with the model. An appropriate conjugation is given by $(Cf)(\boldsymbol{k},\lambda):=-(-1)^{\lambda}\overline{f(-\boldsymbol{k},\lambda)}$.

Let $V\in L^1_{\mathrm{loc}}(\mathbb{R}^3,\mathbb{R})$. Then the *total Hamiltonian* for our model is given by

$$H^V:=\tfrac{1}{2}(-i\nabla_{\boldsymbol{x}}-\varphi(\boldsymbol{G}))^2-\boldsymbol{\sigma}\cdot\varphi(\boldsymbol{F})+\mathrm{d}\Gamma(\omega)+V.$$

A priori it is interpreted as a quadratic form with dense domain $C_0^\infty(\mathbb{R}^3)\otimes\mathbb{C}^2\otimes\mathscr{C}[\mathfrak{d}_C]$, where $\mathfrak{d}_C:=\mathfrak{d}\cap\mathfrak{h}_C$. If this form is bounded from below, then we denote the unique self-adjoint operator representing its closure again by H^V. The SDE's discussed below involve the *generalized fiber Hamiltonians* defined, for fixed $\boldsymbol{\xi},\boldsymbol{x}\in\mathbb{R}^3$, by

$$\widehat{H}^V(\boldsymbol{\xi},\boldsymbol{x}):=\mathbb{1}_{\mathbb{C}^2}\otimes\widehat{H}^V_{\mathrm{scal}}(\boldsymbol{\xi},\boldsymbol{x})-\boldsymbol{\sigma}\cdot\varphi(\boldsymbol{F}_{\boldsymbol{x}}),$$

$$\widehat{H}^V_{\mathrm{scal}}(\boldsymbol{\xi},\boldsymbol{x}):=\tfrac{1}{2}\boldsymbol{v}(\boldsymbol{\xi},\boldsymbol{x})^2-\tfrac{i}{2}\varphi(\mathrm{div}\boldsymbol{G}_{\boldsymbol{x}})+\mathrm{d}\Gamma(\omega)+V(\boldsymbol{x}),$$

where $\boldsymbol{v}(\boldsymbol{\xi},\boldsymbol{x}):=\boldsymbol{\xi}-\mathrm{d}\Gamma(\boldsymbol{m})-\varphi(\boldsymbol{G}_{\boldsymbol{x}})$. They act in $\widehat{\mathscr{H}}$ and $\mathscr{F}$, respectively. Essentially well-known, simple estimations reveal that they are closed on the domain $\widehat{\mathcal{D}}$, where

$$\widehat{\mathcal{D}}:=\mathcal{D}(\mathrm{d}\Gamma(\boldsymbol{m})^2)\cap\mathcal{D}(\mathrm{d}\Gamma(\omega)),\quad \|\psi\|^2_{\widehat{\mathcal{D}}}:=\|\psi\|^2+\big\|\big(\tfrac{1}{2}\mathrm{d}\Gamma(\boldsymbol{m})^2+\mathrm{d}\Gamma(\omega)\big)\psi\big\|^2;\quad (2)$$

cf. App. B of [3] and the references given there. If $\mathrm{div}\boldsymbol{G}_{\boldsymbol{x}}=0$, then they are self-adjoint on $\widehat{\mathcal{D}}$. Moreover, $\mathbb{C}^2\otimes\mathscr{C}[\mathfrak{d}_C]$ (resp. $\mathscr{C}[\mathfrak{d}_C]$) is an operator core.

Example 2.2. In the situation of Ex. 2.1 the SDE's involving $\widehat{H}^0(\boldsymbol{\xi},\boldsymbol{0})$ are of direct physical interest, as H^0 is unitarily equivalent to the direct integral $\int_{\mathbb{R}^3}^{\oplus}\widehat{H}^0(\boldsymbol{\xi},\boldsymbol{0})\mathrm{d}\boldsymbol{\xi}$.

3. Annihilation-preservation-creation processes and SDE's

In what follows, the interval I is either $[0,\mathcal{T}]$ with $0<\mathcal{T}<\infty$ or $[0,\infty)$ and we fix a stochastic basis $\mathbb{B}:=(\Omega,\mathfrak{F},(\mathfrak{F}_t)_{t\in I},\mathbb{P})$ satisfying the usual assumptions [9]. If $\mathscr{K}$ is a Hilbert space, then $\mathsf{S}_I(\mathscr{K})$ denotes the space of *continuous* $\mathscr{K}$-valued semi-martingales w.r.t. $\mathbb{B}$. The bold letter $\boldsymbol{B}$ denotes a $\mathbb{B}$-Brownian motion in $\mathbb{R}^3$ (with covariance matrix $\mathbb{1}$) and $\boldsymbol{X}\in\mathsf{S}_I(\mathbb{R}^3)$ is a solution of the Itô equation

$$\boldsymbol{X}_\bullet=\boldsymbol{q}+\boldsymbol{B}_\bullet+\int_0^\bullet\boldsymbol{\beta}(s,\boldsymbol{X}_s)\,\mathrm{d}s,\quad (3)$$

for some $\mathfrak{F}_0$-measurable $\boldsymbol{q}:\Omega\to\mathbb{R}^3$, under the following convention:

- If $I=[0,\mathcal{T}]$, then $\boldsymbol{\beta}(t,\boldsymbol{x})=\frac{\boldsymbol{y}-\boldsymbol{x}}{\mathcal{T}-t}$, so that $\boldsymbol{X}$ is a Brownian bridge from $\boldsymbol{q}$ to $\boldsymbol{y}$.
- If $I=[0,\infty)$, then $\boldsymbol{\beta}:I\times\mathbb{R}^3\to\mathbb{R}^3$ is continuous and $\boldsymbol{\beta}(t,\cdot)$ is globally Lipschitz continuous with Lipschitz constant L_t, where $[0,\infty)\ni t\mapsto L_t$ is non-decreasing.

Moreover, we always assume that $\mathbb{P}\{V(\boldsymbol{X}_\bullet)\in L^1_{\mathrm{loc}}(I)\}=1$.

Definition 3.1. We define a process of isometries $\iota_t \colon \mathfrak{h} \to L^2(\mathbb{R}^4 \times \mathbb{Z}_2)$, $t \in \mathbb{R}$,

$$\iota_t f(k_0, \boldsymbol{k}, \lambda) := \pi^{-1/2} e^{-itk_0 - i\boldsymbol{m}\cdot(\boldsymbol{X}_t - \boldsymbol{X}_0)} \omega(\boldsymbol{k})^{1/2} (\omega(\boldsymbol{k})^2 + k_0^2)^{-1/2} f(\boldsymbol{k}, \lambda),$$

a.e. $(k_0, \boldsymbol{k}, \lambda) \in \mathbb{R}^4 \times \mathbb{Z}_2$. As usual ι^τ is ι stopped at τ, and we set $w_{\tau,t} := \overline{w}^*_{\tau,t}$ with

$$\overline{w}_{\tau,t} := (\iota_t^\tau)^* \iota_t = e^{-(t-\tau)\omega - i\boldsymbol{m}\cdot(\boldsymbol{X}_t - \boldsymbol{X}_\tau)} 1_{t>\tau} + 1_{t\leqslant\tau} : \Omega \to \mathscr{B}(\mathfrak{h}_C), \quad t, \tau \in I.$$

In the case $\boldsymbol{m} = \boldsymbol{0}$, the isometry ι_t has been introduced by Nelson [11].

Definition 3.2 (Basic processes). *Let $\tau \in I$. Then we $\mathbb{P}$-a.s. define, for all $t \in I$,*

$$K_{\tau,t} := \int_0^t 1_{s>\tau}\, \iota_s \boldsymbol{G}_{\boldsymbol{X}_s}\, \mathrm{d}\boldsymbol{X}_s + \frac{1}{2}\int_0^t 1_{s>\tau}\, \iota_s(\mathrm{div}\boldsymbol{G}_{\boldsymbol{X}_s} - i\boldsymbol{m}\cdot\boldsymbol{G}_{\boldsymbol{X}_s})\,\mathrm{d}s,$$

as well as $K_t := K_{0,t}$, $U^-_{\tau,t} := (\iota_t^\tau)^* K_{\tau,t}$, $U^-_t := U^-_{0,t}$, $U^+_t := \iota_t^* K_t$, and

$$u^V_{\boldsymbol{\xi},t} := \frac{1}{2}\|K_t\|^2 + \int_0^t V(\boldsymbol{X}_s)\,\mathrm{d}s - i\boldsymbol{\xi}\cdot(\boldsymbol{X}_t - \boldsymbol{X}_0), \quad \boldsymbol{\xi} \in \mathbb{R}^3.$$

The process K defined above appears in [4]. In the next definition $\exp\{\cdots\}$ abbreviates exponential series of operators which are strongly convergent on $\mathscr{C}[\mathfrak{h}]$.

Definition 3.3 (Annihilation-preservation-creation process). *We set*

$$\begin{aligned} W^V_{\boldsymbol{\xi},t}\,\zeta(h) &:= e^{-u^V_{-\boldsymbol{\xi},t}} \exp\{ia^\dagger(U^+_t)\}\,\Gamma(w_{0,t})\,\exp\{ia(U^-_t)\}\zeta(h) \\ &= e^{-u^V_{-\boldsymbol{\xi},t} - \langle U^-_t | h\rangle}\zeta(w_{0,t}\,h + U^+_t), \quad t \in I,\ h \in \mathfrak{d}_C, \end{aligned} \tag{4}$$

which, by linear extension, uniquely defines linear operators $W^V_{\boldsymbol{\xi},t}$ on $\mathscr{C}[\mathfrak{d}_C]$.

Proposition 3.1. *Let $h \in \mathfrak{d}_C$. Then $W^V_{\boldsymbol{\xi}}\zeta(h) \in \mathsf{S}_I(\mathscr{F})$ and, $\mathbb{P}$-a.s.,*

$$W^V_{\boldsymbol{\xi},\bullet}\zeta(h) = \zeta(h) - i\int_0^\bullet \boldsymbol{v}(\boldsymbol{\xi},\boldsymbol{X}_s) W^V_{\boldsymbol{\xi},s}\zeta(h)\mathrm{d}\boldsymbol{X}_s - \int_0^\bullet \widehat{H}^V_{\mathrm{scal}}(\boldsymbol{\xi},\boldsymbol{X}_s) W^V_{\boldsymbol{\xi},s}\zeta(h)\mathrm{d}s$$

on $[0, \sup I)$. Moreover, $\mathbb{P}$-a.s., all operators $W^V_{\boldsymbol{\xi},t}$, $t \in I$, extend uniquely to elements of $\mathscr{B}(\mathscr{F})$ – again denoted by the same symbols – such that $\|W^V_{\boldsymbol{\xi},t}\| \leqslant e^{-\int_0^t V(\boldsymbol{X}_s)\mathrm{d}s}$.

Proof. The expression in the second line of (4) is essentially an exponential function of basic processes. Moreover, it is possible to show that $U^\pm \in \mathsf{S}_I(\mathfrak{h}_C)$, $u^V_{\boldsymbol{0}} \in \mathsf{S}_I(\mathbb{R})$, and to represent these processes as stochastic integrals w.r.t. the process $(t, \boldsymbol{X}_t)_{t\in I}$. The Itô formula and the rules of stochastic calculus for *real* Hilbert spaces proved in [10] apply in this situation and eventually lead to the asserted SDE. In the computations we exploit that $\langle U^+|\boldsymbol{G}_{\boldsymbol{X}}\rangle$, $\langle U^+|\mathrm{div}\boldsymbol{G}_{\boldsymbol{X}}\rangle$, and $\langle U^+|i\boldsymbol{m}\cdot\boldsymbol{G}_{\boldsymbol{X}}\rangle$ are real as a consequence of (1). □

Remark 3.1. The adjoint of $W^0_{\mathbf{0},t}$ (with $\boldsymbol{m} = \mathbf{0}$) appears in Hiroshima's FK formula [4] for the scalar semi-group where it is represented as $W^{0\,*}_{\mathbf{0},t} = \Gamma(\iota_0^*)e^{-i\varphi(K_t)}\Gamma(\iota_t)$. As noted in [4], the isometries ι_t are not strongly differentiable w.r.t. t, which seems to prevent an application of Itô's formula at first sight. However, the application of Itô's formula to the more regular representation of $W^0_{\mathbf{0},t}$ as an annihilation-preservation-creation process is fairly straightforward.

Next, let $t\Delta_n := \{(s_1,\dots,s_n) \in \mathbb{R}^n : 0 \leqslant s_1 \leqslant \dots \leqslant s_n \leqslant t\}$. If $t_1,\dots,t_n \in \mathbb{R}$ and $\mathcal{A} \subset [n] := \{1,\dots,n\}$, then we set $t_{\mathcal{A}} := (t_{a_1},\dots,t_{a_m})$ where $\mathcal{A} = \{a_1,\dots,a_m\}$, $a_1 < \dots < a_m$, and analogously for a multi-index $\alpha \in [3]^n = \{1,2,3\}^n$.

Definition 3.4 (Time-ordered integral processes). *Let $t, t_1,\dots,t_n \in I$, $\alpha \in [3]^n$, and $\mathcal{A},\mathcal{B} \subset [n]$. We put $\mathscr{L}^{\alpha_\varnothing}_t(t_\varnothing) := \mathscr{R}_{\alpha_\varnothing}(t_\varnothing) := \mathbb{1}$ and, if $\mathcal{A} \neq \varnothing \neq \mathcal{B}$,*

$$\mathscr{L}^{\alpha_{\mathcal{A}}}_t(t_{\mathcal{A}}) := \prod_{a\in\mathcal{A}} \{a^\dagger(w_{t_a,t}F_{\alpha_a,\boldsymbol{X}_{t_a}}) + i\langle U^-_{t_a,t}|F_{\alpha_a,\boldsymbol{X}_{t_a}}\rangle\},$$

$$\mathscr{R}_{\alpha_{\mathcal{B}}}(t_{\mathcal{B}}) := \prod_{b\in\mathcal{B}} \{a(\overline{w}_{0,t_b}F_{\alpha_b,\boldsymbol{X}_{t_b}}) + i\langle F_{\alpha_b,\boldsymbol{X}_{t_b}}|U^+_{t_b}\rangle\},$$

on $\mathscr{C}[\mathfrak{d}_C]$. If $\mathcal{C} \subset [n]$ with $\#\mathcal{C} \in 2\mathbb{N}_0$, then we further set $\mathscr{I}_{\alpha_\varnothing}(t_\varnothing) := 1$ and

$$\mathscr{I}_{\alpha_{\mathcal{C}}}(t_{\mathcal{C}}) := \sum_{\substack{\mathcal{C}=\cup\{c_p,c'_p\}\\ c_p<c'_p}} \prod_{p=1}^{\#\mathcal{C}/2} \langle F_{\alpha_{c'_p},\boldsymbol{X}_{t_{c'_p}}}|w_{t_{c_p},t_{c'_p}}F_{\alpha_{c_p},\boldsymbol{X}_{t_{c_p}}}\rangle, \quad \text{if } \mathcal{C} \neq \varnothing.$$

Here the sum runs over all possibilities to split $\mathcal{C}$ into disjoint subsets $\{c_p, c'_p\} \subset \mathcal{C}$ with $c_p < c'_p$, $p = 1,\dots,\#\mathcal{C}/2$. Writing $dt_{[n]} := \mathrm{d}t_1\dots\mathrm{d}t_n$, we finally define

$$\mathbb{W}^{V,(n)}_{\boldsymbol{\xi},t}\psi := \sum_{\alpha\in[3]^n} \sigma_{\alpha_n}\dots\sigma_{\alpha_1} \otimes \sum_{\substack{\mathcal{A}\cup\mathcal{B}\cup\mathcal{C}=[n]\\ \#\mathcal{C}\in 2\mathbb{N}_0}} \int_{t\Delta_n} \mathscr{I}_{\alpha_{\mathcal{C}}}(t_{\mathcal{C}})\,\mathscr{L}^{\alpha_{\mathcal{A}}}_t(t_{\mathcal{A}})W^V_{\boldsymbol{\xi},t}\mathscr{R}_{\alpha_{\mathcal{B}}}(t_{\mathcal{B}})\,\psi\,\mathrm{d}t_{[n]},$$

for $\psi \in \mathbb{C}^2 \otimes \mathscr{C}[\mathfrak{d}_C]$, $n \in \mathbb{N}$, and $\mathbb{W}^{V,(0)}_{\boldsymbol{\xi},t} := \mathbb{1}_{\mathbb{C}^2} \otimes W^V_{\boldsymbol{\xi},t}$. Here the second sum runs over all disjoint *partitions of $[n]$, where $\mathcal{A}$, $\mathcal{B}$, or $\mathcal{C}$ may be empty and $\#\mathcal{C}$ is even.*

We can now state our fundamental existence and uniqueness result for the SDE associated with the generalized fiber Hamiltonian; recall that $\widehat{\mathcal{D}}$ is normed by (2).

Theorem 3.1. (a) *$\mathbb{P}$-a.s., all operators $\mathbb{W}^{V,(n)}_{\boldsymbol{\xi},t}$ with $n \in \mathbb{N}_0$ and $t \in I$ extend uniquely to elements of $\mathscr{B}(\hat{\mathscr{H}})$, which are henceforth again denoted by the same symbols. Furthermore, $\mathbb{P}$-a.s., the limit $\mathbb{W}^V_{\boldsymbol{\xi},t} := \lim_{N\to\infty}\sum_{n=0}^N \mathbb{W}^{V,(n)}_{\boldsymbol{\xi},t}$ exists in $\mathscr{B}(\hat{\mathscr{H}})$ locally uniformly in $t \in I$, and $\|\mathbb{W}^V_{\boldsymbol{\xi},t}\| \leqslant e^{\mathfrak{c}t-\int_0^t V(\boldsymbol{X}_s)\mathrm{d}s}$ with $\mathfrak{c} \propto \sup_{\boldsymbol{x}}\|\omega^{-1/2}\boldsymbol{F}_{\boldsymbol{x}}\|^2$.*

(b) *Let $\eta : \Omega \to \widehat{\mathcal{D}}$ be $\mathfrak{F}_0$-measurable. Then $\mathbb{W}^V_{\boldsymbol{\xi}}\,\eta \in \mathsf{S}_I(\hat{\mathscr{H}})$ and, up to indistinguishability, $\mathbb{W}^V_{\boldsymbol{\xi}}\,\eta$ is the unique element of $\mathsf{S}_I(\hat{\mathscr{H}})$ whose paths belong $\mathbb{P}$-a.s. to*

$C(I, \widehat{\mathcal{D}})$ *and which* $\mathbb{P}$*-a.s. solves*

$$\mathscr{X}_\bullet = \eta - i \int_0^\bullet \boldsymbol{v}(\boldsymbol{\xi}, \boldsymbol{X}_s)\, \mathscr{X}_s \, \mathrm{d}\boldsymbol{X}_s - \int_0^\bullet \widehat{H}^V(\boldsymbol{\xi}, \boldsymbol{X}_s)\, \mathscr{X}_s \, \mathrm{d}s \quad on\ [0, \sup I). \tag{5}$$

If V is bounded and continuous, then we infer the following from Thm. 3.1:

- The existence of a stochastic flow for the system comprised of (3) and (5).
- The existence of a family of transition operators associated with the flow, which enjoys the Feller and Markov properties.
- A Blagoveščensky-Freidlin type theorem, i.e., the existence of strong solutions.

4. Feynman-Kac representations

The construction of $\mathbb{W}_{\boldsymbol{\xi}}^V$ depends on the choice of $\boldsymbol{X}$. Since we are dealing with different choices of $\boldsymbol{X}$ at the same time in this section, we indicate this by adding the variable $[\boldsymbol{X}]$ to the symbol $\mathbb{W}_{\boldsymbol{\xi}}^V$.

Theorem 4.1 (FK formula: fiber case). *Assume that* $\boldsymbol{G}$ *and* $\boldsymbol{F}$ *do not depend on* $\boldsymbol{x}$*, and let* $\boldsymbol{\xi} \in \mathbb{R}^3$ *and* $t \geqslant 0$*. Then* $e^{-t\widehat{H}^0(\boldsymbol{\xi},\boldsymbol{0})} = \mathbb{E}[\mathbb{W}_{\boldsymbol{\xi},t}^0[\boldsymbol{B}]^*]$.

Remark 4.1. In the earlier literature FK formulas for fiber Hamiltonians have been deduced from the FK formula for the total Hamiltonian by inserting suitable peak functions [5, 6, 8]. By a standard procedure, based on Thm. 3.1, we are able to avoid this detour and treat the fiber case directly. For non-zero $\boldsymbol{F}$, Thm. 4.1 is new as the earlier formulas involved an additional regularization procedure [6, 8]; compare Rem. 4.2 (a) below.

For $\boldsymbol{x}, \boldsymbol{y} \in \mathbb{R}^3$, we set $\boldsymbol{B}^{\boldsymbol{x}} := \boldsymbol{x} + \boldsymbol{B}$ and let $\boldsymbol{b}^{t;\boldsymbol{y},\boldsymbol{x}} \in \mathsf{S}_{[0,t]}(\mathbb{R}^3)$ denote the Brownian bridge from $\boldsymbol{y}$ to $\boldsymbol{x}$ in time $t > 0$, defined as the solution of $\boldsymbol{b}_\bullet = \boldsymbol{B}_\bullet^{\boldsymbol{y}} + \int_0^\bullet \frac{\boldsymbol{x} - \boldsymbol{b}_s}{t-s} \mathrm{d}s$.

Theorem 4.2 (FK formula). *Assume that* $V = V_+ - V_-$*, where* $V_\pm : \mathbb{R}^3 \to [0, \infty)$ *such that* $V_+ \in L^1_{\mathrm{loc}}(\mathbb{R}^3)$ *and* V_- *is* $-\frac{1}{2}\Delta$*-form bounded with form bound* $b \leqslant 1$*. Then* $\Psi \in \mathcal{Q}(H^{V_+}) \subset \mathscr{H} = L^2(\mathbb{R}^3, \hat{\mathscr{H}})$ *implies* $\|\Psi(\cdot)\| \equiv \|\Psi(\cdot)\|_{\hat{\mathscr{H}}} \in \mathcal{Q}(-\frac{1}{2}\Delta + V_+)$ *with*

$$\big\langle \|\Psi(\cdot)\| \big| (-\tfrac{1}{2}\Delta + V_+) \|\Psi(\cdot)\| \big\rangle \leqslant \big\langle \Psi \big| \big(H^{V_+} + \mathfrak{c} \sup_{\boldsymbol{x}} \|\omega^{-1/2} \boldsymbol{F}_{\boldsymbol{x}}\|^2\big) \Psi \big\rangle. \tag{6}$$

In particular, V_- *is* H^{V_+}*-form bounded with form bound* $\leqslant b$*, so that* H^V *has a distinguished self-adjoint realization defined via quadratic forms (which is again denoted by* H^V*). For all* $t > 0$*,* $\Psi \in \mathscr{H}$*, and a.e.* $\boldsymbol{x} \in \mathbb{R}^3$*, we then have*

$$(e^{-tH^V}\Psi)(\boldsymbol{x}) = \mathbb{E}\big[\mathbb{W}_{0,t}^V[\boldsymbol{B}^{\boldsymbol{x}}]^* \Psi(\boldsymbol{B}_t^{\boldsymbol{x}})\big] = \int_{\mathbb{R}^3} \frac{e^{-|\boldsymbol{x}-\boldsymbol{y}|^2/2t}}{(2\pi t)^{3/2}} \mathbb{E}\big[\mathbb{W}_{0,t}^V[\boldsymbol{b}^{t;\boldsymbol{y},\boldsymbol{x}}]\big] \Psi(\boldsymbol{y}) \mathrm{d}\boldsymbol{y}. \tag{7}$$

Proof. Let $V \in C(\mathbb{R}^3)$ be bounded. Employing (inter alia) the Markov property of the stochastic flow we check that $(T_t\Psi)(\boldsymbol{x}) := \mathbb{E}[\mathbb{W}^V_{\mathbf{0},t}[\boldsymbol{B}^{\boldsymbol{x}}]^*\Psi(\boldsymbol{B}^{\boldsymbol{x}}_t)]$ defines a strongly continuous semi-group of self-adjoint operators $T_t \in \mathscr{B}(\mathscr{H})$. Then we use (5) to verify that the generator of this semi-group agrees with H^V on a suitable core of H^V. This yields the first identity in (7) and the second one follows from Thm. 3.1 and well-known relations for Brownian motions and Brownian bridges. Invoking standard extension procedures we then get (6) and (7) for general V. □

Remark 4.2. **(a)** In the scalar case ($\boldsymbol{F} = \mathbf{0}$), and under slightly more restrictive conditions on $\boldsymbol{G}$, the first equality in (7) has been proved by F. Hiroshima in [4] while the second one is new. For a spin-1/2 electron, the sesqui-linear form of the semi-group has been represented earlier only as a *limit* of expectation values of certain *regularized* FK integrands involving a Poisson jump process to account for the spin degrees of freedom [6]. Hence, both identities in (7) are new for non-zero $\boldsymbol{F}$.

(b) The assumptions of Thm. 4.2 imply $\mathbb{P}\{V(\boldsymbol{B}^{\boldsymbol{x}}_{\bullet}) \in L^1_{\mathrm{loc}}([0,\infty))\} = 1$, for a.e. $\boldsymbol{x}$.

(c) In Thm. 4.2, the integral $\mathbb{E}\big[\mathbb{W}^V_{\mathbf{0},t}[\boldsymbol{b}^{t;\boldsymbol{y},\boldsymbol{x}}]\big] \in \mathscr{B}(\hat{\mathscr{H}})$ is defined only for a.e. $(\boldsymbol{x},\boldsymbol{y}) \in \mathbb{R}^6$. The condition on V of Thm. 4.3 implies its existence for *all* $\boldsymbol{x},\boldsymbol{y} \in \mathbb{R}^3$.

(d) The condition on V_- in Thm. 4.2 ensures the validity of the FK formula for the Schrödinger operator with potential V_- [12], which is what we really use in the proof.

Let us announce the next theorem as a first application. Its proof combines arguments from [1] with certain weighted estimates on $\mathbb{W}^0_{\mathbf{0},t}$ inferred from Thm. 3.1. Here the weights are given by unbounded functions of $\mathrm{d}\Gamma(\omega)$.

Theorem 4.3. *Assume that $V = V_+ - V_-$, where $V_- \geqslant 0$ belongs to the Kato class and $V_+ \geqslant 0$ is in the local Kato class. Then the following map is continuous w.r.t. the norm topology on $\mathscr{B}(\hat{\mathscr{H}})$,*

$$(0,\infty) \times \mathbb{R}^3 \times \mathbb{R}^3 \ni (t,\boldsymbol{x},\boldsymbol{y}) \longmapsto \mathbb{E}\big[\mathbb{W}^V_{\mathbf{0},t}[\boldsymbol{b}^{t;\boldsymbol{y},\boldsymbol{x}}]\big] \in \mathscr{B}(\hat{\mathscr{H}}).$$

Remark 4.3. **(a)** All results presented above hold true for several electrons as well.

(b) Suitable analogs of all results stated here are available for Nelson's model.

Acknowledgements

BG has been financially supported by the SFB 647: Raum-Zeit-Materie. OM has been partially supported by the Lundbeck Foundation, the Villum Foundation, and by the European Research Council under the European Community's Seventh Framework Program (FP7/2007–2013)/ERC grant agreement 202859.

Bibliography

[1] K. Broderix, D. Hundertmark, and H. Leschke, Continuity properties of Schrödinger semigroups with magnetic fields, *Rev. Math. Phys.* **12**, 181–225 (2000).

[2] R. P. Feynman, Mathematical formulation of the quantum theory of electromagnetic interaction, *Phys. Rev. (Ser. II)* **80**, 440–457 (1950).

[3] B. Güneysu, O. Matte, and J. S. Møller, Stochastic differential equations for models of non-relativistic matter interacting with quantized radiation fields, *Preprint*, http://arXiv.org/abs/1402.2242, 80 pp. (2014).

[4] F. Hiroshima, Functional integral representation of a model in quantum electrodynamics, *Rev. Math. Phys.* **9**, 489–530 (1997).

[5] F. Hiroshima, Fiber Hamiltonians in non-relativistic quantum electrodynamics, *J. Funct. Anal.* **252**, 314–355 (2007).

[6] F. Hiroshima and J. Lőrinczi, Functional integral representations of the Pauli-Fierz model with spin 1/2, *J. Funct. Anal.* **254**, 2127–2185 (2008).

[7] R. L. Hudson and K. R. Parthasarathy, Quantum Ito's formula and stochastic evolutions, *Commun. Math. Phys.* **93**, 301–323 (1984).

[8] J. Lőrinczi, F. Hiroshima, and V. Betz, *Feynman-Kac-type theorems and Gibbs measures on path space*, Studies in Mathematics, vol. 34 (Walter de Gruyter & Co., Berlin-Boston, 2011).

[9] M. Métivier, *Semimartingales. A course on stochastic processes*, de Gruyter Studies in Mathematics, vol. 2 (Walter de Gruyter & Co., Berlin-New York, 1982).

[10] M. Métivier and J. Pellaumail, *Stochastic integration* (Academic Press, New York, 1980).

[11] E. Nelson, The free Markoff field, *J. Funct. Anal.* **12**, 211–227 (1973).

[12] J. Voigt, Absorption semigroups, their generators, and Schrödinger semigroups, *J. Funct. Anal.* **67**, 167–205 (1986).

Estimates for numbers of negative eigenvalues of Laplacian for Y-type chain of weakly coupled ball resonators

A. S. Melikhova

Department of Higher Mathematics, ITMO University, Saint Petersburg, 197101, Russia
E-mail: alina.s.melikhova@gmail.com

This article is devoted to the spectral problem for a branched chain of weakly coupled ball resonators. The solution for such a problem in the framework of theory of extensions is obtained. The spectral equations are derived for two types of the resonators coupling (δ-coupling and δ'-coupling).

Keywords: Spectrum, theory of extensions, self-adjoint operator, transfer-matrix.

1. Introduction

Over the last few years a number of systems that have not previously existed has been obtained, for example: quantum dots, graphene, fullerene, *etc.* One of these newly created nanostructures is a chain of coupled ball quantum resonators. For example, there are a lot of articles [1–3] devoted to describing the process of creation and properties of nano peapod, that had been obtained experimentally. In the present paper we suggest solvable model describing the spectral properties of an electron in such nanosystem with Y-type branched geometry. This work incorporates ideas of two models: the model of chain-type quantum graph [4, 5] and the model of point-like aperture [6–8].

As for the quantum graph, our ideas are close to that of [5]. In this paper the authors investigated the spectrum of the Hamiltonian for the quantum graph G of branched geometry (it is shown at Fig. 1(a)). They examine the problem with free Schrödinger operator on the graph edges and δ-coupling at the vertices.

In present paper we suggest to consider a branched chain of weakly coupled ball resonators. In the model the resonators are coupled through point-like windows: with δ-coupling condition in contact points (see Sec. 3) and with δ'-coupling condition (see Sec. 4). The model Hamiltonian is constructed in the framework of the operator extensions theory. This approach is effective for the description of the

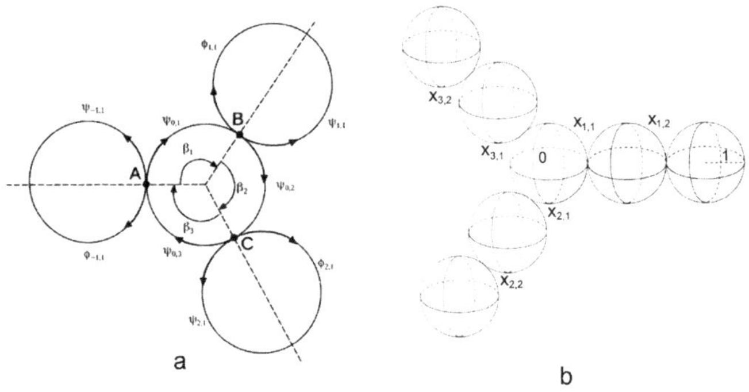

Figure 1. Systems with branched geometry: (a) quantum graph, (b) chains of resonators.

Laplacian perturbed on the set of zero measure [9], [10] and for description of electron transport in the systems of quantum waveguides and resonators [7, 8]. The model of such type allows us to use one-dimensional transfer-matrix technique for the infinite chain of the resonators due to their point-like coupling. Due to this, the spectral equation can be obtained in an explicit form. The main goal of the paper is to examine the spectrum of the model Hamiltonian.

2. Model discussion

First of all, let us describe the geometry of the Y-type system. It is formed by identical balls of unit radii. Also, we consider that there are a ball with number 0 (a "central ball") and three semi-infinite chains of balls attached to the central ball (see Fig. 1(b)). Each ball has two indices: the number of the branch and the number of the ball in this branch. Let us assume for simplicity that this system has symmetry with respect to the rotation by the angles $2\pi/3$, $4\pi/3$ and that it can be laid on the plane.

The quantum spin-less non-relativistic particle is considered living in our branched chain. It is supposed that it's wavefunction $\psi(\mathbf{x})$ satisfies the Schrödinger equation: $\hat{H}\psi(\mathbf{x}) = \lambda\psi(\mathbf{x})$, where the Hamiltonian $\hat{H} = -\Delta$ is the orthogonal sum of the Neumann Laplacians for all balls (it is assumed that the system of units is such that $\hbar = 2m = e = 1$). It is also assumed that neighbour balls are connected by a pinhole. To explain in more details how the extensions theory allows one to switch on the interaction between balls, let us consider the case of two balls, Ω_0, Ω_1, coupled at the point $\mathbf{x}_1$. In this case $-\Delta = -\Delta_0 \oplus (-\Delta_1)$ in $L_2(\Omega_0) \oplus L_2(\Omega_1)$. Let one restrict the operator $-\Delta$ on the set of all smooth functions from $D(-\Delta)$ vanishing at the point $\mathbf{x}_1$. Its closure $-\Delta_0$ is the symmetric operator with the deficiency indices $(2, 2)$. The deficiency elements corresponding to a regular value λ_0 of the spectral parameter are as follows: $(G_0(\mathbf{x}, \mathbf{x}_1, \lambda_0),\ 0)^T$, $(0,\ G_1(\mathbf{x}, \mathbf{x}_1, \lambda_0))^T$, where G_0, G_1 are the Green's functions for balls Ω_0, Ω_1, correspondingly. Self-adjoint extensions of $-\Delta_0$ are restrictions of the adjoint operator $-\Delta_0^*$. To describe self-adjoint extensions we will use the approach based on the von Neumann formulas: $D(-\Delta_0^*) = D(-\Delta_0^F) \dotplus N_{\lambda_0}$, where $-\Delta_0^F$ is the Friedrichs extension of $-\Delta_0$, N_{λ_0} is the deficiency subspace corresponding to real regular value λ_0 ($\lambda_0 < 0$). One can see that in accordance with this representation, a function U, $U \in D(-\Delta_0^*)$, has the following form: $U = \binom{U_0}{U_1} = \binom{U_0^0 + a_0^- G_0(\mathbf{x},\mathbf{x}_1,\lambda_0)+b_0^-}{U_1^0+a_1^+ G_1(\mathbf{x},\mathbf{x}_1,\lambda_0)+b_1^+}$, where $\binom{U_0^0}{U_1^0} \in D(-\Delta_0)$, $U_0^0(\mathbf{x}_1) = U_1^0(\mathbf{x}_1) = 0$; a_0^-, a_1^+, b_0^-, b_1^+ are some coefficients. To construct a self-adjoint extension $-\widetilde{\Delta_0}$ of $-\Delta_0$ it is necessary to establish some linear relation between these coefficients.

For infinite chain of balls one can construct the model Hamiltonian in the same way. Firstly, one restricts the operator $-\Delta$ on the set of all smooth functions from $D(-\Delta)$ that are equal to zero at the points $\mathbf{x}_{i,j}$. Its closure $-\Delta_0$ is the symmetric operator with infinite deficiency indices. The deficiency elements corresponding to the point λ for particular balls are the Green's functions $G(\mathbf{x}, \mathbf{x}_{i,j}, \lambda)$ for these balls. Secondly, to obtain the self-adjoint extension $-\widetilde{\Delta_0}$ one has to establish a linear dependence between coefficients $a_{i,j}^{\pm}$, $b_{i,j}^{\pm}$ (that refer to pinholes $\mathbf{x}_{i,j}$ similarly to the case of two connected resonators and coefficients $a_0^{\pm}$, $b_0^{\pm}$ referring to $\mathbf{x}_0$). The operator $-\widetilde{\Delta_0}$ will be self-adjoint if, for example, the following system is satisfied:

$$\begin{cases} a_{i,j}^+ = -a_{i,j-1}^-, \\ b_{i,j}^+ = b_{i,j}^-. \end{cases} \tag{1}$$

In the following two sections we will discuss a spectral problem for described system with two types of coupling: δ-coupling (Sec. 3), δ'-coupling (Sec. 4).

3. δ-coupling

Let us discuss the case of δ-coupling condition with coupling constant α, $\alpha \in \mathbf{R}$ (see [4, 11]). This condition imposes a little bit different to (1) dependence between our coefficients:

$$\begin{cases} a_{i,j}^+ = -a_{i,j-1}^- \\ b_{i,j}^+ - b_{i,j}^- = -\alpha\, a_{i,j-1}^- \end{cases}, \tag{2}$$

where

$$b_{i,j}^+ = a_{i,j}^+ \lim_{\mathbf{x}\to\mathbf{x}_{i,j}} \left(G(\mathbf{x}, \mathbf{x}_{i,j}, \lambda) - G(\mathbf{x}, \mathbf{x}_{i,j}, \lambda_0)\right) + a_{i,j}^- G(\mathbf{x}_{i,j+1}, \mathbf{x}_{i,j}, \lambda), \tag{3}$$

$$b_{i,j-1}^- = a_{i,j-1}^- \lim_{\mathbf{x}\to\mathbf{x}_{i,j}} \left(G(\mathbf{x}, \mathbf{x}_{i,j}, \lambda) - G(\mathbf{x}, \mathbf{x}_{i,j}, \lambda_0)\right) + a_{i,j-1}^+ G(\mathbf{x}_{i,j-1}, \mathbf{x}_{i,j}, \lambda), \tag{4}$$

$$b_{i,0}^- = a_{i,0}^- \lim_{\mathbf{x}\to\mathbf{x}_{i,1}} \left(G(\mathbf{x}, \mathbf{x}_{i,1}, \lambda) - G(\mathbf{x}, \mathbf{x}_{i,1}, \lambda_0)\right) + \sum_{n\neq i} a_{n,0}^- G(\mathbf{x}_{n,1}, \mathbf{x}_{i,1}, \lambda). \tag{5}$$

Let us obtain relation between the coefficients $a_{i,j}^{\pm}$ and $a_{i,j-1}^{\pm}$ refer to balls number i,j and $i,j-1$, correspondingly, by employing formulas (2)–(5). So, one can arrive at the following relation:

$$\begin{aligned}\begin{pmatrix} a_{i,j}^{+} \\ a_{i,j}^{-} \end{pmatrix} &= \begin{pmatrix} 0 & -1 \\ \frac{G(\mathbf{x}_{i,j-1},\mathbf{x}_{i,j},\lambda)}{G(\mathbf{x}_{i,j+1},\mathbf{x}_{i,j},\lambda)} & \frac{-\alpha+\lim\limits_{\mathbf{x}\to\mathbf{x}_{i,j}}(G(\mathbf{x},\mathbf{x}_{i,j},\lambda)-G(\mathbf{x},\mathbf{x}_{i,j},\lambda_0))}{G(\mathbf{x}_{i,j+1},\mathbf{x}_{i,j},\lambda)} \end{pmatrix} \begin{pmatrix} a_{i,j-1}^{+} \\ a_{i,j-1}^{-} \end{pmatrix} \\ &= \mathbf{M}_{i,j} \begin{pmatrix} a_{i,j-1}^{+} \\ a_{i,j-1}^{-} \end{pmatrix},\end{aligned} \tag{6}$$

where $\mathbf{M}_{i,j}$ is a transfer-matrix. Such values of λ that lead to the transfer-matrix having eigenvalues with unit modulus belong to the continuous spectrum. Taking into account (6), one can get the following condition for λ to belong to the continuous spectrum:

$$|X/G| \leq 2, \tag{7}$$

where $X = -\alpha + 2 \lim\limits_{\mathbf{x}\to\mathbf{x}_{i,j}} (G(\mathbf{x}, \mathbf{x}_{i,j}, \lambda) - G(\mathbf{x}, \mathbf{x}_{i,j}, \lambda_0))$, $G = G(\mathbf{x}_{1,1}, \mathbf{x}_{1,2}, \lambda)$.

As for the point spectrum, the corresponding λ should give the eigenvalue μ (with the eigenvector ν) of the transfer-matrix satisfying $|\mu| < 1$ and, moreover, the vector $\left(a_{i,j}^{+}\ a_{i,j}^{-}\right)^T$ (for any i, j) should be collinear to ν. To obtain the model operator eigenvalue we need the decay of the solution at infinity along each branch. Hence, there is a linear dependence between the coefficients $a_{i,1}^{\pm}$ and $a_{i,0}^{-}$. It leads to the following equation:

$$(X - \mu G)^3 - 3(X - \mu G)A^2 + 2A^3 = 0, \tag{8}$$

where $A = G(\mathbf{x}_{2,1}, \mathbf{x}_{1,1}, \lambda)$; and solutions of (8) belong to the discrete spectrum if:

$$|X/G| > 2. \tag{9}$$

It should be also noted that equation (8) is a homogeneous equation. And after more detailed analysis of expressions (7)–(9) we can state the following

Theorem 3.1. *The essential spectrum for Y-branched chain of weakly coupled ball resonators (with δ-coupling) consists of the continuous spectrum having band structure (the bands are given by equation (7)) and eigenvalues of infinite multiplicity which are given by the eigenvalues of the Neumann Laplacian for the ball corresponding to the eigenfunctions vanishing at the both opposite points of the ball ($\mathbf{x}_{i,j}$ and $\mathbf{x}_{i,j-1}$). There exist values of the model parameters α, λ_0 such that the model Hamiltonian has at most one negative eigenvalue λ.*

4. δ'-coupling

In the case of δ'-coupling condition with coupling constant β, cf. [11, 12], the following system for $a_{i,j}^{\pm}$, $b_{i,j}^{\pm}$ is satisfied:

$$\begin{cases} a_{i,j}^{+} + a_{i,j-1}^{-} = \beta b \\ b_{i,j}^{+} = b_{i,j}^{-} = b \end{cases}. \tag{10}$$

Similarly to Sec. 3, by using condition (10) and expressions (3)–(5), one can obtain the transfer-matrix $\mathbf{M}'$ for this case in the following form:

$$\mathbf{M}'_{i,j} = \begin{pmatrix} \beta G\left(\mathbf{x}_{i,j-1}, \mathbf{x}_{i,j}, \lambda\right) & -X' \\ X' \frac{G(\mathbf{x}_{i,j-1}, \mathbf{x}_{i,j}, \lambda)}{G(\mathbf{x}_{i,j}, \mathbf{x}_{i,j+1}, \lambda)} & X' \frac{1+X'}{G(\mathbf{x}_{i,j}, \mathbf{x}_{i,j+1}, \lambda)} \end{pmatrix}, \tag{11}$$

where $X' = 1 - \beta \lim_{\mathbf{x} \to \mathbf{x}_{i,j}} \left(G\left(\mathbf{x}, \mathbf{x}_{i,j}, \lambda\right) - G\left(\mathbf{x}, \mathbf{x}_{i,j}, \lambda_0\right)\right)$. Also, as explicit form (11) of transfer-matrix for this case is known, one can obtain conditions on continuous and discrete spectrum correspondingly as follows:

$$|(\beta^2 G^2 - X^2 + 1)/(\beta G)| \leq 2, \tag{12}$$

$$|(\beta^2 G^2 - X^2 + 1)/(\beta G)| > 2. \tag{13}$$

By using expressions (10), (3)–(5) one can obtain the main (and also homogeneous) equation on the spectrum for the system with δ'-coupling: $D^3 - 3DA^2 + 2A^3 = 0$, where $D = \left(\mu'(1 - (X')^2) - \beta G X'\right) / \left(\beta(\mu' X' + \beta G)\right)$ and μ' is an eigenvalue of matrix (11).

Analysis of this equation and inequalities (12), (13) gives us

Statement 4.1 *The essential spectrum for Y-branched chain of ball resonators with δ'-coupling consists of the continuous spectrum having band structure (the bands are given by* (12)*) and eigenvalues of infinite multiplicity as in the previous case. There exist values of the model parameters β, λ_0 such that the model Hamiltonian has at most four negative eigenvalues λ.*

Acknowledgments

This work was partially financially supported by the Government of the Russian Federation (grant 074-U01), by Ministry of Science and Education of the Russian Federation (GOSZADANIE 2014/190, Project 14.Z50.31.0031), by grants of the President of Russia (state contracts 14.124.13.2045-MK and 14.124.13.1493-MK).

Bibliography

[1] Y. Wang, H. J. Zhang, L. Lu, L. P. Stubbs, C. C.Wong, J. Lin. Designed Functional Systems from Peapod-like Co@Carbon to Co3O4@Carbon Nanocomposites. *ACS Nano* **4** (8) (2010) 4753–4761.

[2] Y. Yang, L. Li, W. Li. Plasmon Absorption of Au-in-CoAl2O4 Linear Nanopeapod Chains. *J. Phys. Chem.* C **117** (27) (2013) 14142-14148.

[3] E. Abou-Hamad, Yo. Kim, A. V. Talyzin, Ch. Goze-Bac, D. E. Luzzi, A. Rubio, T. Wagberg. Hydrogenation of C60 in Peapods: Physical Chemistry in Nano Vessels. *J. Phys. Chem.* C **113** (2009) 8583–8587

[4] P. Duclos, P. Exner, O. Turek. On the spectrum of a bent chain graph. *J. Phys. A: Math. Theor.* **41** (2008), 415206.

[5] I. Yu. Popov, P. I. Smirnov. Spectral problem for branching chain quantum graph. *Phys. Lett. A.* 2013, **377** (6), 439-442.

[6] I. Yu. Popov. The resonator with narrow slit and the model based on the operator extensions theory. *J. Math. Phys.* **33** (11) (1992) 3794-3801.

[7] I. Yu. Popov, S. L. Popova. The extension theory and resonances for a quantum waveguide. *Phys. Lett. A.* **173** (1993) 484-488.

[8] I. Yu. Popov. The extension theory and the opening in semitransparent surface. *J. Math. Phys.* **33** (5) (1992) 1685-1689.

[9] S. Albeverio, F. Gesztesy, R. Hoegh-Krohn, H. Holden. Solvable models in quantum mechanics. (New York: Springer-Verlag, 1988).

[10] B. S. Pavlov. Operator extensions theory and explicitly solvable models. *Russ. Math. Surv.* **42**, (1987) 127-168.

[11] Yu. Golovaty. Schrödinger operators with $(\alpha\,\delta' + \beta\,\delta)$-like potentials: norm resolvent convergence and solvable models. *arXiv:1201.2610v2[math.SP]*.

[12] P. Kurasov, J. Larson. Spectral Asymptotics for Schrödinger Operators with Periodic Point Interactions. *J. Math. Anal. Appl.* **266** (2002), 127–148.

On thermodynamical couplings of quantum mechanics and macroscopic systems

A. Mielke

Weierstraß-Institut für Angewandte Analysis und Stochastik,
Mohrenstraße 39, 10117 Berlin, Germany
E-mail: alexander.mielke@wias-berlin.de

Quantum mechanics can be formulated as a Hamiltonian system in terms of the Liouville equation for the density matrix. Dissipative effects are modeled via coupling to a macroscopic system, where the coupling operators act via commutators. Following Öttinger (2010) we use GENERIC to construct thermodynamically consistent evolution equations $\dot{q} = \mathbb{J}(q)\mathrm{D}\mathcal{E}(q) + \mathbb{K}(q)\mathrm{D}\mathcal{S}(q)$ as a sum of a Hamiltonian and a gradient-flow contribution, which satisfy a particular non-interaction condition. We give three applications: First, we consider a finite-dimensional quantum system that is coupled to a finite number of simple and finite heat baths, each of which is described by a scalar temperature variable. Second, we model quantum system given by a one-dimensional Schrödinger operator connected to a one-dimensional heat equation on the left and on the right. Finally, we consider thermo-opto-electronics, where the Maxwell-Bloch system of optics is coupled to the energy-drift-diffusion system for semiconductor electronics.

2010 AMS Mathematics Subject Classification: 37N20, 47N50, 80A99, 81V70, 78A35.

Keywords: GENERIC, density matrix, Hamiltonian systems, Onsager systems, canonical correlation, heat reservoirs, Maxwell-Bloch equation, thermo-opto-electronics.

1. Introduction

A fundamental problem in nano-science is the consistent coupling of quantum mechanics with effects on larger scales, see [1] for the modeling of lasers based on quantum-dot layers. One is interested in combining quantum and continuum mechanical models with dissipative effects such that the fundamental axioms of thermodynamics and quantum mechanics are still satisfied. We follow [2, 3] in modeling the coupling between classical dissipative systems and reversible quantum systems formulated in terms of the Liouville equation for the density matrix. Our basis is the theory of GENERIC systems, which stands for the acronym General Equations for Non-Equilibrium Reversible Irreversible Coupling and which was developed earlier under the name metriplectic system, see [4] and [5, Sect. 15.4] for the history.

A GENERIC system is a quintuple $(\mathcal{Q}, \mathcal{E}, \mathcal{S}, \mathbb{J}, \mathbb{K})$ with state space $\mathcal{Q}$, total energy $\mathcal{E} : \mathcal{Q} \to \mathbb{R}$ and total entropy $\mathcal{S} : \mathcal{Q} \to \mathbb{R}$. The evolution equation reads

$$\dot{q} = \mathbb{J}(q)\mathrm{D}\mathcal{E}(q) + \mathbb{K}(q)\mathrm{D}\mathcal{S}(q), \tag{1}$$

which displays an additive split into a Hamiltonian part $\mathbb{J}(q)\mathrm{D}\mathcal{E}(q)$, where $\mathbb{J}$ is a Poisson structure, and a dissipative part $\mathbb{K}(q)\mathrm{D}\mathcal{S}(q)$, where the Onsager operator $\mathbb{K}$ satisfies $\mathbb{K} = \mathbb{K}^* \geq 0$. GENERIC systems are thermodynamically correct in the sense that the total energy $\mathcal{E}$ is preserved while the total entropy $\mathcal{S}$ is nondecreasing. This is guaranteed by the central non-interaction conditions

$$\textbf{(NIC)} \qquad \mathbb{J}(q)\mathrm{D}\mathcal{S}(q) \equiv 0 \quad \text{and} \quad \mathbb{K}(q)\mathrm{D}\mathcal{E}(q) \equiv 0. \tag{2}$$

The evolution of the quantum system on a Hilbert space $\boldsymbol{H}$ will be described via the density matrix $\rho \in \mathfrak{R} := \{\, \rho \in \mathcal{L}^1(\boldsymbol{H}) \mid 0 \leq \rho = \rho^*, \ \operatorname{tr} \rho = 1 \,\}$ and the Liouville equation $\dot{\rho} = \jmath[\rho, H]$, where $\jmath := \mathrm{i}/\hbar$ and $H = H^*$ is the Hamiltonian. Following [2, 3] we couple this quantum system to the dissipative system $\dot{z} = \mathbb{K}_{\mathrm{di}}(z)\mathrm{D}\mathcal{S}_{\mathrm{di}}(z)$ using observables $Q_m = Q_m^*$ as coupling operators. In Section 2.4 we show that

$$\begin{aligned} \dot{\rho} &= \jmath[\rho, H] - \sum_{m=1}^{M} \Big(\big[Q_m \,,\, k_{\mathrm{B}}[Q_m, \rho] + \mathcal{C}_\rho[Q_m, \mathrm{B}_m(z)\mathrm{D}\mathcal{S}_{\mathrm{di}}(z)] \big] \Big), \\ \dot{z} &= \mathbb{K}_{\mathrm{di}}(z)\mathrm{D}\mathcal{S}_{\mathrm{di}}(z) + \sum_{m=1}^{M} \mathrm{B}_m^*(z) \big[Q_m \,,\, k_{\mathrm{B}}[Q_m, \rho] + \mathcal{C}_\rho[Q_m, \mathrm{B}_m(z)\mathrm{D}\mathcal{S}_{\mathrm{di}}(z)] \big] \end{aligned} \tag{3}$$

with $\mathbb{K}_{\mathrm{di}}\mathrm{D}\mathcal{E}_{\mathrm{di}} \equiv 0$ and $\mathrm{B}_m\mathrm{D}\mathcal{E}_{\mathrm{di}} \equiv H$ for $m = 1, ..., M$ is a GENERIC system for

$$\mathcal{E}(\rho, z) = \operatorname{tr}(\rho H) + \mathcal{E}_{\mathrm{di}}(z) \quad \text{and} \quad \mathcal{S}(\rho, z) = -k_{\mathrm{B}} \operatorname{tr}(\rho \log \rho) + \mathcal{S}_{\mathrm{di}}(z).$$

The crucial mathematical structure in our theory is the so-called *canonical correlation operator* $\mathcal{C}_\rho : \mathcal{L}^\infty(\boldsymbol{H}) \to \mathcal{L}^1(\boldsymbol{H})$ given by

$$\mathcal{C}_\rho A := \int_0^1 \rho^s A \rho^{1-s} \,\mathrm{d}s \quad \text{and} \quad \mathcal{C}_\rho^{-1} B = \int_0^\infty (\rho{+}sI)^{-1} B (\rho{+}sI)^{-1} \,\mathrm{d}s,$$

see [2, 3, 6] for its justification via fluctuation theory. The inverse $\mathcal{C}_\rho^{-1}$ is called the Bogoliubov-Kubo-Mori metric, cf. [7, 8]. There is a strong relation between the von Neumann entropy and $\mathcal{C}_\rho$ encoded in the fundamental commutator relations

$$\big[\mathcal{C}_\rho A, \log \rho\big] = \big[A, \rho\big] = \mathcal{C}_\rho\big[A, \log \rho\big], \tag{4}$$

for all $A = A^*$ and $\rho \in \mathfrak{R}$. Recall $\mathrm{D}_\rho \mathcal{S} = -k_{\mathrm{B}} \log \rho$ and note that $\mathrm{D}_\rho^2 \mathcal{S} = -k_{\mathrm{B}} \mathcal{C}_\rho^{-1}$, see [8, 9]. The continuity of $\rho \mapsto \mathcal{C}_\rho$ is proved in [10, Prop. 21.2], see (22).

Section 2 gives the basic features of GENERIC, introduces the relevant notation for quantum systems, and describes the general principles of coupling in terms of commutators $[Q_m, \cdot]$ with coupling operators and the canonical correlation operator. Section 3 describes the finite-dimensional results in [10] where $\dim \boldsymbol{H} < \infty$ and $z = (\theta_1, ..., \theta_M) \in \,]0, \infty[^M$.

In Section 4 we discuss a quantum dot described by a one-dimensional Schrödinger operator that is coupled to a left and a right reservoir described by a heat equation. There we also address the question of the existence of non-equilibrium steady states, which arises if the specific heat of the reservoirs tends to infinity. We also conjecture suitable formulas for the flux through the quantum dot if the temperature of the two reservoirs are different.

Section 5 deals with thermo-opto-electronics, which couples the Maxwell-Bloch system for polarizable crystals to the semiconductor equations for temperature and drift-diffusion for charge carriers. The former is a Hamiltonian system (cf. [11, 12]), while the latter was shown to be a gradient system in [13]. See [14] for the more details.

2. Coupling of quantum and dissipative mechanics

We first introduce the notion of GENERIC systems and then specify to the case where the underlying Hamiltonian system is given by a quantum mechanical system. In Section 2.4 we introduce the Kubo-Mori metric on density matrices, which provides a canonical way of defining dissipation in quantum systems.

2.1. *The GENERIC framework*

The framework of GENERIC was introduced by Morrison [4] under the name *metriplectic systems*, see [5, Sec. 15.4] for an outline of these early developments. In [15, 16] Öttinger and Grmela introduced the name GENERIC to emphasize the thermodynamical modeling aspects that were relevant for their applications in fluid mechanics. A more mathematical formulation is given in [17], where applications to thermoplasticity are developed. A survey on the modeling aspects of GENERIC and various applications will be presented in [18]. Our subsequent discussion is mainly based on the quantum mechanical papers [2, 3].

A GENERIC system is a quintuple $(\mathcal{Q}, \mathcal{E}, \mathcal{S}, \mathbb{J}, \mathbb{K})$, where the functionals $\mathcal{E}$ and $\mathcal{S}$ on the state space $\mathcal{Q}$ denote the total energy and the total entropy, respectively. Moreover, $\mathcal{Q}$ carries two geometric structures, namely a Poisson structure $\mathbb{J}$ and a dissipative Onsager structure $\mathbb{K}$, i.e., the operators $\mathbb{J}(q)$ and $\mathbb{K}(q)$ map the cotangent space $\mathrm{T}_q^*\mathcal{Q}$ into the tangent space $\mathrm{T}_q\mathcal{Q}$. The evolution of the system is given as the sum of the Hamiltonian part $\mathbb{J}(q)\mathrm{D}\mathcal{E}(q)$ and the gradient-flow $\mathbb{K}(q)\mathrm{D}\mathcal{S}(q)$, namely

$$\dot{q} = \mathbb{J}(q)\mathrm{D}\mathcal{E}(q) + \mathbb{K}(q)\mathrm{D}\mathcal{S}(q). \tag{5}$$

The basic conditions on the geometric structures $\mathbb{J}$ and $\mathbb{K}$ are the symmetries

$$\mathbb{J}(q) = -\mathbb{J}(q)^* \text{ and } \mathbb{K}(q) = \mathbb{K}(q)^* \tag{6}$$

and the structural properties

$$\begin{aligned} &\mathbb{J} \text{ satisfies Jacobi's identity,} \\ &\mathbb{K}(q) \text{ is positive semi-definite, i.e., } \langle \xi, \mathbb{K}(q)\xi \rangle \geq 0. \end{aligned} \tag{7}$$

Here, Jacobi's identity for $\mathbb{J}$ holds, if for all $\eta_j \in \mathrm{T}_q^*\mathcal{Q}$ we have

$$\langle \eta_1, \mathrm{D}\mathbb{J}(q)[\mathbb{J}(q)\eta_2]\eta_3 \rangle + \langle \eta_2, \mathrm{D}\mathbb{J}(q)[\mathbb{J}(q)\eta_3]\eta_1 \rangle + \langle \eta_3, \mathrm{D}\mathbb{J}(q)[\mathbb{J}(q)\eta_1]\eta_2 \rangle = 0.$$

The central condition states that the energy functional does not contribute to dissipative mechanisms and that the entropy functional does not contribute to reversible dynamics, which are the following *non-interaction conditions*:

$$\textbf{(NIC)} \qquad \forall q \in \mathcal{Q}: \quad \mathbb{J}(q)\mathrm{D}\mathcal{S}(q) = 0 \quad \text{and} \quad \mathbb{K}(q)\mathrm{D}\mathcal{E}(q) = 0. \tag{8}$$

Of course, the structure of GENERIC is geometric in the sense that it is invariant under coordinate transformations, see [17, 18]. The first observation is that (7) and (8) imply energy conservation and entropy increase:

$$\frac{\mathrm{d}}{\mathrm{d}t}\mathcal{E}(q(t)) = \langle \mathrm{D}\mathcal{E}(q), \dot{q} \rangle = \langle \mathrm{D}\mathcal{E}(q), \mathbb{J}\mathrm{D}\mathcal{E} + \mathbb{K}\mathrm{D}\mathcal{S} \rangle = 0 + 0 = 0, \tag{9}$$

$$\frac{\mathrm{d}}{\mathrm{d}t}\mathcal{S}(q(t)) = \langle \mathrm{D}\mathcal{S}(q), \dot{q} \rangle = \langle \mathrm{D}\mathcal{S}(q), \mathbb{J}\mathrm{D}\mathcal{E} + \mathbb{K}\mathrm{D}\mathcal{S} \rangle = 0 + \langle \mathrm{D}\mathcal{S}, \mathbb{K}\mathrm{D}\mathcal{S} \rangle \geq 0. \tag{10}$$

Of course, to guarantee energy conservation and positivity of the entropy production one needs much less than the two conditions (7) and (8).

However, the *maximum entropy principle* really relies on (8). It states that a maximizer q_* of $\mathcal{S}$ subject to the constraint $\mathcal{E}(q) = E_0$ is an equilibrium of (5), the so-called thermodynamic equilibrium for the given energy. Indeed, if q_* maximizes $\mathcal{S}$ given $\mathcal{E}(q) = E_0$, we obtain a Lagrange multiplier $\lambda_* \in \mathbb{R}$ with $\mathrm{D}\mathcal{S}(q_*) = \lambda_* \mathrm{D}\mathcal{E}(q_*)$. Since $\mathrm{D}\mathcal{S}(q) \neq 0$ for all q (e.g. by $\partial_\theta S > 0$), we have $\lambda_* \neq 0$ and conclude $\mathbb{J}(q_*)\mathrm{D}\mathcal{E}(q_*) = \frac{1}{\lambda_{\mathrm{eq}}}\mathbb{J}(q_*)\mathrm{D}\mathcal{S}(q_*) = 0$ and $\mathbb{K}(q_*)\mathrm{D}\mathcal{S}(q_*) = \lambda_{\mathrm{eq}}\mathbb{K}(q_*)\mathrm{D}\mathcal{E}(q_*) = 0$, where we have used the NIC (8). Vice versa, every steady state q_* of (5) satisfies

$$\mathbb{J}(q_*)\mathrm{D}\mathcal{E}(q_*) = 0 \text{ and } \mathbb{K}(q_*)\mathrm{D}\mathcal{S}(q_*) = 0. \tag{11}$$

Thus, in a steady state there cannot be any balancing between reversible and irreversible forces: both have to vanish independently. To show (11) we simply recall the entropy production relation (10), which implies $\langle \mathrm{D}\mathcal{S}(q_*), \mathbb{K}(q_*)\mathrm{D}\mathcal{S}(q_*) \rangle = 0$ for any steady state. Since $\mathbb{K}(q_*)$ is positive semidefinite, this implies the second identity in (11). The first identity then follows from $\dot{q} \equiv 0$ in (5).

2.2. *Quantum mechanics*

The quantum mechanical system is described by states in a complex Hilbert space $\boldsymbol{H}$ with scalar product $\langle \cdot | \cdot \rangle$ and a Hamiltonian (operator) $H : D(H) \to \boldsymbol{H}$, which is assumed to be selfadjoint and semi-bounded, namely

$$\exists\, h_{\min} \in \mathbb{R}\ \forall\, \psi \in D(H): \quad \langle H\psi | \psi \rangle \geq h_{\min} \|\psi\|^2.$$

The associated Hamiltonian dynamics is given via the time-dependent Schrödinger equation $\dot{\psi} = -\jmath H\psi$, where $\jmath = \frac{\mathrm{i}}{\hbar}$, which has the solution $\psi(t) = \mathrm{e}^{-\jmath t H}\psi(0)$.

We denote by $\mathcal{L}^p(\boldsymbol{H})$ the Banach space of compact operators A from $\boldsymbol{H}$ into itself with norm $\|A\|_p := \big(\sum_{j=1}^{\infty} \sigma_j(A)^p\big)^{1/p}$, where σ_j is the jth singular value of A, see [19, Ch. III]. Moreover, $\mathcal{L}^\infty(\boldsymbol{H})$ is the set of bounded linear operators. On $\mathcal{L}^1(\boldsymbol{H})$ the trace operator $\mathrm{tr} : \mathcal{L}^1(\boldsymbol{H}) \to \mathbb{C}$, $A \mapsto \sum_{j=1}^{\infty}\langle A\phi_j|\phi_j\rangle$ is well-defined, where $\{\,\phi_j \mid j \in \mathbb{N}\,\}$ is an complete orthonormal system in $\boldsymbol{H}$. Using the dual pairing

$$\langle\!\langle\, A \,\|\, B \,\rangle\!\rangle := \mathrm{tr}(AB^*), \text{ where } \mathrm{tr}(\psi\otimes\overline{\phi}) = \langle\psi|\phi\rangle,$$

we see that $\mathcal{L}^2(\boldsymbol{H})$ is a Hilbert space. By $\mathcal{L}^p_{\mathrm{S}}(\boldsymbol{H})$ we denote the subspace of symmetric bounded linear operators in $\mathcal{L}^p(\boldsymbol{H})$. The coupling of a quantum system to a macroscopic one can be described using the density matrices

$$\rho \in \mathfrak{R} := \{\,\rho \in \mathcal{L}^1(\boldsymbol{H}) \mid \rho = \rho^* \geq 0,\ \mathrm{tr}\,\rho = 1\,\} \subset \mathcal{L}^1_{\mathrm{S}}(\boldsymbol{H}).$$

Each $\rho \in \mathfrak{R}$ has the representation $\rho = \sum_{j=1}^{\infty} r_j\,\psi_j\otimes\overline{\psi}_j$, where $r_j \geq 0$, $\sum_1^\infty r_j = 1$, and $\{\,\psi_j \mid j \in \mathbb{N}\,\}$ is an orthonormal set. The evolution of ρ is given via the

$$\textit{Liouville equation:}\quad \dot{\rho} = \jmath[\rho, H], \quad \text{where } [\rho, H] := \rho H - H\rho. \tag{12}$$

Below we will use the *von Neumann entropy*

$$\mathcal{S}_{\mathrm{qm}}(\rho) = -k_{\mathrm{B}}\langle\!\langle\, \rho \,\|\, \log\rho \,\rangle\!\rangle = -k_{\mathrm{B}}\,\mathrm{tr}(\rho\log\rho) = -k_{\mathrm{B}}\sum_{j=1}^{\infty} r_j\log r_j, \tag{13}$$

which remains constant along the solutions ρ of the Hamiltonian system (12).

Defining $\mathcal{Q} = \mathfrak{R}$, $\mathcal{E}_{\mathrm{qm}}(\rho) = \mathrm{tr}(\rho H)$, $\mathbb{J}_{\mathrm{qm}}(\rho) = \jmath[\rho, \square]$ (where "$\square$" indicates the place to insert the argument), and $\mathbb{K}_{\mathrm{qm}} \equiv 0$, we see that (12) is given by the purely Hamiltonian GENERIC system $(\mathfrak{R}, \mathcal{E}_{\mathrm{qm}}, \mathcal{S}_{\mathrm{qm}}, \mathbb{J}_{\mathrm{qm}}, 0)$.

2.3. *General coupling of QM to dissipative models*

We consider an additional variable $z \in \mathcal{Z}$ which is driven by a gradient flow

$$\dot{z} = \mathbb{K}_{\mathrm{di}}(z)\mathrm{D}\mathcal{S}_{\mathrm{di}}(z), \text{ where } \mathbb{K}_{\mathrm{di}}(z) = \mathbb{K}_{\mathrm{di}}(z)^* \geq 0, \tag{14}$$

which we call *Onsager systems*, because of Onsager's fundamental contributions in [20, Part II, (1.11)], where the symmetry $\mathbb{K}_{\mathrm{di}}(z) = \mathbb{K}_{\mathrm{di}}(z)^*$ was derived. See also [21, eq. (2-1) to (2-4)], where it is stated that (14) is the universal form for many physical systems close to (local) equilibrium such as Ohm's law, Fick's law, heat flow, etc. We assume that (14) preserves an energy $\mathcal{E}_{\mathrm{di}}$ via the condition $\mathbb{K}_{\mathrm{di}}(z)\mathrm{D}\mathcal{E}_{\mathrm{di}}(z) \equiv 0$.

Following [2, 3] we couple the quantum and the Onsager system via GENERIC by setting $\mathcal{Q} = \mathfrak{R} \times \mathcal{Z}$ and using an additive split of the energy and the entropy:

$$\mathcal{E}(\rho, z) = \langle\!\langle\, H(z) \,\|\, \rho \,\rangle\!\rangle + \mathcal{E}_{\mathrm{di}}(z) \quad\text{and}\quad \mathcal{S}(\rho, z) = -k_{\mathrm{B}}\langle\!\langle\, \rho \,\|\, \log\rho \,\rangle\!\rangle + \mathcal{S}_{\mathrm{di}}(z).$$

In the general case, the Hamiltonian H may depend on the dissipative variable z (see e.g. (32) and (35)), but for the moment we assume that H is independent of z. We obtain the thermodynamic driving forces

$$\begin{pmatrix}\widetilde{\mu}\\ \widetilde{\zeta}\end{pmatrix}=\mathrm{D}\mathcal{E}(\rho,z)=\begin{pmatrix}H\\ \mathrm{D}_z\mathcal{E}_{\mathrm{di}}(z)\end{pmatrix},\quad \begin{pmatrix}\mu\\ \zeta\end{pmatrix}=\mathrm{D}\mathcal{S}(\rho,z)=\begin{pmatrix}\mathrm{D}_\rho\mathcal{S}_{\mathrm{qm}}(\rho)\\ \mathrm{D}_z\mathcal{S}_{\mathrm{di}}(z)\end{pmatrix}=\begin{pmatrix}-k_{\mathrm{B}}\log\rho\\ \mathrm{D}\mathcal{S}_{\mathrm{di}}(z)\end{pmatrix}.$$

Since z is a totally dissipative variable, the coupled Poisson structure $\mathbb{J}$ reads

$$\mathbb{J}(\rho,z)=\begin{pmatrix}\mathbb{J}_{\mathrm{qm}}(\rho) & 0\\ 0 & 0\end{pmatrix}=\begin{pmatrix}\jmath[\rho,\square] & 0\\ 0 & 0\end{pmatrix}, \tag{15}$$

and using $[\rho,\log\rho]\equiv 0$ we see that $(\mathrm{NIC})_1$ (i.e. $\mathbb{J}\mathrm{D}\mathcal{S}\equiv 0$) is satisfied.

The Onsager operator $\mathbb{K}$ of the coupled system is the sum of $\mathbb{K}_{\mathrm{di}}$ and a dissipative coupling between ρ and z. We write it in terms of the dual entropy-production potential $\mathcal{R}^*(\rho,z,\mu,\zeta)=\frac{1}{2}\langle\binom{\mu}{\zeta},\mathbb{K}(\rho,z)\binom{\mu}{\zeta}\rangle$. Using $\mathbb{K}\geq 0$ it is easy to see

$$(\mathrm{NIC})_2 \quad\Longleftrightarrow\quad \mathcal{R}^*(\rho,z,\mathrm{D}\mathcal{E}(\rho,z))=\mathcal{R}^*\big(\rho,z,(H,\mathrm{D}\mathcal{E}_{\mathrm{di}}(z))\big)\equiv 0.$$

As in [2, 3] we assume that the quantum system can only couple to the dissipative system via observables, which are given by commutators with symmetric bounded linear operators $Q(z)\in\mathcal{L}^\infty_{\mathrm{S}}(\boldsymbol{H})$. Hence, a possible form for $\mathcal{R}^*$ is

$$\mathcal{R}^*(\rho,z,\mu,\zeta)=\frac{1}{2}\langle\zeta,\mathbb{K}_{\mathrm{di}}(z)\zeta\rangle+\frac{1}{2}\sum_{m=1}^{M}\big|\!\big|\!\big|[Q_m(z),\mu-\mathrm{B}_m(z)\zeta]\big|\!\big|\!\big|^2_{\mathfrak{C}_m(\rho,z)},$$

where $\mathrm{B}_m(z)$ is a linear operator from $\mathrm{T}^*_z\mathcal{Z}$ to self-adjoint operators on $\boldsymbol{H}$ such that $\mathrm{B}_m(z)\mathrm{D}\mathcal{E}_{\mathrm{di}}(z)\equiv H$. The norms $|\!|\!|\cdot|\!|\!|_{\mathfrak{C}}$ are given via

$$|\!|\!|A|\!|\!|_{\mathfrak{C}}:=\langle\!\langle\,\mathfrak{C}A\,\|\,A\,\rangle\!\rangle^{1/2}=\|\mathfrak{C}^{1/2}A\|_2,$$

where the super-operators $\mathfrak{C}_m(\rho,z)$ are symmetric and positive semidefinite on $\mathcal{L}^2(\boldsymbol{H})$ (where super-operators are linear operators on $\mathcal{L}^p(\boldsymbol{H})$). Using the notation $\mathfrak{K}^Q_{\mathfrak{C}}A:=[Q^*,\mathfrak{C}[Q,A]]$ we can write the Onsager operator $\mathbb{K}$ in the form

$$\mathbb{K}(q)=\begin{pmatrix}0 & 0\\ 0 & \mathbb{K}_{\mathrm{di}}(z)\end{pmatrix}+\sum_{m=1}^{M}\begin{pmatrix}\mathfrak{K}^{Q_m}_{\mathfrak{C}_m} & -\mathfrak{K}^{Q_m}_{\mathfrak{C}_m}\mathrm{B}_m(z)\\ -\mathrm{B}^*_m(z)\mathfrak{K}^{Q_m}_{\mathfrak{C}_m} & \mathrm{B}^*_m(z)\mathfrak{K}^{Q_m}_{\mathfrak{C}_m}\mathrm{B}_m(z)\end{pmatrix}. \tag{16}$$

The above assumptions guarantee that $(\mathfrak{R},\mathcal{E},\mathcal{S},\mathbb{J},\mathbb{K})$ provides a GENERIC system. The evolution equations for $q=(\rho,z)$ give the coupled system

$$\begin{aligned}\dot\rho&=\jmath[\rho,H]-\sum_{m=1}^{M}\mathbb{K}^{Q_m(z)}_{\mathfrak{C}_m(\rho,z)}\Big(k_{\mathrm{B}}\log\rho+\mathrm{B}_m(z)\mathrm{D}\mathcal{S}_{\mathrm{di}}(z)\Big),\\ \dot z&=\mathbb{K}_{\mathrm{di}}(z)\mathrm{D}\mathcal{S}_{\mathrm{di}}(z)+\sum_{m=1}^{M}\mathrm{B}^*_m(z)\,\mathbb{K}^{Q_m(z)}_{\mathfrak{C}_m(\rho,z)}\Big(k_{\mathrm{B}}\log\rho+\mathrm{B}_m(z)\mathrm{D}\mathcal{S}_{\mathrm{di}}(z)\Big).\end{aligned} \tag{17}$$

2.4. *The Kubo-Mori metric*

In [3, 6, 10] it is argued that the dissipative super-operators $\mathfrak{C}_m(\rho, z)$ should be the *canonical correlation operator* $\mathcal{C}_\rho$, which is defined for all $\rho \in \mathfrak{R}$ via

$$\mathcal{C}_\rho : \begin{cases} \mathcal{L}^\infty(\boldsymbol{H}) \to & \mathcal{L}^1(\boldsymbol{H}), \\ \quad A \quad \mapsto & \int_0^1 \rho^s A \rho^{1-s}\,\mathrm{d}s. \end{cases} \tag{18}$$

By $\rho = \rho^*$ one has $(\mathcal{C}_\rho A)^* = \mathcal{C}_\rho(A^*)$, and if $\rho = \sum r_j \psi_j \otimes \overline{\psi}_j$, then

$$\mathcal{C}_\rho A = \sum_{j,k=1}^{\dim \boldsymbol{H}} \Lambda(r_j, r_k) \langle A\psi_k | \psi_j \rangle \psi_j \otimes \overline{\psi}_k \text{ and } \langle\!\langle\, \mathcal{C}_\rho A \,\|\, A \,\rangle\!\rangle = \sum_{j,k=1}^{\dim \boldsymbol{H}} \Lambda(r_j, r_k) \big| \langle A\psi_k | \psi_j \rangle \big|^2 \tag{19}$$

with $\Lambda(a, b) = \int_0^1 a^s b^{1-s}\,\mathrm{d}s$. In particular, $\mathcal{C}_\rho$ is symmetric and positive semidefinite:

$$\langle\!\langle\, \mathcal{C}_\rho A \,\|\, B \,\rangle\!\rangle = \langle\!\langle\, A \,\|\, \mathcal{C}_\rho B \,\rangle\!\rangle \quad \text{and} \quad \langle\!\langle\, \mathcal{C}_\rho A \,\|\, A \,\rangle\!\rangle \geq 0.$$

Hence, $\mathcal{C}_\rho$ induces the scalar product $\langle A, B \rangle_\rho := \langle\!\langle\, \mathcal{C}_\rho A \,\|\, B \,\rangle\!\rangle$, which is called the *canonical correlation between A and B for the given state* ρ in [3, 22].

For $\rho > 0$ and $\dim \boldsymbol{H} < \infty$ the operator $\mathcal{C}_\rho$ is invertible, namely

$$\mathcal{G}_\rho A := \mathcal{C}_\rho^{-1} A = \int_0^\infty (\rho{+}sI)^{-1} A (\rho{+}sI)^{-1}\,\mathrm{d}s, \tag{20}$$

which can be derived from (19) using $1/\Lambda(a, b) = \int_0^\infty \big((a{+}s)(b{+}s)\big)^{-1}\,\mathrm{d}s$. The super-operator $\mathcal{G}_\rho$ defines the *Bogoliubov-Kubo-Mori metric* on the set of density matrices (cf. [8, 23]). For us, the most important fact is the connection to the von Neumann entropy $\mathcal{S}_{\mathrm{qm}}$, see (13). In fact, $\mathcal{G}_\rho$ can be identified with the Hessian of $\mathcal{S}_{\mathrm{qm}}$ up to the factor $-k_{\mathrm{B}}$, namely $\langle\!\langle\, A \,\|\, \mathrm{D}^2 \mathcal{S}_{\mathrm{qm}}(\rho) B \,\rangle\!\rangle = -k_{\mathrm{B}} \langle\!\langle\, A \,\|\, \mathcal{G}_\rho B \,\rangle\!\rangle$.

In particular, we will heavily use the *fundamental commutator identities*

$$\big[\mathcal{C}_\rho A, \log \rho\big] = \big[A, \rho\big] = \mathcal{C}_\rho \big[A, \log \rho\big], \tag{21}$$

which go back to [24]. They play an important role in the field of dissipative effects in quantum mechanics and manifest the relation between the von Neumann entropy $\mathcal{S}_{\mathrm{qm}}$ and the canonical correlation operator $\mathcal{C}_\rho$. See [10, Prop. 21.1] for a simple proof and [9, Lem. 3.1], where a general calculus for operator functions is developed. (These identities generalize the differential identity $v\nabla(\log v) = \nabla v$, which is crucial for the entropic gradient formulation of the Fokker-Planck equation in [25], see [10, Sect. 21.7] for a comparison of the different gradient structures.)

In addition, [10, Prop. 21.2] provides the continuity of $\mathcal{C}_\rho$ in the form

$$\|\mathcal{C}_{\rho_1} A - \mathcal{C}_{\rho_2} A\|_1 \leq \omega\Big(\tfrac{\|\rho_1 - \rho_2\|_1}{\|\rho_1\|_1 + \|\rho_2\|_1}\Big) \big(\|\rho_1\|_1 + \|\rho_2\|_1\big) \|A\|_\infty \quad \text{with } \omega(\nu) = 2\tfrac{1-\nu}{|\log \nu|}. \tag{22}$$

We emphasize that $\mathfrak{R} \ni \rho \mapsto \mathcal{C}_\rho \in \mathrm{Lin}\big(\mathcal{L}^\infty_{\mathrm{S}}(\boldsymbol{H}), \mathcal{L}^1_{\mathrm{S}}(\boldsymbol{H})\big)$ is continuous, if $\mathfrak{R}$ is equipped with the norm of $\mathcal{L}^1_{\mathrm{S}}(\boldsymbol{H})$.

If we specialize the coupled system (17) by choosing all $\mathfrak{C}_m$ equal to $\mathcal{C}_\rho$, as suggested in [3], we can use the identities (21) giving $\mathfrak{K}^Q_{\mathcal{C}_\rho}\log\rho = \big[Q,[Q,\rho]\big]$, where the right-hand side is continuous. Thus, the singularity in $\log\rho$ and the nonlinearity of $\mathcal{C}_\rho$ interact in a nice way to give rise to a linear and hence continuous term. Hence, (17) takes the form (where we dropped the dependence of Q_m on z for simplicity):

$$\begin{aligned}\dot\rho &= \jmath\big[\rho,H\big] - \sum_{m=1}^{M}\Big(\big[\,Q_m\,,\,k_{\mathrm B}[Q_m,\rho]+\mathcal{C}_\rho[Q_m,G_m(z)]\,\big]\Big),\\ \dot z &= \mathbb{K}_{\mathrm{di}}(z)\mathrm{D}\mathcal{S}_{\mathrm{di}}(z)+\sum_{m=1}^{M}\mathrm{B}^*_m(z)\big[\,Q_m\,,\,k_{\mathrm B}[Q_m,\rho]+\mathcal{C}_\rho[Q_m,G_m(z)]\,\big],\end{aligned} \tag{23}$$

where $G_m(z):=\mathrm{B}_m(z)\mathrm{D}\mathcal{S}_{\mathrm{di}}(z)$. We see the analogy to the Lindblad equations

$$\dot\rho = \jmath[\rho,H] - k_{\mathrm B}\textstyle\sum_{m=1}^{M}\big[Q_m,[Q_m,\rho]\big],$$

where the double bracket terms generate the dissipation, see e.g. [6, 26, 27].

The most important feature of this system is that the singular term $\log\rho$ in the right-hand side has disappeared. Under suitable further assumptions the right-hand side forms a continuous vector field, which wasn't the case for (17).

3. A finite-dimensional model with simple heat baths

In this section we consider a finite-dimensional version of the general coupled system developed above. We will summarize the results obtained in [10, 28].

This model is a special case of system (23) by assuming that $\boldsymbol{H}=\mathbb{C}^n$ and that the dissipative variable z consists of the absolute temperatures $\theta=(\theta_1,\dots,\theta_M)\in\,]0,\infty[^M=:\mathcal{Z}_M$ of the simple, finite heat baths, where "simple" means that the state of the heat bath is determined by a single value $\theta_j>0$ for the absolute temperature. Thus, (23) reduces to an ODE with dimension n^2-1+m. While the Poisson structure $\mathbb{J}$ on $\mathcal{Q}:=\mathfrak{R}_n\times\mathcal{Z}_M$ is given by (15), we still have to specify the energy, the entropy, and the Onsager operator. We assume that the mth heat bath has the constant specific heat $c_m>0$, and thus has finite energy. Then we have

$$\mathcal{E}(\rho,\theta)=\mathrm{tr}(\rho H)+\sum_{m=1}^{M}c_m\theta_m \quad\text{and}\quad \mathcal{S}(\rho,\theta)=-k_{\mathrm B}\,\mathrm{tr}(\rho\log\rho)+\sum_{m=1}^{M}c_m\log(\theta_m/\theta_0),$$

where $\theta_0>0$ is an arbitrary reference temperature. Choosing $\mathfrak{C}_m=\mathcal{C}_\rho$, the Onsager operator $\mathbb{K}$ in (16) is fixed by specifying $\mathrm{B}_m:\mathbb{R}^M\to\mathbb{C}^{n\times n}_{\mathrm S}$. We take $\mathrm{B}_m(\tau_1,\dots,\tau_M)=\frac{\tau_m}{c_m}H$, which highlights the fact that Q_m couples the quantum system to the mth heat bath. Here τ_m denotes the thermodynamically conjugate driving force to θ_m. Using $\mathrm{D}_{\theta_m}\mathcal{E}(\rho,\theta)=c_m$ we also find $\mathrm{B}_m(z)\mathrm{D}_z\mathcal{E}(\rho,z)=H$ as desired for $(\mathrm{NIC})_2$. Hence, we define the dual entropy-production potential $\mathcal{R}^*$ via

$$\mathcal{R}^*(\rho,\theta;\mu,\tau)=\tfrac12\textstyle\sum_{m=1}^{M}\big|\!\big|\!\big|[Q_m,\mu-\frac{\tau_m}{c_m}H]\big|\!\big|\!\big|^2_{\mathcal{C}_\rho}+\tfrac12\tau\cdot\kappa(\theta)\tau.$$

A typical choice for $\kappa(\theta) \in \mathbb{R}_{\mathrm{S}}^{M\times M}$ is given by

$$\tau \cdot \kappa\tau = \sum_{m=1}^{M-1} \sum_{l=m+1}^{M} \widetilde{\kappa}_{ml}(\theta)\big(\tfrac{\tau_m}{c_m} - \tfrac{\tau_l}{c_l}\big)^2, \tag{24}$$

where $\widetilde{\kappa}_{ml}(\theta) \geq 0$ gives the direct heat transfer between the heat baths m and l.

Setting $\mathbb{K}(\rho,\theta) = \mathrm{D}^2_{\mu,\tau}\mathcal{R}^*(\rho,\theta,0,0)$ the quintuple $(\mathfrak{R}_n \times \mathcal{Z}_M, \mathcal{E}, \mathcal{S}, \mathbb{J}, \mathbb{K})$ defines a GENERIC system which generates the following coupled system:

$$\dot{\rho} = \jmath[\rho, H] - \sum_{m=1}^{M} \big[\, Q_m\,,\, k_{\mathrm{B}}[Q_m,\rho] + \frac{1}{\theta_m}\mathcal{C}_\rho[Q_m,H]\,\big], \tag{25a}$$

$$\dot{\theta}_m = \frac{1}{c_m}\sum_{l=1}^{M} \kappa_{ml}(\theta)\frac{c_l}{\theta_l} \;+\; \frac{1}{c_m}\langle\!\langle\, k_{\mathrm{B}}[Q_m,\rho] + \frac{1}{\theta_m}\mathcal{C}_\rho[Q_m,H] \,\|\, [Q_m,H]\,\rangle\!\rangle. \tag{25b}$$

Choosing $\kappa_{ml}(\theta) = \widehat{\kappa}_{ml}\theta_m\theta_l$ makes the first term in (25b) linear. This choice is used in [29] to model linear interface conditions in heterostructures using gradient structures. Using the form (24) we see that the first term in (25b) can be written in terms of the differences $\frac{1}{\theta_m} - \frac{1}{\theta_l} = (\theta_l - \theta_m)/(\theta_m\theta_l)$.

Since $\dim \boldsymbol{H} = n < \infty$ the right-hand side of (25) is analytic in the interior of $\mathcal{Q} = \operatorname{int}\mathfrak{R}_n \times]0,\infty[^M$, while it is continuous on $\mathcal{Q}$. Hence, solutions starting in the interior are unique as long as they stay in the interior. Even for this ODE model the global existence theory is not completely trivial. First, we observe that all temperatures stay bounded. Indeed, using $H \geq h_0 I$ and $\operatorname{tr}\rho = 1$ we have

$$c_m\theta_m(t) \leq \sum_{l=1}^{M} c_l\theta_l(t) \leq \mathcal{E}(\rho(t),\theta(t)) - h_0 = \mathcal{E}(\rho(0),\theta(0)) - h_0 =: e_*.$$

Using the entropy we find a lower bound via

$$\begin{aligned}\mathcal{S}(\rho(0),\theta(0)) &\leq \mathcal{S}(\rho(t),\theta(t)) \leq k_{\mathrm{B}}\log n + \textstyle\sum_{l=1}^{M} c_l \log(\theta_l(t)/\theta_0) \\ &\leq k_{\mathrm{B}}\log n + \textstyle\sum_{l\neq m} c_l \log\big(e_*/(c_l\theta_0)\big) + c_m \log(\theta_m(t)/\theta_0).\end{aligned}$$

Since $\mathfrak{R}_n$ is compact, we find that the solution $(\rho(t),\theta(t))$ stays in a compact set. Moreover, the vector field defined in (25) is uniformly continuous on this compact set by the continuity (22) of $\rho \mapsto \mathcal{C}_\rho$. Hence, a solution could only cease to exist by leaving the compact domain through the boundary of $\mathfrak{R}_n$. However, using the method developed for [10, Thm. 21.3] it can be shown that this is not possible. Hence global existence follows, see [28] for more details.

In [10, Sect. 21.6] the case of one heat bath is analyzed further. In that case the single temperature $\theta = \theta_1$ can be eliminated by using energy conservation $\mathcal{E}(\rho,\theta) = E_0$ in the form $\theta = (E_0 - \operatorname{tr}(\rho H))/c$. Allowing for L coupling operators Q^l connecting to one simple heat bath we arrive at the following equation for ρ alone:

$$\dot{\rho} = \jmath[H,\rho] - k_{\mathrm{B}}\sum_{l=1}^{L}\big[Q^l,[Q^l,\rho]\big] + \frac{c}{E_0 - \operatorname{tr}(\rho H)}\sum_{l=1}^{L}\big[Q^l,\mathcal{C}_\rho[Q^l,H]\big]. \tag{26}$$

Global existence of solutions is established in [10, Thm. 21.3]. The question of uniqueness of solutions remains still open, since non-uniqueness may occur, when solutions touch the boundary of $\mathfrak{R}_n$. Additional assumptions guarantee that the thermodynamic equilibrium $\rho_{\rm eq} = \frac{1}{Z(E_0)}{\rm e}^{-\beta(E_0)H}$ is the only equilibrium and that all solutions decay to this equilibrium for $t \to \infty$. Defining

$$\mu(\boldsymbol{Q}) := \inf\{\, f_{\boldsymbol{Q}}(\psi,\phi) \mid |\psi| = |\phi| = 1,\ \langle\psi|\phi\rangle = 0\,\} \text{ where } f_{\boldsymbol{Q}}(\psi,\phi) = \sum_{l=1}^{L} |\langle Q^l\psi|\phi\rangle|^2,$$

the following result was established in [10, Thm. 21.4].

Theorem 3.1. *Assume that*(26) *satisfies* $\mu(\boldsymbol{Q}) > 0$ *and the commutator condition*

$$\Big(A \in \mathcal{L}^2_{\rm S}(\boldsymbol{H}) \ \text{ and } \ \forall\, l = 1,\dots,L\colon [Q^l, A] = 0\Big) \quad\Longrightarrow\quad \exists\, \alpha \in \mathbb{R}\colon A = \alpha I. \qquad (27)$$

Then, all solutions $\rho : [0,\infty[\,\to \mathfrak{R}$ *of* (26) *satisfy* $\rho(t) \to \rho_{\rm eq}$.

4. A quantum dot coupled to heat equations

We consider a one-dimensional system consisting of a left and a right reservoir that are coupled to a quantum system modeling a quantum dot (QD). Our aim is to model the energy exchanges between the QD and the two reservoirs. In particular, energy can flow from one reservoir into the other only through the QD.

4.1. *QD with two reservoirs*

The two reservoirs are contained in the domain $\Omega = \Omega_+ \cup \Omega_-$ with $\Omega_\pm = \pm]\ell_1, \ell_3[$. We use the absolute temperature $\theta : \Omega \to \,]0,\infty[$ to describe the local state of the heat bath. The quantum system is defined on the Hilbert space $\boldsymbol{H} := {\rm L}^2(\Omega_{\rm qm})$ with $\Omega_{\rm qm} := \,]-\ell_2, \ell_2[$, where $0 < \ell_1 \le \ell_2 \ll \ell_3$. We decompose this space into a subspace where the coupling may occur and the space where the quantum system is left free, namely $\boldsymbol{H} = \boldsymbol{H}_{\rm coup} \times \boldsymbol{H}_{\rm free}$ with $\boldsymbol{H}_{\rm coup} = {\rm L}^2(\Omega_{\rm coup})$, $\boldsymbol{H}_{\rm free} = {\rm L}^2(\Omega_{\rm free})$, $\Omega_{\rm free} = [-\ell_1, \ell_1]$, and $\Omega_{\rm coup} = \Omega_{\rm qm} \setminus \Omega_{\rm free}$. For the Hamiltonian operator H we consider the Schrödinger operator $H\psi = -\frac{1}{2{\sf m}}\Psi'' + V\psi$ with a smooth potential V and boundary conditions $\psi'(\pm\ell_2) = 0$. The total energy and entropy are given by

$$\mathcal{E}(\rho,\theta) = {\rm tr}(H\rho) + \int_\Omega c(x)\theta(x)\,{\rm d}x, \quad \mathcal{S}(\rho,\theta) = -k_{\rm B}\,{\rm tr}(\rho\log\rho) + \int_\Omega c(x)\log(\theta(x)/\theta_0)\,{\rm d}x.$$

While the Poisson structure $\mathbb{J}$ is given by the standard Lie-Poisson structure (15), we have to specify the Onsager structure in more detail. It contains the heat conduction as well as the interaction of the quantum system and the heat baths. We define it in terms of the quadratic dual entropy-production potential

$$\mathcal{R}^*(\rho,\theta,\mu,\tau) = \int_\Omega \frac{k(\theta)}{2}\Big|\nabla\big(\frac{\tau}{c}\big)\Big|^2 {\rm d}x + \sum_{m=1}^{M} \Big|\Big|\Big| [Q_m, \mu - {\rm B}_m\big(\frac{\tau}{c}\big)] \Big|\Big|\Big|^2_{\mathcal{C}_\rho},$$

where $Q_m \in \mathcal{L}^\infty(\boldsymbol{H}_{\rm coup})$, $m = 1, ..., M$, are the coupling operators. For $m = 1, ..., M$ we need a linear mapping B_m from $\mathrm{L}^2(\Omega)$ into the self-adjoint operators on $\boldsymbol{H}$ such that $\mathrm{B}_m 1 = H$, e.g. a possible choice would be

$$\mathrm{B}_A(v)\psi := -\frac{1}{2\mathsf{m}}\operatorname{div}\big(\widehat{v}\nabla\psi\big) + \widehat{v}V\psi, \quad \text{where } \widehat{v}(x) = \begin{cases} v(x) & \text{for } x \in \Omega, \\ \text{aff. interpol.} & \text{for } x \in [-\ell_1, \ell_1]. \end{cases}$$

The adjoint operators $\mathrm{B}_m^* : \mathcal{L}^2_{\rm S}(\boldsymbol{H}) \to \mathrm{L}^2(\Omega)$ are defined as usual, i.e. $\operatorname{tr}((\mathrm{B}_m v)A) = \int_\Omega v(\mathrm{B}_m^* A)\,\mathrm{d}x$ for all A.

Obviously, we have $\mathcal{R}^*(\rho, \theta, \mathrm{D}\mathcal{E}(\rho, \theta)) = 0$ as desired and, hence, have a GENERIC system after defining the Onsager operator $\mathbb{K}(\rho, \theta) = \mathrm{D}^2_{(\mu,\tau)}\mathcal{R}^*(\rho, \theta, 0, 0)$. To highlight the interactions between the quantum system and the heat bath we write the equations for the coupled system explicitly:

$$\begin{aligned} \dot\rho &= \jmath[H, \rho] - \sum_{m=1}^{M}\Big(k_{\rm B}\big[Q_m, [Q_m, \rho]\big] + \big[Q_m, \mathcal{C}_\rho[Q_m, \mathrm{B}_m(1/\theta)]\big]\Big), \\ c\dot\theta &= -\operatorname{div}\big(k(\theta)\nabla(1/\theta)\big) + \sum_{m=1}^{M}\mathrm{B}_m^*\big[Q_m, k_{\rm B}[Q_m, \rho] + \mathcal{C}_\rho[Q_m, \mathrm{B}_m(1/\theta)]\big], \\ 0 &= \partial_x\theta(x) \quad \text{for } x \in \partial\Omega = \{-\ell_3, -\ell_1, \ell_1, \ell_3\}. \end{aligned} \tag{28}$$

4.2. *QD with one left and one right coupling*

We now simplify the above model to highlight its structure more clearly. For this we make particular choices, namely $M = 2$ such that $Q_- = Q_1$ couples to the left reservoir and $Q_+ = Q_2$ couples to the right. In particular, we assume $Q_\pm = g_\pm \otimes \overline{g}_\pm$ with $\operatorname{supp}(g_\pm) \subset \pm[\ell_1, \ell_2]$ and $g_\pm \in \mathrm{H}^2_0(\overline{\Omega}_{\rm coup})$ such that $Hg_\pm$ lies in $\mathrm{L}^2(\Omega)$ and $\jmath[Q_\pm, H] \in \mathcal{L}^2_{\rm S}(\boldsymbol{H})$. Finally, we choose $\mathrm{B}_\pm$ as simple as possible, namely

$$\mathrm{B}_\pm(v) = [\![v]\!]_\pm H \quad \text{with } [\![v]\!]_\pm = \tfrac{1}{\ell_2 - \ell_1}\textstyle\int_{\pm[\ell_1, \ell_2]} v(y)\,\mathrm{d}y,$$

where in the case $\ell_1 = \ell_2$ the value $[\![v]\!]_\pm$ can be interpreted as a point evaluation $v(\pm\ell_1)$. Hence, B_+ sees the average temperature on the right coupling region, while B_- that on the left. In particular, the adjoints $\mathrm{B}_\pm^*$ are simple as well, namely $\mathrm{B}_\pm^* Q = 1\!\!1_{\pm[\ell_1, \ell_2]}\operatorname{tr}(QH)$, where $1\!\!1_S$ denotes the indicator function for the set S.

With these choices (28) takes the more specific form

$$\begin{aligned} \dot\rho &= \jmath[H, \rho] - \sum_{\alpha \in \{+,-\}}\Big(k_{\rm B}\big[Q_\alpha, [Q_\alpha, \rho]\big] + [\![1/\theta]\!]_\alpha\big[Q_\alpha, \mathcal{C}_\rho[Q_\alpha, H]\big]\Big), \\ c\dot\theta &= \operatorname{div}\big(\tfrac{k(\theta)}{\theta^2}\nabla\theta\big) + \sum_{\alpha \in \{+,-\}}\frac{1\!\!1_{\alpha[\ell_1, \ell_2]}}{\ell_2 - \ell_1}\operatorname{tr}\Big([Q_\alpha, H]\big(k_{\rm B}[Q_\alpha, \rho] + [\![1/\theta]\!]_\alpha\mathcal{C}_\rho[Q_\alpha, H]\big)\Big), \end{aligned} \tag{29}$$

plus the same boundary conditions for θ as in (28). In (29) we nicely see the coupling between the heat equation and the quantum system: the dissipative terms in the quantum system give rise to a source term in the heat equation. The GENERIC structure guarantees that the energy remains constant, while the entropy increases.

4.3. *Non-equilibrium steady states and fluxes*

The above model allows us to study non-equilibrium steady states (NESS) by assuming that the reservoirs are huge. We realize this by making the specific heat c very large, namely $c = 1/\varepsilon$ for $\varepsilon \searrow 0$. For a fixed ε we still expect that solutions (ρ, θ) converge to thermodynamic equilibrium. However, starting from (ρ_0, θ_0), where $\theta_0(x) = \widehat{\theta}_\pm$ for $x \in \pm[\ell_1, \ell_3]$, one expects that ρ decays to an equilibrium on a time-scale of order 1 (relative to ε), while θ only changes with a rate proportional to ε. Hence, we can formulate a nonlinear equation for the NESS $\widehat{\rho}$ as follows:

$$0 = \jmath[H, \widehat{\rho}] - \textstyle\sum_{\alpha\in\{+,-\}} \Big(k_\mathrm{B} \big[Q_\alpha, [Q_\alpha, \widehat{\rho}]\big] + \frac{1}{\widehat{\theta}_\alpha} \big[Q_\alpha, \mathcal{C}_{\widehat{\rho}}[Q_\alpha, H]\big] \Big). \tag{30}$$

Here the NESS $\widehat{\rho}$ is the only unknown, since the temperatures $\widehat{\theta}_+$ and $\widehat{\theta}_-$ are given.

The heat equation is no longer part of the problem, however it allows us to calculate the energy flux through the quantum system. The energies in the left and right reservoirs are $\int_{\pm[\ell_1,\ell_3]} \frac{1}{\varepsilon}\theta(x)\,\mathrm{d}x$. Taking the time derivative of these terms and inserting the heat equation we obtain the flux from the left to the right reservoir by

$$\begin{aligned} &\mathcal{F}(\widehat{\theta}_-, \widehat{\theta}_+) := \mathcal{F}_+(\widehat{\theta}_-, \widehat{\theta}_+) - \mathcal{F}_-(\widehat{\theta}_-, \widehat{\theta}_+) \\ &\text{with } \mathcal{F}_\pm(\widehat{\theta}_-, \widehat{\theta}_+) := \mathrm{tr}\,\Big([Q_\pm, H]\big(k_\mathrm{B}[Q_\pm, \widehat{\rho}] + \tfrac{1}{\widehat{\theta}_\pm}\mathcal{C}_{\widehat{\rho}}[Q_\pm, H]\big) \Big), \end{aligned} \tag{31}$$

where the NESS $\widehat{\rho}$ is a function of $(\widehat{\theta}_-, \widehat{\theta}_+)$. The term div $\big(k\nabla(1/\theta)\big)$ does not contribute to the flux terms $\mathcal{F}_\pm$, because of the no-flux boundary conditions. Moreover, the non-equilibrium flux does not depend on the choice of the operators $\mathrm{B}_\pm$.

For $\widehat{\theta}_\pm = \theta_*$ one obtains the thermodynamical equilibrium $\widehat{\rho} = \rho_\mathrm{eq} = \frac{1}{Z}\mathrm{e}^{-\beta H}$ with $\beta = 1/(k_\mathrm{B}\theta_*)$. Assuming $\widehat{\theta} = \theta_* \pm \delta$ it should be possible to calculate $\widehat{\rho}(\delta) = \rho_\mathrm{eq} + \delta\sigma + O(\delta^2)$ and to find an expansion for the non-equilibrium flux $\mathcal{F}(\theta_* - \delta, \theta_* + \delta) = \delta F_1 + O(\delta^2)$, where we expect that F_1 can be related to the Landauer-Büttiker theory, see e.g. [30].

5. Thermo-opto-electronic models for semiconductors

The coupling of quantum systems to classical opto-electronic systems is common practice in the modeling of laser devices, if the optically active zones are based on quantum wells or quantum dots, see e.g. [1]. Fully quantum mechanical models as in [31] cannot be used because of the macroscopic complexity of real semiconductor devices. Hence, such models have to consist of two subparts that are typically considered independently. On the one hand the drift-diffusion equation for the charged particles (usually free electrons and holes in a semiconductor crystal) and the heat equation form a parabolic system that can be written as a gradient system $\dot{q} = \mathbb{K}(q)\mathrm{D}\mathcal{S}(q)$, see [13]. On the other hand the Maxwell-Bloch system for the electromagnetic fields and the polarization form a hyperbolic system that is mostly treated as an energy conserving Hamiltonian system which contains a quantum model, see [11, 12]. We present the main results of [14], which couples both models via GENERIC.

5.1. *The Maxwell-Bloch equation as Hamiltonian system*

For optical waves in a polarizable crystal one couples the electromagnetic fields to continuously distributed quantum system that are described by a finite-dimensional Hilbert space $\boldsymbol{H} = \mathbb{C}^n$ at each material point x in the crystal $\Omega \subset \mathbb{R}^3$. It is assumed that there is no coupling between the different points, which is a good approximation for certain polarizations of an optically active crystal. To introduce the notation we start with Maxwell's equations in the vacuum $\mathbb{R}^3$ by considering the dielectric displacement $\boldsymbol{D}$ and the magnetic flux $\boldsymbol{B}$ as the basic fields:

$$\dot{\boldsymbol{D}} = \operatorname{curl} \boldsymbol{H}, \quad \dot{\boldsymbol{B}} = -\operatorname{curl} \boldsymbol{E}, \quad \operatorname{div} \boldsymbol{D} = 0 = \operatorname{div} \boldsymbol{B}$$

with conserved total energy $\mathcal{E}_{\mathrm{M}}(\boldsymbol{D}, \boldsymbol{B}) = \int_{\mathbb{R}^3} \frac{1}{2\varepsilon_0}|\boldsymbol{D}|^2 + \frac{1}{2\mu_0}|\boldsymbol{B}|^2 \,\mathrm{d}x$, electric field $\boldsymbol{E} = \mathrm{D}_{\boldsymbol{D}}\mathcal{E}_{\mathrm{M}}(\boldsymbol{D}, \boldsymbol{B}) = \frac{1}{\varepsilon_0}\boldsymbol{D}$, and magnetic field $\boldsymbol{H} = \mathrm{D}_{\boldsymbol{B}}\mathcal{E}_{\mathrm{M}}(\boldsymbol{D}, \boldsymbol{B}) = \frac{1}{\mu_0}\boldsymbol{B}$.

The Maxwell-Bloch system is a combination of the Maxwell equations and an n-level quantum system at each material point $x \in \Omega$ described by the density matrix $\rho(t, x) \in \mathfrak{R}_n := \mathfrak{R} \cap \mathbb{C}^{n\times n}_{\mathrm{Herm}}$. This quantum system contains the material properties of the underlying crystal Ω. Without interaction with the electromagnetic waves it is given via $\dot{\rho} = \jmath[\rho, N_{\mathrm{B}}]$ where $N_{\mathrm{B}} = N_{\mathrm{B}}^* \in \mathbb{C}^{n\times n}$ is the Bloch Hamiltonian. The coupling of ρ to Maxwell's equation occurs via the polarization in the form $\boldsymbol{D} = \boldsymbol{E} + \boldsymbol{P}$ with polarization $\boldsymbol{P}(t, x) = \Gamma\rho(t, x)$ where $\Gamma : \mathbb{C}^{n\times n}_{\mathrm{Herm}} \to \mathbb{R}^3$ is called the dipole-moment operator. Thus we arrive at the nonlinear Maxwell-Bloch system

$$\dot{\boldsymbol{D}} = \operatorname{curl} \boldsymbol{H}, \qquad \dot{\boldsymbol{B}} = -\operatorname{curl}(\boldsymbol{D} - \Gamma\rho), \qquad \dot{\rho} = \jmath[\rho, N_{\mathrm{B}} - \Gamma^*(\boldsymbol{D} - \Gamma\rho)]. \tag{32}$$

Choosing the space $\boldsymbol{Q} = \mathrm{L}^2(\Omega, \operatorname{div})^3 \times \mathrm{L}^2(\Omega, \operatorname{div})^3 \times \mathfrak{R}_n$ and setting $\varepsilon_0 = \mu_0 = 1$ for simplicity, these equations form a Hamiltonian system $(\boldsymbol{Q}, \mathcal{E}, \mathbb{J})$ with the energy $\mathcal{E}(\boldsymbol{D}, \boldsymbol{B}, \rho) = \int_\Omega \frac{1}{2\varepsilon_0}|\boldsymbol{D} - \Gamma\rho|^2 + \frac{1}{2\mu_0}|\boldsymbol{B}|^2 + \operatorname{tr}(N_{\mathrm{B}}\rho)\,\mathrm{d}x$ as follows

$$\frac{\mathrm{d}}{\mathrm{d}t}\begin{pmatrix} \boldsymbol{D} \\ \boldsymbol{B} \\ \rho \end{pmatrix} = \mathbb{J}(\rho)\begin{pmatrix} \mathrm{D}_{\boldsymbol{D}}\mathcal{E} \\ \mathrm{D}_{\boldsymbol{B}}\mathcal{E} \\ \mathrm{D}_\rho\mathcal{E} \end{pmatrix} \quad \text{where } \mathbb{J}(\rho) = \begin{pmatrix} 0 & \operatorname{curl} & 0 \\ -\operatorname{curl} & 0 & 0 \\ 0 & 0 & \jmath[\rho, \square] \end{pmatrix}.$$

In [11, 12] the variables $(\boldsymbol{E}, \boldsymbol{H}, \rho)$ are used which lead to a slightly more involved formulation for the Poisson structure $\mathbb{J}$, but make the analysis easier.

5.2. *Energy-drift-diffusion systems as Onsager system*

As is explained in [13], one can also handle general energy-drift-diffusion systems with reactions as treated in [32] as long as the reaction satisfy a detailed balance condition. The formulation is most easily in terms of the densities of the charge carriers, namely $\boldsymbol{c}(t,x) \in [0,\infty[^I$, and the density $u \in \mathbb{R}$ of the internal energy. Then, the total energy $\mathcal{E}$ and the total entropy $\mathcal{S}$ are given by

$$\mathcal{E}(\boldsymbol{c},u) = \int_\Omega \frac{\varepsilon}{2}|\nabla\phi_{\boldsymbol{c}}|^2 + u\,\mathrm{d}x \quad \text{and} \quad \mathcal{S}(c,u)\,\mathrm{d}x,$$

where the electrostatic potential $\phi_{\boldsymbol{c}}$ is the unique solution of the Poisson equation

$$-\operatorname{div}\big(\varepsilon\nabla\phi\big) = \delta_{\text{dop}} + \boldsymbol{z}\cdot\boldsymbol{c} \text{ in } \Omega, \text{ plus suitable boundary conditions,}$$

where $\boldsymbol{z} = (z_1,..,z_I) \in \mathbb{Z}^I$ is the vector of the charge numbers. We note that the differentials of the these functionals have the form

$$\mathrm{D}\mathcal{E}(\boldsymbol{c},u) = \begin{pmatrix}\phi_{\boldsymbol{c}}\boldsymbol{z}\\ 1\end{pmatrix} \quad \text{and} \quad \mathrm{D}\mathcal{S}(\boldsymbol{c},u) = \begin{pmatrix}\partial_{\boldsymbol{c}}S(\boldsymbol{c},u)\\ \partial_u S(\boldsymbol{c},u)\end{pmatrix} \text{ where } \partial_u S(\boldsymbol{c},u) = 1/\theta.$$

We define the Onsager operator $\mathbb{K}$ in terms of the dual entropy-production potential

$$\mathcal{R}^*(\boldsymbol{c},u;\boldsymbol{\mu},\tau) = \frac{1}{2}\int_\Omega \begin{pmatrix}\nabla\mu - \tau\boldsymbol{z}\otimes\nabla\phi_{\boldsymbol{c}}\\ \nabla\tau\end{pmatrix}:\mathbb{M}(\boldsymbol{c},u):\begin{pmatrix}\nabla\mu - \tau\boldsymbol{z}\otimes\nabla\phi_{\boldsymbol{c}}\\ \nabla\tau\end{pmatrix} + \boldsymbol{\mu}\cdot\mathbb{H}(\boldsymbol{c},u)\boldsymbol{\mu}\,\mathrm{d}x,$$

i.e. $\mathbb{K} = \mathrm{D}^2_{\boldsymbol{\mu},\tau}\mathcal{R}^*$. Here $\mathbb{M}$ gives the mobilities of the particles and the heat conduction, while $\mathbb{H}$ gives the reaction coefficients.

By construction we have $\Psi^*(\boldsymbol{c},u;\mathrm{D}\mathcal{E}(\boldsymbol{c},u)) \equiv 0$, which implies energy conservation for the Onsager system via $\mathbb{K}(\boldsymbol{c},u)\mathrm{D}\mathcal{E}(\boldsymbol{c},u) \equiv 0$. The Onsager evolution $\frac{\mathrm{d}}{\mathrm{d}t}\binom{\boldsymbol{c}}{u} = \mathbb{K}(\boldsymbol{c},u)\mathrm{D}\mathcal{S}(\boldsymbol{c},u)$ is then given in the form

$$\frac{\mathrm{d}}{\mathrm{d}t}\begin{pmatrix}\boldsymbol{c}\\ u\end{pmatrix} + \begin{pmatrix}\operatorname{div}\boldsymbol{J}_{\boldsymbol{c}}\\ \operatorname{div}\boldsymbol{j}_u\end{pmatrix} = -\begin{pmatrix}0\\ \boldsymbol{z}\otimes\nabla\phi_{\boldsymbol{c}}:\boldsymbol{J}_{\boldsymbol{c}}\end{pmatrix} + \begin{pmatrix}\mathbb{H}(\boldsymbol{c},u)\partial_{\boldsymbol{c}}S(\boldsymbol{c},u)\\ 0\end{pmatrix} \tag{33}$$

$$\text{where } \begin{pmatrix}\boldsymbol{J}_{\boldsymbol{c}}\\ \boldsymbol{j}_u\end{pmatrix} = \mathbb{M}(\boldsymbol{c},u)\begin{pmatrix}\nabla\partial_{\boldsymbol{c}}S(\boldsymbol{c},u) - \partial_u S(\boldsymbol{c},u)\boldsymbol{z}\otimes\nabla\phi_{\boldsymbol{c}}\\ \nabla\partial_u S(\boldsymbol{c},u)\end{pmatrix}.$$

The equation for u in (33) contains the work of the charge carriers moving against the electric field $\boldsymbol{E} = \nabla\phi_{\boldsymbol{c}}$, namely $\boldsymbol{E}\cdot(\boldsymbol{z}\cdot\boldsymbol{J}_{\boldsymbol{c}})$, where $\boldsymbol{z}\cdot\boldsymbol{J}_{\boldsymbol{c}}$ is the net flux of all moving charges. This term is needed in the energy balance to compensate for the change in the electrostatic energy $\frac{\varepsilon}{2}|\nabla\phi_{\boldsymbol{c}}|^2$.

5.3. *A GENERIC formulation for thermo-opto-electronics*

To couple the Maxwell-Bloch equation and the energy-drift-diffusion system we use the state variable $q = (\boldsymbol{D},\boldsymbol{B},\rho,\boldsymbol{c},u) : \Omega \to \mathbb{R}^3 \times \mathbb{R}^3 \times \mathfrak{R}_n \times [0,\infty[^I \times \mathbb{R}$.

If $S(\boldsymbol{D},\boldsymbol{B},\rho,\boldsymbol{c},u)$ is the entropy density then $\partial_u S$ gives the inverse $1/\theta$ of the temperature θ. The energy $\mathcal{E}$ and the entropy $\mathcal{S}$ are assumed to have the form

$$\mathcal{E}(q)=\int_\Omega \frac{1}{2}|\boldsymbol{D}-\Gamma\rho|^2+\frac{1}{2}|\boldsymbol{B}|^2+\mathrm{tr}(N_{\mathrm{B}}\rho)+u\,\mathrm{d}x,\quad \mathcal{S}(q)=\int_\Omega S(\rho,\boldsymbol{c},u)\,\mathrm{d}x$$

leading to the thermodynamic driving forces

$$\begin{pmatrix}\boldsymbol{E}\\ \boldsymbol{H}\\ \eta\\ \boldsymbol{\mu}\\ \tau\end{pmatrix}=\mathrm{D}\mathcal{E}(q)=\begin{pmatrix}\boldsymbol{D}-\Gamma\rho\\ \boldsymbol{B}\\ \Gamma^*(\Gamma\rho-\boldsymbol{D})+N_{\mathrm{B}}\\ 0\\ 1\end{pmatrix},\quad \mathrm{D}\mathcal{S}(q)=\begin{pmatrix}0\\ 0\\ \partial_\rho S\\ \partial_{\boldsymbol{c}} S\\ \partial_u S\end{pmatrix}. \tag{34}$$

For the Poisson structure we choose the Maxwell-Bloch operator, which is now augmented by a term $\mathcal{A}$ that accounts for reversible exchanges of heat:

$$\mathbb{J}_{\mathrm{TOE}}(q):=\begin{pmatrix}0&\mathrm{curl}&0&0&0\\ -\mathrm{curl}&0&0&0&0\\ 0&0&\jmath[\rho,\Box]&0&-\mathcal{A}(q)\\ 0&0&0&0&0\\ 0&0&\mathcal{A}(q)^*&0&0\end{pmatrix}$$

with $\mathcal{A}(q)\tau=\jmath\,\tau\,\theta\,[\rho,\partial_\rho S(q)]$ and $\mathcal{A}(q)^*\eta=\theta\,\mathrm{Re}\,\mathrm{tr}\,\big(\jmath\eta[\rho,\partial_\rho S(q)]\big)$, cf. [17] for similar latent heat terms in thermoelasticity. Here $\mathcal{A}$ is constructed in such a way that $(\mathrm{NIC})_1$ holds, namely using $\theta\partial_u S\equiv\mathbb{1}$ we find $\mathbb{J}_{\mathrm{TOE}}(q)\mathrm{D}\mathcal{S}(q)=0$. Hence, the reversible part of the dynamics of the density matrix ρ takes the form

$$\dot\rho=\jmath\big[\rho,\,\mathrm{D}_\rho\mathcal{E}(q)-\theta\mathrm{D}_\rho\mathcal{S}(\rho)\big]=\jmath\big[\rho,\,\mathrm{D}_\rho\mathcal{F}(\rho)\big] \tag{35}$$

which emphasizes the rôle of the free energy $\mathcal{F}=\mathcal{E}-\theta{*}\mathcal{S}$ for the reversible dynamics.

The Onsager operator $\mathbb{K}_{\mathrm{TOE}}$ for thermo-opto-electronic has the additive form:

$$\mathbb{K}_{\mathrm{TOE}}(q)=\mathbb{K}_{\mathrm{react}}(q)+\mathbb{K}_{\mathrm{B}}(q)+\mathbb{K}_{\mathrm{diff}}(q)+\mathbb{K}_{\mathrm{heat}}(q),$$

where each term is of the form $\mathbb{K}_j(q)=\mathbb{A}_j(q)^*\mathbb{L}_j(q)\mathbb{A}_j(q)$ with $\mathbb{L}_j\geq 0$. Hence if $\mathbb{A}_j(q)\mathrm{D}\mathcal{E}(q)\equiv 0$ holds for all j then $(\mathrm{NIC})_2$ holds as well.

(i) For the *reactions* between charge carriers (including stimulated emission) we choose $\mathbb{A}_{\mathrm{react}}(q)(\boldsymbol{E},\boldsymbol{H},\eta,\boldsymbol{\mu},\tau)=\boldsymbol{\mu}$ and $\mathbb{L}_{\mathrm{react}}=\mathbb{H}$. Moreover, $\mathbb{A}_{\mathrm{react}}(q)\mathrm{D}\mathcal{E}(q)\equiv 0$ follows from $\mathrm{D}_{\boldsymbol{c}}\mathcal{E}(q)\equiv 0$. We obtain $\mathbb{K}_{\mathrm{react}}(q)\mathrm{D}\mathcal{S}(q)=\big(0,0,0,\mathbb{H}\mathrm{D}_{\boldsymbol{c}}\mathcal{S}(q),0\big)^\top$.

(ii) To model the *quantum relaxations* we set $N(q):=\mathrm{D}_\rho\mathcal{E}(q)=N_{\mathrm{B}}-\Gamma^*(\boldsymbol{D}-\Gamma\rho)$, define $\mathbb{A}_{\mathrm{B}}(q)(\boldsymbol{E},\boldsymbol{H},\eta,\boldsymbol{\mu},\tau):=\eta-\tau N(q)$, and choose an arbitrary $\mathbb{L}(q)\geq 0$, e.g. $\mathbb{L}(q)\eta=\sum_{m=1}^M\big[Q_m,\mathcal{C}_\rho[Q_m,\eta]\big]$. Using the definition of $N(q)$ and $\mathrm{D}_\tau\mathcal{E}(q)=1$ gives $\mathbb{A}_{\mathrm{B}}(q)\mathrm{D}\mathcal{E}(q)\equiv 0$, and $(\mathrm{NIC})_2$ holds. In particular, we obtain

$$\mathbb{K}_{\mathrm{B}}(q)\mathrm{D}\mathcal{S}(q)=\Big(0,0,\mathbb{L}(q)\big(\partial_\rho S(q)-\tfrac{1}{\theta}N(q)\big),0,-\,\mathrm{tr}\,\big(N(q)\mathbb{L}(\partial_\rho S(q)-\tfrac{1}{\theta}N(q))\big)\Big)^\top.$$

We observe that the dissipative relaxations are driven by the *free entropy* $\mathcal{N}$, namely by $\mathrm{D}_\rho\mathcal{N}(q)=\mathrm{D}_\rho\mathcal{S}(q)-\frac{1}{\theta}\mathrm{D}_\rho\mathcal{E}(q)$. Moreover, the last term can be seen as a heating term due to dissipation in the quantum systems.

(iii) The *drift and the diffusion* of the charge carriers is modeled via the mobility. Since the flux tensor $\boldsymbol{J_c}$ for the densities $\boldsymbol{c}$ should have the form $\boldsymbol{J_c} = \mathbb{M}_c(\nabla\boldsymbol{\mu} - \frac{1}{\theta}\boldsymbol{z}\otimes\boldsymbol{E}) \in \mathbb{R}^{I\times d}$ we define the operator $\mathbb{A}_{\text{mob}}(q)(\boldsymbol{E},\boldsymbol{H},\eta,\boldsymbol{\mu},\tau)^\top := \boldsymbol{z}\otimes\boldsymbol{E} + \nabla\boldsymbol{\mu} - \tau\,\boldsymbol{z}\otimes(\boldsymbol{D}-\Gamma\rho) \in \mathbb{R}^{I\times d}$. Using (34) we find the relations

$$\mathbb{A}_{\text{mob}}(q)\mathrm{D}\mathcal{E}(q) \equiv 0, \qquad \mathbb{A}_{\text{mob}}(q)\mathrm{D}\mathcal{S}(q) = \nabla\partial_{\boldsymbol{c}}S - \frac{1}{\theta}\boldsymbol{z}\otimes(\boldsymbol{D}-\Gamma\rho).$$

Thus, $\mathbb{K}_{\text{mob}}(q) = \mathbb{A}(q)^*\mathbb{M}_c(q)\mathbb{A}(q)$ satisfies $(\text{NIC})_2$ and provides

$$\mathbb{K}_{\text{diff}}(q)\mathrm{D}\mathcal{S}(q) = \begin{pmatrix} \boldsymbol{z}\cdot\mathbb{M}_c(\nabla\boldsymbol{\mu}-\frac{1}{\theta}\boldsymbol{z}\otimes\boldsymbol{E}) \\ 0 \\ 0 \\ -\operatorname{div}\left(\mathbb{M}_c(\nabla\boldsymbol{\mu}-\frac{1}{\theta}\boldsymbol{z}\otimes\boldsymbol{E})\right) \\ (\boldsymbol{z}\otimes\boldsymbol{E}){:}\mathbb{M}_c(\nabla\boldsymbol{\mu}-\frac{1}{\theta}\boldsymbol{z}\otimes\boldsymbol{E}) \end{pmatrix}$$

Component 1 gives the charge flux $\boldsymbol{z}\cdot\boldsymbol{J_c}$, component 4 generates the drift-diffusion terms, and component 5 is the power of charges working against the electric field.

(iv) The heat transfer can be given in the form

$$\mathbb{K}_{\text{heat}}(q)(\boldsymbol{E},\boldsymbol{H},\eta,\boldsymbol{\mu},\tau)^\top = \left(0,0,0,0,-\operatorname{div}\left(\kappa(q)\nabla\tau\right)\right)^\top.$$

Clearly $\mathrm{D}_u\mathcal{E}(q) = 1$ gives $\mathbb{K}_{\text{heat}}(q)\mathrm{D}\mathcal{E}(q) = 0$, which is $(\text{NIC})_2$. In fact, in general we may allow for cross-diffusion between charges and heat:

$$\begin{pmatrix} \boldsymbol{J_c} \\ \boldsymbol{j}_u \end{pmatrix} = \begin{pmatrix} \mathbb{M}_c & \mathbb{M}_{\text{cross}} \\ \mathbb{M}^*_{\text{cross}} & \kappa(q) \end{pmatrix}\begin{pmatrix} \mathbb{A}_{\text{mob}}(q)\mathrm{D}\mathcal{S}(q) \\ \nabla\partial_u S(q) \end{pmatrix}. \tag{36}$$

Now we are able to write the full GENERIC system $\dot q = \mathbb{J}_{\text{TOE}}(q)\mathrm{D}\mathcal{E}(q) + \mathbb{K}_{\text{TOE}}(q)\mathrm{D}\mathcal{S}(q)$ with the building blocks described above. It leads to the following coupled system of partial differential equations

$$\begin{aligned}
\dot{\boldsymbol{D}} &= \operatorname{curl}\boldsymbol{B} + \boldsymbol{z}\cdot\boldsymbol{J_c},\\
\dot{\boldsymbol{B}} &= -\operatorname{curl}\boldsymbol{E},\\
\dot\rho &= \jmath\big[\rho,\, N(q)-\theta\partial_\rho S(q)\big] + \mathbb{L}(q)\big(\partial_\rho S(q)-\tfrac{1}{\theta}N(q)\big),\\
\dot{\boldsymbol{c}} &= -\operatorname{div}\boldsymbol{J_c} + \mathbb{H}(q)\partial_{\boldsymbol{c}}S(\rho,\boldsymbol{c},u),\\
\dot u &= \theta\operatorname{tr}\left(\jmath N(q)[\rho,\partial_\rho S(q)]\right) - \operatorname{div}\boldsymbol{j}_u - \operatorname{tr}\left\{N(q)\mathbb{L}(q)\big(\partial_\rho S(q)-\tfrac{1}{\theta}N(q)\big)\right\} + (\boldsymbol{z}\otimes\boldsymbol{E}){:}\boldsymbol{J_c},
\end{aligned}$$

where $(\boldsymbol{J_c},\boldsymbol{j}_u)$ is given in (36), $\boldsymbol{E} = \boldsymbol{D}-\Gamma\rho$, and $N(q) = \mathrm{D}_\rho\mathcal{E}(q) = N_{\mathrm{B}}-\Gamma^*(\boldsymbol{D}-\Gamma\rho)$.

The first equation (combined with the fourth and using the charge neutrality $\boldsymbol{z}\cdot\mathbb{H} \equiv 0$ of the reactions) is compatible with $\operatorname{div}\boldsymbol{D} = \delta_{\text{dop}} + \boldsymbol{z}\cdot\boldsymbol{c}$ for all $t > 0$, if it is satisfied for $t = 0$. The last equation can be written as

$$\begin{aligned}
\dot u &= -\operatorname{div}\boldsymbol{j}_u + (\boldsymbol{z}\otimes\boldsymbol{E}){:}\boldsymbol{J_c} - \operatorname{tr}\left\{N(q)\big(\jmath[\rho,\mathrm{D}_\rho\mathcal{E}-\theta\mathrm{D}_\rho\mathcal{S}] + \mathbb{L}(q)(\mathrm{D}_\rho\mathcal{S}-\tfrac{1}{\theta}\mathrm{D}_\rho\mathcal{E})\big)\right\}\\
&= -\operatorname{div}\boldsymbol{j}_u + \boldsymbol{E}\cdot\boldsymbol{z}\cdot\boldsymbol{J_c} - \operatorname{tr}(\mathrm{D}_\rho\mathcal{E}(q)\dot\rho),
\end{aligned}$$

which displays the balance of internal energy with the work of the charges against the electric field $\boldsymbol{E}$ and the work of the quantum system. For more details see [14].

Acknowledgments. The research is partially supported by the ERC grant FP7-267802 (AnaMultiScale). AM is grateful for stimulating discussions with Uwe Bandelow, Thomas Koprucki, Hagen Neidhardt, Florian Platzek, and Paul Racec.

Bibliography

[1] T. Koprucki, A. Wilms, A. Knorr and U. Bandelow, *Opt. Quantum Electron.* **42**, 777 (2011).
[2] H. C. Öttinger, *Phys. Rev. A* **82**, p. 052119(11) (2010).
[3] H. C. Öttinger, *Europhys. Lett.* **94**, p. 10006(6) (2011).
[4] P. J. Morrison, *Phys. D* **18**, 410 (1986).
[5] A. M. Bloch, P. J. Morrison and T. S. Ratiu, Gradient flows in the normal and Kähler metrics and triple bracket generated metriplectic systems (ch. 15), in *Recent Trends in Dynamical Systems*, eds. A. Johann et al (Springer, 2013).
[6] H. Grabert, *Z. Phys. B* **49**, 161 (1982).
[7] R. F. Streater, *Rep. Mathem. Physics* **38**, 419 (1996).
[8] P. W. Michor, D. Petz and A. Andai, *Infin. Dimens. Anal. Quantum Probab. Relat. Top.* **3**, 199 (2000).
[9] E. A. Carlen and J. Maas, *arXiv* **1203.5377** (2012).
[10] A. Mielke, Dissipative quantum mechanics using GENERIC (ch. 21), in *Recent Trends in Dynamical Systems*, eds. A. Johann et al (Springer Verlag, 2013).
[11] J.-L. Joly, G. Metivier and J. Rauch, *J. Differential Equations* **166**, 175 (2000).
[12] É. Dumas, *J. Diff. Eqns.* **219**, 484 (2005).
[13] A. Mielke, *Nonlinearity* **24**, 1329 (2011).
[14] U. Bandelow, T. Koprucki and A. Mielke, *In preparation* (2014).
[15] M. Grmela and H. C. Öttinger, *Phys. Rev. E (3)* **56**, 6620 (1997).
[16] H. C. Öttinger and M. Grmela, *Phys. Rev. E (3)* **56**, 6633 (1997).
[17] A. Mielke, *Contin. Mech. Thermodyn.* **23**, 233 (2011).
[18] A. Mielke and M. Thomas, *In preparation* (2014).
[19] I. C. Gohberg and M. G. Krein, *Introduction to the theory of linear nonselfadjoint operators* (AMS Transl. Mathem. Monographs, 1969).
[20] L. Onsager, *Physical Review* **37**, 405 (1931), (part II, 38:2265-2279).
[21] L. Onsager and S. Machlup, *Phys. Rev.* **91**, 1505 (1953).
[22] R. Kubo, M. Toda and N. Hashitsume, *Nonequilibrium Statistical Mechanics*, 2nd edn. (Springer, 1991).
[23] D. Petz, *J. Math. Phys.* **35**, 780 (1994).
[24] R. Kubo, *Rep. Prog. Phys.* **29**, 255 (1966).
[25] F. Otto, *Comm. Partial Differential Equations* **26**, 101 (2001).
[26] G. Lindblad, *Comm. Math. Phys.* **48**, 119 (1976).
[27] A. Arnold, Mathematical properties of quantum evolution equations, in *Quantum transport. Modelling, analysis and asymptotics*, eds. G. Allaire

et al (Berlin: Springer; Firenze: Fondazione CIME Roberto Conti, 2008) pp. 45–109.
[28] A. Mielke and H. Neidhardt, *In preparation* (2014).
[29] A. Glitzky and A. Mielke, *Z. angew. Math. Phys. (ZAMP)* **64**, 29 (2013).
[30] H. D. Cornean, H. Neidhardt and V. Zagrebnov, *Ann. Henri Poincaré* **10**, 61 (2009).
[31] H. Neidhardt, L. Wilhelm and V. A. Zagrebnov, *WIAS preprint 1895* (2013).
[32] G. Albinus, H. Gajewski and R. Hünlich, *Nonlinearity* **15**, 367 (2002).

Almost sure purely singular continuous spectrum for quasicrystal models

C. Seifert

Technische Universität Hamburg-Harburg, Institut für Mathematik,
21073, Hamburg, Germany
E-mail: christian.seifert@tuhh.de
http://www.mat.tuhh.de/home/cseifert

We review recent developments in the spectral theory of continuum one-dimensional quasicrystals, yielding purely singular continuous spectrum for these Schrödinger operators. Allowing measures as potentials we can generalize some results to very singular potentials, including Kronig-Penney type models.

Keywords: Schrödinger operators, singular continuous spectrum, quasicrystals.

1. Introduction

The aim of this short note is to review continuum models for quasicrystals in one dimension. Such models have been studied intensively during the last decades.

Here, we allow for measures μ as potentials, i.e. we are studying the spectral theory of self-adjoint Schrödinger operators of the form

$$H_\mu := -\Delta + \mu$$

in $L_2(\mathbb{R})$. In order to deal with quasicrystal models the potential μ should satisfy certain properties. Quasicrystals can be considered as media between order (i.e., perfect crystals) and disorder (i.e., amorphous media). On the one hand, quasicrystals are not too far from periodic crystals, so μ should be locally close to being periodic, but still globally aperiodic. Furthermore, μ should satisfy a finite local complexity condition (in a suitable sense for measures). On the other hand, quasicrystals share features from disordered media, so the corresponding Hamiltonians may be considered as random operators with some ergodicity assumptions. Thus, one is tempted to expect neither pure-point spectrum nor absolutely continuous spectrum for such models.

Combining results on the absolutely continuous spectrum obtained in [5] (which make use of Remlings Oracle Theorem in [7]) with Gordon-type arguments for singular potentials as in [10] (see also [2, 3, 8] for earlier results) we will conclude generic (in the sense of an ergodic measure) purely singular continuous spectrum for such operators. We also give a scheme how to produce examples by means of discrete subshifts and a suspension type construction.

2. Schrödinger operators with measures as potentials

We say that

$$\mu\colon \{B \subseteq \mathbb{R};\ B \text{ is a bounded Borel set}\} \to \mathbb{R}$$

is a *local measure* if $\mathbb{1}_K\mu := \mu(\cdot \cap K)$ is a signed Radon measure for any compact set $K \subseteq \mathbb{R}$. Then there exist a (unique) nonnegative Radon measure ν on $\mathbb{R}$ and a measurable function $\sigma\colon \mathbb{R} \to \mathbb{R}$ such that $|\sigma| = 1$ ν-a.e. and $\mathbb{1}_K\mu = \mathbb{1}_K\sigma\nu$ for all compact sets $K \subseteq \mathbb{R}$. The *total variation* of μ is defined by $|\mu| := \nu$. Let $\mathcal{M}_{\mathrm{loc}}(\mathbb{R})$ be the space of all local measures on $\mathbb{R}$.

A local measure $\mu \in \mathcal{M}_{\mathrm{loc}}(\mathbb{R})$ is called *uniformly locally bounded* if

$$\|\mu\|_{\mathrm{unif}} := \sup_{a\in\mathbb{R}} |\mu|\left((a, a+1]\right) < \infty.$$

Let $\mathcal{M}_{\mathrm{loc,unif}}(\mathbb{R})$ denote the space of all uniformly locally bounded local measures. The space $\mathcal{M}_{\mathrm{loc,unif}}(\mathbb{R})$ naturally extends $L_{1,\mathrm{loc,unif}}(\mathbb{R})$ to measures.

For $\mu \in \mathcal{M}_{\mathrm{loc,unif}}(\mathbb{R})$ let H_μ be the maximal operator in $L_2(\mathbb{R})$ associated with $-\Delta + \mu$ (in the distributional sense): define

$$\begin{aligned} D(H_\mu) &:= \{u \in L_2(\mathbb{R}) \cap C(\mathbb{R});\ -u'' + u\mu \in L_2(\mathbb{R})\}, \\ H_\mu u &:= -u'' + u\mu. \end{aligned}$$

Then H_μ is self-adjoint. Realizations of $-\Delta+\mu$ can also be defined via form methods (see [5]) or Sturm-Liouville theory (see [1]). As shown in Theorem 3.6 of [10] all these methods lead to the same self-adjoint operator.

3. Spectral properties of quasicystalline models

Let us recall some definitions from [5] concerning notions of finite local complexity for measures.

Definition 3.1. A *piece* is a pair (ν, I) consisting of an interval $I \subseteq \mathbb{R}$ with positive length $\lambda(I) > 0$ (which is then called the *length* of the piece) and a signed (local) measure ν on $\mathbb{R}$ supported on I. We abbreviate pieces by ν^I. A *finite piece* is a piece of finite length. We say ν^I *occurs* in a signed (local) measure μ at $x \in \mathbb{R}$, if $\mathbb{1}_{[x,x+\lambda(I)]}\mu$ is a translate of ν.

The *concatenation* $\nu^I = \nu_1^{I_1} \mid \nu_2^{I_2} \mid \ldots$ of a finite or countable family $(\nu_j^{I_j})_{j\in N}$, with $N = \{1, 2, \ldots, |N|\}$ (for N finite) or $N = \mathbb{N}$ (for N infinite), of finite pieces is defined by

$$I = \left[\min I_1, \min I_1 + \sum_{j\in N} \lambda(I_j)\right],$$

$$\nu = \nu_1 + \sum_{j\in N, j\geq 2} \nu_j\Big(\cdot - \Big(\min I_1 + \sum_{k=1}^{j-1} \lambda(I_k) - \min I_j\Big)\Big).$$

We also say that ν^I is *decomposed* by $(\nu_j^{I_j})_{j\in N}$.

Definition 3.2. Let μ be a signed (local) measure on $\mathbb{R}$. We say that μ has the *finite decomposition property* (f.d.p.), if there exist a finite set $\mathcal{P}$ of finite pieces (called the *local pieces*) and $x_0 \in \mathbb{R}$, such that $\mathbb{1}_{[x_0,\infty)}\mu^{[x_0,\infty)}$ is a translate of a concatenation $v_1^{I_1} \mid \nu_2^{I_2} \mid \ldots$ with $\nu_j^{I_j} \in \mathcal{P}$ for all $j \in \mathbb{N}$. Without restriction, we may assume that $\min I = 0$ for all $\nu^I \in \mathcal{P}$.

A signed (local) measure μ has the *simple finite decomposition property* (s.f.d.p.), if it has the f.d.p. with a decomposition such that there is $\ell > 0$ with the following property: Assume that the two pieces

$$\nu_{-m}^{I_{-m}} \mid \ldots \mid \nu_0^{I_0} \mid \nu_1^{I_1} \mid \ldots \mid \nu_{m_1}^{I_{m_1}} \quad \text{and} \quad \nu_{-m}^{I_{-m}} \mid \ldots \mid \nu_0^{I_0} \mid \mu_1^{J_1} \mid \ldots \mid \mu_{m_2}^{J_{m_2}}$$

occur in the decomposition of μ with a common first part $\nu_{-m}^{I_{-m}} \mid \ldots \mid \nu_0^{I_0}$ of length at least ℓ and such that

$$\mathbb{1}_{[0,\ell)}(\nu_1^{I_1} \mid \ldots \mid \nu_{m_1}^{I_{m_1}}) - \mathbb{1}_{[0,\ell)}(\mu_1^{J_1} \mid \ldots \mid \mu_{m_2}^{J_{m_2}}),$$

where $\nu_j^{I_j}$, $\mu_k^{J_k}$ are pieces from the decomposition (in particular, all belong to $\mathcal{P}$ and start at 0) and the latter two concatenations are of lengths at least ℓ. Then

$$\nu_1^{I_1} = \mu_1^{J_1}.$$

We can formulate a striking spectral consequence for Schrödinger operators with potentials having the s.f.d.p.

Theorem 3.1 (see [6, Theorem 7.1]). *Let $\mu \in \mathcal{M}_{\text{loc,unif}}(\mathbb{R})$ such that μ and $\mu(-(\cdot))$ have the s.f.d.p. and assume that neither μ nor $\mu(-(\cdot))$ are eventually periodic. Then $\sigma_{ac}(H_\mu) = \varnothing$.*

As noted in the introduction Schrödinger operators modelling quasicrystals can be thought of as random operators. To this end, let $\Omega \subseteq \mathcal{M}_{\text{loc,unif}}(\mathbb{R})$ be $\|\cdot\|_{\text{unif}}$-bounded and closed with respect to the vague topology. Then Ω is compact and hence metrizable (see [9, Proposition 4.1.2]). Furthermore, assume that Ω is translation invariant, i.e., we have a group action $\alpha\colon \mathbb{R}\times\Omega \to \Omega$, $\alpha(t,\omega) := \alpha_t(\omega) := \omega(\cdot + t)$

acting on Ω. Note that α is continuous. For $\omega \in \Omega$ define H_ω as above. Note that $(H_\omega)_{\omega\in\Omega}$ is a random operator (see [6, Theorem 3.6]) and it is an ergodic operator family if (Ω, α) is ergodic with ergodic measure $\mathbb{P}$. Recall that (Ω, α) is called minimal if every orbit $\{\alpha_t(\omega);\ t \in \mathbb{R}\}$ is dense in Ω. Both ergodicity and minimality are suitable assumptions for quasicrystalline models.

Proposition 3.1 (see [5, Theorem 5.1]). *Let (Ω, α) be minimal, having the s.f.d.p. (i.e. for every $\omega \in \Omega$: ω and $\omega(-(\cdot))$ has s.f.d.p.), and aperiodic (i.e. there exists $\omega \in \Omega$ which is not periodic). Then $\sigma_{ac}(H_\omega) = \varnothing$ for all $\omega \in \Omega$.*

Proof. Assume that $\{\omega \in \Omega;\ \sigma_{ac}(H_\omega) \neq \varnothing\}$ in nonempty. Then by [5, Theorem 4.1], the set $\{\omega \in \Omega;\ \omega \text{ or } \omega(-(\cdot)) \text{ is eventually periodic}\}$ is nonempty as well. W.l.o.g. assume that ω is periodic for $t \geq t_0$ with period p. By closedness of Ω,

$$\tilde{\omega} := \lim_{t\to\infty} \alpha_t(\omega) = \lim_{t\to\infty} \omega(\cdot + t) \in \Omega$$

and $\tilde{\omega}$ is periodic with period p. By minimality, for any $\omega' \in \Omega$ there exists (t_n) in $\mathbb{R}$ such that $\alpha_{t_n}(\tilde{\omega}) \to \omega'$. Since $\tilde{\omega}$ is periodic and α is continuous, we arrive at

$$\alpha_p(\omega') = \alpha_p\left(\lim_{n\to\infty} \alpha_{t_n}(\tilde{\omega})\right) = \lim_{n\to\infty} \alpha_{t_n}\alpha_p(\tilde{\omega}) = \omega'.$$

So, every $\omega \in \Omega$ must be periodic with the same period, a contradiction. □

Let us now focus on the pure point spectrum. As mentioned in the introduction, potentials modelling quasicrystals are close to periodic potentials. The following condition due to Kaminaga [4] suits very well for "local periodicity".

Definition 3.3. Let $(\Omega, \alpha, \mathbb{P})$ be ergodic. We say that it satisfies condition (K) if there exists (p_n) in $(0, \infty)$ with $p_n \to \infty$ such that

$$G_n := \left\{\omega \in \Omega;\ \mathbb{1}_{[0,p_n]}\omega = \mathbb{1}_{[0,p_n]}\alpha_{p_n}(\omega) = \mathbb{1}_{[0,p_n]}\alpha_{-p_n}(\omega)\right\} \quad (n \in \mathbb{N})$$

satisfies $\limsup_{n\to\infty} \mathbb{P}(G_n) > 0$.

Note that the set G_n contains all measures $\omega \in \Omega$ for which the three pieces of the intervals $[-p_n, 0]$, $[0, p_n]$ and $[p_n, 2p_n]$ are equal, i.e. which are locally close to being periodic.

Lemma 3.1. *Let $(\Omega, \alpha, \mathbb{P})$ be ergodic satisfying (K). Then for $\mathbb{P}$-a.a. $\omega \in \Omega$ we have $\sigma_{pp}(H_\omega) = \varnothing$, i.e., H_ω does not have any eigenvalues $\mathbb{P}$-a.s.*

Proof. Let $\Omega_c := \{\omega \in \Omega;\ \sigma_{pp}(H_\omega) = \varnothing\}$. Note that Ω_c is α-invariant. Ergodicity implies $\mathbb{P}(\Omega_c) \in \{0, 1\}$. By a Gordon type argument, see [10, Corollary 5.5], we have

$$G := \limsup_{n\to\infty} G_n = \bigcap_{n\in\mathbb{N}} \bigcup_{k=n}^{\infty} G_k \subseteq \Omega_c.$$

Hence,

$$\mathbb{P}(\Omega_c) \geq \mathbb{P}(G) = \mathbb{P}(\limsup_{n\to\infty} G_n) \geq \limsup_{n\to\infty} \mathbb{P}(G_n) > 0.$$

We conclude that $\mathbb{P}(\Omega_c) = 1$. □

We can now state our main theorem: the almost sure purely singular continuous spectrum for quasicrystalline models.

Theorem 3.2. *Let $(\Omega, \alpha, \mathbb{P})$ be ergodic and minimal, having the s.f.d.p., aperiodic, and satisfying (K). Then H_ω has purely singular continuous spectrum for $\mathbb{P}$-almost all $\omega \in \Omega$.*

Proof. By Proposition 3.1 we obtain that $\sigma_{ac}(H_\omega) = \varnothing$ for all $\omega \in \Omega$. Lemma 3.1 yields $\sigma_{pp}(H_\omega) = \varnothing$ for $\mathbb{P}$-a.a. $\omega \in \Omega$. Hence, for $\mathbb{P}$-a.a. $\omega \in \Omega$, H_ω has purely singular continuous spectrum. □

This result is as far as we know the most general version; previously known results as stated in [5, Theorem 7.4] only work for $L_{1,\mathrm{loc}}$-potentials.

Example 3.1. Let A be a finite set equipped with the discrete topology. Let $X \subseteq A^{\mathbb{Z}}$ be closed and invariant under the shift $S\colon A^{\mathbb{Z}} \to A^{\mathbb{Z}}$, $Sx(n) := x(n+1)$. Then (X, S) is called a subshift. We use a suspension type consruction. For $a \in A$ choose $\nu_a \in \mathcal{M}_{\mathrm{loc,unif}}(\mathbb{R})$ supported on $[0, l_a]$. For $x \in X$ define $\omega_x \in \mathcal{M}_{\mathrm{loc,unif}}(\mathbb{R})$ by

$$\omega_x := \sum_{n\in\mathbb{N}_0} \delta_{\sum_{k=0}^{n-1} l_{x(k)}} * \nu_{x(n)} + \sum_{n\in\mathbb{N}} \delta_{\sum_{k=-n}^{-1} -l_{x(k)}} * \nu_{x(-n)}.$$

Let $\Omega := \{\alpha_t(\omega_x);\ x \in X,\ t \in \mathbb{R}\}$. As shown in [6, Proposition 8.2] many properties of (X, S) such as ergodicity and minimality may be transferred to (Ω, α). Furthermore, (Ω, α) has the s.f.d.p., if at most one of the ν_a's is a multiple of Lebesgue measure. Furthermore, condition (K) can also be checked by an analogue condition for (X, S), see [5, Proposition 5].

Thus, we can generate many examples including Kronig-Penney type models (where the potentials consist of δ-peaks on a quasicrystalline lattice) by choosing appropriate ν_a's and lengths l_a's.

Bibliography

[1] A. Ben Amor and C. Remling, *Direct and inverse spectral theory of one-dimensional Schrödinger operators with measures.* Integr. equ. oper. theory **52**, 395–417 (2005).

[2] D. Damanik and G. Stolz, *A generalization of Gordon's theorem and applications to quasiperiodic Schrödinger operators.* Electron. J. Diff. Eqns. **55**, 1–8 (2000).

[3] A. Gordon, *On the point spectrum of the one-dimensional Schrödinger operator.* Usp. Math. Nauk 31, 257–258 (1976).

[4] M. Kaminaga, *Absence of point spectrum for a class of discrete Schrödiner operators with quasiperiodic potential.* Forum Math. **8**, 63–69 (1996).

[5] S. Klassert, D. Lenz and P. Stollmann, *Delone measures of finite local complexity and applications to spectral theory of one-dimensional continuum models of quasicrystals.* Discrete Contin. Dyn. Syst. **29**(4), 1553–1571 (2011).

[6] D. Lenz, C. Seifert and P. Stollmann, *Zero measure Cantor spectra for continuum one-dimensional quasicrystals.* J. Differential Equations **256**(6), 1905–1926 (2014).

[7] C. Remling, *The absolutely continuous spectrum of one-dimensional Schrödinger operators.* Math Phys Anal Geom **10**, 359–373 (2007).

[8] C. Seifert, *Gordon type theorem for measure perturbation.* Electron. J. Diff. Eqns. **111**, 1–9 (2011).

[9] C. Seifert, *Measure-perturbed one-dimensional Schrödinger operators – A continuum model for quasicrystals.* Dissertation thesis, Chemnitz University of Technology (2012). url: `http://nbn-resolving.de/urn:nbn:de:bsz:ch1-qucosa-102766`

[10] C. Seifert and H. Vogt, *A weak Gordon type condition for absence of eigenvalues of one-dimensional Schrödinger operators.* Integr. equ. oper. theory **78**(3), 383–405 (2014).

Adiabatic theorems with and without spectral gap condition for non-semisimple spectral values

J. Schmid

Fachbereich Mathematik, Universität Stuttgart,
D-70569 Stuttgart, Germany
jochen.schmid@mathematik.uni-stuttgart.de

We present adiabatic theorems with and without spectral gap condition for operators $A(t) : D(A(t)) \subset X \to X$ in a Banach space X. In the case with spectral gap, the considered spectral values $\lambda(t) \in \sigma(A(t))$ are not required to be semisimple and, in particular, need not be eigenvalues. In the case without spectral gap, the considered eigenvalues $\lambda(t) \in \partial\sigma(A(t))$ are not required to be weakly semisimple. We also discuss adiabatic theorems for operators $A(t)$ with time-dependent domains and an application to operators defined by symmetric sesquilinear forms.

Keywords: Adiabatic theorems with and without spectral gap condition, not necessarily (weakly) semisimple spectral values, time-dependent domains.

1. Introduction

Adiabatic theory for general – as opposed to skew self-adjoint [1–3] – operators is a quite young area of research which emerged in [4] and was developed further in [5–7]. (In the complex time method [8, 9] in the spirit of [10] non-self-adjoint operators appear as well, yet as auxiliary objects.) In general terms, adiabatic theory is concerned with linear operators $A(t) : D(A(t)) \subset X \to X$ in a Banach space X over $\mathbb{C}$, spectral values $\lambda(t)$ of $A(t)$ and projections $P(t)$ in X (for $t \in I := [0,1]$) such that

- $A(t)$ is densely defined closed for every $t \in I$ and the initial value problems

$$x' = A(\varepsilon s)x \quad (s \in [0, \frac{1}{\varepsilon}]) \quad \text{or} \quad x' = \frac{1}{\varepsilon}A(t)x \quad (t \in [0,1])$$

  with initial conditions $x(s_0) = y$ or $x(t_0) = y$ are well-posed on the spaces $D(A(t))$ for every value of the slowness parameter $\varepsilon \in (0, \infty)$ (**Condition 1.1**).

- $P(t)$ is spectrally related with $\lambda(t)$ in some natural sense (specified below) for every $t \in I$ except possibly for some few t (**Condition 1.2**).

What one basically wants to know in adiabatic theory is the following: when – under which additional conditions on A, λ and P – does the evolution U_ε generated by the operators $\frac{1}{\varepsilon}A(t)$ (Condition 1.1) approximately follow the spectral subspaces $P(t)X$ related to the spectral values $\lambda(t)$ of $A(t)$ (Condition 1.2) as the slowness parameter ε tends to 0? Shorter and more precisely: when is it true that

$$(1 - P(t))U_\varepsilon(t)P(0) \longrightarrow 0 \quad (\varepsilon \searrow 0)$$

(with respect to a certain operator topology) for all $t \in I$? *Adiabatic theorems* are, by definition, theorems that give such conditions. We distinguish adiabatic theorems *with spectral gap condition (uniform or non-uniform)* and adiabatic theorems *without spectral gap condition* depending on whether $\lambda(t)$ is isolated in $\sigma(A(t))$ for every $t \in I$ (uniformly or non-uniformly) or not.

We present, in this paper, some new adiabatic theorems for general operators $A(t)$ carrying further the adiabatic theory from [5, 7] mainly in two respects:

- first, by no longer requiring the considered spectral values $\lambda(t)$ of $A(t)$ to be (weakly) semisimple
- second, by no longer requiring the domain of $A(t)$ to be time-independent.

We do not strive for the utmost generality and do not give complete proofs here (which can be found in [11]), but instead try to convey the central ideas.

2. Spectral-theoretic preliminaries

In this preliminary section, we identify a natural notion of spectral relatedness – namely, associatedness in the case with spectral gap and weak associatedness in the case without spectral gap – thereby specifying the vague Condition 1.2 above. Suppose that $A : D(A) \subset X \to X$ is densely defined with $\rho(A) \neq \emptyset$ and $\lambda \in \sigma(A)$. If λ is isolated in $\sigma(A)$, then a projection P is called *associated with A and λ* iff P commutes with A and $A_P := A|_{PD(A)}$ is bounded such that

$$\sigma(A_P) = \{\lambda\} \quad \text{whereas} \quad \sigma(A_{1-P}) = \sigma(A) \setminus \{\lambda\}.$$

Additionally, λ is called *semisimple* iff λ is a pole of $(\,.\, - A)^{-1}$ of order 1. If λ is not necessarily isolated in $\sigma(A)$, then a projection P is called *weakly associated (of order m) with A and λ* iff P commutes with A and A_P is bounded such that

$$A_P - \lambda \text{ is nilpotent (of order } m) \text{ whereas } A_{1-P} - \lambda \text{ is injective with dense range.}$$

Additionally, λ is called *weakly semisimple* iff there is a projection P weakly associated with A and λ of order 1, that is, $(A_P - \lambda)^1 = 0$. We have the following central properties of associatedness (which are completely well-known) and of weak associatedness (which seem to be new).

- If λ is isolated in $\sigma(A)$, then there exists a unique projection P associated with A and λ, and $P = (2\pi i)^{-1} \int_\gamma (z - A)^{-1} \, dz$ (Riesz projection). If P is associated with A and λ, and λ is a pole of $(\,.\, - A)^{-1}$ of order m, then

$$PX = \ker(A - \lambda)^k \quad \text{and} \quad (1 - P)X = \operatorname{rg}(A - \lambda)^k \quad (k \geq m).$$

- In general, there exists no projection P weakly associated with A and λ, but if it exists it is unique. If P is weakly associated with A and λ of order m, then

$$PX = \ker(A - \lambda)^k \quad \text{and} \quad (1 - P)X = \overline{\operatorname{rg}(A - \lambda)^k} \quad (k \geq m).$$

Clearly, associatedness is a completely natural notion of spectral relatedness in the case with spectral gap. In order to see that weak associatedness is equally natural in the case without spectral gap, notice that it is certainly natural to require that A_P be bounded with $\sigma(A_P) = \{\lambda\}$ (in other words, that $A_P - \lambda$ be quasinilpotent) and that $A_{1-P} - \lambda$ be injective – and from this one naturally arrives at weak associatedness by further requiring that $A_P - \lambda$ be nilpotent (instead of only quasinilpotent) and that λ belong to the continuous spectrum of A_{1-P} (instead of the residual spectrum). We conclude this section with the following remarks.

- If A is a spectral operator [12] (with spectral measure P^A) and $\lambda \in \sigma(A)$ such that for some bounded neighborhood σ of λ the bounded spectral operator $A|_{P^A(\sigma)X}$ is of finite type, then the projection weakly associated with A and λ exists and is given by $P^A(\{\lambda\})$. In particular, this is true if A is spectral of scalar type.

- If P is associated with A and λ, then the dual operator P^* is associated with A^* and λ. An analogous implication holds for weak associatedness, provided that A is a semigroup generator and X is reflexive.

3. Adiabatic theorems ...

3.1. ... *with spectral gap condition*

We begin with an adiabatic theorem with spectral gap condition, which treats uniform and non-uniform spectral gaps alike and which uses the convenient terminology of $\lambda(\,.\,)$ falling into $\sigma(A(\,.\,)) \setminus \{\lambda(\,.\,)\}$ at a point t_0 (which means there is a sequence (t_n) converging to t_0 such that $\operatorname{dist}(\lambda(t_n), \sigma(A(t_n)) \setminus \{\lambda(t_n)\}) \longrightarrow 0$ as $n \to \infty$). It is a quite simple generalization of the adiabatic theorem of Abou Salem [5] covering the case of semisimple spectral values. Its proof rests upon solving an appropriate commutator equation.

Theorem 3.1. *Suppose that*

- $A(t) : D \subset X \to X$ *for every* $t \in I$ *generates a contraction semigroup and* $t \mapsto A(t)x$ *is continuously differentiable for every* $x \in D$
- $\lambda(t)$ *for every* $t \in I$ *is a spectral value of* $A(t)$, $\lambda(\,.\,)$ *falls into* $\sigma(A(\,.\,)) \setminus \{\lambda(\,.\,)\}$ *at only countably many points accumulating at only finitely many points, and* $t \mapsto \lambda(t)$ *is continuous*
- $P(t)$ *for every* $t \in I \setminus N$ *is associated with* $A(t)$ *and* $\lambda(t)$ *and* $I \setminus N \ni t \mapsto P(t)$ *extends to a twice strongly continuously differentiable map on the whole of* I, *where* N *denotes the set of those points where* $\lambda(\,.\,)$ *falls into* $\sigma(A(\,.\,)) \setminus \{\lambda(\,.\,)\}$.

If V_ε *denotes the (adiabatic) evolution system for* $\frac{1}{\varepsilon}A + [P', P]$ *on* D, *then*

$$\sup_{t \in I} \|U_\varepsilon(t) - V_\varepsilon(t)\| = O(\varepsilon) \quad or \quad o(1) \quad (\varepsilon \searrow 0),$$

depending on whether $N = \emptyset$ *(uniform spectral gap) or* $N \neq \emptyset$ *(non-uniform gap).*

Sketch of proof. Since the case with non-uniform gap can be reduced by a standard argument to the case of uniform gap, we have only to consider this latter case. Setting

$$B(t) := \frac{1}{2\pi i} \int_{\gamma_t} (z - A(t))^{-1} P'(t) (z - A(t))^{-1} \, dz$$

with cycles γ_t in $\rho(A(t))$ encircling $\lambda(t)$ but not $\sigma(A(t)) \setminus \{\lambda(t)\}$, we easily obtain (central properties of associatedness!) the commutator equation $BA - AB \subset [P', P]$. With this commutator equation, in turn, we can rewrite the difference $V_\varepsilon - U_\varepsilon$ as

$$\begin{aligned} V_\varepsilon(t) - U_\varepsilon(t) &= \varepsilon \, U_\varepsilon(t, \tau) B(\tau) V_\varepsilon(\tau) \Big|_{\tau=0}^{\tau=t} \\ &\quad - \varepsilon \int_0^t U_\varepsilon(t, \tau) \big(B'(\tau) + B(\tau)[P'(\tau), P(\tau)] \big) V_\varepsilon(\tau) \, d\tau, \end{aligned}$$

from which the desired conclusion follows by the boundedness of U_ε and V_ε in ε. □

We refer to [11] for simple examples of the above theorem where the previously known adiabatic theorems [5–7] cannot be applied. In particular, these examples show that the considered spectral values $\lambda(t)$ may well be essential singularities of the resolvent of $A(t)$ and need not even be eigenvalues. A similar – but weaker – theorem is a corollary of an adiabatic theorem of higher order by Joye [6] where the operators $A(t)$ have to be required to analytically depend on t and the considered spectral values are required to be poles of the resolvent.

3.2. *. . . without spectral gap condition*

We now present the main result of this paper: an adiabatic theorem without spectral gap condition for not necessarily weakly semisimple eigenvalues. It generalizes an adiabatic theorem independently obtained by Avron, Fraas, Graf, Grech [7] and by Schmid [13], which covers the case of weakly semisimple eigenvalues. Its proof rests upon solving an appropriate approximate commutator equation.

Theorem 3.2. *Suppose that*

- $A(t) : D \subset X \to X$ *for every* $t \in I$ *generates a contraction semigroup and* $t \mapsto A(t)x$ *is continuously differentiable for every* $x \in D$.
- $\lambda(t)$ *for every* $t \in I$ *is an eigenvalue of* $A(t)$ *such that* $\lambda(t)+\delta e^{i\vartheta(t)} \in \rho(A(t))$ *for every* $\delta \in (0, \delta_0]$ *and* $t \mapsto \lambda(t), e^{i\vartheta(t)}$ *are continuously differentiable*
- $P(t)$ *is weakly associated with* $A(t)$ *and* $\lambda(t)$ *for almost every* $t \in I$,
$$\left\| \big(\lambda(t) + \delta e^{i\vartheta(t)} - A(t)\big)^{-1}(1 - P(t)) \right\| \le \frac{M_0}{\delta} \quad (\delta \in (0, \delta_0]),$$
$\operatorname{rk} P(0) < \infty$ *and* $t \mapsto P(t)$ *is continuously differentiable.*

If X *is reflexive, then*

$$\sup_{t\in I} \|U_\varepsilon(t) - V_\varepsilon(t)\| \longrightarrow 0 \quad (\varepsilon \searrow 0),$$

whenever the (adiabatic) evolution system V_ε *for* $\frac{1}{\varepsilon}A + [P', P]$ *exists on* D. *If* X *is arbitrary, then* $\sup_{t\in I} \|(1 - P(t))U_\varepsilon(t)P(0)\| \longrightarrow 0$ *as* $\varepsilon \searrow 0$.

Sketch of proof. We may assume by a standard approximation argument [3] using mollification that $t \mapsto P(t)$ is twice continuously differentiable (in which case the existence of V_ε is guaranteed). Setting $m_0 := \operatorname{rk} P(0) = \operatorname{rk} P(t)$ (indepent of t!) and

$$B_{\boldsymbol{\delta}} := \sum_{k=0}^{m_0-1} \Big(\prod_{i=1}^{k+1} \overline{R}_{\delta_i}\Big) P'P(\lambda - A)^k + \sum_{k=0}^{m_0-1} (\lambda - A)^k PP' \Big(\prod_{i=1}^{k+1} \overline{R}_{\delta_i}\Big)$$

where $\boldsymbol{\delta} := (\delta_1, \dots, \delta_{m_0})$ and $\overline{R}_\delta := (\lambda + \delta - A)^{-1}(1 - P)$, we obtain from $PX = \ker(A - \lambda)^{m_0}$ (central properties of weak associatedness!) the approximate commutator equation $B_{\boldsymbol{\delta}}A - AB_{\boldsymbol{\delta}} \subset [P', P] - C_{\boldsymbol{\delta}}$, where $C_{\boldsymbol{\delta}} := C_{\boldsymbol{\delta}}^+ - C_{\boldsymbol{\delta}}^-$ with

$$C_{\boldsymbol{\delta}}^+ := \sum_{k=0}^{m_0-1} \delta_{k+1} \Big(\prod_{i=1}^{k+1} \overline{R}_{\delta_i}\Big) P'P(\lambda - A)^k, \quad C_{\boldsymbol{\delta}}^- := \sum_{k=0}^{m_0-1} (\lambda - A)^k PP' \delta_{k+1} \Big(\prod_{i=1}^{k+1} \overline{R}_{\delta_i}\Big).$$

At first glance, this solution $B_{\boldsymbol{\delta}}$ of the approximate commutator equation might seem a bit mysterious, but actually, it can be guessed from the solution B of the commutator equation in the case with spectral gap: indeed, if λ is a pole of $(\,.\,-A)^{-1}$ of order at most m_0, then B can be seen to be equal to $B_{\boldsymbol{\delta}=0}$ with the help of Cauchy's theorem. With the above approximate commutator equation, we can now rewrite the difference $V_\varepsilon - U_\varepsilon$ as

$$V_\varepsilon(t) - U_\varepsilon(t) = \varepsilon\, U_\varepsilon(t,\tau) B_\delta(\tau) V_\varepsilon(\tau)\big|_{\tau=0}^{\tau=t} - \varepsilon \int_0^t U_\varepsilon(t,\tau)\big(B_{\boldsymbol{\delta}}'(\tau) + B_{\boldsymbol{\delta}}(\tau)\cdot$$

$$\cdot\, [P'(\tau), P(\tau)]\big) V_\varepsilon(\tau)\, d\tau + \int_0^t U_\varepsilon(t,\tau) C_{\boldsymbol{\delta}}(\tau) V_\varepsilon(\tau)\, d\tau$$

and with the assumed resolvent estimate we obtain the estimates

$$\|B_{\boldsymbol{\delta}}(t)\| \le c\,(\delta_1 \cdots \delta_{m_0})^{-1}, \quad \|B_{\boldsymbol{\delta}}'(t)\| \le c\,(\delta_1 \cdots \delta_{m_0})^{-(m_0+1)},$$

$$\int_0^1 \|C_{\boldsymbol{\delta}}^{\pm}(\tau)\|\, d\tau \le c \sum_{k=0}^{m_0-1} (\delta_1 \cdots \delta_k)^{-1} \eta^{\pm}(\delta_{k+1})$$

where $\eta^+(\delta) := \int_0^1 \delta \left\|\overline{R}_\delta(\tau) P'(\tau) P(\tau)\right\| d\tau$ and $\eta^-(\delta) := \int_0^1 \delta \left\|P(\tau) P'(\tau) \overline{R}_\delta(\tau)\right\| d\tau$ for $\delta \in (0, \delta_0]$. With $(1-P)X = \overline{\mathrm{rg}(A-\lambda)^{m_0}}$ and $(1-P^*)X^* = \overline{\mathrm{rg}(A^*-\lambda)^{m_0}}$ (weak associatedness for semigroup generators in reflexive spaces carries over to the dual operators!) and with the assumed resolvent estimate one can show that

$$\delta\, \overline{R}_\delta(t), \quad \delta\, \overline{R}_\delta(t)^* \longrightarrow 0 \quad (\delta \searrow 0)$$

w.r.t. the strong operator topology for almost every $t \in I$, and since $\operatorname{rk} P(t)^* = \operatorname{rk} P(t) < \infty$, one even gets $\eta^{\pm}(\delta) \longrightarrow 0$. Choosing now $\boldsymbol{\delta} = \boldsymbol{\delta}_\varepsilon = (\delta_{1\,\varepsilon}, \ldots, \delta_{m_0\,\varepsilon})$ in a suitable way, we arrive at the desired conclusion. □

We refer to [11] for a wealth of simple examples of the above theorem where the previously known adiabatic theorems [7, 13] cannot be applied (finding interesting applied examples will be the subject of future research). In these examples the operators $A(t)$ need not be spectral, but for spectral operators the construction of examples is particularly simple: one can ensure the assumptions – in particular, the reduced resolvent estimate – of the above theorem by choosing $A(t)$ for every t as a spectral operator (with spectral measure $P^{A(t)}$) generating a contraction semigroup and $\lambda(t)$ as an eigenvalue of $A(t)$ such that, apart from regularity conditions,

- $A(t)|_{P^{A(t)}(\sigma(t))D}$ is spectral of scalar type for some punctured neighborhood
  $$\sigma(t) := \sigma(A(t)) \cap \overline{U}_{r_0}(\lambda(t)) \setminus \{\lambda(t)\}$$
  of $\lambda(t)$ in $\sigma(A(t))$ ($r_0 \in (0,\infty) \cup \{\infty\}$) and $\operatorname{rk} P^{A(t)}(\{\lambda(t)\}) < \infty$ for a. e. t,
- the open sector $\{\lambda(t) + \delta e^{i\vartheta} : \delta \in (0, \delta_0), \vartheta \in (\vartheta(t) - \vartheta_0, \vartheta(t) + \vartheta_0)\}$ of radius $\delta_0 \in (0,\infty)$ and angle $2\vartheta_0 \in (0,\pi)$ is contained in $\rho(A(t))$.

In the special case where the eigenvalues $\lambda(t)$ from the above theorem are purely imaginary for every $t \in I$, these eigenvalues are automatically weakly semisimple by the contraction semigroup assumption, and so, in this special case, the adiabatic theorem above reduces to the theorem from [7, 13]. Simple examples [6, 11] show that the contraction semigroup assumption cannot be essentially weakened.

3.3. *...for time-dependent domains*

So far, we have always considered operators $A(t)$ with time-independent domains. It is possible to quite easily extend all the presented adiabatic theorems to operators with time-dependent domains, which in essence is due to the following facts:

- An evolution system $W = (W(t, s))$ for operators with time-dependent domains is right-differentiable w.r.t. the second variable s in some appropriate sense
- A continuous, right-differentiable map $f : [a, b] \to X$ with bounded right derivative $\partial_+ f$ belongs to the Sobolev class $W^{1,\infty}$ and therefore the fundamental theorem of calculus is available

We do not state these extended adiabatic theorems with their precise assumptions (which can be found in [11]), but instead give an application to the special case of skew self-adjoint operators $A(t) = iA_{a(t)}$ defined by symmetric sesquilinear forms $a(t)$ (Schrödinger operators with Rollnik potentials $V(t)$, for instance). Consider the situation, where H^+ is continuously and densely embedded in H (both Hilbert spaces), where $a(t)$ is a symmetric sesquilinear form on H^+ such that, for some $m \in (0, \infty)$,

$$\langle \,.\,, ..\rangle_t^+ := a(t)(\,.\,, ..) + m \langle \,.\,, ..\rangle$$

is a scalar product on H^+ and $\|\,.\,\|_t^+$ is equivalent to $\|\,.\,\|^+$ for every $t \in I$, and where $t \mapsto a(t)(x, y)$ is twice continuously differentiable for all $x, y \in H^+$. We then have the following adiabatic theorem without spectral gap condition, which is a generalization of a theorem of Bornemann [2] where the considered eigenvalues are required to belong to the discrete spectrum. It can be proved in a considerably simpler way than in [2], namely by applying the general adiabatic theorems for time-dependent domains mentioned above.

Theorem 3.3. *Suppose that*

- $A(t) = iA_{a(t)} : D(A(t)) \subset H \to H$ *with* $a(t)$ *as above*
- $\lambda(t)$ *for every* $t \in I$ *is an eigenvalue of* $A(t)$ *such that* $t \mapsto \lambda(t)$ *continuous*
- $P(t)$ *is weakly associated with* $A(t)$ *and* $\lambda(t)$ *for almost every* $t \in I$, $\operatorname{rk} P(0) < \infty$ *and* $t \mapsto P(t)$ *is continuously differentiable.*

Then

$$\sup_{t\in I} \|(1 - P(t))U_\varepsilon(t)P(0)\|, \quad \sup_{t\in I} \|P(t)U_\varepsilon(t)(1 - P(0))\| \longrightarrow 0 \quad (\varepsilon \searrow 0).$$

Bibliography

[1] J. E. Avron, A. Elgart: *Adiabatic theorem without a gap condition.* Commun. Math. Phys. **203** (1999), 445-463.

[2] F. Bornemann: *Homogenization in time of singularly perturbed mechanical systems.* Lecture Notes in Mathematics **1687**. Springer, 1998.

[3] S. Teufel: *Adiabatic perturbation theory in quantum dynamics.* Lecture Notes in Mathematics **1821**. Springer, 2003.

[4] G. Nenciu, G. Rasche: *On the adiabatic theorem for non-self-adjoint Hamiltonians.* J. Phys. A: Math. Gen. **25** (1992), 5741-5751.

[5] W. Abou Salem: *On the quasi-static evolution of nonequilibrium steady states.* Ann. Henri Poincaré **8** (2007), 569-596.

[6] A. Joye: *General adiabatic evolution with a gap condition.* Commun. Math. Phys. **275** (2007), 139-162.

[7] J. E. Avron, M. Fraas, G. M. Graf, P. Grech: *Adiabatic theorems for generators of contracting evolutions. Commun. Math. Phys.* **314** (2012), 163-191.

[8] J.-T. Hwang, P. Pechukas: *The adiabatic theorem in the complex plane and the semiclassical calculation of non-adiabatic transition amplitudes.* J. Chem. Phys. **67** (1977), 4640-4653.

[9] A. Joye, H. Kunz, C.-E. Pfister: *Exponential decay and geometric aspect of transition probabilities in the adiabatic limit.* Ann. Phys. **208** (1991), 299-332.

[10] L. D. Landau, E. M. Lifshitz: *Course of theoretical physics 3: quantum mechanics (non-relativistic theory).* 3rd edition. Pergamon, 1977.

[11] J. Schmid: *Adiabatic theorems with and without spectral gap condition for non-semisimple spectral values.* arXiv: 1401.0089 (2013).

[12] N. Dunford, J. T. Schwartz: *Linear operators III.* Wiley, 1971.

[13] J. Schmid: *Adiabatensätze mit und ohne Spektrallückenbedingung.* Diplomarbeit, Universität Stuttgart. arXiv: 1112.6338 (2011).

An eigenvalue counting theorem with applications to random Schrödinger operators

D. Schmidt

Department of Mathematics, Virginia Tech,
Blacksburg, Virginia 24061, USA
E-mail: dfs@vt.edu
www.vt.edu

We present an estimate on the number of eigenvalues of a matrix in a small interval, phrased in terms of the properties of a submatrix. We then apply this result to prove a theorem on the eigenvalue statistics of certain random Schrödinger operators. Specifically, we establish Wegner, Minami, and higher-order estimates for a class of random block operators modeling quasiparticles in the Bogoliubov-de Gennes equation from the theory of dirty superconductors.

1. Abstract Result

We begin with an abstract result on matrix theory, which we will later apply to the study of random Schrödinger operators. Specifically, our first result gives a necessary condition and a separate sufficient condition for a matrix to have at least m eigenvalues in the interval $(-\epsilon, \epsilon)$. (Note that most of the proofs below are only sketched, or omitted altogether. A more complete presentation may be found in [1].)

To fix the notation, let A be an invertible Hermitian $M \times M$ matrix with inverse A^{-1}, and let I_M be the $M \times M$ identity matrix. Let $G(x, y)$ denote the matrix element in the (x, y) position of A^{-1}, also known as the Green function of A. Let us denote by $\mathcal{C}_\epsilon(A)$ the number of eigenvalues (counted according to algebraic multiplicity) of A in the interval $(-\epsilon, \epsilon)$.

We wish to find the criteria under which A has at least m eigenvalues in the interval $(-\epsilon, \epsilon)$. As a start, recall the following well known result from matrix analysis:

$$\mathcal{C}_\epsilon(A) > 0 \iff \|A^{-1}\| > \frac{1}{\epsilon}.$$

We will want to rewrite the second statement from this equivalence. Since $\|B\|_{\max} \leq \|B\| \leq M\|B\|_{\max}$ for any $M \times M$ matrix B with

$$\|B\|_{\max} = \max_{x,y} |B(x, y)|,$$

we can immediately rewrite this result as a relationship between the existence of a small eigenvalue of A, and the size of a single entry of A^{-1}:

$$\mathcal{C}_\epsilon(A) > 0 \Rightarrow \text{There exists a pair } \{x,y\} \text{ such that } |G(x,y)| > \frac{1}{M\epsilon}; \tag{1}$$

$$|G(x,y)| > \frac{1}{\epsilon} \text{ for some pair } \{x,y\} \Rightarrow \mathcal{C}_\epsilon(A) > 0\,. \tag{2}$$

Our first theorem will be an analogous criterion for the existence of not 1, but m eigenvalues in the interval $(-\epsilon, \epsilon)$. Accordingly, this theorem will use properties of an $m \times m$ submatrix of A^{-1}, rather than the 1×1 submatrix (i.e. the Green function at (x,y)) used above.

Theorem 1.1. *Let $A = A^*$ be an $M \times M$ invertible matrix. Let $A[\alpha, \beta]$ denote the submatrix of A with rows indexed by index subset α and columns indexed by index subset β. Consider the following two assertions:*

(1) $\mathcal{C}_\epsilon(A) \geq m$;

(2) There exist index subsets

$$\alpha_m = \{i_1, \ldots, i_m\}, \quad \beta_m = \{j_1, \ldots, j_m\}$$

of $\{1, \ldots, M\}$ such that

$$A^{-1}[\alpha_m, \beta_m] A^{-1}[\beta_m, \alpha_m] > \frac{K^2}{\epsilon^2} I_N[\alpha_m, \alpha_m] \ \ \textit{for some } K > 0,$$

Then (1) *implies* (2) *with*

$$K = \frac{C_m}{M}, \quad C_m = \frac{1}{m!\, 2^{m-1}}$$

Conversely, (2) *with* $K = 1$ *implies* (1).

It will be convenient later to work with principal submatrices $A^{-1}[\gamma_m]$ of A^{-1}. (Here the notation $A^{-1}[\gamma_m]$ represents the submatrix of A^{-1} where both rows and columns are indexed by the index subset γ_m.) After two applications of Theorem 1.1, we have the following:

Theorem 1.2. *Let $A = A^*$ be an $M \times M$ invertible matrix. Consider the following two assertions:*

(1) $\mathcal{C}_\epsilon(A) \geq m$;

(2) There exists an index subset $\gamma_m = \{i_1, \ldots, i_{2m}\}$ of $\{1, \ldots, M\}$ such that for any subset $\gamma \supset \gamma_m$ for which the matrix $A^{-1}[\gamma]$ is invertible

$$\mathcal{C}_{\epsilon/K}\left(\left(A^{-1}[\gamma]\right)^{-1}\right) \geq m \ \ \textit{for some } K > 0.$$

Then (1) *implies* (2) *with*

$$K = \frac{C_m}{M}, \quad C_m = \frac{1}{m!\, 2^{m-1}}$$

Conversely, (2) *with* $K = 1$ *implies* (1).

2. Quasiparticle Model

The setting for our application of Theorem 1.2 is the study of random Schrödinger operators, and more specifically the distribution of eigenvalues for such operators. The prototypical example of such a model would be the Anderson model, which represents a single electron in a disordered solid. The solid itself is represented by a lattice $\mathcal{V}$ with randomly arranged ions at the lattice points. The charge of each ion generates a "single-site potential" and typically these charges are i.i.d. random variables. In this context, the matrix we need to work with represents a random Hamiltonian. Because of the randomness, we will no longer have a deterministic criterion for the number of small eigenvalues. Instead, we will compute an estimate on the probability that the Hamiltonian has at least m small eigenvalues. If $m = 1$, this is called a Wegner estimate, and for $m = 2$ it is called a Minami estimate. The Wegner estimate is especially important in proofs of electron localization (or nonspreading of wave packets).

Eigenvalue statistics for single-electron Hamiltonians have already been extensively studied, but fewer results are known for Hamiltonians with correlated random variables or with interacting charge carriers. For now, we are interested in a certain model with an interacting electron-hole pair, which arises from the Bogoliubov-de Gennes equations in the mean-field approximation of BCS theory. Previous work [2, 3] has already established a Wegner estimate for this model, under the assumption that all single-site potentials are positive. We will give a second proof here that allows sign-indefinite single-site potentials, and we will also prove a weakened version of the Minami and higher-order estimates for the same model. (The problem of extending such results to the sign-indefinite case has arisen before in other contexts. See [4], for example.)

Our model comes from the Bogoliubov-de Gennes equation, which is a version of the time-independent Schrödinger equation:

$$\begin{bmatrix} H_1 & B \\ B^* & -H_1 \end{bmatrix} \begin{bmatrix} \psi_+ \\ \psi_- \end{bmatrix} = E \begin{bmatrix} \psi_+ \\ \psi_- \end{bmatrix}, \tag{3}$$

The 'particle' and 'hole' components ψ_+ and ψ_- of the quasi-particle state belong to the single-particle Hilbert space $\mathcal{H}$.

The matrix above has a 2×2 block form with $N \times N$ blocks, where $N = |\Lambda|$ is the volume of a finite box $\Lambda \subset \mathcal{V}$ to which we restrict the Hamiltonian. It will be advantageous to reorder the basis so that instead we have an $N \times N$ block matrix with 2×2 blocks. In this case each diagonal block will depend only on the random variables at a single lattice site, but will involve both particle and hole components of the wave function. (The off-diagonal blocks will be deterministic.) Our Hamiltonian is now of the following form:

$$(H_\omega(g)\psi)(x) = (H_0\psi)(x) + gA_\omega(x)\psi(x), \tag{4}$$

where $g > 0$ is a coupling constant, ω is a random variable representing the ion configuration, x is a lattice site, and the $A_\omega(x)$ blocks have the form:

$$A_\omega(x) = \begin{bmatrix} u_x & v_x \\ v_x & -u_x \end{bmatrix} \tag{5}$$

In what follows, we will suppress the dependence on the coupling constant g and refer to the Hamiltonian simply as H_ω.

3. Main Theorem

We now come to our main application theorem, an estimate on the probability that the Hamiltonian described above has at least m eigenvalues in the interval $(-\epsilon, \epsilon)$. We will eventually rephrase this as a question about the determinant of a modified Hamiltonian. Ultimately, we would like to prove this estimate by using an inductive argument on the number of blocks in the matrix. The inductive step will require a connection between the eigenvalue statistics of a matrix with L blocks, and the eigenvalue statistics for a matrix with $L+1$ blocks. To find this connection, we use the concept of a Schur complement. If H is a block matrix:

$$H = \left[\begin{array}{c|c} A & B \\ \hline C & D \end{array}\right]$$

then the Schur complement of the block D is defined to be:

$$H/D := A - BD^{-1}C$$

We now have the Schur determinant formula, which for our quasiparticle model can be written as:

$$det(H - E) = det((H - E)/A_\omega(x))det(A_\omega(x)) \tag{6}$$

Note that if $H - E$ has $L + 1$ diagonal blocks, then the Schur complement $(H - E)/A_\omega(x)$ has only L. This gives the desired relationship which will make the inductive step of our argument possible. We will also need an inductive hypothesis, which would most naturally be stated as a condition on the determinant of an individual block. The assumption in the following theorem provides such a condition. (It may initially appear that our assumption involves the determinant of the wrong kind of block, but this is because we will actually be rearranging the Hamiltonian later. See section 4)

Notation: Note that in the case of the quasiparticle model described above, we will have blocks of size 2×2, but since this result can potentially be extended to other cases, we allow $k \times k$ blocks in the statement of the theorem. Also, m and $M = kN$ are used in the same way here as in Theorems 1.1 and 1.2.

Theorem 3.1. *For an integer n, let S be a given set of $2nk$ distinct integers. Assume there exists an $\alpha > 0$ such that, for any integer $a \in S$, any $\epsilon \in [0,1]$ and arbitrary Hermitian $k \times k$ matrix J we have the bound*

$$\mathbb{P}\left(\left|\det\left((A_\omega(x) - a)^{-1} + (J + a)^{-1}\right)\right| \leq \epsilon\right) \leq K\epsilon^\alpha$$

Then we have

$$\mathbb{P}\left(\mathcal{C}_\epsilon(H_\omega - E) \geq m\right) \leq C\left(-ln(N(\epsilon/g)^\alpha)N(\epsilon/g)^\alpha\right)^m$$

for any $E \in \mathbb{R}$, for any $\epsilon \in [0, \min(2^{-k}, N^{-1/\alpha})]$ and for all $m \leq n$. Here the constant C depends on k, α and m but not on N or ϵ.

Note: For the case $m = 1$ we can prove a stronger estimate, without the log factor. Thus we have a Wegner estimate in the strictest sense of the term.

4. Outline of the Proof of Theorem 3.1

Ultimately, we want $\mathcal{C}_\epsilon(H_\omega - E) \geq m$ to imply an inequality of the form

$$|det(H_\omega - E)| \leq C\epsilon^m \tag{7}$$

Given this implication, we will be able to calculate the probability of the event (7) to get the desired bounds for Theorem 3.1. Our motive in rewriting the problem this way is that the determinant is an easy object to work with—specifically, it lends itself to inductive calculations by way of Equation 6. However, it is not immediately possible to use this determinant. The first problem is that the number of factors in the determinant (i.e. the number of eigenvalues of $(H_\omega - E)$, counting algebraic multiplicity) is equal to kN, so that C will turn out to have a strong N-dependence. We can fix this by applying Theorem 1.1, which allows us to replace the matrix $H - E$ with a smaller matrix whose size does not depend on N. Specifically, we define $\tilde{H}_\omega = \left((H_\omega - E)^{-1}[\gamma]\right)^{-1}$, where γ is as in Theorem 1.2. $\tilde{H}_\omega$ is then a $kL \times kL$ matrix, where L depends on m but not on N. (Note that there are two distinct uses of the Schur complement in this work. One is to reduce the number of blocks from N to L, and the other is to allow an induction on L, as described in the next section.)

This solves one problem but creates another: the new matrix comes from a Schur complement of a block in the original matrix, and the process of taking a Schur complement may increase the norm of the matrix (and therefore the size of the eigenvalues) by an arbitrarily large factor. As a simple illustration, consider the matrix:

$$H = \left[\begin{array}{c|c} A & B \\ \hline B* & D \end{array}\right] = \left[\begin{array}{ccc|c} 0 & 0 & 0 & 1 \\ 0 & 0 & 0 & 0 \\ 0 & 0 & 0 & 0 \\ \hline 1 & 0 & 0 & \eta \end{array}\right] \tag{8}$$

If we assume $\eta > 0$, then $||H|| = \frac{1}{2}\left(\eta + \sqrt{\eta^2 + 4}\right)$, so for small η, $||H|| \approx 1$. The Schur complement of D, however, has norm $||H/D|| = \frac{1}{\eta}$, which can be arbitrarily large for small η. In light of this, our second task is to control the norm of the modified Hamiltonian. Again, we do this by replacing our first modified Hamiltonian $\tilde{H}_\omega$ with yet another matrix, which still has similar eigenvalue statistics. The following theorem shows that this is possible.

Theorem 4.1. *Let $B_{1,2}$ be a pair of Hermitian $L \times L$ matrices with $\|B_1\| \leq 1$. Consider the matrices*

$$B = B_1 + B_2, \quad \hat{B} = (B_1 - aI_L)^{-1} + (B_2 + aI_L)^{-1}$$

where $a \in \mathbb{R}$. Then there exists an integer $a \in [-L-3, -3] \cup [3, L+3]$ (which depends on B_2 but not on B_1) and $\epsilon_0 > 0$ (which depends only on L) such that for any $\epsilon < \epsilon_0$

$$\max\left(\left\|(B_1 - aI_L)^{-1}\right\|, \left\|(B_2 + aI_L)^{-1}\right\|\right) \leq \frac{1}{2}; \tag{9}$$

$$\mathcal{C}_{\epsilon/(225L^4)}\left(\hat{B}\right) \leq \mathcal{C}_\epsilon(B) \leq \mathcal{C}_{7L^2\epsilon}\left(\hat{B}\right). \tag{10}$$

When we apply this theorem to $\tilde{H}_\omega$, the B_1 and B_2 matrices will correspond to the hopping term and the potential, respectively. In this case, the potential is literally a block from the original potential matrix, but the "hopping term" has by now been replaced by a random matrix. Thus we have a second modified Hamiltonian $\hat{H}_\omega$, which has at most $kL - m$ large eigenvalues, each of which has absolute value at most 1.

5. Estimating the Probability

We now have $\mathcal{C}_\epsilon(H_\omega - E) \geq m \Rightarrow |\det \hat{H}_\omega| \leq C\epsilon^m$, for a constant C which does not depend on N or ϵ. The proof of Theorem 3.1 now comes down to verifying the appropriate bound on the probability of the event $\{\omega \in \Omega : |\det \hat{H}_\omega| \leq C\epsilon^m\}$, which will follow from the condition below:

$$\mathbb{P}\left(|\det \hat{H}_\omega| \leq \epsilon\right) \leq (2K\alpha)^L \ln^L(\epsilon^{-1})\epsilon^\alpha, \quad \epsilon \in [0, \epsilon_0] \tag{11}$$

where L is the number of diagonal blocks in the matrix $\hat{H}_\omega$. To prove (11), we use induction on L. For $L = 1$ the result follows from the assumption from Theorem 3.1. Suppose that (11) holds for matrices with L diagonal blocks. We wish to establish the induction step, i.e. (11) for $L+1$ diagonal blocks. We can evaluate $det(\hat{H}_\omega)$ using the Schur determinant formula. Namely, for a site x in the lattice, let us denote by $\hat{H}_\omega^{(x)}$ the restriction of $\hat{H}_\omega$ to the site x. Then

$$\det \hat{H}_\omega = \det \hat{H}_\omega^{(x)} \det(\hat{H}_\omega / \hat{H}_\omega^{(x)}) \tag{12}$$

by Schur's determinant formula. Both determinants on the right-hand side are random, but the first one depends only on the random variable associated with $A_\omega(x)$, a fact which we will exploit momentarily. We note now that the Schur complement $\hat{H}_\omega / \hat{H}_\omega^{(x)}$ is by itself also of the same form as the original $\hat{H}_\omega$ matrix (with fewer blocks). Note that the hopping term in $\hat{H}_\omega / \hat{H}_\omega^{(x)}$ might depend on $A_\omega(x)$, but not on other random variables $\{A_\omega(y)\}$. By the induction hypothesis, we have

$$\mathbb{P}\left(|\det(\hat{H}_\omega / \hat{H}_\omega^{(x)})| \leq \epsilon\right) \leq (2K\alpha)^L \ln^L(\epsilon^{-1})\epsilon^\alpha, \quad \epsilon \in [0, \epsilon_0]. \tag{13}$$

Let $S := \{\omega : \; |\det \hat{H}_\omega| \leq \epsilon\}$, and let

$$F_\omega = |\det \hat{H}_\omega^{(x)}|, \quad G_\omega = |\det(\hat{H}_\omega / \hat{H}_\omega^{(x)})|.$$

We set $Q := \{\omega : \; \min(F_\omega, G_\omega) \leq \epsilon\}$, then by the assumption from Theorem 3.1 and the induction hypothesis

$$\mathbb{P}(Q) \leq \left(K + (2K\alpha)^L \ln^L(\epsilon^{-1})\right) \epsilon^\alpha. \tag{14}$$

On the other hand, we have

$$\chi(S \setminus Q) = \int_\epsilon^1 \chi(sG_\omega \leq \epsilon)\delta(F_\omega - s)ds.$$

Taking expectations on both sides and using (13), we obtain

$$\begin{aligned}
\mathbb{E}\chi(S \setminus Q) &\leq \mathbb{E}\int_\epsilon^1 ds\, \delta(F_\omega - s)\mathbb{E}\left(\chi(sG_\omega \leq \epsilon) \,\middle|\, A_\omega(x)\right) \\
&\leq (2K\alpha)^L \epsilon^\alpha \mathbb{E}\int_\epsilon^1 \frac{\ln^L(\frac{s}{\epsilon})\delta(F_\omega - s)}{s^\alpha} ds \\
&= (2K\alpha)^L \epsilon^\alpha \mathbb{E}\, \frac{\ln^L(\epsilon^{-1}F_\omega)\chi(1 > F_\omega > \epsilon)}{(F_\omega)^\alpha} \\
&\leq (2K\alpha)^L \epsilon^\alpha \ln^L(\epsilon^{-1})\mathbb{E}\, \frac{\chi(1 > F_\omega > \epsilon)}{(F_\omega)^\alpha}.
\end{aligned} \tag{15}$$

We may now use the assumption from Theorem 3.1 and the layer cake representation of integrals to get

$$\mathbb{E}\,\frac{\chi(1 > F_\omega > \epsilon)}{(F_\omega)^\alpha} = \int_1^{\epsilon^{-\alpha}} \mathbb{P}\left((F_\omega)^{-\alpha} \geq t\right) dt \leq K \int_1^{\epsilon^{-\alpha}} \frac{1}{t} dt$$
$$= K\alpha \ln(\epsilon^{-1}) \tag{16}$$

Combination of (14), (15), and (16) yields the induction step.

Acknowledgment

This work was supported in part by NSF grant DMS–1210982.

Bibliography

[1] A. Elgart and D. Schmidt, Eigenvalue statistics for random block operators, arXiv 1306.3459, (2013).
[2] M. Gebert and P. Müller, Localization for random block operators, Oper. Theory Adv. Appl. **232**, 229–246 (2013).
[3] W. Kirsch, B. Metzger and P. Müller, Random block operators, J. Stat. Phys. 143, 1035–1054 (2011).
[4] M. Tautenhahn and I. Veselić, Minami's Estimate: Beyond Rank One Perturbation and Monotonicity, Ann. Henri Poincaré *Online First* (2013).

System of fermions with zero-range interactions

A. Teta

Dipartimento di Matematica, Sapienza Università di Roma,
P.le A. Moro 5, 00185 Roma, Italy
E-mail: teta@mat.uniroma1.it

We discuss the stability problem for a system of N identical fermions with unit mass interacting with a different particle of mass m via zero-range interactions in dimension three. We find a stability parameter $m^*(N) > 0$, increasing with N, such that the Hamiltonian of the system is self-adjoint and bounded from below for $m > m^*(N)$.

Keywords: Point interactions, fermions.

1. Introduction

A system of n quantum particles in $\mathbb{R}^3$ with two-body zero-range interactions is described by the formal Hamiltonian

$$\mathcal{H} = -\sum_{i=1}^{n} \frac{1}{2m_i} \Delta_{\mathbf{x}_i} + \sum_{\substack{i,j=1 \\ i<j}}^{n} \mu_{ij}\, \delta(\mathbf{x}_i - \mathbf{x}_j), \tag{1}$$

where $\mathbf{x}_i \in \mathbb{R}^3$, $i = 1, \ldots, n$, m_i is the mass, $\Delta_{\mathbf{x}_i}$ is the Laplacian relative to $\mathbf{x}_i$, and $\mu_{ij} \in \mathbb{R}$ are coupling constants. We set $\hbar = 1$. In recent years these class of Hamiltonians have been extensively used in the physical literature to describe the dynamics of ultra-cold quantum gases (see e.g. [2]). From the mathematical point of view, the definition of $\mathcal{H}$ as a self-adjoint operator in $L^2(\mathbb{R}^{3n})$ is usually given in the following way. Let us consider the operator

$$\dot{\mathcal{H}}_0 = -\sum_{i=1}^{n} \frac{1}{2m_i} \Delta_{\mathbf{x}_i}, \qquad D(\dot{\mathcal{H}}_0) = C_0^\infty(\mathbb{R}^{3n} \setminus \cup_{i<j}\{\mathbf{x}_i = \mathbf{x}_j\}) \tag{2}$$

where $\{\mathbf{x}_i = \mathbf{x}_j\}$ denotes the hyperplane characterized by the coincidence of the coordinates $\mathbf{x}_i$ and $\mathbf{x}_j$. The operator (2) is symmetric but not self-adjoint. Then, any self-adjoint extension of $\dot{\mathcal{H}}_0$, different from the free Hamiltonian, is by definition a Hamiltonian of n quantum particles in $\mathbb{R}^3$ with two-body zero-range interactions. Roughly speaking, such Hamiltonians act as the free Hamiltonian in $\mathbb{R}^3 \setminus \cup_{i<j}\{\mathbf{x}_i = \mathbf{x}_j\}$ and the elements of the domain satisfy a sort of generalized

boundary condition on each hyperplane $\{\mathbf{x}_i = \mathbf{x}_j\}$. The explicit construction of the self-adjoint extensions is not trivial and a general characterization is not known, except in the simpler case $n = 2$. Indeed, for $n = 2$ and extracting the center of mass motion, the domain of each extension consists of $\psi \in L^2(\mathbb{R}^3) \cap H^2(\mathbb{R}^3 \setminus \{0\})$ satisfying the boundary condition at the origin

$$\psi(\mathbf{x}) = \frac{q}{|\mathbf{x}|} + \alpha q + o(1) \tag{3}$$

for $|\mathbf{x}| \to 0$, where $\mathbf{x}$ is the relative coordinate, $q \in \mathbb{C}$ and $\alpha \in \mathbb{R}$ is the strength of the interaction characterizing each extension (see [1] for details). In the general case $n > 2$, proceeding by analogy, one introduces the so-called Skornyakov-Ter-Martirosyan (STM) extension H_α, defined on elements of $L^2(\mathbb{R}^{3n}) \cap H^2(\mathbb{R}^{3n} \setminus \cup_{i<j}\{\mathbf{x}_i = \mathbf{x}_j\})$ satisfying the boundary condition

$$\psi(\mathbf{x}_1, \ldots, \mathbf{x}_n) = \frac{q_{ij}}{|\mathbf{x}_i - \mathbf{x}_j|} + \alpha q_{ij} + o(1) \tag{4}$$

for $|\mathbf{x}_i - \mathbf{x}_j| \to 0$, where q_{ij} are functions defined on $\{\mathbf{x}_i = \mathbf{x}_j\}$ and α is the strength of the interaction. As a matter of fact, the STM extension H_α is a symmetric operator but, in general, it is not self-adjoint. This happens, for instance, in the cases of three identical bosons and three different particles. In such cases, it is known that any self-adjoint extension of H_α is unbounded from below (see [8] and [10]). Such instability effect is known in the literature as Thomas effect.

On the other hand, there is an important physical situation where one can expect absence of the Thomas effect, i.e., the case of a two groups of particles made of identical fermions. This is due to the fact that the antisymmetry constraint makes the zero-range interaction ineffective for fermions of the same type. In this generality the stability problem is open. Here we consider the particular case of N identical fermions, with unit mass, interacting via zero-range interactions with a different particle of mass m.

In the physical literature (see e.g. [2–5]) the case $N = 2$ and $N = 3$ has been studied using analytical and numerical arguments. In both cases it is found that for m larger than a critical mass, depending on N, the STM extension H_α is self-adjoint and bounded from below, while for m smaller the Thomas effect occurs.

Rigorous proofs of the above results for $N = 2$ have been obtained in [12–14], while in [11] a stability result for $N < 5$ and m sufficiently large is found.

In the next section we discuss a stability result recently obtained in [6] (see also [9]) valid for any N. In particular we find a stability parameter $m^*(N) > 0$ (see definition below) such that stability holds for $m > m^*(N)$. The result has been obtained exploiting a quadratic form method along the line developed in [7].

2. Result

In order to explain the result, we introduce the following function

$$\Lambda(m, N) = \frac{2}{\pi}(N-1)(m+1)^2\left[\frac{1}{\sqrt{m(m+2)}} - \arcsin\left(\frac{1}{m+1}\right)\right] \tag{5}$$

It is easy to see that, for each N, the function $\Lambda(m, N)$ is positive, decreasing with m and satisfies $\lim_{m\to 0}\Lambda(m,N) = \infty$, $\lim_{m\to\infty}\Lambda(m,N) = 0$. Therefore we can introduce the following definition.

Definition 2.1. (Stability parameter $m^*(N)$). For any N fixed, $m^*(N)$ is the unique solution of the equation

$$\Lambda(m, N) = 1 \tag{6}$$

Notice that $m^*(N)$ is increasing with N. Moreover the condition $\Lambda(m, N) < 1$ is equivalent to $m > m^*(N)$, which is precisely our stability condition for the system expressed in the following theorem.

Theorem 2.1. *For $m > m^*(N)$ the STM extension H_α is self-adjoint and bounded from below. In particular H_α is positive for $\alpha \geq 0$ and*

$$\inf \sigma(H_\alpha) \geq -\frac{\alpha^2}{4\pi^4(1-\Lambda(m,N))} \tag{7}$$

for $\alpha < 0$.

We remark that the result is optimal for $N = 2$, in the sense that, according to the previously mentioned results, for $m < m^*(2)$ the three-particle system exhibits the Thomas effect. On the other hand, for $N > 2$ our stability condition is only sufficient and, in order to improve the result, the role of the antisymmetry must be more carefully taken into account. The numerical value of $m^*(N)$ can also be easily computed. For example

$m^*(2) = 0.0735$, $m^*(3) = 0.1890$, ... , $m^*(8) = 0.9473$, $m^*(9) = 0.1215$, ...

We notice that in the special case of equal masses we find stability up to $N = 8$.

The proof of the theorem can be found in [6]. Here we only outline the method based on the quadratic form naturally associated to H_α. Limiting ourselves to the simpler case $N = 2$, we introduce relatives coordinates $\mathbf{y}_1 = \mathbf{x}_1 - \mathbf{x}_0$, $\mathbf{y}_2 = \mathbf{x}_2 - \mathbf{x}_0$, where $\mathbf{x}_1$, $\mathbf{x}_2$ are the coordinates of the two fermions and $\mathbf{x}_0$ the coordinate of the different particle. Moreover we denote by $\mathbf{k}_1$, $\mathbf{k}_2$ the corresponding coordinates in the Fourier space. Neglecting the center of mass motion, the Hilbert space is

$L^2_f(\mathbb{R}^6)$, where the subscript f denotes the restriction to antisymmetric functions, and the free Hamiltonian in the Fourier space reads

$$\widehat{H_0} = \mathbf{k}_1^2 + \mathbf{k}_2^2 + \frac{2}{m+1}\mathbf{k}_1 \cdot \mathbf{k}_2 \tag{8}$$

The quadratic form associated to H_0 is $F_0(u) = (u, H_0 u)$ and it is closed and positive on the domain $D(F_0) = H^1_f(\mathbb{R}^6)$. The quadratic form associated to H_α turns out to be a singular perturbation of F_0, defined on a domain larger than $H^1_f(\mathbb{R}^6)$, which can be derived using a renormalization procedure (explained in [6]). The procedure leads to the following definition

Definition 2.2. For any $\alpha \in \mathbb{R}$, the quadratic form $\mathcal{F}_\alpha$, $D(\mathcal{F}_\alpha)$ is given by

$$D(\mathcal{F}_\alpha) = \left\{ u \in L^2_f(\mathbb{R}^6) \,|\, u = w + \mathcal{G}\xi,\ |\nabla w| \in L^2_f(\mathbb{R}^6),\ \xi \in H^{1/2}(\mathbb{R}^3) \right\} \tag{9}$$

$$\mathcal{F}_\alpha(u) = F_0(w) + 2\left(\Phi(\xi) + \alpha\,\|\xi\|^2\right) \tag{10}$$

where $\mathcal{G}\xi$ in the Fourier space is

$$\widehat{\mathcal{G}\xi}(\mathbf{k}_1, \mathbf{k}_2) = \frac{\hat{\xi}(\mathbf{k}_1) - \hat{\xi}(\mathbf{k}_2)}{\mathbf{k}_1^2 + \mathbf{k}_2^2 + \frac{2}{m+1}\,\mathbf{k}_1 \cdot \mathbf{k}_2} \tag{11}$$

and

$$\Phi(\xi) = 2\pi^2 \sqrt{\frac{m(m+2)}{(m+1)^2}} \int d\mathbf{p}\, |\mathbf{p}| |\hat{\xi}(\mathbf{p})|^2 + \int d\mathbf{p} d\mathbf{q}\, \frac{\overline{\hat{\xi}(\mathbf{p})}\hat{\xi}(\mathbf{q})}{\mathbf{p}^2 + \mathbf{q}^2 + \frac{2}{m+1}\mathbf{p} \cdot \mathbf{q}} \tag{12}$$

We remark that $\mathcal{G}\xi$ is locally in $L^2_f(\mathbb{R}^6)$ and $|\nabla \mathcal{G}\xi| \notin L^2_f(\mathbb{R}^6)$. We also notice that in [6] the definition of the quadratic form is given in a slightly different form, but it is easily seen that the two definitions are equivalent.

The proof of the theorem is essentially based on the analysis of the above quadratic form. In particular, we show that for $m > m^*(2)$ the form is closed and bounded from below and that the associated self-adjoint and bounded from below operator coincides with the STM extension H_α.

The key point of the proof is to show that $\Phi(\xi) \geq 0$ for $m > m^*(2)$. Since such a result can be obtained using elementary methods, in the remaining few lines we summarize the main steps.

- We introduce the expansion in spherical harmonics $\hat{\xi}(\mathbf{p}) = \sum_{lm} \xi_{lm}(p) Y_l^m(\theta, \phi)$ and we have

$$\Phi(\xi) = 2\pi^2 \sqrt{\frac{m(m+2)}{(m+1)^2}} \sum_{lm} \int_0^\infty dp\, p^3 |\xi_{lm}(p)|^2$$
$$+2\pi \sum_{lm} \int_0^\infty dp \int_0^\infty dq\, p^2 \overline{\xi_{lm}(p)}\, q^2 \xi_{lm}(q) \int_{-1}^1 dy \frac{P_l(y)}{p^2 + q^2 + \frac{2}{m+1} pqy} \tag{13}$$

- Using the map $\mathcal{M} : \xi_{lm} \to \xi^\sharp_{lm}$ defined as

$$\xi^\sharp_{lm}(k) = \frac{1}{\sqrt{2\pi}} \int_{\mathbb{R}} dx\, e^{-ikx} e^{2x} \xi_{lm}(e^x) \tag{14}$$

and performing explicit integrations, one obtains

$$\Phi(\xi) = \sum_{lm} \int_{\mathbb{R}} dk \left(2\pi^2 \sqrt{\frac{m(m+2)}{(m+1)^2}} + S_l(k) \right) |\xi^\sharp_{lm}(k)|^2 \tag{15}$$

where

$$S_l(k) = 2\pi^2 \int_{-1}^{1} dy\, P_l(y) \frac{\sinh\left(k \arccos \frac{y}{m+1}\right)}{\sin\left(\arccos \frac{y}{m+1}\right) \sinh(\pi k)} \tag{16}$$

- By an explicit study of the function S_l one sees that

$$S_l(k) \geq 0 \qquad \text{for } l \text{ even} \tag{17}$$

$$S_l(k) \geq S_1(k) \geq S_1(0) \qquad \text{for } l \text{ odd} \tag{18}$$

for any $k \in \mathbb{R}$, where

$$S_1(0) \equiv -4\pi(m+1)\sqrt{m(m+2)} \left[\frac{1}{\sqrt{m(m+2)}} - \arcsin\left(\frac{1}{m+1}\right) \right] < 0 \tag{19}$$

- Taking into account of the above bounds one has

$$\begin{aligned}\Phi(\xi) &\geq \sum_{lm} \int_{\mathbb{R}} dk \left(2\pi^2 \sqrt{\frac{m(m+2)}{(m+1)^2}} + S_1(0) \right) |\xi^\sharp_{lm}(k)|^2 \\ &= 2\pi^2 \sqrt{\frac{m(m+2)}{(m+1)^2}} \sum_{lm} \int_{\mathbb{R}} dk\, (1 - \Lambda(m,2)) |\xi^\sharp_{lm}(k)|^2 \end{aligned} \tag{20}$$

where $\Lambda(m,2)$ has been defined in (5). Since $\Lambda(m,2) < 1$ is equivalent to $m > m^*(2)$, formula (20) implies positivity of $\Phi(\xi)$ for $m > m^*(2)$.

Bibliography

[1] S. ALBEVERIO, F. GESZTESY, R. HOEGH-KROHN, H. HOLDEN, *Solvable Models in Quantum Mechanics* (Springer-Verlag, New-York, 1988).

[2] E. BRAATEN, H.W. HAMMER, Universality in few-body systems with large scattering length, *Phys. Rep.* **428**, 259 (2006).

[3] Y. CASTIN, C. MORA, L. PRICOUPENKO, Four-Body Efimov Effect for Three Fermions and a Lighter Particle, *Phys. Rev. Lett.* **105**, 223201 (2010).

[4] Y. CASTIN, F. WERNER, Unitary gas in an isotropic harmonic trap: symmetry properties and applications, *Phys. Rev. A* **74**, 053604 (2006).

[5] Y. CASTIN, F. WERNER, Unitary quantum three-body problem in a harmonic trap, *Phys. Rev. Lett.* **97**, 150401 (2006).

[6] M. Correggi, G. Dell'Antonio, D. Finco, A. Michelangeli, A. Teta, Stability for a system of N fermions plus a different particle with zero-range interactions, *Rev. in Math. Phys.* **24**, no. 5, 1250017 (2012).

[7] G. Dell'Antonio, R. Figari, A. Teta, Hamiltonians for Systems of N Particles Interacting through Point Interactions, *Ann. Inst. H. Poincaré Phys. Théor.* **60**, 253 (1994).

[8] L. Faddeev, R.A. Minlos, On the point interaction for a three-particle system in Quantum Mechanics, *Soviet Phys. Dokl.* **6**, 1072 (1962).

[9] D. Finco, A. Teta, Quadratic Forms for the Fermionic Unitary Gas Model, *Reports on Math. Phys.* **69**, no. 2, 131 (2012).

[10] A.M. Melnikov, R.A. Minlos, On the Pointlike Interaction of Three Different Particles, *Adv. Soviet Math.* **5**, 99 (1991).

[11] R.A. Minlos, On point-like interaction between n fermions and another particle, *Moscow Math. Journal* **11**, 113 (2011).

[12] R.A. Minlos, On point-like interaction between three particles (two fermions and another particle), *ISRN Math. Phys.* **2012**, 230245 (2012).

[13] R.A. Minlos, M.K. Shermatov, On Pointlike Interaction of Three Particles, *Vestnik Mosk. Univ. Ser. Math. Mekh.* **6**, 7 (1989).

[14] M.K. Shermatov, Point Interaction between Two Fermions and a Particle of Different Nature, *Theo. Math. Phys.* **136**, 1119 (2003).

List of registered participants

1.	Achleitner, F.	TU Wien franz.achleitner@tuwien.ac.at
2.	Aftalion, A.	Ecole Polytechnique Paris amandine.aftalion@uvsq.fr
3.	Anapolitanos, I.	Universität Stuttgart anapoli@mathematik.uni-stuttgart.de
4.	Anikevich, A.	ITMO University, Saint Petersburg shadowpoos@gmail.com
5.	Arnold, Andreas	Karlsruher Institut für Technologie arnold@kit.edu
6.	Arnold, Anton	TU Wien anton.arnold@tuwien.ac.at
7.	Bach, V.	TU Braunschweig v.bach@tu-bs.de
8.	Bachmayr, M.	RWTH Aachen bachmayr@igpm.rwth-aachen.de
9.	Ballesteros, M.	TU Braunschweig m.ballesteros@tu-bs.de
10.	Barseghyan, D.	Friedrich-Alexander-Universität Erlangen-Nürnberg barseghyan@math.fau.de
11.	Benedikter, N.	Universität Bonn niels.benedikter@hcm.uni-bonn.de
12.	Biskup, M.	UCLA Los Angeles biskup@math.ucla.edu
13.	Boytsev, A.	ITMO University Saint Petersburg boitsevanton@gmail.com
14.	Branding, V.	TU Wien volker@geometrie.tuwien.ac.at
15.	Brauns, F.	LMU München fbrauns@gmail.com
16.	Butko, Y.	Bauman Moscow State Technical University yanabutko@yandex.ru
17.	Cacciapuoti, C.	Hausdorff Center for Mathematics Bonn cacciapuoti@hcm.uni-bonn.de
18.	Cances, E.	Ecole des Ponts ParisTech cances@cermics.enpc.fr

19.	Cenatiempo, S.	Universität Bonn cenatiempo@iam.uni-bonn.de
20.	Černý, J.	Universität Wien jiri.cerny@univie.ac.at
21.	Cornean, H.	Aalborg University cornean@math.aau.dk
22.	Davies B.	King's College London e.brian.davies@kcl.ac.uk
23.	Demuth, M.	TU Clausthal demuth@math.tu-clausthal.de
24.	De Nittis, G.	Universität Erlangen-Nürnberg denittis.giuseppe@gmail.com
25.	Dereziński, J.	Warsaw University Jan.Derezinski@fuw.edu.pl
26.	Dittrich, J.	Nuclear Physics Institute Řež dittrich@ujf.cas.cz
27.	Drabkin, M.	Universität Erlangen-Nürnberg drabkin@mi.uni-erlangen.de
28.	Duch, P.	Jagiellonian University, Kraków pawel.duch@uj.edu.pl
29.	Dybalski, W.	ETH Zürich dybalski@ma.tum.de
30.	Ehrlacher, V.	cole des Ponts ParisTech ehrlachv@cermics.enpc.fr
31.	Engelmann, M.	Aarhus University matthias@imf.au.dk
32.	Erdös, L.	IST Austria, LMU München lerdos@mathematik.uni-muenchen.de
33.	Esteban, M. J.	University Paris-Dauphine esteban@ceremade.dauphine.fr
34.	Exner, P.	Doppler Institute Prague exner@ujf.cas.cz
35.	Finco, D.	Universit Telematica Internazionale Uninettuno Roma d.finco@uninettunouniversity.net
36.	Gebert, M.	LMU München gebert@math.lmu.de
37.	Gesztesy, F.	University of Missouri Columbia gesztesyf@missouri.edu
38.	Gontier, D.	Ecole des Ponts ParisTech david.gontier@cermics.enpc.fr
39.	Griesemer, M.	Universität Stuttgart m.griesemer@googlemail.com

40. Gruber, M. Leibniz Universität Hannover
gruber@math.uni-hannover.de
41. Güneysu, B. Humboldt-Universität Berlin
gueneysu@math.hu-berlin.de
42. Hainzl, Chr. Universität Tübingen
christian.hainzl@uni-tuebingen.de
43. Hamza, E. Institute for Advanced Studies Cairo
eman.hamza07@gmail.com
44. Hanauska, F. TU Clausthal
hanauska@math.tu-clausthal.de
45. Hansmann, M. TU Chemnitz
marcel.hansmann@mathematik.tu-chemnitz.de
46. Hempel, R. TU Braunschweig
r.hempel@tu-bs.de
47. Hirokawa, M. Okayama University
hirokawa@math.okayama-u.ac.jp
48. Hoecker-Escuti, F. TU Chemnitz
hoecker@math.univ-paris13.fr
49. Huang, X. eGain Communications Foster City
HuangZeng@yahoo.com
50. Ichinose, W. Shinshu University Matsumoto
ichinose@math.shinshu-u.ac.jp
51. Jensen, A. Aalborg University
matarne@math.aau.dk
52. Jex, M. TU Prague
jexmicha@fjfi.cvut.cz
53. Kalvoda, T. TU Prague
tomas.kalvoda@fit.cvut.cz
54. Kerner, J. Universität Stuttgart
JoachimKerner@web.de
55. Khrabustovskyi, A. Karlsruher Institute für Technologie
andrii.khrabustovskyi@kit.edu
56. Kirsch, W. FernUniversität Hagen
werner.kirsch@fernuni-hagen.de
57. König, W. TU Berlin and WIAS Berlin
koenig@wias-berlin.de
58. Kordyukov, Y. Russian Academy of Sciences Ufa
yurikor@matem.anrb.ru
59. Korotyaev, E. St. Petersburg State University
korotyaev@gmail.com
60. Kostenko, A. Universität Wien
duzer80@gmail.com

61.	Kovarik H.	Università degli studi di Brescia hynek.kovarik@ing.unibs.it
62.	Kriz, J.	University of Hradec Kralove jan.kriz@uhk.cz
63.	Laptev, A.	Imperial College London a.laptev@imperial.ac.uk
64.	Lipovsky, J.	University of Hradec Kralove jiri.lipovsky@uhk.cz
65.	Lobanov, I.	ITMO University Saint Petersburg lobanov.igor@gmail.com
66.	Lotoreichik, V.	TU Graz vladimir.lotoreichik@gmail.com
67.	Lu, J.	Duke University Durham jianfeng@math.duke.edu
68.	Malamud, M.	IAMM Donetsk mmm@telenet.dn.ua
69.	Manko, St.	TU Prague stepan.manko@gmail.com
70.	Matte, O.	Aarhus University matte@imf.au.dk
71.	Melikhov, I.	ITMO University Saint Petersburg ivan.melikhov@gmail.com
72.	Mendl, Chr.	TU München mendl@ma.tum.de
73.	Mielke, A.	HU Berlin and WIAS Berlin mielke@wias-berlin.de
74.	Minakov, O.	TU Prague minakov.ilt@gmail.com
75.	Mine, T.	Kyoto Institute of Technology mine@kit.ac.jp
76.	Møller, J. Sch.	Aarhus University jacob@imf.au.dk
77.	Mugnolo, D.	Universität Ulm delio.mugnolo@uni-ulm.de
78.	Naaijkens, P.	Leibniz Universität Hannover pieter.naaijkens@itp.uni-hannover.de
79.	Nakano, F.	Gakushuin University Tokyo fumihiko@math.gakushuin.ac.jp
80.	Narita, M.	Okinawa National College of Technology narita@okinawa-ct.ac.jp
81.	Naumkin, I.	Universidad Nacional Autonoma de Mexico ivannaumkinkaikin@gmail.com

82.	Neidhardt, H.	WIAS Berlin hagen.neidhardt@wias-berlin.de
83.	Niikuni, H.	Doshisha University Kyoto hniikuni@mail.doshisha.ac.jp
84.	Novak, R.	Nuclear Physics Institute Řež r.novak@ujf.cas.cz
85.	Pankrashkin, K.	Universite Paris-Sud 11 Orsay konstantin.pankrashkin@math.u-psud.fr
86.	Pedra, W.	Johannes Gutenberg-Universität Mainz pedra@mathematik.uni-mainz.de
87.	Popov, A.	ITMO University, Saint Petersburg irin-a@yandex.ru
88.	Popov, I. Yu.	ITMO University Saint Petersburg popov1955@gmail.com
89.	Post, O.	Durham University olaf.post@dur.ac.uk
90.	Pusztai, B. G.	University of Szeged gpusztai@math.u-szeged.hu
91.	Racec, P.	WIAS Berlin paul.racec@wias-berlin.de
92.	Rasmussen, M. G.	Aalborg University morteng@math.aau.dk
93.	Raykov, G.	Pontificia Universidad Catlica de Chile graikov@mat.puc.cl
94.	Schilling, Chr.	ETH Zrich schillin@phys.ethz.ch
95.	Schlein, B.	Universität Bonn benjamin.schlein@hcm.uni-bonn.de
96.	Schmid, J.	Universität Stuttgart jochen.schmid@mathematik.uni-stuttgart.de
97.	Schmidt, D.	Virginia Tech Blacksburg dfs@vt.edu
98.	Schneider, R.	TU Berlin schneidr@math.TU-Berlin.de
99.	Schulz-Baldes, H.	Universität Nürnberg-Erlangen schuba@mi.uni-erlangen.de
100.	Seifert, Chr.	TU Hamburg christian.seifert@tuhh.de
101.	Seri, M.	University College London m.seri@ucl.ac.uk
102.	Siegl, P.	University of Bern petr.siegl@math.unibe.ch

103.	Sigal I. M.	University of Toronto im.sigal@utoronto.ca
104.	Skibsted, E.	Aarhus University skibsted@imf.au.dk
105.	Stollmann, P.	TU Chemnitz P.Stollmann@mathematik.tu-chemnitz.de
106.	Stoltz, G.	CERMICS-Ecole des Ponts ParisTech stoltz@cermics.enpc.fr
107.	Stürzer, D.	TU Wien dominik.stuerzer@tuwien.ac.at
108.	Tater, M.	Nuclear Physics Institute, Řež tater@ujf.cas.cz
109.	Teta, A.	"La Sapienza" Universita' di Roma teta@mat.uniroma1.it
110.	Truc, F.	Universit Grenoble I Francoise.Truc@ujf-grenoble.fr
111.	Tušek, M.	TU Prague tusekmat@fjfi.cvut.cz
112.	Ueberschaer, H.	IPhT, CEA Saclay, Paris henrik.uberschar@cea.fr
113.	Ueltschi, D.	University of Warwick daniel@ueltschi.org
114.	Veselić, I.	TU Chemnitz ivan.veselic@mathematik.tu-chemnitz.de
115.	Villegas-Blas, C.	Universidad Nacional Autonoma de Mexico villegas@matcuer.unam.mx
116.	Vogt, H.	TU Hamburg hendrik.vogt@tu-harburg.de
117.	Warzel, S.	TU München warzel@ma.tum.de
118.	Watanabe, S.	Gunma University Maebashi shuwatanabe@gunma-u.ac.jp
119.	Weder, R.	IIMAS-UNAM Mexico City weder@unam.mx
120.	Weich, T.	Philipps Universität Marburg weich@mathematik.uni-marburg.de

Author index